POISONOUS PLANTS: GLOBAL RESEARCH AND SOLUTIONS

Poisonous Plants: Global Research and Solutions

Edited by

Kip E. Panter

Terrie L. Wierenga

James A. Pfister

USDA-ARS Poisonous Plant Research Laboratory
Logan, Utah 84341, USA

www.cabi.org

CABI Publishing is a division of CAB International

CABI Publishing
CAB International
Wallingford
Oxfordshire OX10 8DE
UK

Tel: +44 (0)1491 832111
Fax: +44 (0)1491 833508
E-mail: cabi@cabi.org
Website: www.cabi-publishing.org

CABI Publishing
875 Massachusetts Avenue
7th Floor
Cambridge, MA 02139
USA

Tel: +1 617 395 4056
Fax: +1 617 354 6875
E-mail: cabi-nao@cabi.org

A catalogue record for this book is available from the British Library, London, UK.

Library of Congress Cataloging-in-Publication Data

International Symposium on Poisonous Plants (7th : 2005 : Utah State University)
Poisonous plants : global research and solutions / edited by Kip E. Panter, Terrie L. Wierenga, James A. Pfister.
p. cm.
Includes bibliographical references and index.
ISBN 978-1-84593-273-2 (alk. paper)
1. Livestock poisoning plants--Toxicology--Congresses. 2. Livestock poisoning plants--Congresses. 3. Poisonous plants--Congresses. 4. Poisonous plants--Toxicology--Congresses. 5. Plant toxins--Physiological effect--Congresses. I. Panter, Kip Eugene, 1951- II. Wierenga, Terrie L. III. Pfister, James Alan. IV. Title.
SF757.5I56 2005
636.089'5952--dc22   2007003872

ISBN   978 1 84593 273 2

Printed and bound in the UK from copy supplied by the editors by Cromwell Press, Trowbridge.

Contents

Reproduction

Treatment

CNS

Kidney, GI, Immunology

Management

Diagnosis

Chemistry

Mycotoxins

Herbal Products

Preface

This book contains manuscripts highlighting poisonous plant and natural toxin research from around the world. The manuscripts are peer reviewed, edited and formatted to conform with style, scientific integrity and professional standards. These research results were presented at the 7th International Symposium on Poisonous Plants (ISOPP7) held on the Utah State University campus, Logan, Utah, USA, June 6–10, 2005. This book represents the 7th symposium in a continuing series begun in 1977; planning for ISOPP8 is underway. Historically, past symposia have brought together scientists and researchers representing the multi-disciplinary approach to solving poisonous plant problems. This symposium was no different, with botanists, chemists, biochemists, veterinarians, toxicologists, range scientists and other animal and human health researchers from around the world coming together to renew friendships, build collaborations, share research findings and discuss new techniques and approaches in this global arena of scientific research and regulatory surveillance of poisonous plants and natural toxins.

Poisonous plants are responsible for large economic losses to livestock producers throughout the world. Research, both fundamental and applied, presented at this symposium provided a brief summary of the vast intellectual resources dedicated to solving the many issues surrounding poisonous plants and their toxins. The chapters in this book represent most aspects of plant poisonings and mycotoxicoses including: (1) chemical isolation and identification; (2) better understanding of biochemical mechanisms of action on organ systems; (3) treatments and management recommendations to reduce losses; (4) discussions of animal behavior leading to ingestion of poisonous plants; (5) food safety for humans and feed safety for animals; (6) regulatory issues concerning human health and safety; and (7) others.

The first section of this book contains dedications to recognize the lifelong efforts of Dr Richard F. Keeler, Dr Selwyn Everist and Dr Claude Culvenor on poisonous plant research. The dedicatory presentations and representative plaques were presented by Dr Lynn F. James, Dr Alan Seawright and Dr Steven Colegate respectively on the first day of the symposium. The editors congratulate these great scientists for this recognition and also thank the presenters for their contribution to the opening session of this symposium.

The next symposium will be held in João Pessoa, Paraíba in northeastern Brazil in 2009 under the chairmanship of Dr Franklin Riet-Correa. The organizers of ISOPP7 wish Franklin and colleagues the very best in planning and organizing ISOPP8.

Thanks to each of you for a great ISOPP7 and we hope to see you in beautiful northeastern Brazil for ISOPP8 in 2009.

The Editors

Kip E. Panter, Terrie L. Wierenga and James A. Pfister
USDA Agricultural Research Service,
Poisonous Plant Research Lab
Logan, UT, USA

Acknowledgements

The 7th International Symposium on Poisonous Plants (ISOPP7) was sponsored through a joint effort by the USDA-Agricultural Research Service, Poisonous Plant Research Laboratory, Logan, Utah and The Utah Agricultural Experiment Station, Utah State University, Logan, Utah. The editors wish to thank the Conference Services at Utah State University including Carrie Milligan, Jamie Rupp, Vivian Johnson and the many others that provided services and assistance in the venues, entertainment, receptions, banquets, transportation, and technology for posters and presentations; the line drawings of specific plants were provided by Holly Broome-Hyer, USU graphic artist. We also thank the organizing committee at the Poisonous Plant Research Lab including Pat Bradfield, Chris Spall, Ed Knoppel, Joseph Jacobson, Kermit Price, Scott Larsen, and Drs Dale Gardner, Stephen Lee, and Bryan Stegelmeier. The committee thanks Dr Michael Ralphs, Dr Stan Welsh and those assisting in organizing the field trip. We appreciate Danny Hansen, Al Maciulis, Rex Probst and Jason Tuttle for transportation on the field trip and to various social activities. We thank Dr Lynn F. James, Research Leader at the Poisonous Plant Lab, for his support and encouragement during the organization and hosting of the symposium. The editors appreciate all the participants and session chairs of the symposium and the great contribution each and every one makes in keeping this symposium active and interesting. We appreciate Dr Franklin Riet-Correa and his organizing committee for accepting responsibility to host ISOPP8 in 2009. The venue for the meeting is João Pessoa, Paraíba state in northeastern Brazil. We look forward to an exciting and fruitful meeting.

Claude Culvenor

On the occasion of his 80th Birthday

Dr Culvenor has a long history of association with the International Symposia on Poisonous Plants, being involved at the inception along with Lynn James, Richard Keeler, Selwyn Everist and Alan Seawright amongst others.

Claude Culvenor was awarded his PhD by Melbourne University in 1948 for his synthetic organic chemistry research with Professor W. Davies, and his D. Phil. by Oxford University in 1950 for his work with Sir Robert Robinson on the structural elucidation of quinamine, a minor *Cinchona* alkaloid.

Returning to Australia, Dr Culvenor joined the Commonwealth Scientific and Industrial Research Organisation (CSIRO) to begin his long and fruitful association with pyrrolizidine alkaloids. Initial investigations centred around the animal health issues associated with *Heliotropium europaeum*, *Crotalaria retusa* and *Senecio jacobaea*. Outcomes of the research included the application of partition chromatography to isolate individual pyrrolizidine alkaloids for structural identification and toxicity studies. There followed pioneering work into the mode of action of pyrrolizidine alkaloid poisoning and in 1968, the publication of the reference book '*The Pyrrolizidine Alkaloids: Their Chemistry, Pathogenicity and Other Biological Properties*' with Lionel Bull and Allan Dick.

Dr Culvenor became the target of herbalists and organic gardeners when he publicised the fact of pyrrolizidine alkaloids in comfrey (*Symphytum* spp.), the leaves of which were being recommended as herbal medicines and used as a salad plant. With the weight of evidence provided by Dr Culvenor, including the demonstration of liver damage in rats exposed to the comfrey pyrrolizidine alkaloids, the sale of comfrey was soon banned in his home state of Victoria in Australia and the more widespread use declined.

About 1962, during a brief hiatus in his early pyrrolizidine alkaloid research, Dr Culvenor investigated the toxicity problems associated with *Phalaris aquatica* (formerly *tuberosa*). His identification of the staggers-inducing tryptamine alkaloids eventually led to the targeted development of phalaris cultivars low in these compounds. Present day research into continuing, sporadic toxicity problems with phalaris is closely dependent on the early research of Dr Culvenor.

With the same human food safety interest that he demonstrated in the case of comfrey, Dr Culvenor demonstrated high levels of pyrrolizidine alkaloids, especially echimidine, in honey derived from *Echium plantagineum*. That this work with honey is still going on more than thirty years later is a testament to the foresight of Dr Culvenor. However, at the time, Dr Culvenor's reputation and credibility became a target for the honey industry. As with the personal attacks by herbalists a decade or so earlier, this criticism was also unfair. With no vested interest to protect, Dr Culvenor's sole concern was for human food safety but based on credible scientific data. To this latter end, despite the fact of echimidine-like alkaloids in honey, in an effort to properly assess the risk presented by echimidine in honey Dr Culvenor suggested to the American FDA and the WHO Cancer Group in Lyon, France that they specifically test echimidine for carcinogenicity. The suggestion was rejected on the basis that three other, similar pyrrolizidine alkaloids had already been shown to be carcinogenic so where was the need to test yet another!

In 1975 and 1982 Dr Culvenor sat on two WHO working groups constituted to reach a consensus on the carcinogenicity of pyrrolizidine alkaloids. In 1986 Dr Culvenor chaired a similar WHO working group convened in Tashkent to review the hazard to humans of pyrrolizidine alkaloids present in herbal medicines, contaminated cereal grains and other foods. The findings were consolidated into a milestone final report by Professor Tandon, University of New Delhi, and Dr Culvenor; it was published by the WHO in 1988. It includes appendices detailing the then known sources of toxic pyrrolizidine alkaloids.

From the early 1970s, Dr Culvenor led a highly skilled, multi-disciplinary research group within CSIRO Division of Animal Health in the investigation of plant-associated intoxication problems. Foremost amongst these were lupinosis and Annual Ryegrass Toxicity that eventually led to the identification of the causative toxins and ongoing work that continues today. During this time Dr Culvenor also co-published a reference book on the Phytochemical Survey of Australian plants that produce bioactive secondary metabolites.

After retirement, Dr Culvenor was appointed an Honorary Visiting Professor at La Trobe University where he continued, and continues, to pursue interests in toxic plants especially those containing pyrrolizidine alkaloids. He has also acted as a consultant to some organisations, reviewing aspects of plant chemistry, and, together with Dr Collins of Monash University, compiled a database on the chemistry of Australian plants for the years 1940–2000 available on a CD entitled '*PhytoChem Australia*' published in 2003 by CSIRO Publishing.

The Poisonous Plants Research Group formed by Dr Culvenor still persists within CSIRO, albeit smaller and with mostly different researchers. However, the current Group is gratefully aware of its heritage and the achievements under the guidance and leadership of Dr Culvenor.

Dr Steven Colegate

Selwyn Everist

Selwyn Everist was the leader of the Australian group attending the first Poisonous Plant Symposium, which was held at Logan in July, 1977. He was at that time the Queensland Government Botanist and Secretary of the Queensland Poisonous Plants Committee. This committee had been established in 1940 by the Queensland Department of Agriculture and Stock to coordinate and advise on investigations of poisonous plant mortalities in livestock in that State.

In the 1930s, Selwyn as a young botanist had been located in central western Queensland to study the regeneration and characteristics of native pastures and of fodder trees harvested for feeding to livestock in times of drought. In this appointment he was also called upon to assist in local veterinary investigation of mortalities in livestock caused by poisonous plants. In 1951 he was appointed part-time to the Faculty of the Queensland Veterinary School as the lecturer in agrostology. As final year veterinary students at the time, we were inspired by his enthusiasm for poisonous plants investigations. Later in 1964, he and I collaborated in teaching the new toxicology course in the BVSc curriculum.

In 1975, following a visit to Brisbane by Lynn James, Selwyn obtained support from the Australian government for the joint US Australia Poisonous Plants Symposium, subsequently held in Logan in 1977. Immediately after this meeting, Selwyn traveled to England, where he suffered a heart attack from which he eventually succumbed in October 1981.

Selwyn made very significant contributions to our knowledge of poisonous plants in Australia, publishing the first edition of his definitive book '*The Poisonous Plants of Australia*' in 1974. Following the 1977 Logan meeting, which was a watershed occasion for both the US and Australian groups, he went to work on the second edition of this book. He received an advance copy of the book the day before he died.

Selwyn's contribution both to science and to veterinary science was subsequently recognized by his award of a PhD by the University of Queensland and a Fellowship of the Australian College of Veterinary Scientists. His absence through recent years from the Queensland Poisonous Plants Committee was greatly felt, as was his infectious enthusiasm for all poisonous plants issues. It has always been a pity that his untimely death in 1981 prevented his making contributions to the later International Poisonous Plants Symposia.

A.A. Seawright
Emeritus Professor

Richard Keeler

Dr Alan Seawright (l) and Dr Richard Keeler (r)

Dr Richard Keeler was recognized at the 7th ISOPP meeting held at Logan, Utah for his outstanding contribution to research on the effects of poisonous plants on livestock. Dick spent the early part of his life (8½ years) in Mexico, where he became acquainted with life in a rural community that involved cattle, horses, sheep, etc. His father was president of the Juarez Academy, which was located in that area. Dick grew up with a background where an education was essential and much sought after. This and a mother who insisted proper grammar was essential probably accounts for his excellence as an author. The family moved to Palo Alto, California, where his father advanced his education. They then moved to Grantsville, Utah, where his father taught school. Grantsville was another rural community where there was a real presence of cattle, horses and sheep and the people that raised them. The family then moved to Draper, Utah. It was while living in this area Dick was old enough to seek some employment, which he obtained from the local poultry and dairy producers. It was here he learned the fine art of the use of a shovel and a pitch fork and hard labor as he cleaned chicken coops, cow barns, etc. He also learned to mow, bale, and haul hay and more of the hard labor bit. He went to Brigham Young University on a scholarship to play basketball. It was about this time (1950) he was invited to go to Korea as a representative of the United States government. I believe the position there was in the front lines. He returned from Korea, and shortly thereafter got married, and enrolled at BYU. The girl he married was from a farm family and he got further experience in swathing and baling hay and hard work. He then turned his full attention to academics and graduated in 1954 with a BS degree in soil biochemistry. With his BS degree in one hand and wife in the other, Dick went to Ohio State University at Columbus, where he graduated with a MS and PhD in soil biochemisty.

Dr Keeler accepted a job as biochemist in the Veterinary Research Department at Montana State University in Bozeman, where he became acquainted with animal diseases and the procedures for diagnosis, etc. Here he was able to associate with Dr Hadley Marsh, who not only knew about poisonous plants from a research point of view but also from practical experience. Dr Keeler's experience with poisonous plants started at this point. Dick was a very apt student and he never stopped learning. Dick and Hadley shared the propensity to say things as they saw them with 'no punches pulled'. From Montana State University Dick went to the USDA, ARS National Animal Disease Center in Ames, Iowa as a chemist. It was here at a meeting in 1965 he met Dr Wayne Binns, who was Director of the USDA, ARS Poisonous Plant Research Laboratory (PPRL). Dr William Buck, DVM, who had spent a year at the PPRL, introduced Dr Keeler to a problem under investigation at this laboratory. Scientists at PPRL had demonstrated that when ewes grazed a range plant (*Veratrum californicum)* on the 14th day of pregnancy they gave birth to lambs with a cyclopean-type malformation. Dr Keeler had started work on identifying the teratogen in *V. californicum*. At this time he met Dr Binns and Dr Howard Johnson, Director of the ARS Animal Disease and Parasite Division, who was attending a meeting at Ames. While there Dr Binns convinced Dr Johnson that Dr Keeler should be transferred to the ARS-PPRL in Logan, Utah, where he could continue this research on isolating the toxin in *Veratrum* in a more effective way. While at PPRL Dr Keeler conducted research on a broad array of plant toxins, which added a great deal to our knowledge about plants poisonous to livestock. Two of these will be mentioned, with the impact of these two areas still being significant today.

1. Identification of the toxin cyclopamine in *V. californicum* which is responsible for a cyclopean-type malformation in lambs born to ewes that have grazed this plant on the 14th day of gestation. Cyclopamine is currently under intense investigation as a treatment of certain cancers such as pancreatic and prostate cancer in people.
2. Certain species of lupine, when grazed by cows between 40 and 70 days of gestation, cause calves to be born with crooked legs and cleft palates. Dr Keeler identified the toxin in the lupine responsible for the malformations. This information was key to the development of a model (PPRL) which is being used for the development of intrauterine surgery techniques to correct cleft palates in humans.

Characteristics that make Dr Richard Keeler an outstanding scientist:

- A man of integrity. Without integrity science is meaningless; in this matter Dick was absolute.
- A man of vision. Truly the men of history who have stood out as leaders have been men of vision. Their imaginations created in their respective fields images of the goals which they sought.
- An inquisitive and productive mind.
- Research that was well planned and thought out so it would provide data that was of high quality and worthy of publication – no false starts.
- Fully aware of why he was doing the research.
- An excellent author and able to give clear expression to his thoughts and research findings.
- Did research that mattered and made a difference.
- Always kept in mind those for whom his research was done.

- His own man.
- Internationally recognized for his research work.

Dick has never forgotten the value of hard physical labor. I spent many days with him somewhere out on the range gathering plants to be used in research. We picked truck loads of plant. He never minded getting dirt on his hands; it washes off.

We of the 7th ISOPP express our appreciation for his contribution to our understanding of livestock poisoning by plants.

Lynn F. James

OVERVIEW

Chapter 1

Toxic Plants for Livestock in Brazil: Economic Impact, Toxic Species, Control Measures and Public Health Implications

F. Riet-Correa[1], R.M.T. Medeiros[1], C.H. Tokarnia[2], and J. Döbereiner[2]

[1]*Centro de Saúde e Tecnologia Rural, Universidade Federal de Campina Grande, Campus de Patos, Paraíba, PB 58700-000, Brazil;* [2]*Projeto Sanidade Animal Embrapa/UFRRJ, Seropédica, RJ 23890-000, Brazil*

Economic Impact and Toxic Species

Livestock poisoning by plants in Brazil and Uruguay has been a serious problem since the Spanish and Portuguese pioneers first grazed cattle on natural pastures in these countries. Economic losses due to livestock poisoning by plants can be defined as direct and indirect losses. Direct losses are caused by death of animals, lower reproductive performance (abortions, infertility, malformations), reduced production of surviving animals and other alterations due to transitory illness, subclinical diseases with decreased production of milk, meat or wool, and increased susceptibility to other diseases due to suppressed immunological response. Indirect losses include the cost of controlling poisonous plants in the pastures, management measures to prevent poisoning such as use of fences and alternative grazing, reduction in the value of the forage due to delayed or deferred grazing, reduced land value, cost of replacement of the breeding stock lost by plant poisoning, and the cost associated with diagnosis and treatment of the affected animals (Riet-Correa *et al.* 1993; James 1994; Riet-Correa and Medeiros 2001). The economic loss by plant poisoning is difficult to estimate because reliable data are not available for many components of the cost, but losses due to deaths are easy to measure where data from diagnostic laboratories indicates the different causes of death.

In Rio Grande do Sul, southern Brazil, it is estimated that 5% of the cattle are lost annually to all causes. With a cattle population of 13 million, these losses represent 650,000 deaths per year. Data from the Regional Diagnostic Laboratory at Pelotas University from 1978 to 1998 showed that an average of 10.6% of the cases diagnosed in cattle were due to plant poisoning, varying annually from 5% to 25% (Riet-Correa and Medeiros 2001). Data from the Diagnostic Laboratory of Santa Maria University showed that about 14% of the deaths in cattle are caused by toxic plants (Claudio Barros, 2000, personal communication). Assuming that in Rio Grande do Sul, 10% to 14% of deaths in cattle per year are due to toxic plants, one can estimate that the annual death rate due to toxic plants in this State varies from 64,000 to 90,000 animals (Riet-Correa and Medeiros 2001).

In the State of Santa Catarina, southern Brazil, with a cattle population of 2,960,343 head, mortality is also about 5%. Data from the University of Santa Catarina Veterinary Diagnostic Laboratory showed that deaths caused by toxic plants represent, from 1998 to 2000, 13.9% of these deaths, with an annual variation from 8.5% to 24.3% (Aldo Gava, University for the Development of Santa Catarina, 2001, personal communication). The mean annual cattle mortality due to toxic plants is estimated to be 20,574 head.

In the state of Bahia, northeastern Brazil, where there are about 10 million head, 11.3% of all cattle deaths are due to toxic plants. This represents an annual loss of 56,500 head (Luciano Figueiredo, Federal University of Bahia, 2004, unpublished data).

Because no reliable data on the losses of cattle due to toxic plants is available in other Brazilian States, the losses of cattle due to toxic plants are difficult to estimate for the whole country. However, assuming that the annual losses due to toxic plants are similar to those in Rio Grande do Sul, Santa Catarina, and Bahia (10.6% to 14% of the cases diagnosed), that at least 5% of the cattle die annually and Brazil has a total cattle population of 195 million, toxic plants cause the loss of 975,000 to 1.365 million cattle each year. Considering an average price of US$200 per head, then approximately US$195 to 273 million are lost annually from cattle deaths caused by toxic plants. Furthermore, these values, estimated mainly with data from southern Brazil, are probably underestimates, as cattle mortality in many other regions is higher than 5% annually. It is known that the largest numbers of deaths by plant poisoning occurs in the northern region, followed by the northeastern and central-western regions (Tokarnia *et al.* 2002).

Data from the Regional Diagnostic Laboratory at Pelotas University show that in Rio Grande do Sul, toxic plants cause 7.2% of sheep deaths. Considering total deaths of 15% to 20% and a population of 5 million sheep, losses from toxic plants in this State are estimated at 54,000 to 72,000 sheep annually (Riet-Correa and Medeiros 2001).

There are few data related to indirect losses including reproductive losses due to abortion, infertility and malformations, yet malformation in sheep, goats and cattle is very common, especially in northeastern Brazil. In recent studies in perinatal mortality, 10% of the neonatal deaths in kid goats (Medeiros *et al.* 2005) and 23% of lambs (Nóbrega *et al.* 2005) were due to malformations. The goat and sheep populations are 10,160,737 and 6,717,980 animals, respectively, in this region. The annual losses due to malformation can be estimated at 273,120 kids and 259,582 lambs. These data were estimated assuming 56% of females were in a reproductive phase, a birth rate of 1.6 for goats and 1 for sheep, and a perinatal mortality of 30% of lambs and kids (Medeiros *et al.* 2005; Nóbrega *et al.* 2005). It was reported recently that some of these malformations are caused by *Mimosa tenuiflora* (jurema-preta), a very common plant in the semiarid region of northeastern Brazil (Riet-Correa *et al.*, Chapter 74, this volume).

In Brazil there are many known toxic plants and an ever growing list of newly recognized toxic species are reported each year. There are 113 toxic species belonging to 64 genera that have been reported to cause spontaneous intoxication in livestock. A list of these plants with their major clinical signs or system affected, economic impact, and active principles based on Tokarnia *et al.* (2000, 2002) and Méndez and Riet-Correa (2001) is given in Tables 1–9. These tables also include other recently diagnosed toxic species such as *Vicia villosa* (Fighera and Barros 2004) *Marsdenia* spp, *Ipomoea riedelii, Ipomoea sericophylla, Centratherum brachylepis, Leucaena leucocephala, Echinochloa polystachya, Pennisetum purpureum, Mimosa tenuiflora* (Riet-Correa *et al.*, Chapter 74, this volume), *Aspidosperma pyrifolium* (Medeiros *et al.* 2004), *Aeschynomene indica* (Riet-Correa *et al.*

2003), *Erythroxylum argentinum* (Barros *et al.* 2004), *Erythroxylum deciduum* (Seitz *et al.* 2004), *Dodonea viscosa* (Colodel *et al.* 2003), *Trema micrantha* (Traverso *et al.* 2003), *Eupatorium tremulum* (Lucioli *et al.* 2004), and *Melia azedarach* (Timm and Riet-Correa 1997).

In Rio Grande do Sul, southern Brazil, *Senecio* spp. were responsible for 50.7% of the cattle deaths caused by toxic plants from 1978 to 1998 (Riet-Correa and Medeiros 2001). In sheep in this State, poisoning by *Nierembergia veitchii* represents 77.8% of the losses due to poisonous plants (Riet-Correa and Medeiros 2001). *Ateleia glazioviana,* a tree that causes abortion, cardiac fibrosis, and *status spongiosus* of the central nervous system in cattle, is a very important poisonous plant in Santa Catarina and northeastern Rio Grande do Sul. *Pteridium aquilinum* is one of the more worrying Brazilian plants, causing an acute pancytopenic hemorrhagic syndrome, chronic hematuria, and squamous cell carcinomas of the upper digestive tract. It causes severe losses in cattle in the south and southeast of Brazil, but also causes poisoning in the other Brazilian regions. In the state of Santa Catarina, 244 (7.16%) out of 3,407 necropsies on cattle from 1987 to 2001 were diagnosed as *P. aquilinum* poisoning. One hundred and twenty-two of those cases were of the hemorrhagic form, 103 were cases of tumors in the upper digestive tract, and 19 cases were of chronic hematuria. Annual losses due to poisoning in this State are estimated at 10,657 cattle (Gava *et al.* 2002).

In the state of Bahia, during 1985 to 2003, *Thiloa glaucocarpa* was responsible for 39% of all deaths due to plant poisoning, *Palicourea marcgravii* for 27% of deaths, and *Mascagnia rigida* for 24% (Luciano Figueiredo, Federal University of Bahia, 2004, unpublished data). *Thiloa glaucocarpa* is also very important in other states of the Brazilian semiarid region.

Palicourea marcgravii, which contains fluoroacetate, is probably the most important Brazilian toxic plant for cattle. It causes severe losses in the whole country, except in the three states of the southern region and in the state of Mato Grosso do Sul. Three other plants that cause 'sudden death' are important: *Arrabidaea bilabiata* and *Arrabidaea japurensis* in the northern Amazon region (Tokarnia *et al.* 1979), and *Mascagnia rigida* in the northeast of Brazil. Tokarnia *et al.* (1990) estimated that plants causing 'sudden death' are responsible for about 50% of the losses due to toxic plants in the whole country.

Severe economic losses are caused in cattle and sheep, mainly in the southeastern and central-western regions, by *Brachiaria decumbens*, *B. brizantha*, *B. ruziziensis* and *B. humidicola,* responsible for photosensitization. Another clinical picture caused by *B. decumbens* is progressive wasting with yellow liver and multifocal accumulation of foamy macrophages in the liver, spleen, lymph nodes, and intestinal submucosa (Riet-Correa *et al.* 2000). Brazilian samples of *B. decumbens* contain steroidal saponins, which are associated with photosensitization and deposition of crystalloid material within the biliary system (Cruz *et al.* 2000). Ultrastructurally, foamy macrophages found in the liver, lymph nodes, gut and spleen show negative images of crystals enveloped partially or totally by membranes. These images represent crystals absorbed from the food and carried to the lymphatic circulation (Driemeier *et al.* 1998). A negative correlation between weight gains and the number of foamy macrophages in the liver was reported in cattle grazing *B. decumbens*, suggesting that important economic losses occur in animals without clinical signs when grazing in these pastures (Fioravante 1999).

In the states of Rio de Janeiro and Espírito Santo, southeastern Brazil, the most important toxic plant is *Cestrum laevigatum* (Tokarnia *et al.* 2002). This plant causes liver necrosis in cattle, buffalo, and goats.

Control Measures

Research with toxic plants in Brazil has been concerned mainly with the identification of hazardous plants and definition of the clinical signs, pathology and some aspects of the epidemiology of the poisoning. Few efforts have been made to identify the toxins in these plants and to define the biochemical mechanisms of their actions.

Prevention and control of plant poisoning in Brazil have been based mainly on using knowledge of the plants and the animals to manage them. Such measures include: (1) the prevention of overgrazing, the use of livestock not susceptible to the toxic plants present, and avoidance of hungry and/or thirsty animals on pastures with toxic plants; (2) the use of fences to isolate areas with toxic plants; (3) the elimination of toxic plants by spraying, weeding, ploughing, burning or mowing; (4) the use of clean grass seeds free of contaminating weed seeds; (5) avoidance of contamination by toxic plants during hay or silage production; (6) in the case of poisoning by *Baccharis coridifolia*, the use of techniques that induce aversion (Riet-Correa *et al.* 1993); and (7) in the case of *Senecio* spp., the use of sheep for biological control (Soares *et al.* 2000).

Table 1. Plants causing sudden death[a].

Plant	Family	Toxin	Species affected	Importance[b]
Palicourea marcgravii	Rubiaceae	Fluoracetate	Cattle, buffaloes	xxxx
Palicourea aeneofusca	Rubiaceae	Unknown	Cattle	xx
Palicourea juruana	Rubiaceae	Unknown	Cattle, buffaloes	xx
Palicourea grandiflora	Rubiaceae	Unknown	Cattle	x
Arrabidaea bilabiata	Bignoniaceae	Fluoroacetate	Cattle, buffaloes	xxx
Arrabidaea japurensis	Bignoniaceae	Unknown	Cattle, buffaloes	xxxx
Pseudocalymma elegans	Bignoniaceae	Unknown	Cattle	xx
Mascagnia rigida	Malpighiaceae	Cromonas	Cattle, goats	xxxx
Mascagnia elegans	Malpighiaceae	Unknown	Cattle	x
Mascagnia pubiflora	Malpighiaceae	Cromonas	Cattle, goats, sheep	xx
Mascagnia aff. *rigida*	Malpighiaceae	Unknown	Cattle	xx
Mascagnia exotropica	Malpighiaceae	Unknown	Cattle	xx

[a] Acute death usually occurs after exercise

[b] x = few outbreaks reported; xx = the intoxication occurs sporadically in one state or region or occurs frequently but without causing death; xxx = the intoxication is an important cause of death in one state or causes moderate number of deaths in different states; xxxx = an important cause of death or severe economical losses in at least one Brazilian region (southern, southeastern, central-western, northeastern and northern).

The efficiency of such prevention and control of plant poisoning in Brazil has been limited, and the more important plants mentioned earlier still have a severe economical impact. One of the main goals of future research projects on toxic plants in Brazil should be to develop new control techniques. To control some of these poisonings it is necessary to develop research programs regarding the different aspects of the plant poisoning. First of all, it would be very helpful to identify the active principles of many Brazilian toxic plants and to define their biochemical mechanisms of action. As mentioned by Molyneux *et al.* (1994), the isolation and characterization of toxic natural products is the first step in preventing livestock losses due to poisonous plants. Actually the toxic principle is unknown in at least 51 species of the 113 toxic plants reported in Brazil. This knowledge would help in the treatment of poisoned animals and could help to develop more efficient techniques for the control of poisonous plants, including vaccination, biologic control, microbial detoxification in the rumen, conditioned food aversion, toxic binding agents, and use of plant breeding and selection programs to obtain non-toxic or less toxic forages or crops.

Table 2. Plants causing liver necrosis.

Plant	Family	Toxin	Species affected	Importance
Cestrum parqui	Solanaceae	Carboxyatractyloside	Cattle	x
Cestrum intermedium	Solanaceae	Unknown	Cattle	xx
Cestrum laevigatum	Solanaceae	Unknown	Cattle, goats, buffaloes	xxxx
Cestrum corymbosum	Solanaceae	Unknown	Cattle	xx
Xanthium cavanillesii	Asteraceae	Carboxyatractyloside	Cattle	x
Vernonia mollissima	Asteraceae	Unknown	Cattle	Xx
Vernonia rubricaulis	Asteraceae	Unknown	Cattle	Xx
Sessea brasiliensis	Solanaceae	Unknown	Cattle	xx
Dodonea viscose	Sapindaceae	Unknown	Cattle	x
Trema micrantha	Ulmaceae	Unknown	Goats	x

Table 3. Plants causing hepatic fibrosis.

Plant	Family	Toxin	Species affected	Importance
Crotalaria retusa	Leguminosae Papilionoideae	Pyrrolizidine alkaloids	Horses, cattle, sheep	xxx (horses); x (other species)
Crotalaria juncea		Pyrrolizidine alkaloids	Horses	x
Senecio. brasiliensis *Senecio selloi* *Senecio cisplatinus,* *Senecio heterotrichius* *Senecio leptolobus,* *Senecio oxyphyllus* *Senecio tweediei*	Asteraceae	Pyrrolizidine alkaloids	Cattle, sheep, horses	xxxx
Echium plantagineum	Boraginaceae	Pyrrolizidine alkaloids	Cattle	x
Tephrosia cinerea	Leguminosae Papilionoideae	Unknown	Sheep	x

Table 4. Plants causing secondary photosensitization.

Plant	Family	Toxic principle	Species affected	Importance
Brachiaria decumbens *B. humidicola* *B. brizantha* *B. ruziziensis*	Poaceae	Litogenic saponins	Cattle, sheep, goats, buffaloes, horses	xxxx
Lantana camara, *Lantana tiliaefolia* *Lantana glutinosa,*	Verbenaceae	Lantadene A Lantadene B		x
Myoporum laetum	Myoporaceae	Furanoid sesquiterpenes	Sheep, cattle	x

Table 5. Plants causing digestive signs.

Plant	Family	Toxic principle	Species affected	Importance
Baccharis coridifolia	Asteraceae	Macrocyclic trichothecenes	Cattle, sheep, horses	xx
Baccharis megapotamica	Asteraceae	Macrocyclic trichothecenes	Cattle	x
Baccharidastrum triplinervium	Asteraceae	Unknown	Cattle	x
Nierembergia hippomanica	Solanaceae	Unknown	Cattle	x
Phytolacca decandra	Phytolaccaceae	Unknown	Sheep	x
Sisyrinchium platense	Iridaceae	Unknown	Sheep, cattle	x
Plumbago scandens	Plumbaginaceae	Unknown	Goats, cattle	x
Trifolium spp (bloat)	Leguminosae Papilionaceae		Cattle	xx
Centratherum brachylepis	Asteraceae	Unknown	Cattle, goats, sheep	x
Stryphnodendron obovatum	Leguminosae Mimosoideae	Triterpenoid sapogenins	Cattle	xxx
Ricinus communis (seeds)	Euphorbiaceae	Toxoalbumins	Cattle	x
Eupatorium tremulum	Asteraceae	Unknown	Cattle	x

Table 6. Plants causing nervous signs.

Plant	Family	Toxic principle	Species affected	Importance
Solanum fastigiatum var. *fastigiatum*	Solanaceae	Unknown	Cattle	xx
Ipomoea carnea subsp. *fistulosa*	Convolvulaceae	Swainsonine	Goats, sheep, cattle	xx
Ipomoea sericophylla	Convolvulaceae	Swainsonine	Goats	x
Ipomoea riedelii	Convolvulaceae	Swainsonine	Goats	x
Sida carpinifolia	Malvaceae	Swainsonine	Goats, sheep, horses, cattle	xx
Ipomoea asarifolia	Convolvulaceae	Unknown	Sheep, goats, cattle	xxx
Phalaris angusta	Gramineae	Unknown	Cattle	x
Paspalum dilatatum *P. notatum* *P. vaginatum*	Gramineae	Penitren A (*Claviceps paspali*)	Cattle	xx
Halimium brasiliense	Cistaceae	Unknown	Sheep	x
Equisetum spp.	Equisetaceae	Thiaminase	Horses	x
Prosopis juliflora	Leguminosae Mimosoideae	Unknown	Cattle, goats	xx
Marsdenia spp (two species)	Asclepiadaceae	Unknown	Cattle, sheep	x
Ricinus communis (leaves)	Euphorbiaceae	Alkaloid (ricin)	Cattle	x
Erythroxylum argentinum *E. deciduum*	Erythroxylaceae	Unknown	Sheep	x
Aeschynomene indica	Leguminosae Papilionoideae	Unknown	Pigs	x

The possibility to prevent fluoroacetate poisoning in cattle by use of ruminal bacteria genetically manipulated by the transference of a gene encoding fluoroacetate dehalogenase has been investigated. Recently in Australia the inoculation into the rumen of strains of the bacteria *Butyrivibrio fibrisolvens* with a gene encoding fluoroacetate dehalogenase from *Moraxella* species reduced the clinical signs of fluoroacetate poisoning by degrading this compound (Gregg *et al.* 1994). In Brazil at least 12 plant species cause 'sudden death' but only *Palicourea marcgravii* and *Arrabidaea bilabiata* are known to contain a fluorinated organic compound. The use of microbial detoxification for the prevention of 'sudden deaths' caused by plants containing fluorinated organic compounds should be one of the main objectives for future research on Brazilian toxic plants. Another aim is the necessity to control *Brachiaria* spp. poisoning by the use of pastures with low saponin content. The knowledge of variation of the saponin content of the plants, and the factors determining such content and the occurrence of the poisoning, will be important to establish management measures to avoid the poisoning. The probable correlation between the saponin content and the toxicity will be a suitable way to develop less toxic varieties of *Brachiaria*.

Table 7. Plants causing nephrosis.

Plant	Family	Toxic principle	Species affected	Importance
Amaranthus hybridus *A. quitensis* *A. blitum* *A. spinosus,* *A. retroflexus* *A. viridis*	Amaranthaceae	Unknown	Cattle, pigs	x
Thiloa glaucocarpa	Combretaceae	Tannins	Cattle	xxxx
Setaria anceps var. Kazungula	Poaceae	Oxalates	Cattle	x

Table 8. Cyanogenic plants.

Plant	Family	Toxic principle	Species affected	Importance
Manihot esculenta	Euphorbiaceae	Fasculonatin Lotaustralin	Cattle, sheep, goats	xxx
Manihot glaziovii, *M. piauhyensis,* *Manihot* spp	Euphorbiaceae	Unknown cyanogenic compounds	Cattle, goats	xxx
Piptadenia macrocarpa, *P. viridiflora*	Leguminosae Mimosoideae	Unknown cyanogenic compounds	Cattle	x
Prunus sphaerocarpa	Rosaceae	Amygdalin	Cattle	xx
Cynodon dactylon	Poaceae	Unknown cyanogenic compounds	Cattle	x
Sorghum spp	Poaceae	Cyanide glycosides	Cattle	x

Table 9. Plants affecting other systems.

	Family	Toxic principle	Species affected	Importance
Calcinogenic plants				
Solanum malacoxylon	Solanaceae	Calcitriol	Cattle	xxx
Nierembergia veitchi	Solanaceae	Calcitriol-like	Sheep	xxx
Plant causing hemolytic anemia				
Brachiaria radicans	Gramineae	Unknown	Cattle	xx
Ditaxis desertorum	Euphorbiaceae	Unknown	Cattle	x
Indigofera suffruticosa	Leguminosae Papilionoideae	Unknown	Cattle	xx
Plants causing segmentar muscular necrosis				
Senna occidentalis	Leguminosae Caesalpinoideae	Unknown	Cattlle, horses, pigs	x
Plants causing primary photosensitization				
Ammi majus	Umbelliferae	Furocoumarins	Cattle	x
Mutagenic plants				
Pteridium aquilinum	Polypodiaceae	Ptaquiloside, quercetin	Cattle	xxxx
Plants causing abortion, cardiac fibrosis and *status spongiosus* of the nervous system				
Ateleia glazioviana	Leguminosae Papilionoideae	Unknown	Cattle, sheep	xxxx
Tetrapterys acutifolia *T. multiglandulosa*	Malpighiaceae	Unknown	Cattle	xxx
Plants causing abortion				
Aspidosperma pyrifolium	Apocynaceae	Unknown	Goats	xx
Estrogenic plants				
Trifolium subterraneum	Leguminosae Papilionoideae	Isoflavones	Cattle	x
Teratogenic plants				
Mimosa tenuiflora	Leguminosae Mimosoideae	Unknown	Goats, sheep, cattle	xxxx
Plants causing photosensitivity, digestive signs and abortion				
Enterolobium contortisiliquum *E. gummiferum* *E. timbouva*	Leguminosae Mimosoideae	Saponins	Cattle	xx
Stryphnodendron obovatum	Leguminosae Mimosoideae	Unknown	Cattle	xx
Plants causing nitrite poisoning				
Echinochloa polystachya	Poaceae	Nitrates and nitrites	Cattle	x
Pennisetum purpureum	Poaceae	Nitrates and nitrites	Cattle	x
Plants causing lymphoid tissue necrosis				
Riedeliella graciliflora	Leguminosae Papilionoideae	Unknown	Cattle	x
Polygala klotzschii	Polygalaceae	5-methoxy-pophyllotoxin	Cattle	x
Plants affecting the skin and adnexa				
Leucaena leucocephala	Leguminosae Mimosoideae	Mimosine	Sheep, goats	x

Table 9 continued

Plants causing granulomatous inflammation				
Vicia villosa	Leguminosae Papilionoideae	Unknown	Cattle	xx
Plants causing nervous and digestive signs and muscular lesions				
Melia azedarach	Meliaceae	4-tetranortriter-penes	Pigs	x

Natural food aversion seems to be an important control mechanism because ruminants and horses raised in areas where *Baccharis coridifolia* occurs do not ingest the plant. *Brachiara coridifolia* is a toxic plant from Uruguay, Argentina and southern Brazil that causes poisoning in cattle, sheep and horses transferred from areas free of the plant to areas where the plant exists. Animals raised in areas where the plant occurs do not ingest it, except for lambs which are occasionally affected in their first month of age. Food aversion, induced by different ways, has been used by farmers to prevent poisoning by *B. coridifolia* in animals that are transferred from areas free of the plant (Barros 1998).

Another important issue to be considered in the control of plant poisoning in South America is the development of good reporting systems to improve the information on diseases affecting livestock. We also must acquire better information regarding the phenology of the more important toxic plants; this knowledge will be necessary for correct management practices to prevent poisoning. In a recent paper the study of the phenology of *Senecio* spp. in southern Brazil provided important information for the control of seneciosis in that region (Karam *et al.* 2002).

Public Health Implications

An important aspect of the ingestion of toxic plants by animals is that toxins can be transferred to humans by way of the consumption of milk, meat, eggs, or other animal products. There are few descriptions of poisoning in humans by the consumption of animal products, related to toxic plants. It is well known that the consumption of milk from cows grazing on *Eupatorium rugosum*-infested pastures causes a disease known as 'milk sickness' that can produce death in humans (James *et al.* 1994). Other milk transfer plant toxins are pyrrolizidine alkaloids. No human disease has been reported from the consumption of these alkaloids from milk, but we must consider the risk of their mutagenic and teratogenic effects. In Brazil, milk from goats fed with *Crotalaria spectabilis* was toxic to rats (Medeiros *et al.* 1999), and litters of rats fed *C. spectabilis* or monocrotaline were also poisoned via milk (Medeiros and Górniak 1995; Medeiros *et al.* 1998). Ptaquiloside, the carcinogenic toxic principle of *Pteridium aquilinum*, one of the most important toxic plants in Brazil, passes through the milk and can cause lesions in lactating rats, mice and calves (James *et al.* 1994). Swainsonine, a toxin present in *Sida carpinifolia* (Colodel *et al.* 2002), *Ipomoea carnea* (Haraguchi *et al.* 2003), *Ipomoea riedelli* and *Ipomoea sericophyla* (Riet-Correa *et al.*, Chapter 74, this volume) from Brazil, is excreted in milk and can cause lesions in suckling calves and lambs (James *et al.* 1994). The observation of photosensitization in calves younger than 30 days old from cows grazing *Brachiaria decumbens* strongly suggests that the steroidal saponins contained in this plant are also excreted in toxic amounts in the milk (Lemos *et al.* 1998). In developed countries, the risk of human poisoning by natural toxins in milk is very remote, because blending of milk associated with processing and marketing dilutes any toxin that may enter milk products (James *et al.* 1994). However, in Brazil, where many people still drink unprocessed milk from small farms, the risk of poisoning is still present. In this aspect, goats, common in several Brazilian regions, would be a more likely source of contaminated milk than dairy cattle (Cheeke 1998). Aflatoxin can also pass into milk from cows fed contaminated peanut meal.

An example of the presence of residues of plant toxins in meat is a case of secondary poisoning reported in dogs fed with meat from horses poisoned by *Indigofora linnaei* containing the toxic amino acid indospecine (Hegarty *et al.* 1988). In Australia, dogs were apparently poisoned by the consumption of bone marrow from pigeons that ingested seeds of Fluoroacetate–containing plants. In that country, some native animals in contact with plants containing high levels of fluoroacetate acquired a high degree of tolerance to this substance (Seawright 1994). In Brazil, *Palicourea marcgravii*, which contains fluoroacetate, is the most important toxic plant, and another 11 species of the genera *Palicourea*, *Mascagnia*, *Arrabidaea* and *Pseudocalymma*, which also cause 'sudden death' (Tokarnia *et al.* 1990), probably contain fluoroacetate-derived compounds. In many areas of Brazil some people eat meat from cattle that die suddenly. Another example of secondary poisoning is that caused by the bufadienolides cardiac glycosides in humans and dogs consuming meat from poisoned cattle (Kellerman *et al.* 1990).

Acknowledgements

The authors wish to thank Drs Claudio S. L. Barros from the University of Santa Maria, Aldo Gava from the University of Santa Catarina, Ricardo Lemos from the University of Mato Grosso do Sul, and Luciano Figueiredo from the Federal University of Bahia, for providing information for the preparation of the manuscript and for its revision. This paper was financially supported by the Programa de Apoio a Núcleos de Excelência (Pronex-Grant 001/04, CNPq, FAPESQ, MCT).

References

Barros CSL (1998). Livestock poisoning by *Baccharis coridifolia.* In *Toxic Plants and Other Natural Toxicants* (T Garland and AC Barr, eds), pp. 569–572. CAB International, Wallingford, Oxon, UK.

Barros RR, Teixeira FR, Oliveira FN, Rissi DR, Rech RR, and Barros CSL (2004). Poisoning in sheep from the ingestion of fruits from *Erythroxylum argentinum. Veterinary and Human Toxicology* 46:173–175.

Cheeke PR (1998). *Natural Toxicants in Feeds, Forages, and Poisonous Plants*, 2nd ed. Interstate Publishers, Danville, Illinois.

Colodel EM, Gardner D, Zlotowsky P, and Driemeier D (2002). Identification of swainsonine as a glycoside inhibitor responsible for *Sida carpinifolia* poisoning. *Veterinary and Human Toxicology* 44:177–178.

Colodel EM, Traverso SD, Seitz AL, Correa A, Oliveira FN, Driemeier D, and Gava A (2003) Spontaneous poisoning by *Dodonea viscosa* (Sapindaceae) in cattle. *Veterinary and Human Toxicology* 45:147–148.

Cruz C, Driemeier D, and Pires VS (2000). Isolation of steroidal sapogenins implicated in experimentally induced cholangiopathy of sheep grazing *Brachiaria decumbens* in Brazil. *Veterinary and Human Toxicology* 42:142–145.

Driemeier D, Barros SS, Peixoto PV, Tokarnia CH, Dobereiner J, and Brito MF (1998). Estudo histológico, histoquímico e ultra-estrutural de fígados e linfonodos de bovinos com presença de macrófagos espumosos ('foam cells'). *Pesquisa Veterinária Brasileira* 18:29–34.

Fighera RA and Barros CSL (2004). Systemic granulomatous disease in Brazilian cattle grazing pasture containing vetch (*Vicia* spp). *Veterinary and Human Toxicology* 46:62–66.

Fioravante MC (1999). Incidencia, avaliação clínica, laboratorial e anatomopatológica da intoxicação subclínica por esporidesmina em bovinos. PhD thesis. Faculdade de Medicina veterinária e Zootecnia, UNESP, Botucatu, in CD-ROM.

Gava A, Neves DS, Gava D, Saliva TM, Schild AL, and Riet-Correa F (2002). Bracken fern (*Pteridium aquilinum*) poisoning in cattle in southern Brazil. *Veterinary and Human Toxicology* 44:362–365.

Gregg K, Cooper CL, Schafer DJ, Sharpe H, Beard CE, Allen G, and Xu J (1994). Detoxification of the plant toxin fluoroacetate by a genetically modified rumen bacterium. *Bio/Technology* 12:1361–1365.

Haraguchi M, Gorniak SL, Ikeda K, Minami H, Kato A, Watson AA, Nash R, Molyneux R, and Asano N (2003). Alkaloidal components in the poisonous plant, *Ipomoea carnea* (Convolvulaceae). *Journal of Agricultural and Food Chemistry* 51:4995–5000.

Hegarty M, Kelly WR, McEwen D, Williams OJ, and Cameron R (1988). Hepatotoxicity to dogs of horse meat contaminated with indospecine. *Australian Veterinary Journal* 65:337–340.

James LF (1994). Solving poisonous plant problems by a team approach. In *Plant Associated Toxins* (SM Colegate and PR Dorling, eds), pp. 1–6. CAB International, Wallingford, Oxon, UK.

James LF, Panter KE, Molyneux RJ, Stegelmeier BL, and Wagstaff DJ (1994). Plant toxicants in milk. In *Plant Associated Toxins* (SM Colegate and PR Dorling, eds), pp. 83–88. CAB International, Wallingford, Oxon, UK.

Karam FSC, Méndez MC, Jarenkow JA, and Riet-Correa F (2002). Fenologia de quatro espécies tóxicas de *Senecio* (Asteraceae) na região Sul do Rio Grande do Sul. *Pesquisa Veterinária Brasileira* 22(1):33–39.

Kellerman TS, Coetzer JAW, and Naude TW (1990). *Plant Poisonings and Mycotoxicosis of Livestock in Southern Africa*, pp.103–105. Oxford University Press, Cape Town.

Lemos RAA, Nakazato L, and Pozo Del CF (1998). Intoxicação por *Brachiaria* sp. In *Principais Enfermidades de Bovinos de Corte do Mato Grosso do Sul*. (RAA Lemos, ed), pp.299–306. Universidade Federal do Mato Grosso do Sul, Campo Grande.

Lucioli J, Reis RN, and Gava A (2004). Aspectos clínicos e patológicos da intoxicação por *Eupatorium tremulum* (Asteraceae) em bovinos. *Pesquisa Veterinária Brasileira* 24 (suppl):38.

Medeiros J, Tabosa IM, Simões SV, Nóbrega J, Vasconcelos JS, and Riet-Correa F (2005). Mortalidade perinatal em cabritos no semi-árido da Paraíba. *Pesquisa Veterinária Brasileira* 25:201–206.

Medeiros RMT and Górniak SL (1995). Efeitos da administração de sementes de *Crotalaria spectabilis* e monocrotalina (MCT) na ração de ratas em lactação, no desenvolvimento físico de seus filhotes. *Revista da Sociedade Brasileira de Toxicologia* (Supl). Abstracts of the Congresso Brasileiro de Toxicologia, Ribeirão Preto, p.296.

Medeiros RMT, Górniak SL, and Guerra JL (1998). Comparative effects of prenatal and postnatal monocrotaline effects in rats. In *Toxic Plants and Other Natural Toxicants* (T Garland and AC Barr, eds), pp. 312–316. CAB International, Wallingford, Oxon, UK.

Medeiros RMT, Górniak SL, and Guerra JL (1999). Effects of milk from goat fed *Crotalaria spectabilis* seeds on growing rats. *Brazilian Journal of Veterinary Research and Animal Science* 36(2):1–9.

Medeiros RMT, Neto SAG, Riet-Correa F, Schild AL, and Sousa NL (2004). Mortalidade embrionária e abortos em caprinos causados por *Aspidosperma pyrifolium*. *Pesquisa Veterinária Brasileira* 24 (suppl.):42.

Mendez MC and Riet-Correa F (2001). Plantas tóxicas e micotoxicoses. In *Doenças de Ruminantes e Eqüinos*, Vol. 2 (F Riet-Correa, AL Schild, MC Méndez, and RAA Lemos, eds), pp. 219–299. Varela, São Paulo.

Molyneux RJ, James LF, Ralphs MH, Pfister JA, Panter KE, and Nash RJ (1994). Polyhydroxy alkaloids glycosidase inhibitors from poisonous plants of global distribution: analysis and identification. In *Plant Associated Toxins* (SM Colegate and PR Dorling, eds), pp. 107–112, CAB International, Wallingford, Oxon, UK.

Nóbrega J, Medeiros J, Nóbrega R, Simões SV, Tabosa IM, Vasconcelos JS, and Riet-Correa F (2005). Mortalidade perinatal em caprinos no semi-árido da Paraíba. *Pesquisa Veterinária Brasileira* 25(4):201–206.

Riet-Correa F and Medeiros RMT (2001). Intoxicações por plantas no Brasil e no Uruguai: importância econômica, controle e riscos para a saúde pública. *Pesquisa Veterinária Brasileira* 21(1):38–42.

Riet-Correa F, Mendez MC, and Schild AL (1993). *Intoxicações por Plantas e Micotoxicoses em Animais Domésticos*. Editorial Hemisfério Sur, Montevideo.

Riet-Correa G, Riet-Correa F, Schild AL, and Driemeier D (2000). Wasting and death in cattle associated with a chronic intoxication by *Brachiaria Decumbens*. *Veterinary and Human Toxicology* 44(3):179–180.

Riet-Correa F, Timm C, Barros SS, and Summers B (2003). Symmetrical focal degeneration in the cerebellar and vestibular nuclei in swine caused by ingestion of *Aeschynomene indica* seeds. *Veterinary Pathology* 40:311–316.

Seitz AL, Schmitz M, Borba MR, Raymundo DG, Driemeier D, Emmel XE, and Colodel EM (2004). Intoxicação espontânea pelos frutos de *Erythroxylum deciduum* (Erythroxylaceae) em ovinos no município de Lagoa Vermelha, RS. *Pesquisa Veterinária Brasileira* 24 (suppl.):68.

Seawright AA (1994). Toxic plants residues in milk. In *Plant Associated Toxins* (SM Colegate and PR Dorling, eds), pp. 77–82. CAB International, Wallingford, Oxon, UK.

Soares MP, Riet-Correa F, Méndez MC, Rosa FG, and Carreira EG (2000). Controle biológico de *Senecio* spp. com pastoreio de ovinos. Abstracts of the II Reunión Argentina de Patologia Veterinaria, Faculdad de Ciencias Veterinarias, Universidad Nacional del Nordeste, Corrientes, Argentina, p.79.

Timm C and Riet-Correa F (1997). Plantas tóxicas para suínos. *Ciencia Rural* 27:521–528.

Tokarnia CH, Döbereiner J, and Silva MF (1979). *Plantas Tóxicas da Amazônia a Bovinos e outros Herbíivoros*. INPA, Manaus.

Tokarnia CH, Peixoto PV, and Dobereiner J (1990). Poisonous plants affecting heart functions of cattle in Brazil. *Pesquisa Veterinária Brasileira* 10:1–10.

Tokarnia CH, Döbereiner J, and Peixoto PV (2000). *Plantas Tóxicas do Brasil.* Editora Helianthus, Rio de Janeiro.

Tokarnia CH, Döbereiner J, and Peixoto PV (2002). Poisonous plants affecting livestock in Brazil. *Toxicon* 40:1635–1660.

Traverso SD, Correa AM, Pescador CA, Colodel ED, Cruz CEF, and Driemeier D (2003). Intoxicação natural por *Trema micrantha. Ciência Rural* 33:133–136.

Chapter 2

Major Poisonous Plants of the Western United States

A. P. Knight

Department of Clinical Sciences, College of Veterinary Medicine and Biomedical Sciences, Colorado State University, Fort Collins, CO 80523, USA

Introduction

In 1898 Chesnut named 28 plants and 2 mushrooms in North America that were poisonous to man and animals (Table 1). One hundred years later these same plant species and many additional toxic plants continue to be a problem in the western United States, causing significant economic losses to the livestock industry. The economic impact of poisonous plants in 17 western states in 1978 was estimated to be in excess of $107 million annually (Nielsen 1978). Despite significant advances in the understanding of plant toxicology, the economic impact of plant poisoning remains substantial, and continues to grow for a variety of reasons. The desire to own land and live in rural environments has meant an ever increasing influx of people into western rangelands. In many areas this has resulted in increased livestock grazing pressure and therefore increased consumption of toxic plant species. Many of the new small acreage owners are unfamiliar with range management and the role of toxic plants and consequently have to learn from costly mistakes.

A good example of this is the recurrence of congenital cyclopia as a result of new small acreage owners allowing sheep to graze western false hellebore (*Veratrum californicum*), a plant that ranchers and early researchers recognized as the cause of the problem, and avoided grazing sheep on the plants in the first month of pregnancy (Binns *et al.* 1963).

Significant Poisonous Plants

Kingsbury (1964) listed over 1000 toxic plants, and more recently Burrows and Tyrl (2001) have greatly expanded the list of toxic plants in a comprehensive review of toxic plants in North America. Although there is wide diversity in toxic plants and the toxins they contain, the economic impact on the livestock industry is attributed to relatively few plants. Schuster (1978) compiled a list of 30 toxic plant genera of the southwest and western states that had significant impact on livestock production, and this list can be shortened further to those indigenous plants that have major impact on livestock because of their wide geographic distribution (Table 2).

Table 1. Thirty poisonous plants of 1898 (Chesnut 1898).

Common Name	Botanical Name
Fly amanita	*Amanita muscaria*
Death cap	*Amanita phalloides*
American false hellebore	*Veratrum viride*
Poke weed	*Phytolacca decandra*
Corn cockle	*Agrostemma githago*
Dwarf larkspur	*Delphinium tricorne*
Wyoming larkspur	*Delphinium geyeri*
Purple larkspur	*Delphinium menziesii*
Black cherry	*Prunus serotina*
Wooly locoweed	*Astragalus mollissimus*
Stemless locoweed	*Astragalus lambertii*
Rattlebox	*Crotolaria sagittalis*
Caper spurge	*Euphorbia lathyris*
Snow on the mountain	*Euphorbia marginata*
Poison ivy	*Rhus radicans*
Poison oak	*Rhus diversiloba*
Poison sumac	*Rhus vernix*
Red buckeye	*Aesculus pavia*
Water hemlock	*Cicuta maculata*
Oregon water hemlock	*Cicuta vegans*
Poison hemlock	*Conium maculatum*
Broad-leaf laurel	*Kalmia latifolia*
Narrow-leaf laurel	*Kalmia angustifolia*
Great laurel	*Rhododendron maximum*
Stagger bush	*Pieris mariana*
Branch ivy	*Leucothoe catesbei*
Jimson weed	*Datura stramonium*
Black nightshade	*Solanum nigrum*
Bittersweet	*Solanum dulcamara*
Sneeze weed	*Helenium autumnale*

Table 2. Toxic plants (indigenous species) having the greatest economic impact on livestock in the western United States.

Common name	Genus
Death camas	*Zigadenus*
Larkspur	*Delphinium*
Locoweeds	*Astragalus & Oxytropis*
Lupine	*Lupinus*
Milkweed	*Asclepias*
Oak	*Quercus*
Ponderosa pine	*Pinus*
Sneezeweed	*Dugaldia*
Water hemlock	*Cicuta*
Choke cherry	*Prunus*
Selenium obligate accumulator plants	*Astragalus, Stanleya, Xylorrhiza*

Members of the *Astragalus* and *Oxytropis* genera (locoweeds) have long been considered the most economically important poisonous plants in North America (Marsh 1919; James and Nielsen 1988; Ralphs and James 1999). However, climatic conditions can have major impact on locoweed populations to the extent that the locoweeds completely disappear from large areas in times of sustained drought (Ralphs *et al.* 2003). The ability for locoweed seeds to retain viability in the soil for at least 50 years ensures that the locoweeds will return when conditions are suitable for growth (Ralphs and Cronin1987).

Other plants that have had and continue to have significant effects on livestock production in the western United States include *Delphinium* and *Lupinus* species. Both of these plant species are well adapted to growing in a vast geographic area where they cause significant death loss and reproductive failure in cattle in particular (Panter and Keeler 1993; Panter *et al.* 1998; Pfister *et al.* 1999). In some years there are significant reproductive losses in cattle that browse on pine needles (*Pinus ponderosa*) especially in the late winter when other forages may be scarce (Panter *et al.* 1992).

Sporadically livestock losses can be significant in areas where *Zigadenus* spp. (death camas; Panter *et al.* 1987), *Asclepias* spp. (milkweeds), *Quercus* spp. (oak), *Dugaldia/Helenium* spp. (sneeze weed/bitterweeds), and *Cicuta* spp. (water hemlock; Panter *et al.* 1988) are prevalent.

Plant nitrate poisoning is a perennial problem associated with livestock consuming weeds or crop plants. The drought conditions prevalent in the western States from 1998 to 2004 has contributed to increased nitrate poisoning especially where producers graze or feed their cattle on abandoned croplands.

Selenium Accumulators

Selenium poisoning continues to be a detriment to horses and other livestock raised in many parts of the Rocky Mountain states rich in seleniferous soils. Many plants in these areas accumulate toxic levels of selenium that result in chronic selenium poisoning characterized by defective keratin formation, resulting in hair loss, cracked hooves, lameness and loss of weight (Olson 1978). In drought conditions the uptake of selenium by plants increases and predisposes animals to chronic selenium toxicity.

Noxious Weeds

Unfortunately, introduced noxious weeds are adding to the number of toxic plants encountered by livestock (Table 3). These introduced noxious weeds not only cause economic loss through death and disease, but by their nature are invasive and displace useful forages thus increasing the cost of livestock production. Factored into the overall costs is the expense of controlling or eradicating the noxious weeds using herbicides, mowing, biological controls, etc. Weeds such as *Centaurea solstitialis* (yellow star thistle) and *Acroptilon repens* (Russian knapweed) are on the noxious weed lists in six states because of their invasiveness. *Halogeton glomeratus* has become permanently established in the dry alkaline soils of the western states and is a continual risk to sheep and cattle in these areas (Lincoln and Black 1980). Similarly *Cynoglossum officinale* (hounds tongue) has made the noxious weed lists in Colorado, Wyoming and Nevada.

Table 3. Introduced noxious plants toxic to livestock.

Common name	**Botanical Name**
Black henbane	*Hyoscyamus niger*
Jimson weed	*Datura stramonium*
St John's wort	*Hypericum perforatum*
Halogeton	*Halogeton glomeratus*
Poison hemlock	*Conium maculatum*
Hounds tongue	*Cynoglossum officinale*
Russian knapweed	*Acroptilon repens*
Yellow star thistle	*Centaurea solstitialis*
Tansy ragwort	*Senecio jacobaea*
Kochia weed	*Kochia scoparia*
Flix weed	*Descurainia sophia*

Yet another group of plants that are having a significant effect upon livestock production in the western United States are those that are not necessarily toxic but which are highly invasive. The Arizona, California, Colorado, New Mexico, Texas, Utah and Wyoming noxious weed lists are extensive, with non-native species such as leafy spurge (*Euphorbia esula*), salt cedar (*Tamarix* spp.), Canada thistle (*Cirsium arvense*), spotted knapweed (*Centaurea maculata*), and diffuse knapweed (*Centaurea diffusa*) being of primary concern because they competitively eliminate useful plant species. Plants such as salt cedar are particularly problematic as they consume considerable quantities of water and therefore deprive arid areas of precious water resources.

References

Binns W, James LF, Shupe JL, and Everett G (1963). A congenital cyclopian type malformation in lambs induced by maternal ingestion of a range plant, *Veratrum californicum*. *American Journal of Veterinary Research* 24:1164–1175.

Burrows GE and Tyrl RJ (2001) *Toxic Plants of North America*. Iowa State University Press, Ames.

Chesnut VK (1898). Thirty poisonous plants of the United States. *USDA Farmer's Bulletin* No. 86:1–32.

James LF and Nielsen DB (1988). Locoweeds: assessment of the problem on western rangelands. In *The Ecology and Economic Impact of Poisonous Plants on Livestock Production* (LF James, MH Ralphs, and DB Nielsen, eds), pp. 171–180. Westview Press, Boulder, Colorado.

Kingsbury JM (1964). *Poisonous Plants of the United States and Canada*. Prentice-Hall Inc., Englewood Cliffs, New Jersey.

Lincoln SD and Black B (1980). Halogeton poisoning in range cattle. *Journal of the American Veterinary Medical Association* 176(8):717–718.

Marsh CD (1919; revised 1936). The locoweed disease. *USDA Farmer's Bulletin* 1054.

Nielsen DB (1978). The economic impact of poisonous plants on the range livestock industry in 17 western states. *Journal Range Management* 31:325–328.

Olson OE (1978). Selenium in plants as a cause of livestock poisoning. In *Effects of Poisonous Plants on Livestock* (RF Keeler, KR Van Kampen, and LF James, eds), pp. 121–133. Academic Press, New York.

Panter KE, Ralphs MH, Smart RA, and Duelke B (1987). Death camas poisoning in sheep: a case report. *Veterinary and Human Toxicology* 29(1):45–48.

Panter KE, Keeler RF, and Baker DC (1988). Toxicoses in livestock from the hemlocks (*Conium* and *Cicuta* spp.). *Journal of Animal Science* 66(9):2407–2413.

Panter KE, James LF, and Molyneux RJ (1992). Ponderosa pine needle-induced parturition in cattle. *Journal of Animal Science* 70(5):1604–1608.

Panter KE and Keeler RF (1993). Quinolizidine and piperidine alkaloid teratogens from poisonous plants and their mechanism of action in animals. *Veterinary Clinics of North America – Food Animal Practice* 9(1):33–40.

Panter KE, Gardner DR, and Molyneux RJ (1998). Teratogenic and fetotoxic effects of two piperidine alkaloid–containing lupines (*L. formosus* and *L. arbustus*) in cows. *Journal of Natural Toxins* 7(2):131–140.

Pfister JA, Gardner DR, Panter KE, Manners GD, Ralphs MH, Stegelmeier BL, and Schoch TKl (1999). Larkspur (*Delphinium* spp.) poisoning in livestock. *Journal of Natural Toxins* 8:81–94.

Ralphs MH and Cronin EH (1987). Locoweed seed in soil: density, longevity, germination and viability. *Weed Science* 35:792–795.

Ralphs MH and James LF (1999). Locoweed grazing. *Journal of Natural Toxins* 8(1):47–51.

Ralphs MH, Pfister JA, Welsh SL, Graham JD, Purvines J, Jensen DT, and James LF (2003). Locoweed population cycles. *Rangelands* 25:14–18.

Schuster JL (1978). Poisonous plant management problems and control measures on U.S. rangelands. In *Effects of Poisonous Plants on Livestock* (RF Keeler, KR Van Kampen, and LF James, eds), pp. 23–34. Academic Press, New York.

Chapter 3

Locoweeds of North America: Taxonomy and Toxicity

S.L. Welsh[1], M.H. Ralphs[2], K.E. Panter[2], J.A. Pfister[2], and L.F. James[2]

[1]*Stanley L. Welsh Herbarium, Brigham Young University, Provo, UT 84601 USA;* [2]*USDA-ARS Poisonous Plant Research Laboratory, Logan UT 84341, USA*

Introduction

Marsh, Clawson, and Eggleston (1930) commented that 'A large number of plants have been called 'locoweeds.'' Previously Marsh (1929) had stated that, 'Without any doubt the most destructive of all poisonous plants are those going under the general name of loco.' And further, 'A great deal of interest attaches to these plants because of their wide distribution and the large number of animals that have been poisoned by them, including cattle, horses, goats, and sheep, also because of the difficulty of actually proving the existence of a poisonous principle in the plants themselves.' Even in the late 20th and early 21st centuries, work continues on the 'loco' plants. In contemporary works the species of plants involved are considered as belonging to two allied genera, *Astragalus* and *Oxytropis*. The first of those genera, *Astragalus* is very large, with some 354 species, 198 varieties, and 552 total taxa in North America north of Mexico (Welsh 2005). Should the 47 species and 9 varieties from Mexico be added, the totals would be 401 species and 207 varieties, or a total of 608 taxa, an addition of 54 taxa since the 1964 publication of Barneby. Species are present in all of the contiguous United States and Alaska, and in all of the Canadian provinces. *Oxytropis* is much smaller, with some 57 taxa in only 22 species, including one introduced from the Old World (Welsh 2001). Species of *Oxytropis* are confined mainly to western North America, with a few of the taxa extending in disjunct across North America to eastern Canada.

Only a few of the great number of species involved grow in sufficient numbers as to constitute important components of a given flora, and among them several have been identified as toxic to livestock. In Alaska and northwestern Canada the boreal oxytrope, *Oxytropis borealis* DC., and the variable field oxytrope, *O. campestris* (L.) DC. var. *varians* (Rydberg) Barneby, grow in large populations on gravel bars (Welsh 1967, 1974) and sometimes along gravelly roadsides, where they are grazed by bears with evidently little if any indication of poisoning. And many of the species evidently lack poisonous compounds, or if they contain such substances, they are not those associated with the syndromes of so-called 'locoism.' Indeed, some are considered of value as grazing plants, especially some of the small number of species introduced from the Old World.

Marsh (1929) and associates (Marsh *et al.* 1930) had worked to understand incidents of poisoning in the Great Plains; specifically they dealt with the species of *Astragalus* and *Oxytropis* on the high plains of the United States, especially with *A. mollissimus*, the woolly loco, and *Oxytropis sericea* Nuttall ex Torrey & Gray and *O. lambertii* Pursh. Two species, *i.e.*, *O. lambertii* and *O. sericea*, were cited by James *et al.* (1989) as poisonous due to the presence of the indolizidine alkaloid, swainsonine. The name 'swainsonine' is derived from the Australian genus *Swainsonia*, which has been noted to result in a similar series of toxic syndromes in that country. There is, however, more than a single syndrome of pathological effects exhibited by livestock. The symptomology of swainsonine poisoning was sometimes obscured or misdiagnosed due to the presence of nitro-compounds or selenium, or perhaps both present along with swainsonine.

Livestock poisoning in the western US has resulted in tremendous economic losses due to *O. lambertii* and *O. sericea*. Despite a century of research the problem of poisoning persists, and will continue into the future as long as domestic livestock graze native ranges on which plants of those two genera grow abundantly.

Modern listing of *Astragalus* and *Oxytropis* species known to contain the physiologically disruptive chemical, swainsonine, are those by James *et al.* (1991), Molyneux *et al.* (1991), Smith *et al.* (1992), Fox *et al.* (1998), and Ralphs *et al.* (2002), which includes, in addition to the species investigated by Marsh and his associates, several others that have been investigated subsequently. The list of species known to contain the indolizidine alkaloid swainsonine is very small compared to the total number of species in North America.

Astragalus Species Reported as Toxic

Some 20 species of *Astragalus* have been noted as poisonous due to the presence of the alkaloid swainsonine in their tissues (Ralphs *et al.* 2002). These are: *Astragalus allochrous*, *A. asymmetricus*, *A. bisulcatus*, *A. didymocarpus*, *A. drummondii*, *A. emoryanus*, *A. humistratus*, *A. lentiginosus*; *A. lonchocarpus*, *A. missouriensis*, *A. mollissimus*, *A. nothoxys*, *A. oxyphysus*, *A. praelongus*, *A. pubentissimus*, *A. purshii*, *A. pycnostachyus*, *A. tephrodes*, *A. thurberi* and *A. wootonii*. Undoubtedly there are many more species that contain swainsonine, but they have not been called to the attention of investigators due to lack of reports of poisoning. The species reported by Ralphs *et al.* (2002) occur widely in the American West from the high plains to California, and in Canada as well. Investigations of the toxic nature of the plants and of their physiologically active ingredients have been based in large part on actual poisoning episodes. Such episodes occur generally where the plant species is an integral part of a local flora in large numbers, numbers that allow consumption of large quantities of plant material of a given entity.

Other toxic substances in various species of the genus, sometimes in association with swainsonine, include the element selenium and various nitro-containing substances, sometimes in combination (accounting for duplication of some species names in the two lists). Among those known to accumulate selenium are *Astragalus albulus*, *A. beathii*, *A. bisulcatus* (three varieties), *A. crotalariae*, *A.cutleri*, *A. debequaeus*, *A. eastwoodiae*, *A. flavus* (three varieties), *A. grayii*, *A. moencoppenis*, *A. nelsonianus*, *A. oocalycis*, *A. pattersoni*, *A. pectinatus*, *A. saurinus*, *A. sophoroides*, and *A. toanus*. Possible selenium accumulators, those that occur solely on seleniferous substrates but do not smell strongly of that element, include *A. ampullarius*, *A. ampullarioides*, *A. asclepiadoides*, and *A. woodruffii*.

Some of these species have caused severe losses in livestock. Syndromes involved an acute selenium poisoning, a chronic selenium poisoning known as alkali disease, and another chronic selenium poisoning known as blind staggers (James *et al.* 1991).

Many species of *Astragalus* also contain nitro-compounds, perhaps as many as half of them (Williams and Barneby 1977), and are, hence, potentially poisonous to livestock. A very small number of those so noted are the following: *A. emoryanus*, *A. tetrapterus* (Marsh 1929), *A. pterocarpus, A. atropubescens, A. canadensis*, *A. falcatus*, *A. miser,* and *A. whitneyi*. All of those listed here are native to the southwestern or western states except for *A. falcatus*, which was introduced from the Old World for use in reclamation and later demonstrated to be poisonous. Poisoning by members of the genus *Astragalus* is not restricted to those from North America alone.

Astragalus is a huge genus, one of the largest in the world, with an estimated 1200 to 2000 species with circumboreal distribution, and with extensions to South America, equatorial Africa, and subtropical India. Major centers of occurrence include central Asia, Iran, and Turkey in the Old World, and western North America in the New World. There are about 390 in North America, some 355 north of Mexico (including a few introduced Old World species), and about 100 in South America. The genus is complex because of its size, the intricacy of its taxa, the morphological diversity, chromosome number, and edaphic specialization.

Astragalus L.

Interpretations of *Astragalus* have been in progress since the time of Linnaeus' (1753) *Species Plantarum*, which included descriptions of two American species, *A. canadensis* and *A. carolinianus*. Additional species were added by Michaux (1803) in his *Flora Boreali-Americana*. Still others became known following Pursh's *Flora Americae-Septentrionalis* (1814) based in part on collections by Lewis and Clark and Bradbury, and in part by Thomas Nuttall. Perhaps the most important early work on American *Astragalus* and the related genus *Phaca* was that by Hooker (1831) in *Flora Boreali-Americana* (1829–1834). Hooker listed some 29 species, with 18 of them being described for the first time, based mainly on collections taken in western Canada by John Richardson, David Douglas, and Thomas Drummond. *Phaca* was considered to include all species with unilocular pods, while in *Astragalus* the pods were bilocular. This distinction, thought at the time to be of fundamental importance, broke down upon examination of pods of *Astragalus* species then being discovered in North America, and *Phaca* was laid to rest by Asa Gray in his revision of the genus (Gray 1864).

An important historical work on American plants is Torrey and Gray=s (1838) *Flora of North America*, which not only included *Astragalus* and *Phaca*, but also treated two related new genera proposed by Thomas Nuttall, i.e. *Homalobus* and *Kentrophyta*. Total number of species in the *Flora* was 83, with more than half of them described for the first time, 34 of them from collections by Nuttall on his 1834 journey across the country with the Wyeth expedition, and later on his explorations along the coast of California.

From 1838 onward the province of North American species of *Astragalus* was dominated by Americans, not by the Europeans as it had been previously. The next giant stride in understanding North American species was presented by Gray (1864) in his *Revision of the North American Species of Astragalus and Oxytropis*. That work formed the basis upon which subsequent North American treatments in both *Astragalus* and *Oxytropis* are based in small to large part. Gray=s revision included some 111 species of *Astragalus* placed in 27 sections. The sections were arranged in order beginning with the fully bilocular species through the subbilocular ones into the truly unilocular groups. His employment of fruit as the main diagnostic criterion would be followed by subsequent workers through most the following century. Those workers subsequent to Gray placed emphasis on the pod and its texture, compression, inflation, and intrusion of the dorsal suture as a septum. In Eurasian treatments, emphasis during the century following Gray was on the leaf, stipule, flower, trichome, and duration (whether annual or perennial). European workers followed the two-parted *Species Gerontogeae* of Alexander von Bunge (1868, 1869), a much regarded monograph of the *Astragalus* species of the Old World.

Work on the genus passed from Asa Gray to his able colleague Sereno Watson, who had served as botanist on King=s Exploration of the Fortieth Parallel (Watson 1871). On that survey he spent two seasons collecting in the American West, mainly in Nevada and Utah, and had become familiar with many of the species of *Astragalus* as they grew in the field, and he subsequently named several of them as new. His summary of the genus (Appendix. Synopses of Genera. *Astragalus*. 1871; 435–446) in the Botany of King's exploration comprised some 128 species, placed within the 27 sections noted previously by Gray, but with discussion of six obscure additional species.

Sheldon (1894), working from the University of Minnesota and far removed from the great geographical mass of the genus, made an attempt at the genus *Astragalus*, which is perhaps best known as being inept. His list includes some 313 species, some of which were described by him and are genuine taxa still recognized into the 21st century. Perhaps the best that can be said of Sheldon=s list is its stimulus to another 19th century worker whose devotion to the genus lapped into the first three decades of the 20th as well, Marcus E. Jones. Jones (1923), age 41 in 1893, indicated that because of Sheldon=s work his mind was made up to undertake a revision of the North American species. By then Jones had some 14 years of experience with the genus as he had encountered it in his travels through *Astragalus*-rich country in the American West and Baja California.

The arrival of Marcus Eugene Jones in the American West in February 1880 presaged a lifetime of dedication to the collection and naming of plants, which ended at his death on 3 June 1934 in an automobile accident in San Bernardino, California. By the 1890s, Jones had come to know the plants better than his eastern, herbarium-bound botanical contemporaries, who were by then either gone from the scene or would be soon. In 1894, under a financial arrangement with the Smithsonian Institution, Jones set out to survey the region around the Grand Canyon for its botanical novelties. Subsequently, he spent the winter of 1894–95 at the Smithsonian identifying the results of his labors from the previous collecting season, wherein he had taken a vast number of specimens. His contribution to the understanding of *Astragalus* (Jones 1923) appeared following his removal from Utah to California in the early 1920s.

Overlapping Jones' time on the stage of North American plant exploration was a young botanist, Per Axel Rydberg, who was originally from Sweden and trained as a mining engineer. Rydberg as a student of botany was situated initially at the University of Nebraska at Lincoln, where he studied with Ernest A. Bessey, who developed a system of classification for the flowering plants with considerable merit. Rydberg published a paper on the flora of the Sand Hills of Nebraska (Rydberg 1895b), and was engaged in a revision of the Rosales for the Nebraska State Flora (Rydberg 1895a). These first approaches to *Astragalus* were conservative, and perhaps better described as 'Grayian' in both specific and generic concepts. From 1895 onward he authored a long series of papers mainly in the

Bulletin of the Torrey Botanical Club, devoted in some or large part, to the genus *Astragalus*, culminating in his summary monograph of 27 narrowly proscribed genera within the inclusive *Fabaceae: Galegeae. Subtribe 10. Astragalaneae* in the *North American Flora* (Rydberg 1929); that attempt at the flora of North America, never to be completed, was centered at the New York Botanical Garden. In that work, Rydberg regarded the North American species within a large number of segregate genera (see above) and narrowly defined species often based on trivial features that resulted in an untenable system of classification and species based often on minutiae, which must be designated as 'Rydbergian.'

Arriving in the United States in the late 1930s were two Englishmen, Harry Dwight Dillon Ripley and his companion Rupert Charles Barneby (Crase 2004), who established residence in southern California in the late 1930s, and at once began to travel through the western states collecting plants. *Astragalus* had come to the attention of Barneby in the early 1930s when the duo had been vacationing in Spain (Barneby, personal communication). There Barneby had encountered a half-dozen or so species of this vast genus. The plants were individually attractive and diverse morphologically. Upon his return to London, Barneby visited the great herbarium at Kew and had an opportunity to view the fantastic *Index Kewensis*. He turned to the bibliographic citations for the species of *Astragalus,* and was amazed at the pages of names, not only in the first pair of volumes, which was published in 1895, but in the supplements as well. In one of the latter he encountered the name *Astragalus asclepiadoides*, and wondered if there could be a member of this genus that looked like a milkweed. He viewed that splendid plant on a spring day in 1942 in the Uinta Basin or in the San Rafael Swell of Utah and perhaps in both places. There followed a great many days spent in the field searching for elusive members of this genus. Barneby developed an uncanny ability to find local endemics and to present the results in a clear and scholarly manner over the next two decades.

The two works dealing with species of *Astragalus* by Jones (1923) and Rydberg (1929), and the summary treatments of Gray (1864) and Watson (1871) were to be the ashes upon which would arise the phoenix of Barneby's own treatment of *Astragalus*, and his unique system of classification. Both Jones and Rydberg were gone from the scene shortly prior to the arrival of Ripley and Barneby; Jones died in 1934 and Rydberg in 1931.

By the late 1940s and 1950s, Barneby was thoroughly involved with a series of publications on the genus, which culminated in his own monumental opus (Barneby 1964), the amazing *Atlas of North American Astragalus*.

Oxytropis Species Reported as Toxic

Four species of *Oxytropis*, i.e., *O. besseyi* Rydberg, *O. campestris* (L.) DC., *O. lambertii* Pursh, and *O. sericea* Nuttall ex Torrey & Gray, are cited by Ralphs *et al.* (2002) as toxic or potentially so due to the presence of swainsonine. 'Of the locoweeds, *O. sericea* is probably the most destructive, not because it is more toxic but because it grows over a larger geographic region than do other locoweed plants' (James *et al.* 1991). The quote by James *et al.* did not take into account the very broad distribution of several other taxa included under the general heading of locoweeds, but certainly *O. lambertii* and *O. sericea* have broad distributions. Phases of *O. sericea* extend from southeastern Yukon south to Idaho, Utah, New Mexico, and Oklahoma, where it is especially common in lands grazed by domestic livestock. *O. lambertii* is common from the prairie provinces of Canada south to Arizona, New Mexico, and Oklahoma, and east as far as Minnesota and Iowa.

Oxytropis campestris, known as field or yellow oxytrope or locoweed, is the most widely distributed species in the genus in North America. It is known from a relatively large number of variants that extend from Alaska south to Washington, Idaho, Utah and Colorado, and across Canada east to Labrador, Wisconsin, and Maine.

Red or Bessey=s loco, *O. besseyi* (Rydberg) Blankinship, occurs in Wyoming, western Montana, and barely into Utah, with a disjunction in northeastern Nevada. The species is locally common in Wyoming, where it forms, at times, a conspicuous part of the vegetation.

Oxytropis *de Candolle*

The genus *Oxytropis* in North America has long been a fascination of American botanists, having been revised in the 19th century by Asa Gray (1884), and in the 20th century by Rupert Barneby (1952), who recognized far fewer taxa than appeared to be the case upon examination of taxonomic literature. He rightly relegated to synonymy most of the names within the genus as it was known at mid-century.

The earliest-named, strictly American species is *O. lambertii* Pursh, named in 1813, is based on an 1811 collection by Bradbury from along the Missouri River near the Nebraska–South Dakota line. Beginning in the latter part of the second decade of the 19th century the British attempted to find a route through the ice-clogged passages of the Canadian Arctic Archipelago and channels adjacent to continental North America. Initial work was under the direction of Captain William Edward Parry (Parry 1821). Specimens from the Parry expeditions are deposited at both K (Kew) and BM (British Museum of Natural History), with some of the private herbaria now represented better at one, some at the other of the two herbaria (Welsh 1994).

John Franklin and John Richardson traveled overland from York Factory on Hudson Bay to Great Slave Lake, thence down the Coppermine River in an expedition lasting from 1819 to 1822 (Franklin 1823). They then worked eastward along the coast of northern Canada until too late in the summer. Their return trip back to their camp at the head of Coppermine led to one of the great tragedies in geographical and botanical exploration. Only nine of the party of 20 men survived the ordeals of hunger, cold, and exposure. Cannibalism and murder formed a part of the saga, with one of the party being shot by naturalist John Richardson (Houston 1984; Burton 1988; Savours 1999). A second Franklin and Richardson expedition explored the region both east and west of the mouth of the Mackenzie River in 1825–27 (Franklin 1828).

Plants from the Parry (Hooker 1825) and from the Franklin–Richardson expeditions were available for study by English botanist William Jackson Hooker (1785–1865). Also available to him were collections by Thomas Drummond, who was with Franklin on the trip westward along the north coast, and those of David Douglas, intrepid English botanical explorer (Houston 1984; Burton 1988; Savours 1999). The earliest revision of *Oxytropis* based on North American materials was that by William Jackson Hooker, whose concepts of the species were set forth in *Flora boreali-americana* (1831, in series from 1829 to 1840).

Welsh (1994) identified the material that formed the basis for interpretations by Richardson (1823), Brown (1823, 1824), and Hooker (1831). Treatments in *A Utah Flora* by Welsh (2003a,b) give indication of the distribution of species of *Oxytropis* in Utah.

The treatment of *Oxytropis* in the *Flora of North America* by Torrey and Gray (1838) followed Hooker's account in nearly all details, but added six new species collected and described by Thomas Nuttall (1786–1859) on his journey across the continent with the Wyeth expedition in 1834. These were the first American species described by a botanist familiar with them in the field. The great diversity among the American species must have overwhelmed Nuttall, who proposed several other species, most of which were not published in Torrey and Gray's monumental work or subsequently. The manuscript names appear on sheets of Nuttall's personal herbarium at the British Museum, and some were later cited as synonyms. The difficulty in comprehending a genus as large and complex as *Oxytropis* is understandable.

Taxonomic problems are disproportionate with the size of the genus (Welsh 1987, 1991, 1994). Plasticity within species, and indistinct lines of separation, have led to a huge number of names in this relatively small genus. Responsibility for many of the synonyms can be traced to workers at the beginning of the 20th century, who viewed any morphological distinction as worthy of naming. Involved in the genus were those botanists renowned for their recognition of plants differing by inconsequential morphological minutia as distinctive taxa.

The taxa within *Oxytropis* tend to lack definitive taxonomic characteristics and tend to intergrade with closely allied taxa, differing in that respect from most species of *Astragalus*, wherein the lines are typically rather sharply drawn. The summary revision by Barneby (1952) analyzed the numerous names accumulated in the previous century and a half of exploration, discovery, and interpretation of the genus.

Conclusions

The genera *Astragalus* and *Oxytropis* each contain several to numerous species of toxic plant species. Those plants with intoxicant properties have led to numerous livestock deaths, amounting to untold amounts of property loss. The first of those genera, *Astragalus*, is very large, and only a small proportion of the species have been investigated as to their potential for intoxication of livestock. Species of *Astragalus* are present in all of the contiguous United States and Alaska, and in all of the Canadian provinces. *Oxytropis* is much smaller. Species of *Oxytropis* are confined mainly to western North America, with a few of the taxa extending across the northern portion of the continent to eastern Canada. Certainly, additional species will be discovered to contain swainsonine, and at least some of the species will be found to cause death loss in domestic animals at least on a local basis.

References

Barneby RC (1952). A revision of the North American species of *Oxytropis* DC. *Proceedings of the California Academy of Sciences IV* 27:177–312.

Barneby RC (1964). Atlas of North American Species of *Astragalus*. *Memoirs of the New York Botanical Gardens* 13:1–1188.

Brown R (1823). *Chloris Melvilliana*. A list of plants collected in Melville Island (latitude 74E–75E N. longitude 110E–112EW.) In the year 1820; by the officers of the voyage of discovery under the orders of Captain Parry. With characters and descriptions of new genera and species by Robert Brown. W. Clowes, London.

Brown R (1824). A supplement to the appendix of Captain Parry's first voyage for the discovery of a North-West passage, in the years 1819–20. Containing an account of the subjects of natural history. IX. Botany. A list of plants collected in Melville Island, by the officers of the expedition; with characters and descriptions of the new species. P. 261–309. In *Journal of a Voyage for the Discovery of a North-West Passage From the Atlantic to the Pacific: Performed in the Years 1819–20 in His Majesty's Ships Hecla and Griper* (WE Parry), p. 166. John Murray, London.

Bunge A von (1868). Generis *Astragali* Species Gerontogeae. Pars Prior. Claves Diagnosticae. Mémoirs de l'Academie Impériale des Sciences de Saint Petersbourg, ser. VII, 11: 1–140.

Bunge A von (1869). Generis *Astragali* Species Gerontogeae. Pars Altera: Species enumeratio. Mémoirs de l'Academie Impériale des Sciences de Saint Petersbourg, ser. VII, 15: 1–254.

Burton P (1988). *The Quest for the North West Passage and the North Pole, 1818–1909*. McClleland and Steward, Toronto, Ontario.

Crase D (2004). *Both: A Portrait in Two Parts*. Pantheon Books, New York.

Fox WE, Allred KW, and Roolson EH (1998). A guide to common locoweeds and milkvetches of New Mexico. *New Mexico Agricultural Experiment Station Circular* 557.

Franklin J (1823). *Narrative of a Journey to the Shores of the Polar Sea in the Years 1819, 20, 21, and 22*. John Murray, London

Franklin J (1828). *Narrative of a Second Expedition to the Shores of the Polar Sea in the Years 1825B1826 and 1827*. John Murray, London

Gray A (1864). A revision and arrangement (mainly by the fruit) of the North American species of *Astragalus* and *Oxytropis*. *Proceedings of the American Academy of Arts and Sciences* 6:188–236.

Gray A (1884). A revision of the North American species of *Oxytropis* DC. *Proceedings of the American Academy of Arts and Sciences* 20:1–7

Hooker WJ (1825). Botanical appendix. In *Appendix to Captain Parry=s Journal of a Second Voyage for the Discovery of a North-West Passage from the Atlantic to the Pacific Performed in His Majesty's Ships Fury and Hecla, in the Years 1821B22B23* (WE Parry), pp. 381–430. John Murray, London.

Hooker WJ (1831). *Flora Boreali-Americana; or, the botany of the northern parts of British America: compiled principally from the plants collected by Dr Richardson & Mr. Drummond on the late northern expeditions, under command of captain Sir John Franklin, R.N., to which are added (by permission of the Horticultural Society of London) those of Mr. Douglas, from North-West America, and other naturalists.* Treuttel & Wurz, London. [Published as fascicles between 1829 and 1840].

Houston CS (1984). *Arctic Ordeal. The Journal of John Richardson Surgeon-Naturalist with Franklin 1820B1822*. McGill-Queen's University Press, Montreal, Quebec, Canada.

James LF, Elbein AD, Molyneux RJ, and Warren CD (eds) (1989). *Swainsonine and Related Glycosidase Inhibitors*. Iowa State University Press, Ames, Iowa.

James LF, Hartley WJ, and Van Kampen KR (1991). Syndromes of *Astragalus* poisoning in livestock. *Journal of the American Veterinary Medical Association* 178:146–150.

Jones ME (1923). *Revision of North-American Species of* Astragalus. Salt Lake City, Utah: by the author.

Linnaeus C (1753). *Species Plantarum*, Vol. 2. Holmiae: Laurentii Salvii.

Marsh CD (1929). Stock-poisoning plants of the range. *USDA Department Bulletin* No. 1245.

Marsh CD, Clawson AB, and Eggleston WW (1930). The locoweed disease. *USDA Farmers' Bulletin* No. 1054. 25 p.

Michaux A (1803). *Flora Boreali-Americana*, Vol. 1. Parisiis: Bibliopola Journaux.

Molyneux RJ, James LF, Panter KE, and Ralphs MH (1991). Analysis and distribution of swainsonine and related polyhydroxy indolizidine alkaloids by thin layer chromatography. *Phytochemical Analysis* 2:125–129.

Parry WE (1821). *Journal of a Voyage for the Discovery of a North-West Passage From the Atlantic to the Pacific: Performed in the Years 1819–20 in His Majesty's Ships Hecla and Griper*. John Murray, London.

Pursh F (1814). *Flora Americae Septentrionalis*, Vol. 2. White, Cochrane, and Company, London.

Ralphs MH, Graham JD, and James LF (2002). A close look at locoweed poisoning on shortgrass prairies. *Rangelands* 24(2):30–34.

Richardson J (1823). Botanical appendix. In *Narrative of a Journey to the Shores of the Polar Sea* (J Franklin). John Murray, London.

Rydbery PA (1895a). *Flora of Nebraska: Part 21: Rosales.* Botanical Seminar of the University of Nebraska.

Rydberg PA (1895b). *Flora of the Sand Hills of Nebraska* (Contributions to the U. S. Nat. Herbarium, Vol. III, No. 3.

Rydberg PA (1929). Galegeae, subtribe 10. Astragalaneae. *North American Flora* 24:251–462.

Savours A (1999). *The Search for the North West Passage*. St. Martins's Press, New York.

Sheldon EP (1894). Preliminary list of the North American species of *Astragalus*. *Minnesota Botanical Studies* 1:115–175.

Smith GS, Allred KW, and Kiehl OE (1992). Swainsonine content of New Mexico locoweeds. *Proceedings of the Western Section of the American Society of Animal Science* 43:405B409.

Torrey J and Gray A (1838). *A Flora of North America*, Vol. 1. Wiley & Putnam, New York.

Watson S (1871). Botany. *In Report of the Geological Exploration of the Fortieth Parallel*, Vol. 5 (C King). Washington, Government Printing Office.

Welsh SL (1967). Legumes of Alaska II: *Oxytropis* DC. *Iowa State Journal of Science* 41:1974.

Welsh SL (1974). *Oxytropis* DC. In *Anderson's Flora of Alaska and Adjacent Parts of Canada*, pp. 275–282. Brigham Young University Press, Provo, Utah.

Welsh SL (1987). *Oxytropis* DC. In *A Utah Flora* (SL Welsh, ND Atwood, S Goodrich, and LC Higgins), pp. 396–398. Great Basin Naturalist Memoirs No. 9.

Welsh SL (1991). *Oxytropis* DC.–Names, basionyms, types, and synonyms–Flora North America Project. *Great Basin Naturalist* 51:377–396.

Welsh SL (1994). North American *Oxytropis* (L.) DC. (Leguminosae) types at the Natural History Museum and Royal Botanic Garden, England, with nomenclatural comments and a new variety. *Great Basin Naturalist* 55:271–281.

Welsh SL (2001). Revision of North American species of *Oxytropis* de Candolle (Leguminosae). E.P.S., Inc., Orem, Utah.

Welsh SL (2003a). *Astragalus*. In *A Utah Flora*, 3rd ed. (SL Welsh, ND Atwood, S Goodrich, and LC Higgins), pp. 354–392. Brigham Young University, Provo, Utah.

Welsh SL (2003b). *Oxytropis* DC. In *A Utah Flora* (SL Welsh, ND Atwood, S Goodrich, and LC Higgins), pp. 409–412. Brigham Young University, Provo, Utah.
Welsh SL (2005). North American Species of *Astragalus* Linnaeus (Leguminosae). Herbarium, M. L. Bean Life Science Museum, Provo, Utah (unpublished manuscripts).
Williams MC and Barneby RC (1977). The occurrence of nitro-toxins in North American *Astragalus* (Fabaceae). *Brittonia* 29:310–326.

GENERAL CASE REPORTS

Chapter 4

Evaluation of Plant Poisoning Reports

D.J. Wagstaff

276 East 100 South, Pleasant Grove, UT 84062, USA

Introduction

Validity of a claim of plant poisoning is determined by evaluation of a primary report, *i.e.*, an eyewitness account of observers qualified by training and experience. This applies equally to field and experimental poisonings. Examples for this presentation were taken from my worldwide poisonous plant database of over 22,000 documents which are being indexed by standard plant names and other factors. Also it is based upon documents published or archived at Cornell University that were located for me by Susanne Whittaker, Reference Librarian.

Uncertainty in a report diminishes confidence that can be placed in it. Anything an author meant to write or should have written is of no value when readers take action. It is virtually impossible to retract the printed word, as illustrated by the aftermath of the hearsay report of a fatal poisoning by poinsettia (Rock 1920). Stewart (1936) in *Ordeal by Hunger* described the arrival of the infamous Lansford W. Hastings into the camp of the ill-fated Donner Party in present-day Wyoming. He urged them to take his new cutoff to California. Hastings was not only a dashing young horseman from the West, he was the author of a book. 'And here was the author himself whose words must be true because they were in print.' Uncritical evaluation of his printed words by the Donners led to the horrors of starvation, murder, and cannibalism. Every effort must be made in recording words of this symposium to prevent an ordeal not of hunger for food but a hunger for truth.

Reporting Elements

There are basic elements of reporting or as Kipling (1902) poetically phrased it, 'I keep six honest serving-men: (They taught me all I knew). Their names are What and Where and When and How and Why and Who.' Neglect of these elements is common in scientific writing and poisonous plant reports are no exception. Examples presented are mainly from authors long dead to lessen sensitivities but deficiencies continue with discomforting regularity and are in all symposia of this series (ISOPP 1–6).

Evaluation occurs at three phases. Scientists evaluate their own manuscripts during the writing process; they evaluate reports of others when acting as reviewers and editors during the publication process; and afterwards in the reading process they evaluate printed reports.

Patently false reports in our field are rare but reports with significant reporting defects are distressingly common. Readers should always be vigilant even when using respected references works. For example, *Dorland's Medical Dictionary* (Friel 1985) erroneously states that loco is a condition caused by selenium.

Sections of a Report

Problems start in the title, the statement of **What**. Use of pretentious terms such as cicerism, coturnism, djenkolism and mycetism confuses readers. Authors demonstrate their awareness of possible confusion when they use parenthetical definitions of these terms. Another burden authors unfairly place upon readers is superfluous wording such as ‘A contribution to the knowledge of …’ (Schultes 1941).

Difficulties are next found in the byline. The set of authors sometimes does not include individuals who made significant contributions. Mention in text of an unnamed practitioner alerts the reader to the existence of a contributing person but does not identify that person, give their qualifications as a qualified observer, nor provide a means to obtain further information. Saying that a plant was identified or a substance was isolated suggests that the authors made the identification or performed the chemical analyses when in fact this is often someone else.

Further problems emerge in the text. One of a general nature is editorial emphasis on grammatical form over scientific substance. Insistence on use of the passive voice leaves Kipling’s servant **Who** virtually unemployed.

Other general problems are obscure claims of doubtful validity. Examples are ubiquitous as exemplified by the statement of Pamell on page 101 of his textbook (1911) that bracken fern (or as he called it *Pteris aquilina*) ‘is said to be poisonous.’ The much-used phrase ‘reported to be poisonous,’ if not supported by a primary report, is misleading in that it suggests a documentary source that may not exist.

A set of more specific issues involves references in text. Correct placement of references is an unappreciated skill. The purpose of a reference is to help the reader obtain further information and then to aid in evaluating the evidence and conclusions of the original report. Some references are incorrect; editors generally do not have the time to check their accuracy.

General Issues

An extremely serious issue is plagiarism, the failure to attribute information to its legitimate source or to use material without permission. This can cause disciplinary action against a journalist but among scientists it is generally not recognized or is ignored. This ethical breach is widespread in poisonous plant reports. It can not be justified by ignorance of acceptable reporting standards nor can it be excused by the pressure to write hastily in a system that rewards quantity of publications over quality. The ease with which material can be copied in this electronic cut and paste age increases the chances for identity of a source to become separated from the material. Notwithstanding difficulties, this unethical conduct even if done in ignorance must cease. Symposium organizers should take a stand against it.

Another issue with references perhaps can not be characterized as being unethical but it is certainly a disservice to readers. This is the practice of referencing secondary sources, e.g. reviews. In preparation for writing an article several years ago about Q Fever, I spoke with R.J. Huebner, an author of an early report on the subject, who said that he had often been misquoted because subsequent authors had not read his actual paper but instead relied on a review that had misinterpreted his report (Huebner, personal communication, 1963). In addition to the mischief of misinterpreting, the inappropriate use of reviews and other secondary material can form a chain of references leading to a dead-end where no primary report is found. Referencing a high quality review, which contains primary references, is at times justified if done with care. An example is a reference to analyses and conclusions of a reviewer. In contrast, using the handiest genuine-appearing reference is an attempt to put up a front of legitimacy for what in reality is sloppy work. Symposium organizers have made a good start to addressing some of these issues in their instructions for manuscript submission.

The most prevalent reporting problem in our field is neglect of Kipling's servant **How**. Readers are frequently not told how a plant was identified. The basis of plant identification such as preservation of a voucher (Wagstaff *et al.* 1999) should always be given despite the lack of such a requirement by most journals.

Detailed Example of Reporting Deficiencies

A more detailed example of documentation deficiencies is taken from a set of five articles by William Hagen and coworkers reporting the first confirmed cases of bracken fern [*Pteridium aquilinum* (L.) Kuhn] poisoning of cattle in North America (Bosshart and Hagan 1920; Hagan 1922, 1925; Hagan and Zeissig 1927; Zeissig 1928). Among the impressive credentials of Hagen are author of the textbook *Infectious Diseases of Domestic Animals*, Dean of Veterinary Medicine at Cornell University, and founding director of the National Animal Disease Laboratory. If one of such sterling scientific reputation could have serious reporting lapses, then it could happen to any of us.

Hagen, after unsuccessful attempts to transmit this newly recognized febrile hemorrhagic enteritis by animal inoculation, postulated intoxication by an environmental agent. Because he seems not to have had any extensive botanical training other than a class in medical botany (Kansas State University Registrar, personal communication, 2005), an unidentified botanist investigated pastures where affected cattle had grazed. However, it is not recorded whether the botanist pointed out bracken to Hagan (1925). Later in the same article Hagan said that bracken was noticed (note the fact-obscuring passive voice) in pastures where affected cattle had been. Based on this he collected samples of the plant and then he and a graduate student, Zeissig, proved its etiologic role through animal feeding studies.

He did not report whether identity of the collected plant was confirmed by a botanist. No basis for plant identification was given until Hagan's fourth report, in which an unsigned unattributed drawing of bracken (then called *Pteridium latuisculum*) together with other botanical information were provided. Later, Zeissig in his thesis published a photograph of bracken in a typical woodland setting.

In different places in the five papers, there is botanical information that Hagan and coworkers likely would not have known without botanical consultation. They neglected to acknowledge project contributors. The unknown botanist might have been Walter Muenscher, instructor in the Veterinary Botany class at that time (*The Register*, 1921). The more serious issue of plagiarism, albeit unintentional, could be raised if the artist of the drawing were not Hagan or Zeissig. Possibly the artist was Helen Hill Craig, who made the drawings for Muenscher's *Poisonous Plants of the United States* (1958) but differences in her drawing of bracken in this text and the unsigned drawing cited above cause some doubt. Hagan and coworkers probably had permission to publish the artist's work but until documentation can be found, their otherwise excellent reports remain under a cloud of uncertainty.

Conclusions

There are other important issues of report evaluation concerning such things as hypothesis generation, experimental design and evidentiary support for conclusions that are beyond the scope of this article. Care should be taken in evaluation of reports in the writing, publishing and reading processes.

References

Bosshart JK and Hagan WA (1920). A fatal undiagnosed disease of cattle in New York State. *Cornell Veterinarian* 10:102–113.

Friel JP (1985). *Dorland's Illustrated Medical Dictionary*, 26th ed. W.B. Saunders, Philadelphia, Pennsylvania.

Hagan WA (1922). Further study of the etiology of an undiagnosed disease of cattle. *Cornell Veterinarian* 12:43–54.

Hagan WA (1925). Bracken poisoning of cattle. *Cornell Veterinarian* 15:326–332.

Hagan WA and Zeissig A (1927). Experimental bracken poisoning of cattle. *Cornell Veterinarian* 17:194–207.

Kipling R (1902). *Just So Stories*. Doubleday, New York.

Muenscher WC (1958). *Poisonous Plants of the United States*, rev. ed. Macmillan, New York.

Pammel LH (1911). *A Manual of Poisonous Plants*. The Torch Press, Cedar Rapids, Iowa.

Rock JF (1920). The poisonous plants of Hawaii. *The Hawaiian Forester and Agriculturalist* 17:59–62.

Schultes RE (1941). A contribution to our knowledge of *Rivea coymbosa*, the narcotic ololiuqui of the Aztecs. *Monograph of the Botanical Museum of Harvard University*.

Stewart GR (1936). *Ordeal by Hunger, The Story of the Donner Party*. Houghton Mifflin, New York.

The Register (1921). Cornell University, Ithaca, New York

Wagstaff DJ, Wiersema JH, and Lellinger DB (1999). Poisonous plant vouchers. *Veterinary and Human Toxicology* 41:162–164.

Zeissig A (1928). *Bracken Poisoning of Cattle*. PhD Thesis, Cornell University, Ithaca, New York.

Chapter 5

Comparative Studies on the Susceptibility of Buffalo and Cattle to Three Poisonous Plants (*Palicourea marcgravii, Palicourea juruana* and *Arrabidaea bilabiata)* in the Amazon Region

C.H. Tokarnia,[1] F. Riet-Correa,[2] J.D. Barbosa,[3] and C.M.C. de Oliveira[3]

[1]Instituto de Zootecnia, Universidade Federal Rural do R. de Janeiro, Seropédica, RJ 23835-000, Brazil; [2]Centro de Saúde e Tecnologia Rural, Universidade Federal de Campina Grande, Campus de Patos, 58700-000, PB, Brazil; [3]Escola de Medicina Veterinária, Campus Castanhal, Universidade Federal do Pará, Rua Maximino Porpino 1000, Castanhal, PA, 68743-080

Introduction

The most important poisonous plants in Brazil are a group of 12 plants that cause 'sudden death syndrome,' mainly during exercise: *Palicourea marcgravii*; *P. aeneofusca*; *P. juruana*; *P. grandiflora*; *Arrabidaea bilabiata*; *A. japurensis*; *Pseudocalyma elegans*; *Mascagnia rigida*; *M. elegans*; *M. pubiflora*; *M.* aff. *rigida* and *Mascagnia* sp. (Tokarnia *et al.* 1979, 1990, 2000, 2002). Fluoroacetate has been identified as the active principle in *P. marcgravii* (Oliveira 1963; Hall 1972; Moraes 1993) and *A. bilabiata* (Krebs *et al.* 1994). In the other plants the active principle has not been identified, but they probably also contain fluoroacetate.

In Brazil, at least 5% of the cattle population die annually, and data from diagnostic laboratories show that from 1978 to 1998, 10.6% to 14% of these losses were due to plant poisoning (Riet-Correa and Medeiros 2001). With a cattle population of 195 million, deaths from different causes represent 9.75 million cattle per year. Assuming that 10% to 14% of those deaths are due to toxic plants, it is estimated that the annual death rate due to toxic plants in Brazil varies from 975,000 to 1.365 million cattle. *Palicourea marcgravii*, which contains fluoroacetate, is probably the most important Brazilian toxic plant for cattle. It causes severe losses in the whole country except in the three states of the southern region and the state of Mato Grosso do Sul in the central-western. (Tokarnia *et al.* 1990, 2000). Tokarnia *et al.* (1990) estimated that plants causing 'sudden death' are responsible for approximately half of the losses from toxic plants in Brazil.

The northern region of Brazil suffers the greatest losses by plant poisoning in cattle. In this region, plant poisoning is the main cause of deaths in adult cattle and also water buffaloes (*Bubalus bubalis*), which are quite numerous in the Amazon region. Almost all these deaths are caused by only two plants: *P. marcgravii*, which occurs on unflooded land ('terra firme'), and *Arrabidaea bilabiata*, which occurs on temporary flooded plains of the Amazon River and its tributaries. *Palicourea juruana* and *P. grandiflora* are two other species of that family that occur on unflooded land of the Amazon region; however, they are much less important. In the state of Roraima (the northernmost point in Brazil) *Arrabidaea japurensis* occurs instead of *A. bilabiata.*

Anecdotal information from the Amazon region suggests that buffalo are more resistant to plant poisoning than are cattle. The smaller number of deaths of buffalo due to the action of poisonous plants can be explained by two reasons. One reason seems to be the difference in the habitat for the two animal species. On unflooded land, the preferential habitat of cattle, the most common poisonous plant is *P. marcgravii*, which is more toxic and is consumed under any circumstances, even when animals are not hungry. In the temporarily flooded areas, the preferential habitat of buffalo, the most common plant is *A. bilabiata*, which is less toxic than *P. marcgravii* and is only consumed when the animals are hungry. For instance, on a farm in the Amazon region where the two plants occurred abundantly in their respective habitats, a farmer commented that when he raised cattle, he always suffered heavy death losses. When he raised buffalo instead of cattle, the number of deaths decreased considerably. On this farm we observed that the paddock was in an unflooded area; when the buffalo grazed, they went to the temporarily flooded areas where *A. bilabiata* existed and remained in that area most of the time. On the other hand, according to the farmer, the cattle preferred to graze on unflooded areas, intermixed with 'capoeiras' (regrowth) and forest, where *P. marcgravii* was common, thus causing many death losses, mainly when the animals were moved (Tokarnia 1978, unpublished data). The second reason for higher incidence may be an acquired resistance (tolerance) in buffalo. One anecdotal report suggests that a farm in a never flooded area had fewer losses of buffalo than cattle (Cunha 1987). This paper reviews the results published in previous papers (Barbosa *et al.* 2003; Oliveira *et al.* 2004; Tokarnia *et al.* 2004) conducted to verify if buffalo are more resistant than cattle to *P. marcgravii*, *A. bilabiata* and *P. juruana*. Additionally, experiments with *A. bilabiata* were performed to verify if there are differences in the toxicity of the plant with respect to young and mature leaves and in different seasons of the year.

Experimental Results

Twenty-seven buffalo and 16 cattle were used, mostly males, 1 to 2 years old. Previously determined doses of fresh leaves of *P. marcgravii*, *A. bilabiata* and *P. juruana* were put into the mouths of the penned animals, which chewed and swallowed them without difficulty. Soon after the administration of the plant, the animals were placed into small pastures and were observed. At intervals of approximately 12 hours over 2 days, they were exercised four times for 15 min, by horse or by foot. *Palicourea marcgravii*, *A. bilabiata* and *P. juruana* were dosed simultaneously at equal dosages to bovines and buffalo.

A non-lethal dose of *P. marcgravii* in water buffalo ranged from 0.5 to 2 g kg^{-1} bw, whereas in cattle the non-lethal dose was < 0.25 g kg^{-1} bw (Figure 1). For *P. juruana* the non-lethal dose in buffalo ranged from 0.25 to 1 g kg^{-1} bw; in contrast the non-lethal dose in cattle was < 0.125 g kg^{-1} bw. For *A. bilabiata* mature leaves (collected in October), a non-lethal dose for buffalo ranged from 1 to 5 g kg^{-1}; a non-lethal dose of mature leaves in cattle was 1 to 2 g kg^{-1} bw. For immature leaves of *A. bilabiata* (collected in May), a non-lethal dose for buffalo ranged from 1 to 2 g kg^{-1}; a non-lethal dose of immature leaves in cattle ranged from 0.5 to 1 g kg^{-1} bw. Nine buffalo and ten cattle showed clinical signs and died. Two buffalo ingested 6 g kg^{-1} bw of *A. bilabiata* and 1 g kg^{-1} *P. juruana*, respectively, and showed mild clinical symptoms and survived. One bovine ingested 1 g kg^{-1} of *A. bilabiata* and also survived after showing mild symptoms.

Clinical and pathological signs were characteristic of fluoroacetate intoxication and were similar in buffalo and cattle. In most animals clinical signs appeared when the animals were exercising. The characteristic signs were loss of balance when getting up, ataxia, and periodic recumbency, staying in lateral recumbence and pedaling in attempts to get up. Tachycardia and labored breathing, sometimes stertorous or with open mouth and protruded tongue, were also observed. Muscular tremors were observed occasionally. One buffalo also had strabismus and nystagmus and another had bruxism. Some animals got up, but fell down again immediately. In some animals these signs appeared without exercise, being characterized by suddenly falling on their side when they were already in sternal recumbency. These severe signs lasted in most cases from 2 to 7 min but one showed clinical signs for up to 40 min before death. Early clinical signs in some of the animals were mild and consisted of reluctance to walk or moving slowly, laying down frequently and for long periods, palpitation, salient pulsing jugular vein and dyspnoea after running. These mild signs lasted from about 30 min to almost 12 hours before severe signs were manifest.

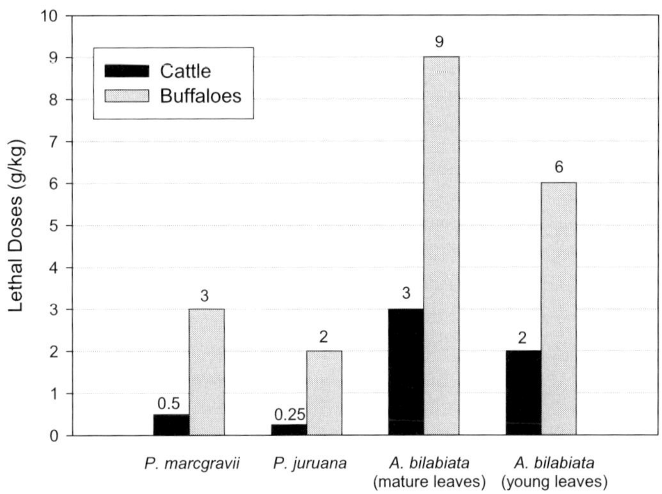

Figure 1. Susceptibility of buffalo and cattle to poisoning by *Palicourea marcgravii, P. juruana* and *Arrabidaea bilabiata*.

Post-mortem findings of lung edema or emphysema were observed only in some animals. Histologically, the kidneys of two buffalo and three cattle had hydropic-vacuolar degeneration of the epithelial cells of the convoluted distal tubules. Another buffalo had incipient necrosis of the epithelial cells (nuclei with condensed chromatin and cytoplasm more eosinophylic) of some urinary tubules in the cortex. Moderate to severe diffuse congestion and edema were observed in the lungs of some animals. One buffalo and one bovine revealed vacuolization of the hepatocytes of the intermediate zone of the hepatic lobule. In one bovine large areas of incipient coagulative necrosis were observed in the myocardium.

Discussion

The experiments show that buffalo are six and eight times more resistant than cattle to the poisonous action of *P. marcgravii* and *P. juruana*, respectively. The experiments performed with *A. bilabiata* in buffaloes during two seasons of the year, October (end of the dry season) and May (end of the rainy period), show that buffalo are susceptible to poisoning by this plant. The experiments performed simultaneously in buffaloes and in cattle at the end of the rainy period (May), with the young as well as with the mature leaves, show that buffalo are three times more resistant than cattle to the poisonous action of *A. bilabiata*. In the experiments done with buffalo in October (end of the dry season), the young leaves were twice as toxic as the mature leaves. In May (end of the rainy season), the young leaves were one and a half times more toxic than the mature leaves in buffalo and cattle.

These data suggest that the lower number of deaths of buffalo than cattle by poisonous plants in the Amazon region is probably due to two factors. First, buffalo are more tolerant of the toxic action of *A. bilabiata* and *P. marcgravii*. Second, the preferential habitat of buffalo coincides with the habitat of *A. bilabiata*, a plant that is less toxic than *P. marcgravii*, whose habitat is preferred by cattle. Thus, losses could be reduced by grazing buffalo instead of cattle in areas where *P. marcgravii* occurs.

There are many possibilities for the higher resistance of buffaloes to poisoning by these plants when compared with that of cattle: (1) a more efficient detoxification by defluorination, which implicates a gluthatione-dependent reaction and occurs mainly in the liver (Cheeke 1998); (2) a lower absorption in the digestive system; (3) bacterial degradation of fluoroacetate in the rumen; or (4) other unknown physiological mechanisms. Ruminal detoxification occurs by the breakdown of fluoroacetate. Australian researchers modified genetically the ruminal bacteria *Butyrivibrio fibrisolvens*, introducing a gene, isolated from *Moraxella* sp., which codifies a dehalogenase and is able to hydrolyze fluroacaetate (Gregg *et al.* 1994). Monofluoroacetic acid is the toxic principle of several poisonous plants in Australia and South Africa. Once ingested, the substance is converted to fluorocitrate, which inhibits aconitase, the enzyme responsible for the breakdown of the citrates, resulting in an interruption of the Krebs cycle. We speculate that buffalo have ruminal bacteria which break down monofluoroacetic acid.

Promising results have been obtained in the prevention of poisoning by fluoroacetate in laboratory animals by the administration of substances that deliver acetate during metabolism. The best results were obtained with glycerol monoacetate (monoacetina) and acetamide (Vickery and Vickery 1973; Kellerman *et al.* 1988). Positive results were obtained by the administration of acetamide to sheep to prevent *Dichapetalum cymosum* poisoning, a South African plant, which contains monofluoroacetic acid (Egyed and Schultz 1986). In buffalo, the metabolic processes may make more acetate available to lessen the effects of monofluoroacetic acid.

References

Barbosa JD, Oliveira CMC, Tokarnia CH, and Riet-Correa F (2003). Comparação da sensibilidade de bovines e búfalos à intoxicação por *Palicourea marcgravii* (Rubiaceae). *Pesquisa Veterinária Brasileira* 23(4):167–172.

Cheeke PR (1998). *Natural Toxicants in Feeds, Forages, and Poisonous Plants*, 2nd ed. Interstate Publishers, Danville, Illinois.

Cunha MR (1987). Búfalo prova sua resistência à erva cafezinho. *Búfalo em Evidência* 1(3):2.

Egyed MN and Schultz RA (1986). The efficacy of acetamide for the treatment of experimental *Dichapetalum cymosum* (gifblaar) poisoning in sheep. *Onderstepoort Journal of Veterinary Research* 53:231–234.

Gregg K, Cooper CL, Schafer DJ, Sharpe H, Beard CE, Allen G, and Xu J (1994). Detoxification of the plant toxin fluoroacetate by a genetically modified rumen bacterium. *Bio/Technology* 12:1361–1365.

Hall RJ (1972). The distribution of organic fluorine in some toxic tropical plants. *New Phytology* 71:855–871.

Kellerman TS, Coetzer JAW, and Naudé TW (1988). *Plant Poisoning and Mycotoxicoses of Livestock in Southern Africa.* Oxford University Press, Cape Town.

Krebs HC, Kemmerling W, and Habermehl G (1994). Qualitative and quantitative determination of fluoroacetic acid in *Arrabidaea bilabiata* and *Palicourea marcgravii* by ^{19}F-NMR spectroscopy. *Toxicon* 32(8):909–913.

Moraes RLF (1993). *Comprovação química e biológica da presença de monofluoroacetato nas folhas de Palicourea marcgravii St. Hil.* Tese de Doutoramento, Fac. Ciênc. Farmacêuticas, Universidad San Paulo.

Oliveira MM (1963). Chromatographic isolation of monofluoroacetic acid from *Palicourea marcgravii* St. Hil. *Experientia* 19:586.

Oliveira CMC, Barbosa JD, Macedo RSC, Brito MF, Peixoto PV, and Tokarnia CH (2004). Estudo comparativo da toxidez de *Palicourea juruana* (Rubiaceae) para búfalos e bovinos. *Pesquisa Veterinária Brasileira* 24(1):27–30.

Riet-Correa F and Medeiros RMT (2001). Intoxicações por plantas em ruminantes no Brasil e no Uruguai: importância econômica, controle e riscos para a saúde pública. *Pesquisa Veterinária Brasileira* 21(1):38–42.

Tokarnia CH, Döbereiner J, and Silva MF (1979). *Plantas Tóxicas da Amazônia a Bovinos e outros Herbíivoros*. INPA, Manaus.

Tokarnia CH, Peixoto PV, and Dobereiner J (1990). Poisonous plants affecting heart functions of cattle in Brazil. *Pesquisa Veterinária Brasileira* 10:1–10.

Tokarnia CH, Döbereiner J, and Peixoto PV (2000). *Plantas Tóxicas do Brasil.* Editora Helianthus, Rio de Janeiro.

Tokarnia CH, Döbereiner J, and Peixoto PV (2002). Poisonous plants affecting livestock in Brazil. *Toxicon* 40:1635–1660.

Tokarnia CH, Barbosa JD, Oliveira CMC, Brito MF, Oliveira RB, and Barbas LAI (2004). Aspectos epidemiológicos e clínico-patológicos comparados da intoxicação por *Arrabidaea bilabiata* (Bignoniaceae) em búfalos e bovinos. *Pesquisa Veterinária Brasileira* 24(2):74–79.

Vickery B and Vickery ML (1973). Toxicity for livestock of organofluorine compounds present in *Dichapetalum* plant species. *Veterinary Bulletin* 43(10):537–542.

Chapter 6

Xanthoparmelia chlorochroa Intoxication in Wapiti (*Cervus canadensis*)

W.E. Cook[1], T.E. Cornish[2], E.S. Williams[2], B. Brown[3], G. Hiatt[3], T.J. Kreeger[4], R.N. Dailey[2], and M.F. Raisbeck[2]

[1]*Wyoming Livestock Board, 2020 Carey Ave, 4th Floor, Cheyenne, WY 82002, USA;* [2]*Wyoming State Veterinary Laboratory, University of Wyoming, 1174 Snowy Range Road, Laramie, WY 82070, USA;* [3]*Wyoming Game and Fish Department, 260 Buena Vista, Lander, WY 82520, USA;* [4]*Wyoming Game and Fish Department, 2362 Highway 34, Wheatland, WY 82201, USA*

Introduction

On February 6, 2004 coyote hunters discovered two recumbent cow elk near Separation Creek (41° 41' N 107° 32' W) on the Red Rim in Carbon County, Wyoming. Believing that the elk had been shot in an attempted poaching, they reported the situation to Wyoming Game & Fish Department (WGFD) personnel. The local WGFD agent investigated the report and found that while it did not appear that the cows had been shot, several more recumbent animals were hidden amongst the sagebrush. During the subsequent 7 weeks, 326 more recumbent or dead elk were discovered. Once recumbent, none recovered, even if provided feed and water. In all an estimated 400–500 animals died from a herd of approximately 600.

Clinical

Most elk were discovered in sternal or lateral recumbency; however, early in the episode WGFD personnel noticed individuals who lagged behind the rest of their band, appeared slightly weak or ataxic and were unable to jump fences. Discolored, reddish-orange urine was found staining the snow. Recumbent elk exhibited normal mentation, vocalizing and struggling to rise when approached. Vision and cranial reflexes appeared normal, at least in animals that had not progressed to lateral recumbency. When approached they focused on the perceived threat (us) and could coordinate their head and eyes to track the threat. Patellar and tricipital reflexes appeared slightly diminished. Appetite was not suppressed; in most cases the recumbent elk had eaten all the vegetation within reach and the few that were offered hay ate it readily.

Morphologic Findings

A total of 18 recumbent elk were subjected to complete (40 tissue) post-mortem workups. All were in fair to good nutritional condition. Elk that had been recumbent for more than a day appeared mildly to moderately dehydrated and had external lacerations and abrasions on the shoulders, neck, head, face, lips, and around the eyes, and early cutaneous ulcers over bony prominences. Most elk had under-filled rumens, some containing visible fragments of lichen. Several elk had multifocal to coalescing pale white/tan streaks in large locomotory skeletal muscles of front and hind limbs, consistent with exertional or nutritional myopathy grossly. Muscle lesions were not observed in any elk recumbent less than 2 days.

No significant histological lesions were observed in sections of brain, spinal cord, or peripheral nerves from any elk to explain recumbency and eventual death. All elk were tested by immunohistochemistry for chronic wasting disease (CWD) (Williams *et al.* 2002) with negative results. In elk with significant gross evidence of myopathy, microscopic examination revealed lesions typical of exertional or possibly nutritional disease, notably myofiber swelling, hyaline change and hypereosinophilia of cytoplasm, nuclear condensation or pyknosis, and in more severe lesions, myofiber fragmentation and/or rupture, with mineralization of myofibers and with infiltration by small numbers of macrophages. In lesions of some chronicity there were attempts at repair and regeneration, with more florid mononuclear inflammation, myofiber nuclear rowing and centralization, early fibrosis, and neovascularization. Similar lesions were lacking in sections of diaphragm from all elk. No consistent significant microscopic lesions were observed in remaining organs or tissues, with the exception of infrequent mild tubular nephrosis in the kidneys of four elk and multifocal, random mild acute hepatocellular necrosis in the livers of three elk.

Clinical Chemistry/Clinicopathology

Every animal from which blood was sampled had elevated creatine kinase (CK). In most cases CK activity was increased *at least* 20-fold over our reference range of 48–129 U/L. All but two elk had elevated serum lactate dehydrogenase (1600– 2744 U/L, ref:692–1295) and aspartate aminotransferase (368–1588 U/L, ref: 34–42) activities. Serum albumin and Ca^{+2} were similarly depressed in most animals. Complete blood counts were performed on 15 animals. All exhibited leukocytosis with neutrophilia. Blood urea nitrogen was slightly elevated (38–58 mg/dL, ref: 31–33) in samples collected during the first 3 weeks of the episode, but was normal in subsequent samples. There were no consistent trends in serum alkaline phosphatase, gamma-glutamyltransferase, magnesium, phosphorous, total protein, globulin, creatinine, sodium, potassium or chloride.

Urine specific gravity ranged from 1.000 to 1.010 in all (7) samples collected. One sample contained a trace of occult blood, but all were negative for leukocytes, nitrite, urobilinogen, bilirubin, hemoglobin, myoglobin, protein, pH, glucose, epithelial cells, casts and bacteria.

Toxicology

Ingesta were analyzed for organophosphate and/or carbamate insecticides with negative results and both blood and brain cholinesterase activities were within normal limits. There was no detectable (MDL <5 ppm) dicamba, MCPA, 2,4-D, 2,4,5-T, 2,4-DB or silvex in the ingesta, nor were there any detectable (MDL < 50 ppm) petroleum hydrocarbons. Kidney samples were analyzed for 1080 (sodium monofluoroacetate) with negative results.

Aqueous humor did not contain significant amounts of nitrate, nitrite or chlorate. Liver and kidney metal (V, Cr, Mn, Fe, Ni, Co, Cu, Zn, As, Se, Mo, Cd, Ba, Tb, Hg, Tl, Pb, Bi) concentrations were within normal limits with the exception of hepatic Se, which was moderately elevated (2–3 ppm wet weight). Brain Na^+, K^+, Ca^{+2} and Mg^{+2} concentrations were within normal limits. Liver tocopherol concentrations were adequate (6–13 ppm).

Rumen contents and urine did not contain detectable atropine, anabasine, coniine, deltaline, nicotine, scopolamine, sparteine, or 4-aminopyridine, and rumen contents were negative for cyanide, vomitoxin, T-2 tetraol, fusarenone-X, 3-acety DON, 15-acety DON, DAS, T-2 triol, T-2 toxin, iso T-2 toxin, scirpenol, nivalenol, 15-acetoxyscirpenol, neosolaniol, HT-2 toxin, acetyl T-2, solanine, solanidine and tomatine. Rumen gas caps contained less than 200 ppm H_2S. Rumen and liver specimens from four animals were evaluated for *Clostridium botulinum* toxin by mouse inoculation. Ingesta was also negative for domoic acid.

Samples of water from the three unfrozen sources in the area were analyzed for Na, Mb, Al, Si, P, S, Cl, K, Ca, Sc, Ti V Cr, Mn, Fe, Co, Ni, Cu, Zn, Ga, Ge, As, Se, Br, Rb, Sr, Zr, Nb, Mo, Ru, Rh, Pd, Ag, Cd, In Sn, Sb, Te, I, Cs, Ba, La, Ce, Pr, Nd, Sm, Eu Gd Tb, Ta W, Os Tp, Au, Hg, Tl, Pb, Bi th, and U. With the exception of Na, Ca, K, Sr and Ba none were present at > 5 ppb. Sodium was relatively high (7000–8000 ppm); other cations were present at 10 ppm or less. Water samples were also provided to mice as their sole water source for 2 weeks without effect.

Experimental

Three recumbent elk were moved from the Red Rim to the Wyoming State Veterinary Laboratory (WSVL) for closer study. Clinicopathologic studies of these animals yielded similar results to samples collected in the field. Treatment with thiamine, lactated Ringers, vitamins A, B, D, E and selenium, calcium gluconate, dextrose, phosphorus, magnesium and potassium, flunixin meglumine and dexamethasone did not result in any improvement. The animals remained unable to rise for 14 days until they were euthanized. Post-mortem lesions were similar to those seen in field cases.

Early in the episode one of us (BB) noted that the elk spent quite a bit of time browsing a small, free-living lichen, later identified as *Xanthoparmelia chlorochroa* (Dr Robert Egan, University of Nebraska, Omaha Lichen Herbarium accession number 15,631). Specimens of ingesta examined microscopically by Dr John Reagor of the Texas Veterinary Medical Diagnostic Lab consisted of as much as 50% of this lichen. To test the hypothesis that the lichen was responsible for signs in the elk, a group of four mice were gavaged with a slurry of finely ground lichen in water twice daily. A similar group received water as a control. After 4 days one mouse was discovered dead. The following day two more were showing reluctance to move and ataxia. Both were euthanized *in extremis.*

Three yearling elk were obtained from northwestern Wyoming and housed at the WGFD Sybille Wildlife Research Unit (41° 45.778' N, 105° 22.605' W). For 3 days, the elk were offered a mix of alfalfa and lichen, with the proportion of alfalfa to lichen decreasing daily. On the 4th day, only lichen was offered, which was weighed every 24 hours to determine the total amount consumed by the three elk. The three elk consumed 1.1–3.6 kg daily. After 4 days on 100% lichen, orange-red urine, similar to that seen in the field, was noted in the pen. After 7 and 10 days respectively, two of three displayed signs of muscular weakness and paresis similar to those seen in field cases. After euthanasia, gross and histological examination of tissues revealed lesions identical to those seen in field cases. The remaining animal ate little, if any, lichen, showed no clinical signs and was returned to a normal diet.

Discussion

Xanthoparmelia chlorochroa is a small, foliose, vagrant lichen common to the high altitude deserts of Wyoming, Idaho and Colorado. It has been described as 'excellent wildlife forage' (Richardson 1996; Brodo *et al.* 2001) although there is a paucity of information about consumption of this species by elk. Most forage studies of elk do not list lichens; presumably they are occasional winter forage, especially when other forage is limited. Marcum (1980) reported that 33% of rumens sampled from a western Montana herd contained an average of 3% lichen (species not specified) by volume. Schwartz and Mitchell (1945) reported that lichens on the trunks and limbs of trees on the Olympic Peninsula were an important part of Roosevelt elk diets, especially at higher elevations in winter.

Even less information is available about the toxic properties of *X. chlorochroa* in mammals. A number of secondary citations refer to work with a lichen identified as '*Parmelia molliuscula*' by O.A. Beath of the Wyoming Agricultural Experiment Station. The latter species does not occur in North America (Esslinger 2002); however, photographs (Beath 1937) of the Wyoming lichen and its reported collection locale (Burzlaff 1950) are consistent with *X. chlorochroa.* According to personal communications with Beath quoted by Kingsbury (1964), feeding experiments were performed with sheep and cattle and 'it was found that *Parmelia* in the amount of 1% of an animal's weight per day for five days is lethal. A single dose of 3.6% of an animal's weight is also lethal.' Clinical signs consisted of incoordination and ascending paresis; however, affected animals continued to eat and drink (Beath 1939). Supposedly, 'no lesions or any characteristic pathology develops' but it is not clear from the reports whether and/or what kind of post-mortem exams were performed.

Beath (1939) reported the active ingredient of *Parmelia* to be usnic acid (2,6-diacetyl-7,9-dihydroxy-8,9b-dimethyl-1,3(2H,9bH)-dibenzo-furandione). Kingsbury (1964) quotes Beath as saying 'typical symptoms were produced in laboratory animals fed an emulsion of usnic acid extracted from the lichen. The residue of lichen remaining after extraction was not toxic.' Beath (Beath 1939; Beath *et al.* 1953) indicated that 'carbon bisulfide is a very satisfactory solvent' for usnic acid and thus was presumably the solvent used to extract the lichen in question for the previous statement; however, in our hands, usnic acid is relatively insoluble in carbon disulfide. Beath's conclusion also ignores the probability that many substances co-extract with usnic acid.

Intravenous sodium usneate (10 mg kg^{-1}) produced signs of augmented metabolism such as hyperventilation, hyperpyrexia, and increased oxygen consumption in anesthetized cats (Soderburg 1953), an observation that seems consistent with its ability to uncouple oxidative phosphorylation *in vitro* (Abo-Khatwa *et al.* 1996; Durazo *et al.* 2004). It is plausible that such a biochemical effect might result in muscle necrosis; however, reports of usnic acid poisoning in rodents and human beings describe hepatic necrosis rather than muscle damage as the predominant toxic effect (Favreau *et al.* 2002; Estes *et al.* 2003; Durazo *et al.* 2004). We thus suspect the active ingredient in *X. chlorochroa* is something other than or in conjunction with usnic acid. Further work is underway to identify the causative agent of muscle necrosis in *X. chlorochroa.*

References

Abo-Khatwa AN, Al-Robai AA, and Al-Jawhari DA (1996). Lichen acids as uncouplers of oxidative phosphorylation of mouse-liver mitochondria. *Natural Toxins* 4:96–102.

Beath OA (1937). Lichen poisoning from the common 'lichen'. *University of Wyoming Extension Bulletin*, p. 1.

Beath OA (1939). Poisonous plants and livestock poisoning. *University of Wyoming Agricultural Experiment Station Bulletin* 231:50–53.

Beath OA, Gilbert CS, Eppson HF, and Rosenfeld I (1953). Poisonous plants and livestock poisoning. *University of Wyoming Agricultural Experiment Station Bulletin* 324:42–46.

Brodo IM, Sharnoff SD and Sharnoff S (2001). *Lichens of North America*, p. 734. Yale University Press, New Haven, Connecticut.

Burzlaff DF (1950). The poisonous lichen, *Parmelia molliuscula. Wyoming Range Management* 28:1–2

Durazo FA, Lassman C, Han SHB, Saab S, Lee NP, Kawano M, Saggi B, Gordon S, Farmer DG, Yersiz H, Goldstein RL, Ghobrial M, and Busuttil RW (2004). Fulminant liver failure due to usnic acid for weight loss. *American Journal of Gastroenterology* 99:950–952.

Esslinger TL (2002). A cumulative checklist for the lichen-forming, lichenicolous and allied fungi of the continental United States and Canada, version #8, posted 17 July 2002 http://www.ndsu.nodak.edu/instruct/ esslinge/chcklst/chcklst7.htm.

Estes JD, Stolpman D, Olyaei A, Corless CL, Ham JM, Schwartz JM, and Orloff SL (2003). High prevalence of potentially hepatotoxic herbal supplement use in patients with fulminant hepatic failure. *Archives Surgery* 138:852–8.

Favreau JT, Ryu ML, Braunstein G, Orshansky G, Park SS, Coody GL, Love LA, and Fong TL (2002). Severe hepatotoxicity associated with the dietary supplement LipoKinetix. *Annals Internal Medicine* 16:590–595.

Kingsbury JM (1964). *Poisonous Plants of the United States and Canada*, pp. 86–87. Prentice Hall, Englewood Cliffs, New Jersey.

Marcum CL (1980). Summer-fall food habits and forage preferences of a western Montana elk herd. In *North American Elk: Ecology, Behavior and Management* (MS Boyce and LD Hayden-Wing, eds), pp. 54–62. University of Wyoming, Laramie.

Richardson DH (1996). Medicinal and other economic aspects of lichens. In *CRC Handbook of Lichenology*, pp. 93–108. CRC Press, Boca Raton, Florida.

Schwartz JE and Mitchell GE (1945). The Roosevelt elk on the Olympic Peninsula, Washington. *Journal of Wildlife Management* 9:295–319.

Söderberg U (1953). A note on the action of usnic acid on anesthesized cats. *Acta Physiologica Scandinavica* 28:202–210.

Williams ES, Miller MW, Kreeger TJ, Kahn RH, and Thorne ET (2002). Chronic wasting disease of deer and elk: A review with recommendations for management. *Journal of Wildlife Management* 66:551–563.

Chapter 7

Poisoning by Day Lily (*Hemerocallis* sp.; Hemerocallidaceae) in Brazilian Cats

T.M. Souza, R.A. Fighera, G.D. Kommers, and C.S.L. Barros

Laboratório de Patologia Veterinária, Universidade Federal de Santa Maria, Santa Maria, RS, 97105-900, Brazil

Introduction

Uremic syndrome described commonly in cats and other species results from three basic mechanisms: (1) decreased blood flow through the kidneys (pre-renal uremia); (2) renal failure (renal uremia); and (3) obstruction to urine flow in the lower urinary pathways (post-renal uremia) (Polzin *et al.* 1992). In cats, renal uremia occurs mainly in association with chronic renal failure, but cases of severe clinical disease due to acute renal failure are not uncommon in the cat; in this species, acute renal failure is mainly due to toxic tubular necrosis (Polzin *et al.* 1992). Although the histological pattern (tubular necrosis) of lesions associated with the toxic injury is easily recognized, the specific cause of the tubular necrosis can be difficult to determine with gross and histopathological examination (Confer and Panciera 2001). Thus, in these cases, a thorough clinical investigation is necessary to establish the specific etiology, which might include antibiotics (e.g. aminoglycosides), antifungal agents (e.g. amphotericin B), chemotherapeutic drugs (e.g. cisplatin), heavy metals (e.g. mercury) and chemical toxicants (e.g. ethylene glycol) (Confer and Panciera 2001).

Since the early 1990s renal failure caused by renal tubular necrosis in cats has been associated with the ingestion of different species of lily, including *Lilium longiflorum* (Easter lily), *L. tigrinum* (tiger lily), *L. lancifolium*, *L. speciosum* (Japanese lily) and *Hemerocallis* spp. (day lily) (Hall 1992; Mullaney *et al.* 1993; Carson *et al.* 1994). In these instances, clinical signs first appear approximately 2 hours after ingestion, renal failure is well established at 48–96 hours after ingestion (Gulledge *et al.* 1997), and death occurs 3–5 days after ingestion (Rumbeiha *et al.* 2004). It is believed that the ingestion of only one to two leaves or of a whole flower can cause the toxicosis (Rumbeiha *et al.* 2004).

Plants belonging to the families Liliaceae and Hemerocallidaceae are frequently found in home gardens and in the interior of houses in the USA (Rumbeiha *et al.* 2004). Recently, several lily species have been introduced into Brazil and they are cultivated and commercialized as ornamental plants. This fact and the peculiar attraction lilies have for cats increase the risk of accidental poisoning by these plants in cats. This paper describes the poisoning by day lily (*Hemerocallis* sp.) in cats in southern Brazil.

Case Report

From 1995 to 2000, six 2- to 4-year-old Brazilian domestic shorthair cats of both sexes, and from five different households, presented with clinical history of apathy (6/6), anorexia (6/6), vomiting (5/6) and polydipsia (1/6). The owners informed us that the cats had ingested day lily several hours prior to the onset of clinical signs. Upon physical exam, three cats were dehydrated and two had dependent subcutaneous edema which extended from the neck to the perineum. Serum creatinine and blood urea nitrogen (BUN) levels determined by spectrophotometry in the blood drawn from two cats had markedly increased creatinine (7.6 and 11.2 mg dL^{-1}) and moderately increased BUN (98 and 250 mg dL^{-1}). All six cats were given supportive therapy with antiemetics and parenteral fluids. Five out of the six affected cats died 24–48 hours after admission to the Veterinary Teaching Hospital of the Universidade Federal de Santa Maria (UFSM) and were necropsied at the Veterinary Pathology Laboratory of the UFSM.

Necropsy findings in all cases included marked perirenal edema. The kidneys were swollen and the cortical cut surface was slightly granular. In two cats there were moderate hydrothorax and ascites. The urinary bladder was empty in all cases. Histologically, there were severe degeneration and necrosis of the epithelial lining of the renal convoluted proximal tubules. Several desquamated epithelial cells with intensely bright eosinophilic cytoplasm and pyknotic nuclei could be observed within the lumina of tubules denuded from their epithelial lining. Hyaline and granular casts were frequently seen both in convoluted tubules and collecting ducts. Occasionally great amounts of cellular debris obstructed the tubules. In the kidneys of all five cats there were varying degrees of tubular epithelial regeneration. Regenerated cells had large nuclei with loose chromatin and one or two large nucleoli; the number of mitosis was variable. In the urinary spaces of most glomeruli there were small amounts of pink proteinaceous material resulting from urine reflux into the urinary space.

Discussion

The clinical signs and necropsy findings of the cases reported here are very similar to those reported for the lily toxicosis in cats and constitute circumstantial evidence that indicates the same diagnosis in our cats. Although the frequency of micturation was not determined in the live cats in this report, necropsy findings of empty urinary bladders are highly suggestive of oliguria or anuria, which are typical clinical signs described in the literature in most cases of lily poisoning in cats (Gulledge *et al.* 1997; Volmer 1999; Brady and Janovitz 2000; Langston 2002; Hadley *et al.* 2003; Rumbeiha *et al.* 2004).

Renal excretion of metabolites (creatinine, BUN) is impaired in lily poisoning in cats (Gulledge *et al.* 1997; Volmer 1999; Brady and Janovitz 2000; Rumbeiha *et al.* 2004) and these changes were observed in the two cats in which the levels of these metabolites were determined in the blood. These laboratory findings and the clinical signs indicated a clinical diagnosis of renal failure. It is interesting that the concentration of creatinine is disproportionately higher than that of BUN. Although there is no apparent explanation for this discrepancy, it was also observed in other reports of lily poisoning in cats (Volmer 1999; Brady and Janovitz 2000). The lack of urine in the urinary bladder in all of these cases compromised the determination of other laboratory changes reported in lily poisoned cats, such as glucosuria and cylindruria (Volmer 1999).

The pathogenesis of renal lesions in cats poisoned by lily is still poorly understood due mainly to current difficulties inherent to experimental studies in this and other domestic species; the toxicosis was only recently reproduced experimentally (Rumbeiha *et al.* 2004). Attempts have failed to experimentally reproduce lily toxicosis in dogs and laboratory animals such as rats and rabbits (Rumbeiha *et al.* 2004). The ingestion of both the leaves and the flowers can induce the intoxication (Gulledge *et al.* 1997; Langston 2002; Rumbeiha *et al.* 2004), and the flowers are by far more toxic (Rumbeiha *et al.* 2004). The aqueous extracts from the plant are toxic (Rumbeiha *et al.* 2004) but the active principle is still unknown (Gulledge *et al.* 1997; Langston 2002; Rumbeiha *et al.* 2004).

In dogs the intoxication by lily-of-the-valley (*Convallaria majalis*), an ornamental species that causes sudden death by cardiogenic shock, was related to glycosides with similar activity to that of digitalis (Moxley *et al.* 1989). Another species of lily of yet another genus (*Veratrum viride*), which is associated with severe cardiac disturbances in human beings, has alkaloids that act to increase sodium transport through lipid membranes (Prince and Stork 2000). In a retrospective study carried out in two Children's Hospitals in Helsinki it was found that the lily-of-the-valley is the most important cause of plant poisoning in children of that particular region of Europe (Lamminpaa and Kinos 1996).

In a recent report (Hadley *et al.* 2003), spontaneous poisoning by four species of day lily (*Hemerocallis dumortierei*, *H. fulvi*, *H. graminea*, and *H. seiboldii*) was described in 22 cats. Thus it is worth noting that although lily poisoning in cats is usually related to ingestion of plants of the Liliaceae family, the toxicosis is also associated with the ingestion of plants of the Hemerocallidaceae family (genus *Hemerocallis*) as is the case of this report.

A factor influencing the high mortality rate in cats of this report may be related to the delay in onset of therapy; it is known that the prognosis is poor when treatment is instituted over 18 hours after ingestion of the plant (Gulledge *et al.* 1997; Volmer 1999; Rumbeiha *et al.* 2004). Furthermore, rapid detoxification occurs when emetics and cathartics are used and they are considered important for the recovery of the patient (Gulledge *et al.* 1997; Volmer 1999; Hadley *et al.* 2003). These agents were not used in the cats in this report. Oliguria and anuria are indicators of a poor prognosis. In a study of acute renal failure caused by lily ingestion in six cats, three cats presenting these signs died (Langston 2002); oliguria and anuria may have contributed to the deaths of the five cats in our report.

An important issue concerning prognosis and that should always be mentioned to the owner of affected cats is that even cats surviving the acute phase of renal failure will probably develop chronic renal failure after months or years. This development was documented in a follow-up study in six poisoned cats that were treated with drugs and hemodialysis (Langston 2002): three surviving cats developed chronic renal failure within 1½ years after the acute crisis.

References

Brady MA and Janovitz EB (2000). Nephrotoxicosis in a cat following ingestion of Asiatic hybrid lily (*Lilium* sp.). *Journal of Veterinary Diagnostic Investigation* 12(6):566–568.

Carson TL, Sanderson TP, and Halbur PG (1994). Acute nephrotoxicosis in cats following ingestion of lily (*Lilium* sp.). In *American Association of Veterinary Laboratory Diagnosticians*, pp. 43. Proceedings of the 37th Annual Meeting, Grand Rapids, Missouri.

Confer AW and Panciera RJ (2001). The urinary system. In *Thomson's Special Veterinary Pathology* (MD McGavin, WW Carlton, and JF Zachary, eds), pp. 235–277. Mosby, St. Louis.

Gulledge L, Boos D, and Wachsstock R (1997). Acute renal failure in a cat secondary to tiger lily (*Lilium tigrinum*) toxicity. *Feline Practice* 25(5/6):38–39.

Hadley RM, Richardson JA, and Gwaltney Brant SM (2003). A retrospective study of day lily toxicosis in cats. *Veterinary and Human Toxicology* 45(1):38–39.

Hall JO (1992). Nephrotoxicosis of Easter lily (*Lilium longiflorum*) when ingested by the cat. In *American College Veterinary Internal Medicine Forum*, pp. 804. Proceedings of the 10th Annual Meeting, San Diego, California.

Lamminpaa A and Kinos M (1996). Plant poisonings in children. *Human & Experimental Toxicology* 15(3):245–249.

Langston CE (2002). Acute renal failure caused by lily ingestion in six cats. *Journal of American Veterinary Medicine Association* 220(1):49–52.

Moxley RA, Schneider NR, Steinegger DH, and Carlson MP (1989). Apparent toxicosis with lily-of-the-valley (*Convallaria majalis*) ingestion in a dog. *Journal of American Veterinary Medicine Association* 195(4):485–487.

Mullaney TP, Slanker MR, and Poppenga RH (1993). Acute nephrotoxicosis in cats following ingestion of lily (*Lilium* sp.). In *North Central Conference of Veterinary Laboratory Diagnosticians*, Proceedings of the 35th Annual Meeting, Madison, Wisconsin.

Polzin D, Osborne C, and O'Brien T (1992). Moléstias dos rins e ureteres. In *Tratado de medicina interna veterinária: moléstias do cão e do gato* (SJ Ettinger, ed), pp. 2047–2137. Manole, São Paulo, São Paulo.

Prince LA and Stork CM (2000). Prolonged cardiotoxicity from poison lily (*Veratrum viride*). *Veterinary and Human Toxicology* 42(5):282–285.

Rumbeiha WK, Francis JA, Fitzgerald SD, Nair MG, Holan K, Bugyei KA, and Simmons H (2004). A comprehensive study of Easter lily poisoning in cats. *Journal of Veterinary Diagnostic Investigation* 16(6):527–541.

Volmer PA (1999). Easter lily toxicosis in cats. *Veterinary Medicine* April:331.

Chapter 8

Naturally Occurring Acute Selenium Toxicosis in Sheep

A.K. Tiwary[1,3], K.E. Panter[2], B.L. Stegelmeier[2], L.F. James[2], and J.O. Hall[3]

[1]CAHFS Laboratory System, SVM, University of California, Davis, CA 95616, USA; [2]USDA-ARS Poisonous Plant Research Laboratory, Logan, UT 84341, USA; [3]Utah Veterinary Diagnostic Laboratory, Utah State University, Logan, UT 84322, USA

Introduction

Selenium (Se) plays a critical role in the health and well-being of humans and animals. Deficiency of Se as well as Se toxicosis in animals has been reported from several regions in the United States (Edmonson *et al.* 1993). Animals residing in Se-deficient areas are often supplemented with Se for optimal growth and production. Due to a relatively low margin of safety, several cases of Se poisoning occur from time to time due to accidental overdosing of injectable Se (sodium selenite with added vitamin E). Natural Se poisoning can also occur in seleniferous areas and areas with high Se contamination, most likely from ingestion of certain plants that can accumulate Se obtained from the soil. Organic Se such as selenomethionine (SeMet) is the major component of Se-accumulating plants (Whanger 2002). In natural settings, chronic selenosis (alkali disease) is more common than acute cases. However, in May 2003, several hundred sheep died acutely after grazing in a Se-rich area in Soda Springs, ID (Fessler *et al.* 2003).

Considerable differences exist between the metabolic pathways of organic and inorganic Se compounds in the body (Ganther 1986). Most of the Se from selenite forms physiologically functional selenoproteins, or is eliminated from the body, whereas a significant portion of the Se from SeMet, in addition to forming functional selenoproteins, is also incorporated into non-functional or structural proteins (Ip 1988). Published literature on acute Se toxicosis predominantly involves sodium selenite (supplementation form of Se). Little is known about such effects of organic Se. The purpose of this investigation was to reproduce the field toxicosis reported from Soda Springs under experimental conditions using organic selenium (SeMet).

Case Report – Incident at Soda Springs, ID

In May 2003, about 1447 sheep (622 ewes and 825 lambs) were moved to an unnamed location designated as site X within a phosphate mining reclamation area. One dead ewe was found at 10:00 PM on the same day, and about three other animals were reported to be visually abnormal. At 7:00 AM the next morning, 127 sheep (25 ewes and 102 lambs) were found dead. The remaining flock was immediately moved away from site X. By 3:00 PM, 245 animals had died (185 lambs + 60 ewes). Animals that succumbed generally isolated themselves from the flock (lagged behind), became recumbent, and died within 1 hour. In most cases there were no noticeable signs of struggling or ataxia until 15 min prior to death. During the last few minutes difficulty in breathing was observed, followed by recumbency. Most animals died with little struggle or paddling. Moving the apparently 'normal' animals predisposed them to manifest clinical signs and death.

Several plant and water samples from site X were taken and submitted for mineral analysis. Blood and serum samples were also collected from seven animals. Tissues from four dead animals (two ewes and two lambs) were submitted to the Washington Animal Disease Diagnostic Laboratory for histopathology.

Materials and Methods

To reproduce organic Se poisoning in a controlled environment, eight crossbred lambs (3 months old) were acclimatized to the local conditions and were given free-choice alfalfa hay and water. In addition, the lambs had access to salt block. Through a naso-gastric tube, two of the eight lambs were administered 6 mg kg^{-1} Se as SeMet. Two other lambs were administered 8 mg kg^{-1} Se as SeMet. The remaining four lambs were controls.

Results and Discussion

Various plant samples collected from site X were identified and analyzed for selenium using ICP/MS (inductively couples plasma/mass spectrometry).

Table 1. Plants growing at site X and their selenium (Se) concentrations.

Plants in pasture	Se (ppm)
#1 Gumweed (*Grindelia* spp.)	5016–6860
#2 Gumweed (*Grindelia* spp.)	67.6–106.3
#1 Eaton's aster (*Aster eatonii*)	5.62–9.17
#2 Eaton's aster (*Aster eatonii*)	3039–4288
#1 Alfalfa (*Medicago sativa*)	64.4–94
#2 Alfalfa (*Medicago sativa*)	54.4–84.3
#3 Alfalfa (*Medicago sativa*)	48.3–70.6
#4 Alfalfa (*Medicago sativa*)	165–204
#1 Grass (unidentified)	13.2–17
#1 Bear berry (unidentified)	34.7–52.6

Analysis performed by diagnostic laboratories at University of Idaho, University of Wyoming, and Utah State University. All values in parts per million (ppm), on a dry weight basis.

Selenium content was quite variable between and within plant species (Table 1). All tested plant samples contained toxic Se concentrations. Dietary Se concentration above 5 ppm is known to cause chronic Se poisoning (alkali disease) in animals (Rosenfeld and

Beath 1964). An oral dose for acute organic Se poisoning is not well known. Also, it was not possible to estimate the individual amount of each of the listed plants consumed by the affected animals. The three water samples (representing all water sources in that area) collected from site X contained 56, 2.1, and 3 ppb (parts per billion) Se, respectively. These water Se concentrations were not considered of toxicological significance in this case. The Se concentration in biological samples collected from several animals at the site of the incident is listed in Table 2.

Table 2. Selenium (Se) concentration in samples from seven selenium-intoxicated sheep.

Animal no.	Blood	Serum	Liver	Kidney	Heart
1	2.8	2.1	–	–	–
2	1.8	2	7.8	8	6.8
3	1.94	1	6.7	9.6	4.9
4	2.1	2	6.7	8.5	4.8
5	1.7	1.7	8.1	8.1	5.4
6	0.7	0.74	–	–	–
7	0.48	0.5	–	–	–

Analysis performed by Veterinary Diagnostic Toxicology Service, University of Idaho. All data in parts per million (ppm) on a wet weight basis.

The field gross necropsy findings performed at site X included congestion of the lungs and petechial hemorrhages in the epicardium. No cause of sudden death was evident from subsequent histological examination of the submitted tissues. However, liver, heart, and lung lesions were noted. Liver changes (centrilobular hepatocyte degeneration) were consistent with hypoxia, or right heart failure. Protozoan cysts (presumed to be *Sarcocystis* spp.) were present in the heart (causing mild myositis and myocarditis) and considered to be an incidental finding. Lungs had multiple focal hemorrhages within the parenchyma.

Clinical signs in the experimental Se dosing were similar to those reported from the field incident. One lamb from each of the two treatments (6 and 8 mg kg^{-1} Se as SeMet) was euthanized at 24 and 36 hours post dosing, respectively, due to development of severe clinical signs. These lambs were dull, ataxic, and exhibited wheezing and respiratory grunts for about an hour, which was followed by recumbency. At this point the lambs were euthanized. When compared to the control lambs, all four lambs experimentally dosed with SeMet had a reduction in feed intake, were depressed, reluctant to move, and had tachypnea. These affected animals tended to stand with their heads down and necks extended taking short rapid shallow breaths (Figure1). A distinct garlic-like smell was also noted in the breath of all clinically affected lambs.

Gross necropsy findings on the two euthanized lambs revealed pulmonary lesions characterized by interlobular edema and accumulation of serosanguinous fluid and foam in the trachea, bronchi and bronchioles. The hearts of these animals were soft and pale with mottled red appearance. The livers were red and mildly swollen. Acute multifocal myocardial necrosis, pulmonary edema, and alveolar vasculitis with hemorrhage were evident on microscopic examination. In the two euthanized lambs, the Se concentrations in liver and kidney for the 6 and 8 mg Se kg^{-1} bw SeMet treatments were 26.0 and 11.2 ppm, and 23.4 and 12.9 ppm, respectively, on a wet weight basis. The remaining two SeMet-dosed lambs survived the clinical illness and apparently recovered by the end of day 3. The 24 hour post-dosing blood and serum samples from the four SeMet treated lambs ranged from 2.3 to 3.8 ppm and from 2.1 to 2.7 ppm, respectively. The four control animals remained clinically healthy throughout the 7-day observation period and had blood and serum Se concentrations within reference ranges for sheep (reference range for Se in blood of sheep = 0.1–0.5 ppm; reference range for Se in serum of sheep = 0.08–0.4 ppm).

Figure 1. Depression and respiratory distress in sheep with experimental selenium intoxication.

The outcome of the experimental SeMet intoxication was similar to the field case from site X, Soda Springs, ID. These clinical changes and lesions were consistent with previously reported acute sodium selenite toxicosis in sheep (Blodgett and Bevill 1987). A considerable difference existed, however, between the liver Se concentration from the field cases (7.8–8.1 ppm), and the liver Se concentration from the experimental case (26.0 and 23.4 ppm). Liver Se concentrations greater than 3–5 ppm on a wet weight basis are consistent with previously reported acute selenite poisonings (reference range for liver selenium in sheep being 0.25–1.5 ppm). Toxic liver Se concentrations associated with acute organic Se poisoning is not known. However, liver Se concentration in animals administered SeMet can be higher than in animals administered sodium selenite because of the greater availability of organic Se for incorporation into proteins (Panter *et al.* 1996). This is also consistent with the findings in the experimental SeMet poisoning reported here.

Most of the ingested selenite is present as Se in specific selenoproteins such as glutathione peroxidase. Additionally, Se from SeMet can deposit directly and non-specifically into proteins as a methionine replacement. This difference in metabolism can partly explain the difference in liver Se concentrations between selenite- and SeMet-intoxicated animals.

Conclusion

Most losses from acute Se poisoning occur within 1 to 3 days of intoxication and are characterized by depression, dyspnea and death. These findings were consistent with observations in the reported field case, and the SeMet-dosed experimental lambs. The clinical signs and pathological lesions seen in both these cases appeared similar to previously reported cases of acute selenite poisonings. The source of excessive Se in the field case was suspected to be the plant material that is known to predominantly contain organic selenium such as SeMet. Even so, there existed a considerable difference in the liver Se concentrations between field deaths and the SeMet-intoxicated experimental lambs, even though the duration of illness was similar in both cases. Since acute selenite poisoning can cause a similar clinical presentation, with liver Se concentrations similar to that reported in the field case, we believe that a considerable portion of the plant Se in this case is likely to be in a non-SeMet form, and possibly inorganic. It would be reasonable to identify the form of Se present in the plant material in order to provide an explanation for the observed difference in the liver Se concentrations.

Acknowledgements

The authors thank Dr Patricia A. Talcott, Dr Tim Baszler and Dr Wayne Cutler for providing valuable information and diagnostic data related to the sheep deaths at Soda springs, ID. USDA-ARS PPRL, Utah Agricultural Experiment Station, and the Utah State Veterinary Diagnostic Laboratory supported for the experimental portion of the study.

References

Blodgett DJ and Bevill RF (1987). Acute selenium toxicosis in sheep. *Veterinary and Human Toxicology* 29:233–236.

Edmonson AJ, Norman BB, and Suther D (1993). Survey of state veterinarians and state veterinary diagnostic laboratories for selenium deficiency and toxicosis in animals. *Journal of the American Veterinary Medical Association* 202:865–872.

Fessler AJ, Moller G, Talcott PA and Exon JH (2003). Selenium toxicity in sheep grazing reclaimed phosphate mining sites. *Veterinary and Human Toxicology* 45:294–298.

Ganther HE (1986). Pathways of selenium metabolism including respiratory excretory products. *Journal of the American College of Toxicology* 5:1–5.

Ip C (1988). Differential effect of dietary methionine on the biopotency of selenomethionine and selenite in cancer chemoprevention. *Journal of the Natural Cancer Institute* 80:258–262.

Panter KE, Hartley WJ, James LF, Mayland HF, Stegelmeier BL, and Kechele PO (1996). Comparative toxicity of selenium from seleno-DL-methionine, sodium selenate, and *Astragalus bisulcatus* in pigs. *Fundamental and Applied Toxicology* 32:217–223.

Rosenfeld I and Beath OA (1964). *Selenium. Geobotany, Biochemistry, Toxicity, and Nutrition*. Academic Press, New York.

Whanger PD (2002). Selenocompounds in plants and animals and their biological significance. *Journal of the American College of Nutrition* 21:223–232.

Chapter 9

Ornamental Plants in Southern Brazil with Toxic Potential for Companion Animals

J.G. Sena Filho, K.A.Q. Pontual, C.P. Ferreira, D.C. Florencio, and H.S. Xavier

Laboratório de Farmacognosia, Departamento de Ciências Farmacêuticas, Universidade Federal de Pernambuco-UFPE; Av. prof. Arthur de Sá s/n, cidade universitária, 50740-521 Recife-PE, Brazil

Introduction

The accumulated body of knowledge about poisonous plants includes cases of accidental poisoning and even intentional use by humans. Some plant toxins have been used for hunting, killing enemies during war or disputes, and to alter individual mental states as hallucinogens. Even so, plants with toxins are often rich in alkaloids, which provide opportunities for benefits to mankind through the toxin's pharmacological activity (Delaveau *et al.* 1985). Plant toxins are products of secondary metabolism, not necessarily essential to plant development or reproduction, but more likely related to plant defense and survival. Some toxins have a bitter flavor and also may induce physiological alterations that reduce the development or survival of herbivores (Osweiller 1998).

The Center for Toxicological Information (CTI) of the state of Rio Grande do Sul, Brazil, from 1998 to 2002 reported 2792 incidents of toxicity involving animals. Of this total 94% occurred in pets and 3% in livestock.

Materials and Methods

Where possible, cases of intoxication of pets were investigated by personal visits to residences, including homes, yards, apartments and common garden areas, to determine the presence of poisonous plants. When present, toxic plants were collected and prepared for botanical identification. After taxonomic classification, the plants involved were grouped by family and toxin.

Results

The plant species most commonly observed in cases of intoxication of pets (*i.e.*, companion animals) were: *Dieffenbachia picta* Shott*; Sansevieria trifasciata*; *Alamanda cathartica* L.; *A. blanchett* Muell. Arg; *Euphorbia milli*; *E. pulcherrima*; *Lantana camara*; *Philodendron bipinnatifidum*; and *Brugmansia suaveolens* (Table 1).

Table 1. Major toxic plants found to be involved in cases of intoxication of companion animals in the state of Rio Grande do Sul, Brazil, during 1998 to 2002.

Plant Family	Scientific name	Toxin	Reference
Apocynaceae	*Alamanda cathartica* L.; *A. blanchett*	Resin latex	Ladeira 1981
Araceae	*Diffenbachia picta* Shott	Calcium oxalate	Becker 1984
	Philodendron bipinnatifidum	Calcium oxalate	Caldas Barbosa *et al.* 1983
Euphorbiaceae	*Euphorbia milli*; *E. pulcherrima*	Toxoalbumin	Schvartsman 1992
Solanaceae	*Brugmansia suaveolens*	Alkaloids	Scavone and Panizza 1981
Verbenaceae	*Lantana camara*	Lantadenes (triterpenoids)	Caldas Barbosa *et al.* 1983

Conclusion

The correct diagnosis of pet intoxication by household and garden plants is extremely important to establish proper treatment and to prevent future incidences of intoxication. The correct identification of the plant causing intoxication is vital so that the veterinarian can understand the likely toxin and prescribe the proper treatment. In this investigation, we visited households with pet intoxications over a 4-year period in order to determine the identity of the plants involved. The veterinarians involved in these cases used this information to build a database of clinical histories about plant poisonings. Hopefully, this information will be useful to educate the public about the potential dangers of some common household plants to companion animals.

Reference

Becker G (1984). *Plants Toxiques*. Grund, Paris.

Caldas Barbosa RCSB, Barbosa Filho JM, Medeiros DF, and Xavier Filho L (1983). *Revisão e Atualização da Literatura em Plantas Tóxicas do Estado da Paraíba*. Universitária UFPB, Brasil.

Delaveau P, Lorrain M, and Mortier F (1985). *Secrets et Vertus des Plantes Médicinales*. Sélection du Reader's Digest, Paris.

Ladeira AM (1981). *Plantas Tóxicas*, Instituto de Botânica, folheto 14, pp 2–17. São Paulo, Brasil.

Osweiller, GD (1998). *Toxicologia Veterinária*. Artes Médicas, Porto Alegre.

Scavone O and Panizza S (1981). *Plantas Tóxicas*. CODAC/USP, São Paulo, Brasil.
Schvartsman S (1992). *Plantas Venenosas e Animais Peçonhentos*. Sarvier, São Paulo, Brasil.

Chapter 10

Research on Lupine-Induced 'Crooked Calf Disease' at the Poisonous Plant Research Laboratory: Past, Present and Future

K.E. Panter[1], L.F. James[1], T.L. Wierenga[1], C.C. Gay[2], E.S. Motteram[2], S.T. Lee[1], D.R. Gardner[1], J.A. Pfister[1], M.H. Ralphs[1], and B.L. Stegelmeier[1]

[1] *USDA-ARS Poisonous Plant Research Laboratory, Logan, UT 84341, USA;* [2]*Field Disease Investigation Unit, College of Veterinary Medicine, Washington State University, Pullman, WA 99164-6610, USA*

Introduction

The *Lupinus* genus contains more than 500 classified species of annual, perennial, or shrub-like species worldwide (Wink *et al.* 1995); over 300 species in North and South America, 150 species in the Intermountain West, USA (Cronquist *et al.* 1989); 95 species in California (Riggins and Sholars 1993) and at least 12 species in Europe and Africa (Wink *et al.* 1995). Morphological taxonomy of lupine remains a difficult task because of lack of uniformity among flowers and pods and extensive hybridization that may occur between certain species. Lupines belong to the Leguminosae family, with alternate palmately compound leaves with 5–17 oblong to lanceolate leaflets. Flowers are terminal and pea-like and can be a variety of colors including blue, purple, white, yellow, reddish or shades of these colors. The plant emerges early in the spring, flowering in early to mid summer and forming seed pods in late summer to early fall. Plant maturity varies from region to region and depends heavily on environmental conditions, a major factor in years when lupine-induced 'crooked calf disease' causes large losses to the cattle industry in the western US.

There are many domestic lupines cultivated and selected for low alkaloid content and used in commercial seed production systems for animal and human foods. These low alkaloid lupines are referred to as 'sweet' or 'semisweet' lupines, whereas wild lupines containing relatively high levels of alkaloids are called 'bitter' lupines. Commercial production of lupines in the US is increasing while in Australia and other countries this protein source has been utilized for many years. Other lupine species have been selected for ornamental purposes because of their large colorful flower racemes. In Texas the state flower is the wild lupine called bluebonnet (*Lupinus texensis* Hook).

In the western US, lupines continue to cause sporadic economic losses to cattle producers. While lupines are considered to be major poisonous plants in range systems throughout the west, they do contain good nutritional qualities and are important forage sources for livestock and wildlife, especially later in the grazing season. However, when grazed excessively over a short period of time they poison and kill livestock in addition to causing the more common 'crooked calf disease' syndrome in pregnant cattle. This syndrome has been extensively studied and much information is known to help ranchers and land managers utilize ranges for maximum production.

This discussion of lupine research, past, present and future in this text will be limited to that involving poisoning in livestock and more specifically the problem of the lupine-induced 'crooked calf disease' in the western US.

Past Research, Historical Perspectives

Lupine poisoning in the western US was first recognized in sheep production systems in the late 1800s or early 1900s as an acute intoxication (Chesnut and Wilcox 1901). Poisoning of sheep in the late 1800s and early 1900s usually occurred when sheep were allowed to graze lupine in the seed pod stage or when fed lupine hay with many seed pods attached. Heavy grazing of immature plants occasionally occurred when sheep and cattle were trailed long distances or unloaded from rail cars into lupine-infested pastures. Bedding sheep near heavy lupine populations or when hungry animals were introduced directly into or near heavy patches of lupine resulted in losses.

Unfortunately, losses similar to this occur still today, although not to the extent as in the past because range bands are much smaller and we now understand lupine as the source of poisoning. Lupines are usually not extensively grazed by sheep through spring and early summer when other forage is plentiful. Cattle and horses may graze early-growth lupine before other more desirable forages emerge. However, losses among cattle and horses are rare. In late fall and early winter, lupines remained somewhat green and succulent after other forage had dried up and were considered a very important late season forage. In the late 1800s and early 1900s lupines not only offered good green forage for grazing but were among the most important native hay crops.

In the late 1800s the practice of feeding lupine hay in the winter to many sheep bands would suggest these ranchers knew lupine hay was a good food source for winter. Lupine has feed value equal to alfalfa hay and seeds are often 30–40% protein. Unfortunately, seed pods also contain 3–5 % total alkaloids and most of these alkaloids are toxic, and if ingested in high enough quantities will cause poisoning and result in acute death.

Harvesting lupine hay in those early days occurred between the first of July and the middle of September. If harvested in September the seed pods would be shattered; seeds were therefore absent and the toxicity low. If hay was harvested in July seed pods would more often be likely to contain toxic levels of alkaloids. While not understood then, it was later determined that most losses resulted when winter hay harvested in July was fed. This proved to be especially true in the winter of 1898–1899, when thousands of sheep died in the Judith Basin of Montana from feeding lupine hay (Chesnut and Wilcox 1901). In one flock alone, 3600 of 7000 sheep died from eating lupine hay. Many sheep producers lost over 50% of their flocks that winter from lupine hay poisoning. The early harvest of lupine hay between July 1 and July 20 and increased seasonal moisture resulted in the 'unusual quantity of seeds' in the hay. While lupine alkaloids had not yet been identified, it became clear to these early investigators that the seed pods were 'rich in the poison'. It is also important to note that seeds of many lupines also contain high levels of protein, which if ingested in moderation are a good source of nutrients.

With improved understanding and the unfortunate experience of heavy losses associated with lupine, the practice of feeding lupine hay was discontinued. Since the turn of the century, large losses in sheep flocks rarely occur, although isolated cases of poisoning and death losses continue to occur (Panter, unpublished cases, 2005).

The association with 'crooked calf disease' was not reported until the late 1950s (Palotay 1959; Wagnon 1960; Binns and James 1961). The 'crooked calf syndrome' was described as a condition in calves born with skeletal malformations including arthrogryposis, scoliosis, kyphosis, torticollis and occasionally cleft palate (Shupe *et al.* 1967 a,b, 1968). Chemically, the quinolizidine and piperidine alkaloid types were identified as the teratogens; however, the primary teratogen was reported to be the quinolizidine, anagyrine (Keeler 1976; Keeler and Panter 1989). Although the piperidine alkaloids are known to cause the same skeletal malformations in cattle, the condition has only been produced in feeding trials in a laboratory setting and has not been reported in the field. Therefore, monitoring lupine species for anagyrine content and concentration remains an important tool in identifying and managing lupine-induced 'crooked calf disease' in cattle.

While lupine-induced losses in sheep have declined, lupines continue to cause large losses to cattle producers, especially cow/calf operators. 'Crooked calf disease' continues to cause major economic losses in the western US. The most recent outbreak occurred in the Scabland Region of Washington State in 1997, where more than 2800 calves in a single county were born with skeletal defects so severe that most required euthanasia. Thirteen ranches were surveyed specifically and 628 of 2210 calves born were deformed (Gay *et al.* 2007). The individual ranch incidence of crooked calves ranged from 0 to 100%. This was an economic catastrophe for this area's cattle producers and associated businesses. This loss followed a similar catastrophic loss to a single producer in central Oregon, in which over 56% of his calf crop was lost (Panter *et al.* 1997) and a later incident in southeastern Idaho in 2003 where 40% of the calves in one herd were destroyed (Panter, unpublished data, 2004).

From 1950 through the 1960s lupine-induced 'crooked calf disease' was emerging as a major problem to cow/calf producers in the northwestern US. Binns, James and Shupe described the condition, reproduced the skeletal malformations and cleft palate in pen studies, and defined the susceptible stages of pregnancy (Binns and James 1961; Shupe *et al.* 1967 a,b, 1968). During the late 1960s Shupe and others identified the most sensitive stages of pregnancy in cattle to be 40 to 70 days gestation potentially extending to day 100 (Shupe *et al.* 1967 a,b). These same researchers described and characterized the skeletal malformations as contracture-type defects and included arthrogryposis, scoliosis, kyphosis, torticollis and associated defects to the rib cage, etc. Bone density and growth appeared to be normal but contractures or mal-alignment occurred at any anatomical location in the fetus where a joint or articulating surface occurred. Shupe *et al.* (1968) identified cleft palate as a condition that occurred with the crooked calf syndrome. Soon after the disease was described the toxins were characterized and a single quinolizidine alkaloid (anagyrine) was determined to be the primary cause of the deformities (Keeler 1976). Later, the piperidine alkaloid ammodendrine was identified in a few lupines in the western US. Ammodendrine was confirmed as a teratogen causing 'crooked calf disease.' While ammodendrine is found in fewer lupine species it is more potent than anagyrine and closely related to teratogens in poison-hemlock and *Nicotiana* spp. (Keeler and Balls 1978; Keeler 1979; Keeler and Panter 1989).

Present Research

From the late 1980's to present most research concentrated on further defining the critical stage of pregnancy when lupine-induced 'crooked calf disease' would occur, defining the fetal response to poisoning, further characterizing the toxins and teratogens and determining plant animal relations to identify when and why cattle graze lupines.

In 1983, Panter (1983) speculated that these malformations were associated with a reduction in fetal movement. Using radio-ultrasound in a goat model it was determined that the teratogenic alkaloids in lupines, poison-hemlock and *Nicotiana* spp. induced a reduction in fetal movement and if severe enough and persistent over the susceptible stages of pregnancy, skeletal malformations would result (Panter *et al.* 1990). Similarly, the cleft palate is induced at the time of palate closure (a narrow period) when the fetus is unable to move, thus preventing the tongue from dropping from the roof of the mouth during a very critical time of pregnancy (Panter and Keeler 1992). The narrow period of cleft palate induction was described to be gestation days 30–41 in pigs (Panter *et al.* 1985), 35–41 in goats (Panter and Keeler 1992) and 40–50 in cattle (Panter *et al.* 1998). A similar susceptible gestation period is speculated to be 50–60 in humans.

Ultrasound studies have demonstrated that first fetal movements in the normal goat become evident at about day 35 of gestation (Panter and Keeler 1992). Interestingly, this coincides with the end of the embryonic period and the beginning of the fetal period when the palatal shelves close and fuse. Further observations suggest that the first fetal movements observed are extension/flexure-type movements of the head and neck. Theoretically this movement is critical as it allows the tongue to drop from the roof of the mouth to permit the palatal shelves to come together at the pre-programmed time of fusion (38–40 days in the goat). In support of this theory, we have induced cleft palate and arthrogryposis in goats using two neuromuscular blockers (succinyl choline and tubocurarine) infused into the amniotic vesicle early in gestation (Panter, unpublished data). In addition to the ultrasound studies, Panter *et al.* (1999) have documented deformities in calves and twin goats of compression malformations of the rib cage, head, neck, spine and legs obviously caused by pressure *in utero* during protracted inhibition of fetal movement. In these cases the fetal positioning at the time of insult could be reconstructed postnatally, demonstrating the cause of certain rib cage deformities, or pressure defects of the spine, neck, limbs etc.

Currently, one quinolizidine alkaloid (anagyrine) and three piperidine alkaloids (ammodendrine, N-methyl ammodendrine, and N-acetyl hystrine) are believed to be teratogens in lupine (Figure 1). Earlier research on teratogenic alkaloids of poison-hemlock and *Nicotiana* spp., which cause the same type of malformations, demonstrated that certain chemical structural requirements were necessary for the alkaloid to induce these contracture-type malformations and cleft palate (Keeler and Balls 1978). While these structural requirements have been accepted for the piperidine alkaloids, there still remains a major question about the teratogenic quinolizidine alkaloid anagyrine. Anagyrine and those lupines containing only quinolizidine alkaloids appear to only be teratogenic in cattle but not sheep and goats. Lupines containing the piperidine teratogens ammodendrine or the N-methyl derivative are teratogenic in cattle, sheep and goats. Research theories of species-specific metabolism and toxicokinetics have been proposed but not yet resolved (Keeler and Panter 1989).

Future Research

In 1999 greater research emphasis was placed on plant/animal relationships in the context of why and when pregnant cattle most prefer lupine plants. This emphasis resulted from the catastrophic losses that occurred in the Scabland Region of eastern Washington State in 1997 and continues today. For the last 6 years grazing studies and pen feeding trials have been conducted in collaboration with the Field Disease Investigation Unit of the Department of Veterinary Medicine at Washington State University. The results of some of these studies are reported in other chapters of this publication.

Because of this collaboration and meetings with livestock producers, important questions remain unanswered for the management of lupine grazing. Some of these questions include: (1) How do environmental factors influence lupine survival, population cycles and toxin profiles; (2) How much lupine and for how long does a pregnant cow need to graze lupine to result in crooked calves; (3) Why do cattle graze lupines and what factors cause cattle to eat lupine when they do; (4) What is the disposition of anagyrine and other teratogens in the fetal compartment; (5) Can a gene marker be identified in livestock as a tool to select for resistance or sensitivity to the effects of lupine; and (6) What improved management strategies can be recommended for ranchers to reduce losses? We believe that answering these research questions and others will not only benefit producers in eastern Washington but will help all ranchers that manage rangelands with poisonous plants.

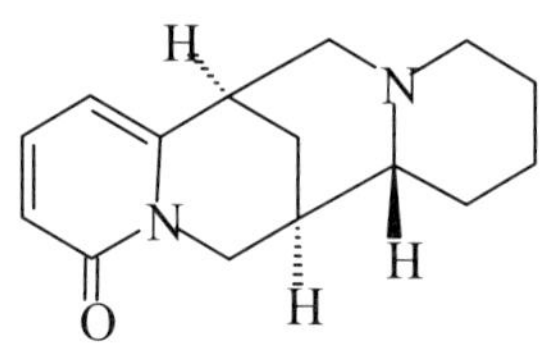

Anagyrine

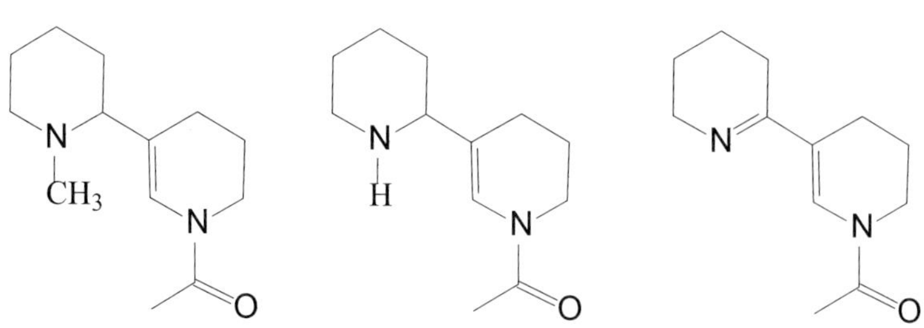

N-methyl ammodendrine Ammodendrine N-acetyl hystrine

Figure 1. Anagyrine, the quinolizidine teratogen, and known piperidine teratogens (N-methyl ammodendrine, ammodendrine, and N-acetyl hystrine).

Recent grazing studies in eastern Washington State determined that cattle graze lupines more readily when grasses become mature, dry and are less available (Ralphs *et al.* 2006). This is not surprising and is expected where cattle graze open ranges. Once grasses are less available cattle rely on forbs and shrubs for nutrients. Lupine is an important source of nutrients, especially when other forages become mature and less available. Lupine is known to be very nutritious, containing 18% protein in its vegetative parts and over 40% protein in the seeds (Panter *et al.* 2001). Research in the Scabland Region determined that cows start to graze lupines early in July when cheatgrass has matured and other forbs were being depleted. This time frame overlaps the critical period of gestation (40–100 days) under current breeding schedules when crooked calf disease is most likely to occur. Most recently, grazing studies conducted during the first week of July using native cattle demonstrated that cattle grazed cheatgrass, wheatgrasses, lupine, skeleton weed and mustards very readily. The cattle grazed lupines beginning the first day of the study and continued to graze it at about 10% of the diet until all lupine was gone. Frequently cows would eat large lupine plants until most of the leaves, flowers, and seed pods were gone then move on to other forages (Panter, personal observation, 2005). While cattle did not seek out the lupine they readily grazed it as they came to the lupine plants during their normal grazing routine.

Lactation did not change a cow's propensity to graze lupine (Pfister, unpublished data, 2005). However, body condition appears to influence the animal's= propensity to graze plants like lupines. This hypothesis is supported by unpublished studies done with cows grazing pine needles, in which cattle in low body condition grazed pine needles earlier and for longer periods of time compared to cows in better body condition. In sheep, body condition was an important factor in disposition of lupine alkaloids, increasing absorption rate and decreasing elimination rate (Lopez-Ortiz *et al.* 2004). Disposition of alkaloids in cattle was not different in cows that had a history of crooked calf disease versus cows with no crooked calf history (Gay *et al.* 2004). Future research is required to compare alkaloid (particularly anagyrine) absorption and elimination in cattle under different scenarios. While absorption, distribution and elimination of the teratogenic alkaloid, anagyrine, in the cow is important, understanding these same parameters in the fetal compartment is critical and the focus of future research.

While anagyrine is believed to be the only quinolizidine alkaloid in lupine responsible for crooked calf disease, there are at least three piperidine alkaloids in lupines with teratogenic activity and six alkaloids in other plant species that will cause the same contracture-type skeletal defects. Research to determine toxic and teratogenic potency of these alkaloids based on chemical structure/activity relationship is ongoing. Undoubtedly there are other alkaloids we don=t know about that are also teratogenic and fit the structural pattern for teratogenicity established by Keeler and Balls (1978). Interestingly, the two teratogenic piperidines found in some lupines have enantiomeric pairs that occur naturally. While we don=t know if one enantiomer is more teratogenic than another, we do know there is a difference in toxicity (Lee *et al.* 2005). Linking teratogenicity to one or both enantiomers is important and is a significant element of future research.

Even though extensive research has been done in the past and present research has answered critical questions, there is still much information needed to further reduce or eliminate losses to cattle producers. Furthermore, collaborative research between the Poisonous Plant Research Lab and biomedical partners at Harvard University Medical School, Lahey Clinic Department of Plastic Surgery, and University of Michigan Medical School is providing leading edge information for improved treatment of cleft palate in children (Weinzweig *et al.* 2002). Using the goat model, fetal surgical treatment of cleft palate and new understandings and techniques to manage cleft palate-related issues postnatally are being investigated.

References

Binns W and James LF (1961). A congenital deformity in calves, similar to 'crooked calf disease,' has been experimentally produced by feeding heifers lupine and lead. *Proceedings of the Western Section of the American Society of Animal Production* 12:LXVI, 1–3.

Chesnut VK and Wilcox EV (1901). The stock poisoning plants of Montana: A preliminary report. *U.S. Department of Agriculture Washington DC. Bulletin* No. 26, pp. 100–110.

Cronquist A, Holmgren AH, Holmgren NH, Reveal JL, and Holmgren PK (1989). *Intermountain Flora, Vascular Plants of the Intermountain West, USA*, V. 3B, pp. 236–268.

Gay CC, Panter KE, Mealey KL, Gay JM, Hjartarson SW, Tibary A, Motteram ES, Wierenga T, and James LF (2004). Comparison of plasma disposition of alkaloids after lupine challenge in cattle that had given birth to calves with lupine-induced arthrogryposis or clinically normal calves. *American Journal of Veterinary Research* 65(11):1580–1583.

Gay CC, Panter KE, Motteram E, Gay JM, Hantz H, Wierenga T, and Platt T (2007). Risk factors for lupine-induced crooked calf disorder in east-central Washington State. *Chapter 27, this publication.*

Keeler RF (1976). Lupine alkaloids from teratogenic and nonteratogenic lupins. III. Identification of anagyrine as the probable teratogen by feeding trials. *Journal of Toxicological and Environmental Health* 1:887–889.

Keeler RF (1979). Congenital defects in calves from maternal ingestion of *Nicotiana glauca* of high anabasine content. *Clinical Toxicology* 15:417–426.

Keeler RF and Balls LD (1978). Teratogenic effects in cattle of *Conium maculatum* and *Conium* alkaloids and analogs. *Clinical Toxicology* 12:49–64.

Keeler RF and Panter KE (1989). Piperidine alkaloid composition and relation to crooked calf disease-inducing potential of *Lupinus formosus*. *Teratology* 40:423–432.

Lee ST, Molyneux RJ, Panter KE, Chang CWT, Gardner DR, Pfister JA, and Garrossian M (2005). Ammodendrine and N-methylammodendrine enantiomers: Isolation, optical rotation, and toxicity. *Journal of Natural Products* 68:681–685.

Lopez-Ortiz SL, Panter KE, Pfister JA, and Launchbaugh KL (2004). The effect of body condition on disposition of alkaloids from silvery lupine (*Lupinus argenteus* Pursh) in sheep. *Journal of Animal Science* 82:2798–2805.

Palotay JL (1959). Crooked calves. *Western Veterinarian* 6:16–20.

Panter KE (1983). Toxicity and teratogenicity of *Conium maculatum* in swine and hamsters. PhD dissertation, University of Illinois, Urbana.

Panter KE, Keeler RF, and Buck WB (1985). Induction of cleft palate in newborn pigs by maternal ingestion of poison-hemlock (*Conium maculatum*). *American Journal of Veterinary Research* 46:1368–1371.

Panter KE and Keeler RF (1992). Induction of cleft palate in goats by *Nicotiana glauca* during a narrow gestational period and the relation to reduction in fetal movement. *Journal of Natural Toxins* 1:25–32.

Panter KE, Bunch TD, Keeler RF, Sisson DV, and Callan RJ (1990). Multiple congenital contractures (MCC) and cleft palate induced in goats by ingestion of piperidine alkaloid-containing plants: Reduction in fetal movement as the probable cause. *Clinical Toxicology* 28:69–83.

Panter KE, Gardner DR, Gay CC, James LF, Mills R, Gay JM, and Baldwin TJ (1997). Observations of *Lupinus sulphureus*-induced 'crooked calf disease.' *Journal of Range Management* 50:587–592.

Panter KE, Gardner DR, and Molyneux RJ (1998). Teratogenic and fetotoxic effects of two piperidine alkaloid-containing lupines (*L. formosus* and *L. arbustus*) in cows. *Journal of Natural Toxins* 7:131–140.

Panter KE, James LF, and Gardner DR (1999). Lupines, poison-hemlock and *Nicotiana* spp: Toxicity and teratogenicity in livestock. *Journal of Natural Toxins* 8(1):117–134.

Panter KE, Mayland HF, Gardner DR, and Shewmaker G (2001). Beef cattle losses after grazing *Lupinus argenteus* (silvery lupine). *Veterinary and Human Toxicology* 43:279–282.

Ralphs MH, Panter KE, Gay C, Motteram E, and Lee ST (2006). Cattle consumption of velvet lupine (*Lupinus leucophyllus*) on the Channel Scablands of eastern Washington. *Rangeland Ecology and Management* 59:204–207.

Riggins R and Sholars R (1993). Lupinus. In: *Jepson Manual, Higher Plants of California*, (JC Hickman, ed.), pp. 622–636. University of California Press, Berkeley.

Shupe JL, Binns W, James LF, and Keeler RF (1967a). Lupine, a cause of crooked calf disease. *Journal of the American Veterinary Medical Association* 151:198–203.

Shupe JL, James LF, and Binns W (1967b). Observations on crooked calf disease. *Journal of the American Veterinary Medical Association* 151:191–197.

Shupe JL, James LF, Binns W, and Keeler RF (1968). Cleft palate in cattle. *Cleft Palate Journal* 1:346–354.

Wagnon KA (1960). Lupine poisoning as a possible factor in congenital deformities in cattle. *Journal of Range Management* 13:89–91.

Weinzweig J, Panter KE, Spangenberger A, Harper JS, McRae R, and Edstrom LE (2002). The fetal cleft palate: III. Ultrastructural and functional analysis of palatal development following *in utero* repair of the congenital model. *Plastic and Reconstructive Surgery* 109(7):2355–2362.

Wink M, Meibner C, and Witte L (1995). Patterns of quinolizidine alkaloids in 56 species of the genus *Lupinus*. *Phytochemistry* 38:139–153.

LIVER

Chapter 11

Intoxication by the Pods of *Enterolobium contortisiliquum* in Goats

T.M.A. Benício, M.J. Nardelli, F.R.B. Nogueira, J.A.S. de Araújo, and F. Riet-Correa

Centro de Saúde e Tecnologia Rural, University of Campina Grande, Campus de Patos, Paraíba, Brazil, 58700-000, PB

Introduction

Many species of *Enterolobium* (Leguminosae-Mimosoideae) are found in Brazil including *E. contortisiliquum*, *E. timbouva, E. gummiferum* and *E. shombungkii* (Lorenzi 1998). *Enterolobium contortisiliquum* (Vell.) Morang (common names: orelha de negro, orelha de macaco, timbaúba, tamboril) is a tree up to 20–35 m high and found in many Brazilian regions, from Pará to Rio Grande do Sul. It flowers in September–November and sets seeds in June–July, but the pods (6–10 cm long) stay on the tree for a longer time. It has characteristic black pods, with a shape similar to a human ear (Figure 1; Lorenzi 1998).

In Brazil, the intoxication by pods of *E. contortisiliquum* has been reported in the states of Bahia (Tokarnia *et al.* 1999), Rio Grande do Sul, and Mato Grosso (Grecco *et al.* 2002). Intoxication by pods of *Enterolobium gummiferum* was reported from the state of Minas Gerais (Deutsch *et al.* 1965), and the intoxication by pods of *Enterolobium timbouva* has been reported from São Paulo (Tokarnia *et al.* 1999) and Mato Grosso do Sul (Lemos *et al.* 1998). Clinical signs observed in spontaneous intoxication by *Enterolobium* spp. are characterized by photosensitization (Lemos *et al.* 1998; Tokarnia *et al.* 1999; Grecco *et al.* 2002) and abortion in cows (Tokarnia *et al.* 1999; Grecco *et al.* 2002). In attempts to produce the intoxication by the administration of *E. contortisiliquum* and *E. timbouva*, the main clinical signs were characterized by anorexia, depression and diarrhea, but not photosensitization (Tokarnia *et al.* 1960, 1999) or abortion (Tokarnia *et al.* 1999). In other experiments the administration of *E. gummiferum* (Deutsch *et al.* 1965) and *E. timbouva* (Lemos *et al.* 1998) caused photosensitization. Mild photosensitization was observed by Grecco *et al.* (2002) in calves that survived the acute digestive signs induced by the administration of *E. contortisiliquum* pods. Both the experimental and the spontaneous intoxication cause liver degeneration and necrosis (Tokarnia *et al.* 1960, 1999; Grecco *et al.* 2002). Anecdotal reports by ranchers in the state of Paraíba have linked cases of diarrhea and anorexia to the ingestion of *Enterolobium* spp. (Riet-Correa *et al.* 2003).

Figure 1. *Enterolobium contortisiliquum* tree and pods (insert).

The intoxication was also produced in guinea pigs with doses of 10 g kg^{-1} bw of pods of *E. contortisiliquum*. The animals died 12–18 hours after administration. The liver and gallbladder were enlarged and severe hemorrhagic lesions were also observed in the gut (Grecco 1999). Histologic lesions of the liver, characterized by severe periportal vacuolation, suggested that those lesions are caused by the same chemical compound that causes photosensitization and liver lesions in cattle. Saponins isolated from *E. gummiferum* were toxic to guinea pigs but the pathology of the intoxication in this species was not studied (Carvalho 1981).

In this paper we report two cases of spontaneous intoxication by the pods of *E. contortisiliquum* in goats in the state of Paraíba, northeastern Brazil, and the experimental reproduction of the disease in this species.

Case Report

The disease occurred in July 2004 in the municipality of Puchinanan, in the semiarid region of the state of Paraíba, when many pods of *E. contortisiliquum* fell to the ground. One male goat restrained beneath a tree of *E. contortisiliquum* ingested the pods for approximately 3 days. After this period, the goat had diarrhea for 3 days and died. At necropsy, large amounts of *E. contortisiliquum* seeds were observed in the digestive tract, and the gut and abomasum were reddish colored. Approximately 2 weeks later another pregnant goat, kept under the same *E. contortisiliquum* tree in the same conditions, aborted after ingestion of the pods. No clinical signs were reported by the farmer in the aborted goat. No significant lesions were observed in the histology of the liver, kidney, and heart of the goat that died nor in the heart, spleen, lung, and liver of the fetus from the goat that aborted.

Experimental Intoxication

Pods were collected under the same tree that caused the spontaneous intoxications. The ground pods were given experimentally to four goats. A one-year-old male and an adult female received 10 g kg^{-1} bw. In the same day of administration, the male goat had anorexia, depression, abdominal pain, continuous head movements, salivation, and bloating. After 24 hours the animal had nasal and ocular discharge and soft feces with mucous. It fully recovered in 5 days. The female adult goat had no clinical signs. Another two adult goats received 15 and 20 g kg^{-1} bw of the pods, respectively. Immediately after administration, both animals showed abdominal pain, salivation, and dyspnoea, and after 4–5 hours, severe diarrhea. Both animals died after a clinical manifestation period of 7–8 hours. At necropsy, the liver was yellowish, and the abomasum and large and small gut were reddish. The main histological lesions were severe enteritis with epithelial necrosis in the gut and a few necrotic hepatocytes in the liver. One goat had diffuse vacuolization of the periportal hepatocytes, but this lesion was probably not associated with the intoxication. Mild vacuolation of hepatocytes was also observed in the other goat. The stratified epithelium of the rumen had severe vacuolation and vesicle formation followed by infiltration of neutrophils with formation of intraepithelial pustules.

Discussion

The history of spontaneous intoxication in goats ingesting pods of *E. contortisiliquum,* and the reproduction of the intoxication with the same signs observed in the goat that died spontaneously, confirmed the occurrence of the intoxication in this species. Previously, the spontaneous and experimental intoxication by pods of *Enterolobium* spp. had been reported only in cattle (Tokarnia *et al.* 1960, 1999; Deutsch *et al.* 1965; Lemos *et al.* 1998; Grecco *et al.* 2002).

In the spontaneous cases, two types of clinical signs were observed: abortion and acute digestive signs. In the experimental reproduction, digestive signs and lesions were similar to that reported in the experimental intoxication by *E. contortisiliquum* in cattle (Tokarnia *et al.* 1960; Grecco *et al.* 2002). Similar digestive signs had been observed in cattle by farmers in Paraíba after the ingestion of *E. contortisiliquum* pods. The main histologic lesions affected the gut, the forestomachs and the liver. Congestion and hemorrhages of the gut were also observed in cattle (Grecco *et al.* 2002) and in guinea pigs (Grecco 1999). Necrosis and degeneration of the liver are also observed in cattle (Tokarnia *et al.* 1960, 1999; Grecco *et al.* 2002) and guinea pigs (Grecco, 1999) and are probably responsible for the photosensitization observed in field cases in cattle. The stratified epithelium of the rumen had vacuolation and vesicle formation followed by infiltration of neutrophils with formation of intraepithelial pustules (Grecco *et al.* 2002).

Abortion had been observed in spontaneous intoxication by *Enterolobium* spp., but was not reproduced experimentally. However, abortion was observed in guinea pigs after the administration of a non-lethal dose of ground pods (Josiane B. Raposo, Federal University of Pelotas, 2002, unpublished data). In cattle, the most frequent clinical sign associated with the intoxication is hepatogenous photosensitization, which was not observed in these field cases in goats.

The toxic dose of *E. contortisiliquum* pods for goats (10–20 g kg^{-1} bw) was similar to the toxic dose for cattle. In cattle the lethal dose varied from 10 to 12.5 g kg^{-1} bw (Tokarnia *et al.* 1960; Grecco *et al.* 2002). Cattle with smaller doses had clinical signs and recovered (Tokarnia *et al.* 1999). Nevertheless, we have seen goats ingesting pods of *E. contortisiliquum* without apparent toxic effect, and some believe that the pods can be used as forage (Braga 1960), especially for sheep (Tokarnia *et al.* 2000). This lack of toxicity may be due to the occurrence of non-toxic plants or to resistance acquired by the animals. Tokarnia *et al.* (1960, 1999) demonstrated that cattle become tolerant to the toxicity of *E. contortisiliquum* and *E. timbouva* pods after the ingestion of successive, increasing doses. The acquisition of resistance to toxic plants is well known with some plants that contain proteins as active principles. The seeds of *Ricinus communis* and *Abrus precatorius* when given in increasing doses confer resistance to much higher amounts of the toxic principles, which are toxic albumins. The treatment by heat of the seeds eliminates the toxicity but maintains the antigenicity (Tokarnia *et al.* 2000).

Acknowledgements

This work was financially supported by the Programa de Apoio a Núcleos de Excelência (PRONEX-Grant Nº 001/04, CNPq, FAPESQ, MCT). We thank Graça Marinho, PhD, University of Campina Grande, Campus of Patos, Patos, PB, for the identification of the plant.

References

Braga R (1960). *Plantas do Nordeste, especialmente do Ceará*, 2nd ed. Imprensa Oficial, Fortaleza.

Carvalho LR (1981). *Estudos químicos e biológicos de uma saponina do Enterolobium gummiferum*. Master thesis, University of São Paulo.

Deutsch J, Döbereiner JV, and Tokarnia CH (1965). Fotossensibilidade hepatogênica em bovinos na intoxicação pela fava de *E. gummiferum. Proceedings, IX Congresso Internacional de Pastagens*, São Paulo, pp. 1279–1282.

Grecco FB (1999). *Intoxication by* Enterolobium contortisiliquum *in cattle and guinea pigs*. Master thesis, Pelotas University.

Grecco FB, Dantas AFM, Riet-Correa F, Leite CGD, and Raposo JB (2002). Cattle intoxication by *Enterolobium controtisiliquum* pods. *Veterinary and Human Toxicology* 44:160–162.

Lemos RA, Purisco L, Nakazato L, Bonilha R, Gattass CBA, and Brum KB (1998). Intoxicação experimental por *Enterolobium timbouva* Mart em bovinos. *Arquivos Instituto Biológico*, São Paulo, 65 (suppl.):109.

Lorenzi H (1998). *Árvores brasileiras, Manual de Identificação e Cultivo de plantas árboreas do Brasil*, 2nd ed. Editora Plantarum, Nova Odessa, São Paulo.

Riet-Correa F, Tabosa IM, Azevedo EO, Medeiros RMT, Simões SVD, Dantas AA, Alves CJ, Nobre VMT, Athayde AC, Gomes AA, and Lima EF (2003). Doenças dos ruminantes e eqüinos no semi-árido da Paraíba. *Semi Árido em Foco*, Patos, 1:2–86.

Tokarnia CH, Canella CFC, and Döbereiner J (1960). Intoxicação experimental pela fava da 'timbaúba' *Enterolobium contortisiliquum* (Vell) Morong em bovinos. *Arquivos Instituto de Biologia Animal* 3:73–81.

Tokarnia CH, Döbereiner J, Dutra IS, Brito IS, Chagas BR, França TN, and Brust LAG (1999). Experimentos em bovinos com as favas de *Enterolobium contortisiliquum* e *Enterolobium timbouva* para verificar propriedades fotossensibilizantes e/ou abortivas. *Pesquisa Veterinária Brasileira* 19:39–45.

Tokarnia CH, Döbereiner J, and Peixoto PV (2000). *Plantas Tóxicas do Brasil.* Editora Helianthus, Rio de Janeiro.

Chapter 12

Poisonous Plants Producing Acute Hepatic Disease in Brazilian Cattle

D.R. Rissi[1], D. Driemeier[2], M.C. Silva[1], R.R Barros[1], and C.S.L. Barros[1]

[1]*Laboratório de Patologia Veterinária, Universidade Federal de Santa Maria, Santa Maria, RS, 97105-900, Brazil;* [2]*Departamento de Patologia Clínica Veterinária, Universidade Federal do Rio Grande do Sul, Porto Alegre, RS, 91540-000, Brazil*

Introduction

Poisonous plants are an important cause of death in Brazilian cattle. Data from our laboratory indicate that approximately 14% of cattle deaths in our region (southern Brazil) are caused by ingestion of poisonous plants. Plants causing acute hepatic lesions in cattle constitute one important group of poisonous plants in Brazil. Acute hepatic failure induced by poisonous plants in cattle runs a course with clinical signs, necropsy and histopathology findings which are very similar regardless of the specific plant that is the cause. Thus, in most cases a field investigation is necessary to establish which plant is causing the deaths. Known poisonous plants that cause spontaneous acute hepatic lesions in cattle in Brazil include *Cestrum parqui* (Riet-Correa *et al.* 1986), *C. corymbosum* var. *hirsutum* (Gava *et al.* 1991), *C. intermedium* (Gava *et al.* 1996), *C. laevigatum* (Döbereiner *et al.* 1969), *Xanthium cavanillesii* (Méndez *et al.* 1998; Driemeier *et al.* 1999), *Vernonia mollissima* (Döbereiner *et al.* 1976; Gava *et al.* 1987), *V. rubricaulis* (Tokarnia and Döbereiner 1982; Brum *et al.* 2002), *Sessea brasiliensis* (Canella *et al.* 1968) and *Dodonea viscosa* (Colodel *et al.* 2003). An acute hepatoxicosis was induced in cattle by the experimental feeding of the green leaves of *Trema micrantha* (Traverso *et al.* 2004). Although the spontaneous disease caused by *T. micrantha* is reported only in goats, cattle from Santa Catarina state (SC) are suspected of being spontaneously affected. This review is intended to be a source of information to serve as an aid in the differential diagnosis of acute liver diseases caused by ingestion of hepatic toxic plants in cattle.

Epidemiology

Under natural conditions cattle from several regions of Brazil are affected by plants of this group (Tokarnia *et al.* 2000). *V. mollissima* and *S. brasiliensis* cause disease in cattle from limited areas in the midwest and south, respectively; although *V. mollissima* is observed in the southern state of Rio Grande do Sul (RS) the disease is not reported in cattle there. *Cestrum corymbosum* var. *hirsutum*, *C. parqui*, *C .intermedium*, *D. viscosa* (Colodel *et al.* 2003) and *X. cavanillesii* (Méndez *et al.* 1998; Driemeier *et al.* 1999) cause disease only in cattle in the southern states. Furthermore, *C. parqui*, *D. viscosa* and *X. cavanillesii* are reported only from RS and *C. corymbosum var. hirsutum* and *Cestrum intermedium* only from the southern state of SC. The disease caused by the ingestion of *Vernonia rubricaulis* is reported from the Pantanal region of the midwestern states of Mato Grosso (MT) and Mato Grosso do Sul (MS). Poisoning by *Cestrum laevigatum* occurs in the southeast, midwest and northeast but in different locations than *S. brasiliensis*, *V. mollissima* and *V. rubricaulis*. Poisoning in cattle by *D. viscosa* is described only from the coastal region of RS (Colodel *et al.* 2003).

Morbidity and mortality rates are variable (5–30%) and lethality approaches 100% (Brum *et al.* 2002), although usually the outbreaks are associated with a dearth of forage and hunger. However, there are exceptions since *C. parqui* toxicosis was reported to occur in good pastures (Riet-Correa *et al.* 1986). *Trema micrantha* is reportedly palatable for cattle (Traverso *et al.* 2004) and *X. cavanillesii* poisoning occurs after flooding of pastures, which induces sprouting of this weed (Méndez *et al.* 1998). Additionally the burrs of *X. cavanillesii* cause the toxicosis when admixed with mechanically harvested grains supplied to cattle (Driemeier *et al.* 1999).

Conditions favoring the appearance of outbreaks of *V. rubricaulis* poisoning in cattle in MS are agricultural management measures that induce sprouting of the plant (burning or jungle clearing of the paddocks) and shipping of cattle from other farms, changing them from one paddock to another on the same farm, and high stocking rate (Brum *et al.* 2002). It is reported that draft oxen after a day of hard labor are especially prone to *C. intermedium* poisoning (Gava *et al.* 1996).

Clinical Signs

Clinical signs in affected cattle appear shortly (12–16 hours) after ingestion and run an acute course (2–72 hours), although longer clinical courses have been reported in the experimental poisoning with *T. micrantha* in cattle (Traverso *et al.* 2004). Main findings include anorexia, apathy, dry muzzle, drooling, tachycardia, tachypnea (Chaulet *et al.* 1990), decreased ruminal movements or ruminal stasis, tenesmus and dehydration; feces are generally dry with blood streaks and covered by a film of mucous (Riet-Correa *et al.* 1986; Tokarnia *et al.* 2000). Signs such as generalized muscle tremors, uncoordinated gait, hyperexcitability, aggression and paralysis of pelvic limbs have been frequently reported (Driemeier *et al.* 1999). In the terminal stages, head pressing (Driemeier *et al.* 1999), seizures, grinding of the teeth (Tokarnia *et al.* 2000), blindness, recumbency and paddling movements with the limbs (Driemeier *et al.* 1999) can be observed; those signs are occasionally erroneously interpreted as part of a primary CNS disease. Serum chemistry reveals increased liver enzymes, γ-glutamyl transferase, alkaline phosphatase and alanine aminotransferase (Loretti *et al.* 1999). Although lethality rates are usually close to 100%, less severely affected cattle can recover (Tokarnia *et al.* 2000).

Pathology

Necropsy findings include ascites (Driemeier *et al.* 1999; Loretti *et al.* 1999) and subcapsular hemorrhages in the liver and the serosal surface of several viscera (Driemeier *et al.* 1999; Colodel *et al.* 2003); the colon and rectal contents are dry, covered by thin film of mucous and tinged by streaks of clotted blood.

Most striking and typical lesions are found in the liver, which is enlarged (Driemeier *et al.* 1999) and has a marked accentuation of the lobular pattern. This pattern can be observed on the capsular surface but is more evident on the cut surface as pale areas alternating with depressed red areas (Tokarnia and Döbereiner 1982; Driemeier *et al.* 1999). The red areas correspond to necrosis and hemorrhage. Other significant findings include distended gall bladder with edematous wall (Tokarnia and Döbereiner 1982) and hydropericardium (Driemeier *et al.* 1999).

Histologically there is a variable degree of coagulative necrosis of the hepatic lobule (Tokarnia and Döbereiner 1982; Gava *et al.* 1987; Chaulet *et al.* 1990; Gava *et al.* 1991; Driemeier *et al.* 1999; Colodel *et al.* 2003). The necrosis is centrilobular; usually the midzonal hepatocytes are vacuolated (hydropic and fat degeneration) and those of the periphery are more or less preserved. In other cases the whole lobule is affected by massive coagulative necrosis (Tokarnia and Döbereiner 1982; Gava *et al.* 1991; Colodel *et al.* 2003). Although the degree of the necrosis within the lobule varies from case to case (and seems dose-dependent) all lobules of one liver tend to be involved equally with exceptions reported in relation to *Vernonia* spp. poisoning (Tokarnia *et al.* 2000), in which the border of the left hepatic lobe of affected cattle is grossly and homogenously yellow (no marked lobular pattern). However, microscopically there is centrilobular necrosis of these homogenous yellow areas but congestion and hemorrhage are lacking (Tokarnia *et al.* 2000; Brum *et al.* 2002). In the field, the diseases reported in cattle poisoned by the plants mentioned above are acute, but chronic lesions (cirrhosis) of the liver were induced in cattle experimentally poisoned by *C. laevigatum* (Döbereiner *et al.* 1969) and *S. brasiliensis* (Chaulet *et al.* 1990). The active principle in *C. laevigatum* (Tokarnia *et al.* 2000), *S. brasiliensis* (Chaulet *et al.* 1990) and *C. corymbosum var. hirsutum* (Gava *et al.* 1991) is cumulative. In the toxicosis produced by the ingestion of *V. mollissima* (Döbereiner *et al.* 1976) and *V. rubricaulis* (Tokarnia and Döbereiner 1982; Brum *et al.* 2002), degenerative lesions of renal tubular epithelia may occur and necrosis of neurons of the telencephalic cortex have been described in cattle experimentally poisoned by *T. micrantha* (Traverso *et al.* 2004). Reasons for the centrilobular (periacinar) distribution of lesions caused by hepatotoxicants include: (1) a relatively lower oxygenation status than that of hepatocytes in more peripheral regions of the hepatic lobule; and (2) a high concentration of cytochrome P450 mixed-function oxidase system enzymes (Loretti *et al.* 1999).

The nervous signs presented by affected cattle are attributed to derangements in the urea cycle (hyperamonemia) due to hepatic failure (Tokarnia *et al.* 2000), a condition known as hepatic encephalopathy. However, the characteristic intramyelinic edema found in hepatic encephalopathy in cattle (Tokarnia *et al.* 2000) was not reported with acute hepatotoxicosis (Brum *et al.* 2002). For the most part, the active principles of these plants are yet unknown. However, saponins are mentioned as active principles for *C. laevigatum* (Tokarnia *et al.* 2000) and *D. viscosa* (Colodel *et al.* 2003), trematoxin for *Trema* spp. (Traverso *et al.* 2004), a cyanogenic glycoside for *D. viscosa* (Colodel *et al.* 2003) and the glycoside carboxyatractyloside for *C. parqui* and *Xanthium* spp. It is known that carboxyatractyloside inhibits oxidative phosphorylation and translocation of adenine nucleotides across the mitochondrial membrane (Loretti *et al.* 1999).

Diagnosis

The differential diagnosis with this clinical picture includes diseases caused by other hepatotoxins and primary diseases of the CNS. The geographic distribution of the poisonous plants helps to determine which plant is involved.

The ingestion of the sawfly (*Perreyia flavipes*) larvae causes clinical disease and lesions in cattle that closely resemble those evoked by the ingestion of poisonous plants causing acute toxic liver disease (Dutra *et al.* 1997). *Microcystis aeruginosa*, the blue-green algae growing as a bloom on lakes and ponds, produces microcystin, a well-documented cause of similar lesions in cattle. This type of poisoning is still unreported in cattle in Brazil, although an unfortunate accidental episode with several deaths due to microcystin toxicosis occurred in human beings in the country (Azevedo *et al.* 2002).

Primary diseases of the CNS (rabies, necrotizing meningoencephalitis by bovine herpesvirus 5 and cerebral babesiosis by *Babesia bovis*) and secondary diseases of the CNS (hepatic encephalopathy) should also be included in the list of differential diagnoses. The centrilobular congestion of the liver (congestive heart failure) imparts a gross nutmeg aspect to the liver. Nutmeg liver is morphologically similar to, although pathogenetically quite distinct from, the marked lobular pattern induced by acute hepatotoxins and should be differentiated from it.

References

Azevedo SMFO, Carmichael WW, Jochimsen EM, Rinehart KL, Lau S, Shaw GR, and Eaglesham GK (2002). Human intoxication by microcystins during renal dialysis treatment in Caruaru – Brazil. *Toxicology* 181–182:441–446.

Brum KB, Purisco E, Lemos RAA, and Riet-Correa F (2002). Intoxicação por *Vernonia rubricaulis* em bovinos no Mato Grosso do Sul. *Pesquisa Veterinária Brasileira* 22:119–128.

Canella CFC, Tokarnia CH, and Döbereiner J (1968). Intoxicação por *Sessea brasiliensis* (Solanaceae) em bovinos. *Arquivos do Instituto Biológico de São Paulo* 31:157–1266.

Chaulet JHF, Peixoto PV, and Tokarnia CH (1990). Intoxicação experimental por *Sessea brasiliensis* em bovinos e caprinos. *Pesquisa Veterinária Brasileira* 10:71–84.

Colodel EM, Traverso, SD, Seitz AL, Correa A, Oliveira FN, and Driemeier D (2003). Spontaneous poisoning by *Dodonea viscosa* (*Salpindaceae*) in cattle. *Veterinary and Human Toxicology* 45:147–148.

Döbereiner J, Tokarnia CH, and Canella CFC (1969). Intoxicação por *Cestrum laevigatum* Schlocht, a causa de mortandades em bovinos no Estado do Rio de Janeiro. *Pesquisa Veterinária Brasileira* 4:165–193.

Döbereiner J, Tokarnia CH, and Purisco E (1976). *Vernonia molissima*, planta tóxica responsável por mortandades de bovinos no sul de Mato Grosso. *Pesquisa Agropecuária Brasileira* 11:49–58.

Driemeier D, Irigoyen LF, Loretti AP, Colodel EM, and Barros CSL (1999). Intoxicação espontânea pelos frutos de *Xanthium cavanillesii* (Asteraceae) em bovinos no Rio Grande do Sul. *Pesquisa Veterinária Brasileira* 19:1–13.

Dutra F, Riet-Correa F, Méndez MC, and Paiva N (1997) Poisoning of cattle and sheep in Uruguay by sawfly (*Perreyia flavipes*) larvae. *Veterinary and Human Toxicology* 39:281–286.

Gava A, Peixoto PV, and Tokarnia CH (1987). Intoxicação experimental por *Vernonia mollissima* (Compositae) em ovinos e bovinos. *Pesquisa Veterinária Brasileira* 7(2):33–41.

Gava A, Stolf L, Pilati C, Neves DS, and Viganó L (1991). Intoxicação por *Cestrum corymbosum* var. *hirsutum* (Solanaceae) em bovinos no Estado de Santa Catarina. *Pesquisa Veterinária Brasileira* 11(3/4):71–74.

Gava A, Stolf L, Varaschin MS, Neves DS, Tigre AP, and Lesmann F (1996). Intoxicação por *Cestrum intermedium* (Solanaceae) em bovinos. *Pesquisa Veterinária Brasileira* 16(4):117–120.

Loretti AP, Bezerra PS, Ilha MRS, and Barros CSL (1999). Intoxicação experimental pelos frutos de *Xanthium cavanillesii* (Asteraceae) em ovinos. *Pesquisa Veterinária Brasileira* 19(2):71–78.

Méndez MC, Santos RC, and Riet-Correa F (1998). Intoxication by *Xanthium cavanillesii* in cattle and sheep in southern Brazil. *Veterinary and Human Toxicology* 40(3):144–147.

Riet-Correa F, Schild AL, Méndez MC, and Pinheiro MH (1986). Intoxicação por *Cestrum parqui* (Solanaceae) em bovinos no Rio Grande do Sul. *Pesquisa Veterinária Brasileira* 6(4):111–115.

Tokarnia CH and Döbereiner J (1982). Intoxicação de bovinos por *Vernonia rubricaulis* (Compositae) em Mato Grosso. *Pesquisa Veterinária Brasileira* 2:143–147.

Tokarnia CH, Döbereiner J, and Peixoto PV (2000). *Plantas Tóxicas do Brasil*. Helianthus, Rio de Janeiro.

Traverso SD, Corrêa AMR, Schmitz M, Colodel EM, and Driemeier D (2004). Intoxicação experimental por *Trema micrantha* (*Ulmaceae*) em bovinos. *Pesquisa Veterinária Brasileira* 24(4):211–216.

Chapter 13

Effects of Dietary Pyrrolizidine (*Senecio*) Alkaloids on Plasma and Liver Vitamin E Distribution in Broiler Chickens

A.L. Lulay[1], S.W. Leonard[2], M.G. Traber[2], M.R. Keller[1], and P.R. Cheeke[1]

[1]*Department of Animal Sciences, Oregon State University, Corvallis, OR 97331, USA;*
[2]*Linus Pauling Institute, Oregon State University, Corvallis, OR 97331, USA*

Introduction

Pyrrolizidine alkaloids (PA) occur in a number of plants such as *Senecio* and *Crotalaria* spp. and cause numerous toxicological problems in humans, poultry and livestock (Mattocks 1986; Cheeke 1988). The PA are metabolized by liver enzymes to reactive metabolites such as pyrroles, which cross-link DNA strands, causing impaired cell division and protein synthesis. These effects lead to a loss of liver function and irreversible liver necrosis. Exposure to dietary PA may modify nutrient metabolism and tissue distribution. For example, an interaction between PA exposure and vitamin A metabolism has been reported (Moghaddam and Cheeke 1989; Huan *et al.* 1992). Rats and broiler chickens fed dietary tansy ragwort (*Senecio jacobaea*) as a source of PA had reduced plasma and liver vitamin A concentrations. This result was attributed to a PA-induced impairment of the hepatic synthesis of retinol-binding proteins, which was confirmed by Huan *et al.* (1993).

Because vitamin E is transported and stored in association with lipoproteins synthesized in the liver, we hypothesize that dietary PA will reduce liver and plasma vitamin E concentrations, as was previously observed with vitamin A. This hypothesis is supported by the observation that red blood cells from PA-fed animals are susceptible to *in vitro* hemolysis (Cheeke 1989), a classic sign of vitamin E deficiency. Thus, the objective of this study was to determine if blood and liver vitamin E concentrations in broiler chickens are influenced by dietary exposure to PA.

Materials and Methods

Fifty-four broiler chicks were separated into six feeding groups of nine birds per group. They were housed under controlled environmental conditions with artificial light, propane heaters and ventilated wood shaving litter floor pens. At one week of age chicks were wing banded and individual weights were recorded. At this time the chicks were started on their diets of: (1) positive control (chick diet); (2) positive control + 500 mg vitamin E kg^{-1}; (3) positive control + 2000 mg vitamin E kg^{-1}; (4) negative control (chick diet + tansy); (5) negative control + 500 mg vitamin E kg^{-1}; and (6) negative control + 2000 mg vitamin E kg^{-1}. Vitamin E was provided as liquid DL-α-tocopherol acetate, which was mixed with corn oil and added as 2% of the diet. Tansy ragwort (*Senecio jacobaea*) flowers were harvested, dried and ground by hammer mill in order to incorporate them into the diet as a source of PA. The tansy ragwort constituted 5% of the diet. The dried material contained approximately 0.18% total PA. The chick diet was a corn soybean meal starter grower diet with 23% crude protein. Chicks were fed their respective diets for a total of 25 days. They were weighed weekly, and at the end of the trial final weights and liver weights were taken as well as blood and liver samples. Blood and liver were analyzed for α-tocopherol using an HPLC method as previously described (Podda *et al.* 1996).

Results and Discussion

Birds fed tansy ragwort had reduced feed intakes, growth rates and liver weights as compared to controls (Table 1). Livers of ragwort-fed birds were shrunken, abnormally shaped and straw colored. Some of the chicks had ascites in the abdominal cavity. Plasma and liver vitamin E concentrations were greatly reduced in the tansy ragwort-fed groups (Table 2, Figures 1 and 2). Total liver vitamin E levels are shown in Figure 3. The reductions in both the liver and plasma vitamin E with dietary PA exposure are likely a consequence of PA-induced inhibition of hepatic synthesis of vitamin E-transporting lipoproteins.

Impairment of vitamin E transport, storage and metabolism may play a role in the pathogenesis of PA toxicity, although there was no evidence that high dietary vitamin E levels protected against PA toxicosis in this study.

Table 1. Effect of dietary tansy ragwort on body and liver weights of broiler chicks.

Dietary tansy ragwort (%)	Dietary vitamin E (mg kg^{-1})	Final body weight (g)	Liver weight (g)	Liver as % body weight
0	0	3431±429 [a]	40±8 [a]	1.2±0.1 [a]
0	500	3668±280 [a]	50±11 [a]	1.4±0.3 [a]
0	2000	3629±302 [a]	45±8 [a]	1.2±0.3 [a]
5	0	1299±330 [b]	12±4 [b]	0.9±0.2 [c]
5	500	1354±352 [b]	12±5 [b]	0.8±0.2 [c]
5	2000	1377±510 [b]	14±6 [b]	1.0±0.3 [b,c]

a different from b (P<0.001)
b different from c (P<0.05)

Conclusion

Dietary consumption of PA results in pronounced reduction in liver and plasma α-tocopherol concentrations in chickens. This effect is probably a consequence of PA-induced inhibition of hepatic synthesis of vitamin E-transporting proteins, as has been observed previously with retinol-binding protein.

Table 2. Effect of dietary tansy ragwort on plasma and liver vitamin E concentrations.

Dietary tansy ragwort (%)	Dietary vitamin E (mg kg^{-1})	Plasma vitamin E (nmol ml^{-1})	Liver vitamin E (nmol g^{-1})	Total liver vitamin E (nmol)
0	0	19±6 [a]	33±10	1332±525[a]
0	500	35±6 [c]	76±22	3864±1408 [c]
0	2000	36±11[e]	111±23	4937±1236 [g]
5	0	3±3 [b]	10±5	119±77 [b]
5	500	12±10 [d]	39±15	453±205 [d]
5	2000	30±11 [f]	69±19	997±545 [h]

a different from b ($\underline{P}$<0.001)
c different from d ($\underline{P}$<0.001)
e different from f (P <0.05)
g different from h ($\underline{P}$<0.001)

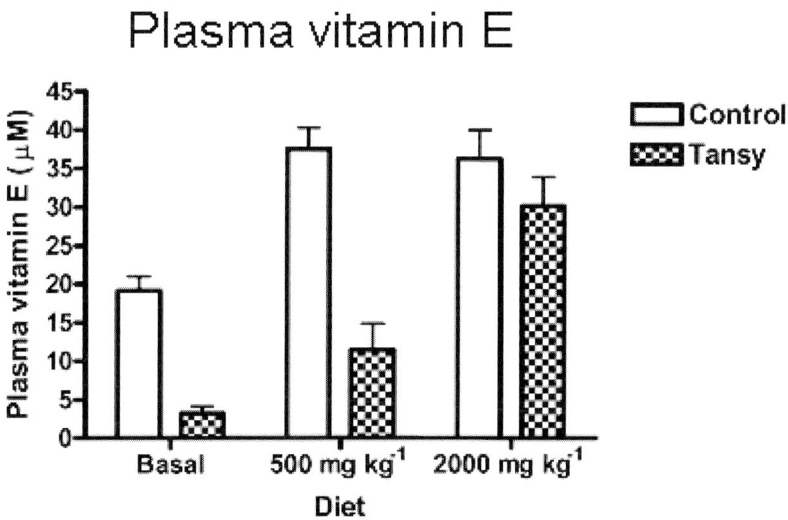

Figure 1. Plasma vitamin E concentrations in chicks fed tansy ragwort or control diets.

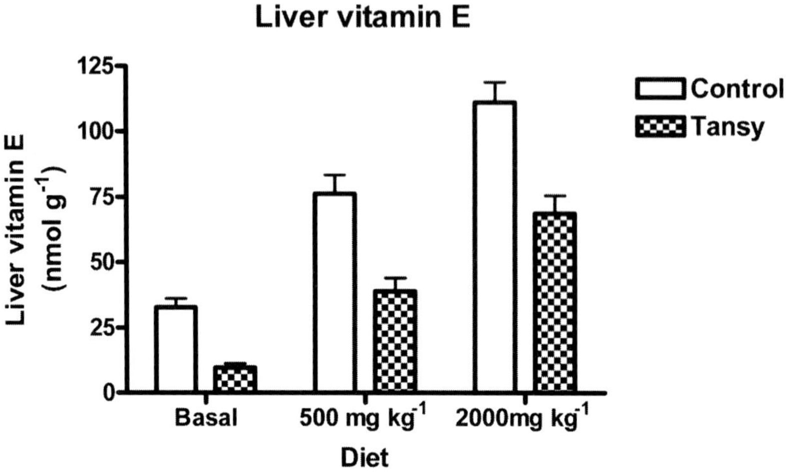

Figure 2. Liver vitamin E concentrations in chicks fed tansy ragwort or control diets.

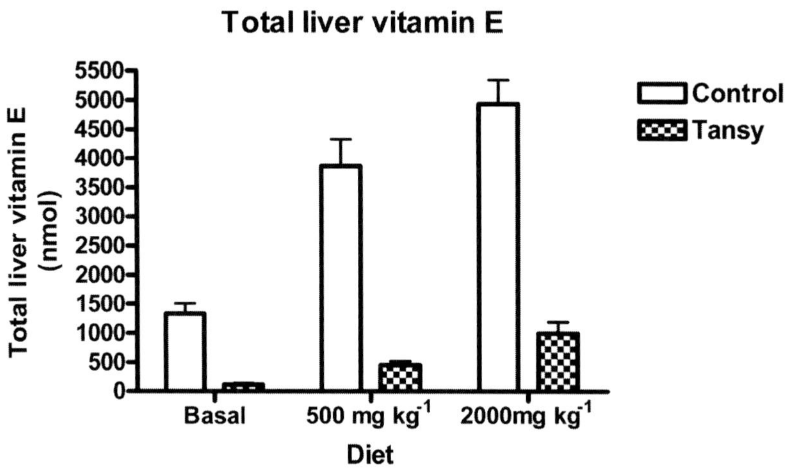

Figure 3. Total liver vitamin E concentrations in chicks fed tansy ragwort or control diets.

References

Cheeke PR (1988). Toxicity and metabolism of pyrrolizidine alkaloids. *Journal of Animal Science* 66:2343–2350.

Cheeke PR (1989). Pyrrolizidine alkaloid toxicity and metabolism in laboratory animals and livestock. In *Toxicants of Plant Origin*, Vol. 1 (PR Cheeke, ed.), pp. 1–22. CRC Press, Boca Raton, Florida.
Huan J, Cheeke PR, Lowry RR, Nakaue HS, Snyder SP, and Whanger PD (1992). Dietary pyrrolizidine (*Senecio*) alkaloids and tissue distribution of copper and vitamin A in broiler chickens. *Toxicology Letters* 62:139–153.
Huan JY, Cheeke PR, and Blaner WS (1993). Modification of vitamin A metabolism by dietary pyrrolizidine (*Senecio*) alkaloids in rats. In *Proceedings, Western Section, American Society of Animal Science* 44:275–278.
Mattocks AR (1986). *Chemistry and Toxicology of Pyrrolizidine Alkaloids*. Academic Press, New York.
Moghaddam MF and Cheeke PR (1989). Effects of dietary pyrrolizidine (*Senecio*) alkaloids on vitamin A metabolism in rats. *Toxicology Letters* 45:149–156.
Podda M, Weber C, Traber MG, and Packer L (1996). Simultaneous determination of tissue tocopherols, tocotrienols, ubiquinols, and ubiquinones. *Journal of Lipid Research* 37(4):893–901.

Chapter 14

Heterologous Expression of a Cytochrome P450 from Turkey Liver That Activates Aflatoxin B_1

R.A. Coulombe, Jr. and S.S.M Yip

Interdepartmental Graduate Toxicology Program and Department of Animal, Dairy, and Veterinary Sciences, Utah State University, Logan, UT 84322-4620, USA

Introduction

Mycotoxins are ubiquitous, mold-produced toxins that contaminate a wide variety of foods and feeds. Ingestion of mycotoxins causes a range of acute and chronic toxic responses in animals and people (Coulombe 1993). One important example is the mycotoxin aflatoxin B_1 (AFB_1) produced by strains of *Aspergillus flavus* and *A. parasiticus*, which is a potent liver toxin and carcinogen in animals (Coulombe 1993). This mycotoxin was discovered in the early 1960s as the etiological agent of 'Turkey X Disease', responsible for widespread deaths in turkeys and other poultry throughout Europe due to contaminated Brazilian peanut meal (Asao *et al.* 1965).

As demonstrated by the 'Turkey X' incident, turkeys are extremely sensitive to the toxic effects of AFB_1, resulting in reductions in growth rate, feed efficiency, increased susceptibility to viral and bacterial diseases, and severe hepatotoxicosis (Klein *et al.* 2002b; Van Vleet *et al.* 2002). For AFB_1 to be toxic, it must first be oxidized by one or more members of an enzyme superfamily called cytochrome P450 (P450) to the reactive and electrophilic AFB_1-8,9-epoxide (AFBO), which binds to DNA, proteins and other critical cellular nucleophiles (Iyer *et al.* 1994). We have shown that the extreme hypersensitivity of turkeys is associated with efficient hepatic P450-mediated metabolic activation (Klein *et al.* 2000, 2002a). Using specific P450 inhibitors, we have previously demonstrated that AFB_1 is metabolized in turkey liver microsomes primarily by a CYP1A and to a lesser extent by a CYP3A homologue (Klein *et al.* 2000).

Because of the presumed importance of the postulated role of the turkey P450 in the sensitivity of this species toward AFB_1, the purpose of this study was to clone, express and partially characterize this gene product from turkey liver. We discovered the presence of a novel P450 from turkey liver that has AFB_1 metabolizing activity. To our knowledge, this is the first such enzyme cloned and sequenced from turkeys.

Materials and Methods

Amplification of P450 cDNA and bacterial expression

Full-length DNA coding for turkey P450 was amplified from turkey liver RNA (Poly(A)Pure mRNA purification kit, Ambion, Austin, TX) using the Rapid Amplification of cDNA ends (RACE) technique. Primers for 3′-RACE were designed from sequence alignments of CYP1A2 in human, rabbit, mouse, rat and a partial clone of chicken using ClustalW (www.ebi.ac.uk). Primers for 5′-RACE were based on the alignments plus the sequences found in 3′-RACE. The gene-specific forward primer sequence for 3′-RACE was 5′-CCCTTC-ACCATCCCCCACAGCAC-3′, and for 5′-RACE was 5′-GCGGCGATGGA-GAAGTTCTTCAGGG-3′. Both reactions used a mixture of two reverse primers (long primer: 5′-CTAATACGACTCACTATAGGGCAAGCAGTGG-TATCAACGCAGAGT-3′ and short primer: 5′-CTAATACGACTCAC-TATAGGGC-3′) provided in the BD SMART RACE cDNA Amplification Kit (BD Biosciences Clontech, Palo Alto, CA). The full-length gene (which we named CYP1A5) was cloned using forward (5′-ATGGGGCCGGAGGAA-GTGATGG-3′) and reverse (5′-GTTTGAGCTCTTCATGGAGA-AGCGTTTC-3′) primers which were designed from the sequence information obtained from 3′- and 5′-RACE. RACE-PCRs were performed using BD Advantage 2 PCR Kit (BD Bioscience Clontech, Palo Alto, CA), and PCR products were cloned into pCR 4-TOPO vector. The final sequence was confirmed by at least three clones in any segment, with at least one sequenced from either direction.

Bicistronic constructs consisting of the coding sequence of the CYP1A5 followed by that of the human NADPH-P450 reductase (hNPR), a redox partner, were constructed from the pCW1A2/hNPR expression vector (Parikh *et al.* 1997). Because N-terminal modifications have been shown to improve bacterial expression of P450s (Sandhu *et al.* 1994; Gillam *et al.* 1995; Shimada *et al.* 1998), we engineered the tCYP1A5 sequence such that it had a G-to-A mutation at the second codon and lacked codons 3 to 31. In addition, the nucleotide sequence of codons 2–8 was modified to enhance P450 protein expression (Barnes *et al.* 1991). The construct also had a PCR-introduced *Nde*I site at the 5′-end and an *Xba*I site at the 3′-end for subcloning; a 6×His tag was also added to the C-terminal to facilitate immunodetection and purification. The expression constructs were created by digesting the original pCW1A2/hNPR vector with *Nde*I and *Xba*I to excise the CYP1A2 coding sequence. The gel purified vector fragment was ligated with the CYP1A5 *Nde*I and *Xba*I digested, PCR-amplified fragment using T4 DNA ligase, with 5:1 insert-to-vector ratio. The resulting ligation mix was transformed into DH5αF'IQ cells. Individual clones were screened by colony PCR and diagnostic restriction digest before sequencing. Sequences in the entire reading frame of both constructs, as well as the 5′- and 3′-junctions of the inserts to the vectors were verified before individual clones were used in expression.

Expression in *Escherichia coli* was performed as described (Jenkins *et al.* 1998). Cells were harvested after 24 hours and levels of P450 expression were determined by reduced CO/reduced difference spectrum (Omura and Sato 1964). Western blots using anti-His tag monoclonal antibody and anti-rat NADPH-P450 reductase polyclonal antibody (which cross-reacts with human NADPH-P450 reductase) were also employed to verify the expression of the recombinant proteins. *E. coli* membrane and cytosol were prepared as described (Jenkins *et al.* 1998). Sample aliquots were stored at −80°C. All kinetic assays were performed with CYP1A5 *E. coli* membranes.

Catalytic properties of expressed P450

O-dealkylation of 7-methoxyresorufin (MROD) was assayed as described (Chang and Waxman 1998). *E. coli* membranes containing 165pmol CYP1A5 were incubated in reaction buffer (50mM Tris, pH 7.5, 25mM $MgCl_2$, 1.6mg/ml BSA) with 0.25mM NADPH and 5μM substrate at 37°C. Selective inhibitors of P450 α-naphthoflavone (CYP1A1, 1A2), 17α-ethynylestradiol (3A4), quinidine (2D), 4-methylpyrazole (2E) and erythromycin (3A1) (Klein *et al.* 2000; Lewis 2001) were evaluated for their potential inhibitory effect on MROD activity catalyzed by CYP1A5-expressing membranes.

Metabolism of AFB_1 in *E. coli* membranes was determined by HPLC using conditions we have described previously (Kelly *et al.* 1997). Because of its short half-life, AFBO was indirectly measured as the GSH-trapped conjugate, which provides an accurate and quantitative measurement (Ramsdell and Eaton 1990). Kinetic data were analyzed by SigmaPlot Enzyme Kinetics Module software (Systat, Point Richmond, CA), and were fit using the Michaelis-Menten equation $[V=V_{max} [S]/([S] + K_m)]$ to calculate the kinetic constants V_{max} and K_m.

Results

Sequence analyses of the entire open reading frame revealed a sequence of 1587bp coding for 528 amino acids, plus 44bp in the 5′-untranslated region (UTR) and 101bp in the 3′-UTR. The predicted molecular weight of the gene product is 60kDa (Figure 1). The final sequence of CYP1A5 was remarkably close to that of the chicken CYP1A5, with 94.7% nucleic acid and amino acid sequence identity. Therefore, we named this turkey P450 gene CYP1A5.

Figure 1 shows the predicted amino acid sequence of the turkey CYP1A5. The N-terminal sequence contained the typical endoplasmic reticulum (ER) anchoring signal in the hydrophobic region, followed by a halt-transfer signal containing several positively charged amino acids (Sabatini *et al.* 1982). The heme-binding cysteine at the active site of CYP1A5 was found at residue 467. The heme-binding motif, usually represented as FXXGXXXCXG, was identified in CYP1A5 as FGLGKRRCIG (amino acids 460–469) (Figure 1). Other potential secondary modification sites found in CYP1A5 included two casein II phosphorylation sites (SVLE, amino acids 60–63; TAVE, amino acids 273–276), an N-linked glycosylation site (NFSI, amino acids 153–156), a tyrosine kinase phosphorylation site (KQSFDPYRY, amino acids 190–198), an amidation site (DGKK, amino acids 494–497), a cAMP/cGMP protein kinase phosphorylation site (KRFS, amino acids 520–523) and a protein kinase C phosphorylation site (SMK, amino acids 523–525). We also identified a 10-amino acid prokaryotic membrane lipoprotein lipid attachment site (IAASPTASSSC, amino acids 156–166) in turkey CYP1A5. The nucleotide sequence for the turkey CYP1A5 gene has been deposited in the GenBank database under GenBank Accession Number AY964644. The amino acid sequence of this protein can be accessed through NCBI Protein Database under NCBI Accession # AAX73011.

The truncated CYP1A5 construct had high levels of P450 expression when compared to that of other P450 isoforms in *E. coli* (Parikh *et al.* 1997; Shimada *et al.* 1998). We routinely obtained 200–300 nmol tCYP1A5 per liter culture, as estimated by reduced CO/reduced difference spectrum using whole cells. After membrane and cytosol isolation, approximately 30% of the tCYP1A5 was recovered in the membrane fraction, and approximately 30–50% was in the cytosol fraction. Wild type (*i.e.*, non-truncated) constructs had no detectable P450 expression as measured by the reduced CO/reduced difference spectrum. The specific CYP1A inhibitor, α-naphthoflavone, completely inhibited MROD activity catalyzed by CYP1A5 from *E. coli* membranes. 17α-ethynylestradiol (3A4) had some, while quinidine (2D), 4-methylpyrazole (2E) and erythromycin (3A1) had little or no inhibitory effect (Table 1).

```
CCCTCTGTCCGTCTCTGCCCAGCTTCAAGGACCTGCAGGTGGCAATGGGGCCGGAGGAAGTGATGGTACAGGTGGGCAGC   80
                                            M  G  P  E  E  V  M  V  Q  V  G  S
CCAGGTCTCATCTCGGCCACCGAGATGCTGGTGGCAGCTGCCACTTTCTGCCTGCTCCTGCTGCTGACCCAGACCCGCCG   160
P  G  L  I  S  A  T  E  M  L  V  A  A  A  T  F  C  L  L  L  L  L  T  Q  T  R  R
GCAGCACACACCCAAGGGGCTGCGCAGACCCCCGGGTCCCCGTGGGCTCCCACTGCTGGGCAGTGTGCTGGAGCTGAGGA   240
 Q  H  T  P  K  G  L  R  R  P  P  G  P  R  G  L  P  L  L  G  S  V  L  E  L  R  K
AGGACCCACACCTGGTGCTCACCCAGATGAGCCGCAAATACGGGGATGTGATGGAGGTGACCATCGGCTCCCGGCCCGTG   320
  D  P  H  L  V  L  T  Q  M  S  R  K  Y  G  D  V  M  E  V  T  I  G  S  R  P  V
GTGGTGCTCAGCGGGCTGGAAACCATCAAGCAAGCCTTGGTGAGGCAAGCAGAAGACTTCATGGGACGCCCCGACCTCTA   400
V  V  L  S  G  L  E  T  I  K  Q  A  L  V  R  Q  A  E  D  F  M  G  R  P  D  L  Y
CAGCTTTCGACACGTTACGGACGGACAGAGCCTGACCTTCAGCACTGACACGGGGGAAGTGTGGAAAGCCCGTAGGAAGC   480
 S  F  R  H  V  T  D  G  Q  S  L  T  F  S  T  D  T  G  E  V  W  K  A  R  R  K  L
TGGCACAGAACGCCCTGAAGAACTTCTCCATCGCTGCCAGCCCCACAGCCTCCTCCAGCTGCCTCCTGGAGGAGCACGTC   560
  A  Q  N  A  L  K  N  F  S  I  A  A  S  P  T  A  S  S  S  C  L  L  E  E  H  V
ACCAACGAGGCCAGCTACCTGGTCACCAAATTCCTGCAGCTGATGGAGGAGAAGCAGAGCTTTGACCCCTACCGCTACAT   640
T  N  E  A  S  Y  L  V  T  K  F  L  Q  L  M  E  E  K  Q  S  F  D  P  Y  R  Y  M
GGTGGTGTCGGTGGCCAACGTCATCTGCGCCATTTGCTTTGGCAAGCGCTACGACCACGACAACCAGGAGCTGCTCAGCG   720
 V  V  S  V  A  N  V  I  C  A  I  C  F  G  K  R  Y  D  H  D  N  Q  E  L  L  S  V
TGGTGAACGTGGTTGAGGAGTTCGGAGATGTGACTGCTGTTGGCAACCCAACTGACTTCATCCCATTGCTCCAGTACCTC   800
  V  N  V  V  E  E  F  G  D  V  T  A  V  G  N  P  T  D  F  I  P  L  L  Q  Y  L
CCCAGCCGCAACATGGATTTATTTCTGGATTTCAACAAGCGATTCATGAAATTGTTGAAGACAGCTGTGGAAGAGCACTA   880
P  S  R  N  M  D  L  F  L  D  F  N  K  R  F  M  K  L  L  K  T  A  V  E  E  H  Y
CGAGACTTTTGACAAGAACAACATCCGAGATGTCACGGACTCCCTCATTGAGCAGTGCATGGAGAAAAAAACCGAAGCCA   960
 E  T  F  D  K  N  N  I  R  D  V  T  D  S  L  I  E  Q  C  M  E  K  K  T  E  A  N
ACAGTGCCACACAGATCCCAAATGAGAAGATCATCAACCTGGTGAATGACATCTTTGGAGCAGGCTTTGACACTGTGACA   1040
  S  A  T  Q  I  P  N  E  K  I  I  N  L  V  N  D  I  F  G  A  G  F  D  T  V  T
ACGGCCCTGTCCTGGAGCCTCATGTACCTTGTGACATACCCCCACATCCAGAAGAAGATTCAGGCAGAGCTGGATCAGAC   1120
T  A  L  S  W  S  L  M  Y  L  V  T  Y  P  H  I  Q  K  K  I  Q  A  E  L  D  Q  T
CATCGGCCGTGAGAGGAGACCACGGCTGTCCGACCGAGGCACGCTGCCCTACACAGAAGCCTTCATCCTGGAGATGTTCC   1200
 I  G  R  E  R  R  P  R  L  S  D  R  G  T  L  P  Y  T  E  A  F  I  L  E  M  F  R
GGCACTCCTCCTTCATGCCCTTCACCATCCCACACAGCACGACCAGGGACACGGTGCTGAATGGCTACTATATCCCAAAG   1280
  H  S  S  F  M  P  F  T  I  P  H  S  T  T  R  D  T  V  L  N  G  Y  Y  I  P  K
GACCGCTGCGTGTTCATCAACCAGTGGCAAGTGAATCACGATGAGAAACTTTGGAAGGATCCACAGGCTTTCAACCCAGA   1360
D  R  C  V  F  I  N  Q  W  Q  V  N  H  D  E  K  L  W  K  D  P  Q  A  F  N  P  E
GCGTTTCCTCAATGCTGAAGGGACTGAAGTGAACAAAGTGGATGCGGAGAAGGTGATGACTTTTGGCCTGGGGAAAAGGA   1440
 R  F  L  N  A  E  G  T  E  V  N  K  V  D  A  E  K  V  M  T  F  G  L  G  K  R  R
GGTGCATTGGGGAAAACATAGGCAAGTGGGAGGTGTTCCTTTTCCTGTCCACATTGCTCCAGCAACTGGAGTTCAGCATC   1520
  C  I  G  E  N  I  G  K  W  E  V  F  L  F  L  S  T  L  L  Q  Q  L  E  F  S  I
CGCGATGGCAAGAAGGCAGACATGACACCTATATACGGACTGTCTGTCAAGCACAAGAGATGTGAGCACTTTCAAGTCAA   1600
R  D  G  K  K  A  D  M  T  P  I  Y  G  L  S  V  K  H  K  R  C  E  H  F  Q  V  K
GAAACGCTTCTCCATGAAGAGCTCAAACTGAGCAGTTTATCCATAGCCCAGATGTTGCCAATTGGGCACCTTTGTGACAG   1680
 K  R  F  S  M  K  S  S  N  *
ACCTGGGTCTGGCATAACCTCACAGAACCATGCAATAAACAAAAGCCATTGA                               1732
```

Figure 1. cDNA and predicted amino acid sequence of turkey CYP1A5. The start and stop codons are bolded, the heme-binding site is underlined. The open reading frame consists of 528 residues.

The CYP1A5 expressed in *E. coli* membranes was especially active in AFB_1 metabolizing activity, forming two principal metabolites, AFBO and AFM_1 at a molar ratio of 1:1.5–2.0 (Figure 2). Metabolite formation for both products conformed to simple Michaelis-Menten kinetics with K_m and V_{max} of AFBO formation being 32 μM and 0.46 nmol min^{-1} $nmol^{-1}$ P450 respectively ($R^2 = 0.87$), whereas those values of AFM_1 formation were 28 μM and 0.86 nmol min^{-1} $nmol^{-1}$ P450 respectively ($R^2 = 0.84$).

Table 1. Effects of specific P450 inhibitors on MROD activity by turkey CYP1A5 heterologously expressed in *E. coli* membrane.

Inhibitor	P450 isoform	MROD Activity* (% of Control)
α-Naphthoflavone	1A1, 1A2	0
17α-Ethynylestradiol	3A4	60
Quinidine	2D	87
4-Methylpyrazole	2E	89
Erythromycin	3A1	104

*values are mean of duplicate determinations

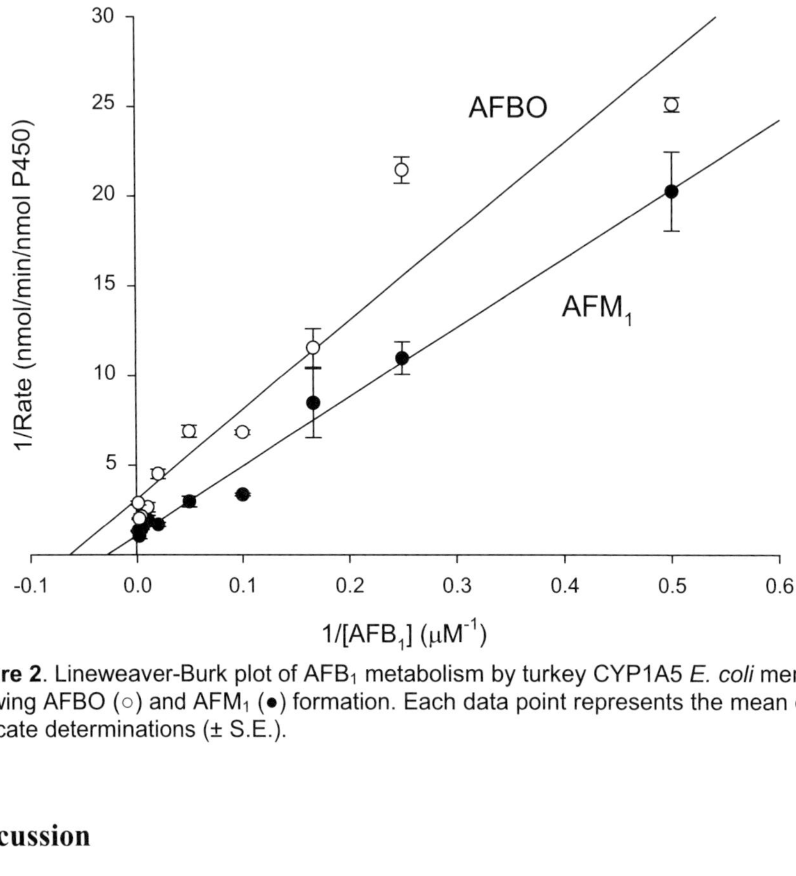

Figure 2. Lineweaver-Burk plot of AFB_1 metabolism by turkey CYP1A5 *E. coli* membrane showing AFBO (○) and AFM_1 (●) formation. Each data point represents the mean of triplicate determinations (± S.E.).

Discussion

Here we describe the cloning and heterologous expression of a CYP1A5 from turkey liver. Membranes from CYP1A5-expressing cells metabolized AFB_1 to form only two metabolites, AFBO and AFM_1. Hepatic formation of AFBO is associated with chronic adverse effects of AFB_1, such as liver cancer, while formation of AFM_1 has been postulated to be an important mediator of shorter-term toxicity (Eaton and Gallagher 1994). The acute toxicity of AFB_1 has more importance in turkeys, which are generally marketed at 16–18 weeks of age. Although AFM_1 is usually considered to be a detoxification product of AFB_1, it is still a potent cytotoxin and genotoxin, active at relatively low doses (Neal *et al.* 1998). It is interesting that the K_m values we determined for CYP1A5 were nearly identical with those of human CYP1A2 cDNA-expressed microsomes (32 vs. 41 μM for AFBO; 28 vs. 36 μM for AFM_1) (Gallagher *et al.* 1996). V_{max} values for AFM_1 formation of CYP1A5 and human CYP1A2 were similar (0.86 vs. 0.92 nmol min^{-1} $nmol^{-1}$ P450), while that for AFBO formation was more than five-fold lower than that measured for human CYP1A2 (0.46 vs. 2.63 nmol min^{-1} $nmol^{-1}$ P450).

In conclusion, we report the identification of a novel P450, which we designated CYP1A5 based on the nucleic acid and amino acid sequence similarities with chicken CYP1A5. The gene, which was heterologously expressed in sufficient quantities in *E. coli*, oxidized AFB_1 to form two metabolites with kinetic properties largely similar to human CYP1A2, the P450 isoform principally responsible for metabolizing AFB_1 in human liver and lung. Because of the importance of these metabolites in the toxicity of AFB_1, we conclude that this P450 probably plays some role in the hypersensitivity of turkeys to AFB_1.

Acknowledgements

The authors wish to thank Professor Gary Yost, University of Utah for his helpful suggestions and review of a draft manuscript. This work was supported by USDA-NRI competitive grant 2002-35204-12294, and by the Utah Agricultural Experiment Station, where this paper is published as number 7717.

References

Asao T, Buechi G, Abdel-Kader MM, Chang SB, Wick EL, and Wogan GN (1965). The structures of aflatoxins B and G. *Journal of the American Chemical Society* 87:882–886.

Barnes HJ, Arlotto MP, and Waterman MR (1991). Expression and enzymatic activity of recombinant cytochrome P450 17 alpha-hydroxylase in *Escherichia coli*. *Proceedings of the National Academy of Sciences U S A* 88:5597–5601.

Chang TKH and Waxman DJ (1998). Enzymatic analysis of cDNA-expressed human CYP1A1, CYP1A2, and CYP1B1 with 7-ethoxyresorufin as substrate. In *Cytochrome P450 Protocols* (IR Phillips and EA Shephard, eds), pp. 103–109. Humana Press, Totowa, NJ.

Coulombe RA, Jr (1993). Biological action of mycotoxins. *Journal of Dairy Science* 76:880–891.

Eaton DL and Gallagher EP (1994). Mechanisms of aflatoxin carcinogenesis. *Annual Review of Pharmacology and Toxicology* 34:135–172.

Gallagher EP, Kunze KL, Stapleton PL, and Eaton DL (1996). The kinetics of aflatoxin B_1 oxidation by human cDNA-expressed and human liver microsomal cytochromes P450 1A2 and 3A4. *Toxicology and Applied Pharmacology* 141:595–606.

Gillam EM, Guo Z, Martin MV, Jenkins CM, and Guengerich FP (1995). Expression of cytochrome P450 2D6 in *Escherichia coli*, purification, and spectral and catalytic characterization. *Archives of Biochemistry and Biophysics* 319:540–550.

Iyer RS, Coles BF, Raney KD, Thier R, Guengerich FP, and Harris TM (1994). DNA adduction by the potent carcinogen aflatoxin B_1: Mechanistic studies. *Journal of the American Chemical Society* 116:1603–1609.

Jenkins CM, Pikuleva I, and Waterman MR (1998). Expression of eukaryotic cytochrome P450 in *E. coli*. In *Cytochrome P450 Protocols* (IR Phillips and EA Shephard, eds), pp. 181–193. Humana Press, Totowa, New Jersey.

Kelly JD, Eaton DL, Guengerich FP, and Coulombe RA, Jr (1997). Aflatoxin B_1 activation in human lung. *Toxicology and Applied Pharmacology* 144:88–95.

Klein PJ, Buckner R, Kelly J, and Coulombe RA, Jr (2000). Biochemical basis for the extreme sensitivity of turkeys to aflatoxin B_1. *Toxicology and Applied Pharmacology* 165:45–52.

Klein PJ, Van Vleet TR, Hall JO, and Coulombe RA, Jr (2002a). Biochemical factors underlying the age-related sensitivity of turkeys to aflatoxin B_1. *Comparative Biochemistry and Physiology Part C: Toxicology & Pharmacology* 132:193–201.

Klein PJ, Van Vleet TR, Hall JO, and Coulombe RA, Jr (2002b). Dietary butylated hydroxytoluene protects against aflatoxicosis in turkeys. *Toxicology and Applied Pharmacology* 182:11–19.

Lewis DF (2001). *Guide to Cytochromes P450 Structure and Function*. Taylor & Francis, London, UK.

Neal GE, Eaton DL, Judah DJ, and Verma A (1998). Metabolism and toxicity of aflatoxins M_1 and B_1 in human-derived in vitro systems. *Toxicology and Applied Pharmacology* 151:152–158.

Omura T and Sato R (1964). The carbon monoxide-binding pigment of liver microsomes. II. Solubilization, purification, and properties. *Journal of Biological Chemistry* 239:2379–2385.

Parikh A, Gillam EM, and Guengerich FP (1997). Drug metabolism by *Escherichia coli* expressing human cytochromes P450. *Nature Biotechnology* 15:784–788.

Ramsdell HS and Eaton DL (1990). Species susceptibility to aflatoxin B_1 carcinogenesis: comparative kinetics of microsomal biotransformation. *Cancer Research* 50:615–620.

Sabatini DD, Kreibich G, Morimoto T, and Adesnik M (1982). Mechanisms for the incorporation of proteins in membranes and organelles. *Journal of Cell Biology* 92:1–22.

Sandhu P, Guo Z, Baba T, Martin MV, Tukey RH, and Guengerich FP (1994). Expression of modified human cytochrome P450 1A2 in *Escherichia coli*: stabilization, purification, spectral characterization, and catalytic activities of the enzyme. *Archives of Biochemistry and Biophysics* 309:168–177.

Shimada T, Wunsch RM, Hanna IH, Sutter TR, Guengerich FP, and Gillam EM (1998). Recombinant human cytochrome P450 1B1 expression in *Escherichia coli*. *Archives of Biochemistry and Biophysics* 357:111–120.

Van Vleet TR, Klein PJ, and Coulombe RA, Jr (2002). Metabolism and cytotoxicity of aflatoxin B_1 in cytochrome p-450-expressing human lung cells. *Journal of Toxicology and Environmental Health A* 65:853–867.

Chapter 15

The Effects of Vaccination on Riddelliine Toxicity in Rats

B.L. Stegelmeier, S.T. Lee, L.F. James, D.R. Gardner, K.E. Panter, M.H. Ralphs, and J.A. Pfister

USDA-ARS Poisonous Plant Research Laboratory, Logan, UT 84341, USA

Introduction

Pyrrolizidine alkaloid (PA)-containing plants are found throughout the world and commonly poison livestock and humans. As many plants are noxious weeds they commonly invade fields and contaminate prepared feeds and grains. Consequently most poisonings occur as a result of contaminated feed and foods (Stegelmeier *et al.* 1999). To better monitor feed and food quality, several PA–protein conjugates were constructed, and class and alkaloid enzymatic specific immunologic detection techniques have been developed (Lee *et al.* 2001). The purpose of this study was to determine if these conjugates might be used as vaccines and see if immunization alters the toxicity when animals are challenged with PA.

Methods

Two groups of 30 rats each were immunized with two riddelliine conjugates (Figure 1). A third group of 30 rats was sham immunized as a control. The first conjugate was linked to fetuin at position 18 with a succinate linker (SAFet). The second was also conjugated to fetuin but the link was made through the nitrogen using a N-crotonate (NCrot) linker. About 0.2 g of the individual fetuin conjugates were mixed with complete Freund's adjuvant and injected subcutaneously into each respective group of vaccinated rats. After 21 days the rats were immunized with a similar amount of booster conjugate mixed with incomplete Freund's. The sham immunized control group was injected with the adjuvants only. Two weeks after the booster all three groups were re-randomized, divided into four groups of ten animals and challenged with 0, 1, 5 and 25 mg riddelliine kg^{-1} bw day^{-1} for 10 days. The dose was divided into two and given as a gavage twice daily. Twelve hours after the final dose all the rats were euthanized and necropsied, and serum and tissues were collected for serum biochemical and histologic analyses. Individual animal response to immunization was determined by measuring titers following previously described techniques (Lee *et al.* 2001). The animal weights, serum biochemistries and titers were analyzed using analysis of variance with P≤0.05 interpreted as being significant (Proc GLM, SAS Institute, Cary NC).

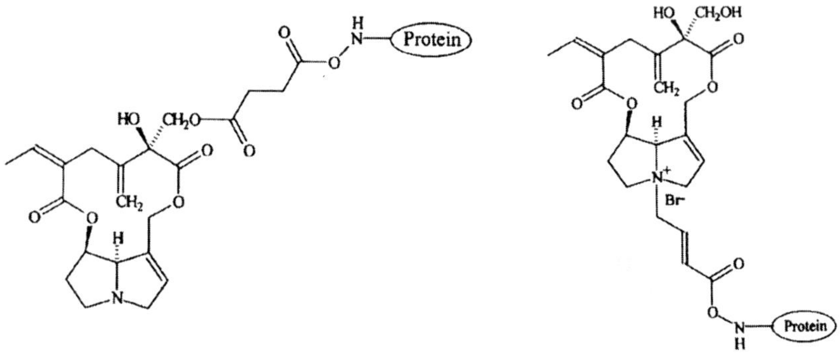

Figure 1. Fetuin-riddelliine conjugates that were used to immunize rats. The left is a riddelliine succinate fetuin conjugate that attaches the protein away from the pyrrolizidine base. The right is a riddelliine N-crotanate fetuin conjugate that attaches the protein to the base nitrogen.

Results and Discussion

All of the rats, including the high dose groups, had minimal response to the riddelliine challenge. Though the high dose animals lost more weight than the other groups, they were in good condition with little clinical evidence of liver disease. However, the serum biochemistries did significantly differ according to dose. High dose animals (25 mg kg^{-1}) had significantly increased ALP, ALT, AST, LDH, bilirubin and bile acid with decreased albumin when compared with the lower dose groups (0, 1 and 5 mg kg^{-1}). However, there were no significant differences between vaccination groups (Table 1).

Table 1. Serum biochemistries of rats that were immunized with two different riddelliine conjugates (SAFe or NaCrot) or only adjuvant (Sham). After the immunization and booster, the rats were challenged with riddelliine for 10 days. The data shown are only from the group dosed with 25 mg kg^{-1}.*

	Units	SAFet	NaCrot	Sham
Urea nitrogen	mg/dL	110±8	73±19	94±6
Albumin	g/dL	1.2±0.2	1.4±0.2	1.2±0.2
Alkaline phosphatase activity	U/L	715±105	842±88	751±80
ALT	U/L	192±30	213±40	200±37
AST	U/L	227±15	285±27	242±47
LDH	U/L	123±38	243±113	163±75
Total bilirubin	mg/dL	4.2±2.2	3.0±1.2	4.2±2.6
Indirect bilirubin	mg/dL	2.2±1.1	1.5±0.5	2.3±1.4
Bile acid	μmol/L	452±86	406±140	403±102

*Data are presented as means±standard deviation.

All of the rats vaccinated with either riddelliine conjugate had poor and variable response to vaccination with few animals developing titers greater than 1:1000. This indicates that the vaccination was not immunogenic in most animals. This is the most likely explanation for the lack of biochemical differences between the vaccination groups. Previous vaccination studies (Lee *et al.* 2003) suggest that biochemical and clinical changes indicative of toxicity are best interpreted when they are adjusted for their immunologic responses or titers. However, in this case, as there were few animals with adequate responses, it was impossible to adjust the response according to titer and do any statistical comparison. As seen in Figure 2 there seemed to be little correlation between titer and serum AST change.

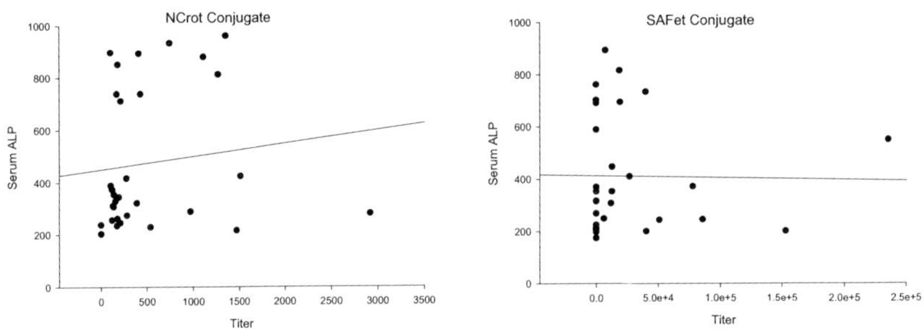

Figure 2. Graphs of serum ALP (alkaline phosphatase) plotted against serum titer in rats that were challenged with 25 mg kg^{-1} riddelliine for 10 days. Notice the wide variation in animal response to the challenge and that neither conjugate produced consistent high titers.

Initial histologic studies showed that the high dose animals all had hepatocellular swelling and vacuolation with small zones of necrosis, increased hepatocellular apoptosis, and congestion with small hemorrhages (Figure 3). Previous riddelliine toxicity studies of similar duration demonstrated extensive hepatocellular necrosis, ovalocyte and biliary proliferation with early fibrosis at much lower doses (Chan *et al.* 2003; Stegelmeier *et al.* 2004). This lack of response to the riddelliine challenge is disappointing. The riddelliine subacute LD_{50} in rats is estimated to be about 20 mg/kg for 10 days (Mattocks 1986; Chan *et al.* 2003). Toxicity is much higher in livestock species with whole plant toxicity studies suggesting lethal doses of 3–8 mg kg^{-1} for 7–10 days for cattle, horses and pigs (Johnson *et al.* 1985; Molyneux *et al.* 1991; Stegelmeier *et al.* 2004). As these challenge doses flank this range, we expected more severe toxicity. These rats were mature and as it has been shown that younger animals are often more susceptible to poisoning, younger animals may respond better and provide a better indicator of vaccination protection (Stegelmeier *et al.* 2004). Additionally animals may need more time after exposure for the disease to progress.

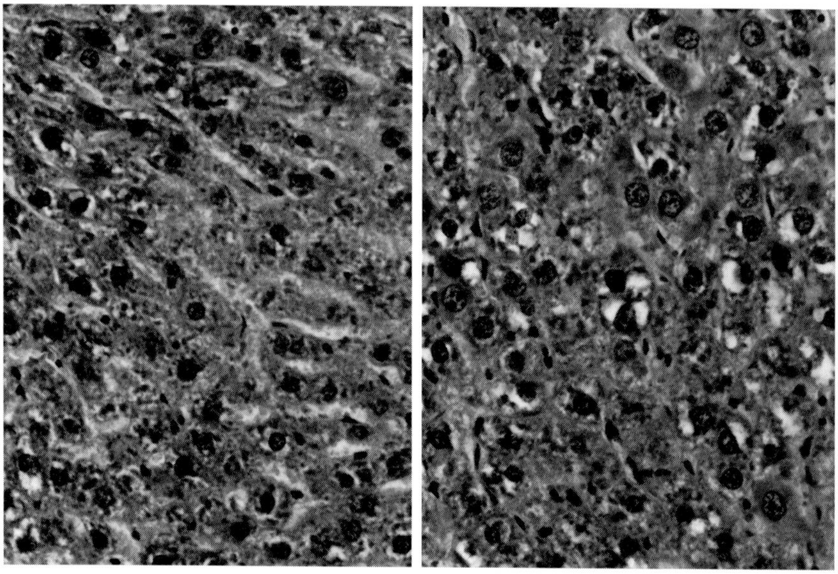

Figure 3. Photomicrograph of the liver from rats that were immunized with the NaCrot–riddelliine conjugate and challenged with 5 mg kg^{-1} riddelliine for 10 days. The liver on the left is from a rat with a high anti-riddelliine titer and the liver on the right is from an animal with a low titer. Notice the low titer liver has more hepatocellular swelling and vacuolation.

When the histologic lesions were grouped according to titer, the liver changes from animals with high titers were slightly different from changes seen in animals from the same group that had low titers. The livers from the high titer animals had less hepatocellular swelling, vacuolation and apoptosis. As only two to three animals per group responded to the vaccination with high titers there were few animals to include in this comparison and these findings are subjective. However, it is encouraging and suggests that additional work is needed to better answer these questions. Studies are currently underway with improved immunization schedules and adjuvant systems. The titers in these animals will be monitored *in vivo* and only animals that respond immunologically will be included in the toxicity challenge.

Summary

These rats had poor and variable response to vaccination with few animals developing titers greater than 1:1000. Although there was no significant biochemical difference between the vaccinated and control animals, rats with high titers appear to have less hepatocellular swelling and necrosis. Additional studies are underway to better develop this model and determine if immunization can alter pyrrolizidine alkaloid toxicity.

References

Chan PC, Haseman JK, Prejean JD, and Nyska A (2003). Toxicity and carcinogenicity of riddelliine in rats and mice. *Toxicology Letters* 144:295–311.

Johnson AE, Molyneux RJ, and Stuart LD (1985). Toxicity of Riddell's groundsel (*Senecio riddellii*) to cattle. *American Journal of Veterinary Research* 46:577–582.

Lee ST, Schoch TK, Stegelmeier BL, Gardner DR, Than KA, and Molyneux RJ (2001). Development of enzyme-linked immunosorbent assays for the hepatoxic alkaloids riddelliine and riddelliine n-oxide. *Journal of Agriculture and Food Chemistry* 49:4144–4151.

Lee ST, Stegelmeier BL, Panter KE, Pfister JA, Gardner DR, Schoch TK, and James LF (2003). Evaluation of vaccination against methyllycaconitine in mice. *Journal of Animal Science* 81:232–238.

Mattocks AR (1986). *Chemistry and Toxicology of Pyrrolizidine Alkaloids*, p. 195. Academic Press, Orlando, Florida.

Molyneux RJ, Johnson AE, Olsen JD, and Baker DC (1991). Toxicity of pyrrolizidine alkaloids from Riddell's groundsel (*Senecio riddellii*) to cattle. *American Journal of Veterinary Research* 52:146–151.

Stegelmeier BL, Edgar JA, Colegate SM, Gardner DR, Schoch TK, Coulombe RA, and Molyneux RA (1999). Pyrrolizidine alkaloid plants, metabolism and toxicity. *Journal of Natural Toxins* 8:95–116.

Stegelmeier BL, James LF, Panter KE, Molyneux RJ, Gardner DR, Schoch TK, and Lee ST (2004). Toxicity of a pyrrolizidine alkaloid, riddelliine in neonatal pigs. In *Poisonous Plants and Related Toxins* (T Acamovic, CS Stewart, and TW Pennycott, eds), pp. 44–49. CAB International, Wallingford, UK.

Chapter 16

Hepatotoxic Pyrrolizidine Alkaloids and Their *N*-Oxides in Honey and Pollen

K. Beales[1], K. Betteridge[3], M. Boppré[2], Y. Cao[1], S.M. Colegate[1], and J.A. Edgar[1]

[1]CSIRO Livestock Industries, Australian Animal Health Laboratory, Private Bag 24, Geelong, Victoria, 3220, Australia; [2]Forstzoologisches Institut, Albert-Ludwigs-Universität, D-79085 Freiburg i.Br., Germany; [3]AgResearch Grasslands, Private Bag 11008, Palmerston North, New Zealand

Introduction

The presence of toxic pyrrolizidine alkaloids and their *N*-oxides in animal feed and in the human food supply has resulted in well-documented, adverse effects and caused some countries to set regulations limiting exposure of humans to these toxins (Edgar *et al.* 2001).

Ingested bee products, such as honey and pollen, especially those derived mainly (if not entirely) from pyrrolizidine alkaloid-producing plants, constitute a direct exposure route for humans. Several studies (Deinzer *et al.* 1977; Culvenor *et al.* 1981; Crews *et al.* 1997) have demonstrated the presence of pyrrolizidine alkaloids in honey but the risk associated with these products remains to be fully quantified.

The potential and real disadvantages associated with using GC-MS or HPLC-MS for the trace analysis of pyrrolizidine alkaloids have been previously discussed (Beales *et al.* 2004b). For this current study of pyrrolizidine alkaloids in honey and pollen, HPLC-ion trap MS was utilized. The pyrrolizidine alkaloid content of over 60 Australian honey samples was quantitated using HPLC-atmospheric pressure chemical ionization (APCI)-MS whilst application of HPLC-electrospray ionization (ESI)-MS to extracts of Australian and New Zealand honeys, and to pollen samples collected from the plants or from bees that foraged the plants, allowed simultaneous quantitation of the pyrrolizidine alkaloids and their *N*-oxides. The latter application of HPLC-ESI-MS for identification and quantitation of pyrrolizidine-*N*-oxides (Betteridge *et al.* 2005; Boppré *et al.* 2005) was enabled by a different selection of solid phase extraction (SPE) media to that used in the analysis of 67 Australian honeys and one New Zealand honey using HPLC-APCI-MS (Beales *et al.* 2004a).

Materials and Methods

Sample collection and preparation

Honey samples were provided by commercial packagers or were purchased at retail outlets. Pure pollen samples were collected directly from the plant by washing individual anthers in *n*-hexane. Bee-harvested pollen was collected by physically removing the pollen baskets of honeybees (*Apis mellifera*).

For HPLC-APCI-MS

Honey samples were diluted with sulfuric acid (0.05 M), centrifuged and applied to polymeric resin-based strong cation exchange (SCX), SPE columns. The captured pyrrolizidine alkaloids and *N*-oxides were eluted from the columns using 20 mL of a solution of water/methanol/concentrated sulfuric acid (135/40/25) and immediately treated with zinc to reduce the *N*-oxides. Subsequent acid/base solvent–solvent extraction provided the pyrrolizidine alkaloids (including those derived from reduction of the *N*-oxides) for analysis (Beales *et al.* 2004a).

For HPLC-ESI-MS

Diluted (0.05 M H_2SO_4) and clarified (centrifugation at 12,000x g) honey samples, or filtered (0.45 µm) aqueous acid (0.05 M H_2SO_4) extracts of pollen samples were applied to silica-based SCX SPE columns. In contrast to the resin-based SPE columns, the captured pyrrolizidine alkaloids and *N*-oxides could be eluted from the columns using ammoniated methanol (methanol saturated with ammonia gas at 0–4°C and stored at room temperature). Immediate evaporation of the ammoniated methanol under a flow of nitrogen and reconstitution of the residue in methanol yielded the analytical sample (Betteridge *et al.* 2005; Boppré *et al.* 2005; Colegate *et al.* 2005).

Sample analysis

Aliquots of the analytical samples were applied to C8 (Beales *et al.* 2004a; Colegate *et al.* 2005) or C18 (Betteridge *et al.* 2005; Boppré *et al.* 2005) reverse phase HPLC columns and eluted with a decreasing gradient of 0.1% formic acid into 0.1% formic acid in acetonitrile. Total ion chromatograms of the column eluate were collected. Structural information was obtained from MS-MS experiments conducted in the second scan of two-scan sequences where a specific ion from the first scan was isolated in the ion trap and further fragmented to provide a mass spectrum of daughter ions.

Identification and quantitation of pyrrolizidine alkaloids/*N*-oxides

Pyrrolizidine alkaloids were identified in the analytical samples by comparison of their HPLC retention times and mass spectra with those of authenticated pyrrolizidine alkaloid standards. Where standard pyrrolizidine alkaloids were not available, the pyrrolizidine character of eluted compounds was indicated by the characteristic fragmentations observed in the MS-MS experiments (Beales *et al.* 2004a; Boppré *et al.* 2005; Colegate *et al.* 2005). For quantitation purposes, reconstructed ion chromatograms (RICs) were generated displaying only the molecular ion adduct ($[M + H]^+$) for specific pyrrolizidine alkaloids/*N*-oxides and the area under the peak integrated. The peak area was adjusted relative to an internal standard and the concentration of the pyrrolizidine alkaloid/*N*-oxide estimated from a calibration curve of a standard, authenticated pyrrolizidine alkaloid/*N*-oxide.

Results

Representative base ion (*m/z* 200–500) chromatograms are shown in Figure 1. The chromatograms in Figure 1 are each normalized to 100% Relative Abundance, which reveals the major components. Minor components (data not shown) were readily identified using RICs that displayed only the molecular ion adduct ($[M + H]^+$) for specific pyrrolizidine alkaloids/oxides.

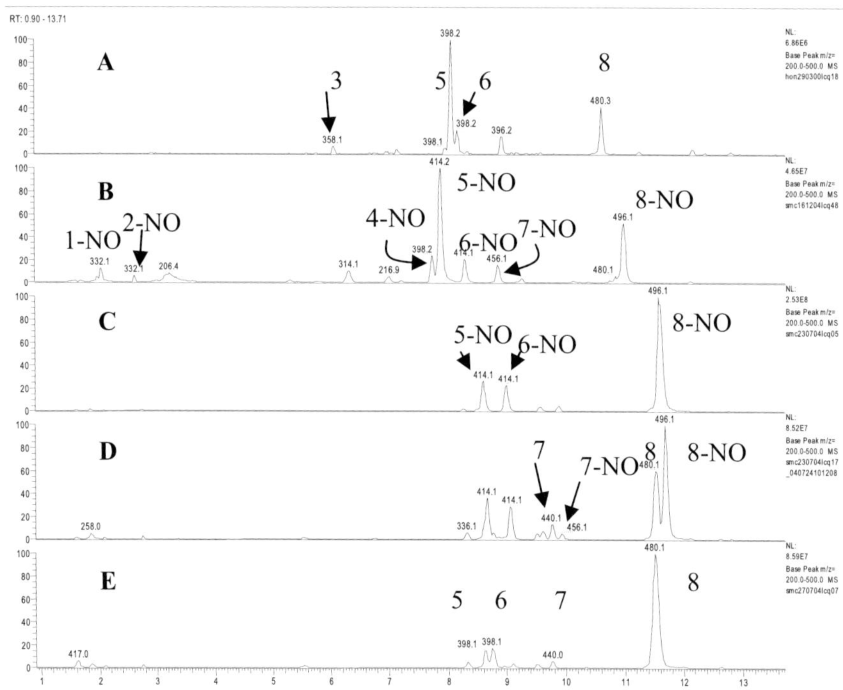

Figure 1. Typical HPLC-ion trap MS base ion (*m/z* 200–500) chromatograms showing: **A.** APCI-MS of pyrrolizidine alkaloids from *Echium vulgare* honey; and ESI-MS of pyrrolizidine alkaloids/*N*-oxides from: **B.** *Echium vulgare* honey; **C.** pure *Echium vulgare* pollen; **D.** *Echium vulgare* pollen collected from honeybees; and **E.** extract from pure pollen (trace C) treated with redox resin. The slight variations in retention times reflect slightly different HPLC conditions. Peak numbers refer to the structures shown in Figure 3.

The 68 honeys analyzed using HPLC-APCI-MS were classified into groups (Figure 2). Thus, 13 samples were attributed to *Echium plantagineum* ('Salvation Jane' or 'Paterson's Curse'), nine samples were listed as an *E. plantagineum* mix, four samples were collected in the vicinity of *Heliotropium amplexicaule* ('Purple Top'), two samples were derived mainly from *H. europaeum* ('Common Heliotrope' or 'Potato Weed') and 35 were attributed to floral sources with no known association with pyrrolizidine alkaloids. In addition, five retail samples of honey were analyzed. Three of these were samples of blended honeys, with no labeling to suggest floral sources; another was described as 'Leatherwood' whilst the last was described as 'Alpine Borage' honey. The latter two are derived from bees foraging on *Eucryphia lucida* and *Echium vulgare* respectively.

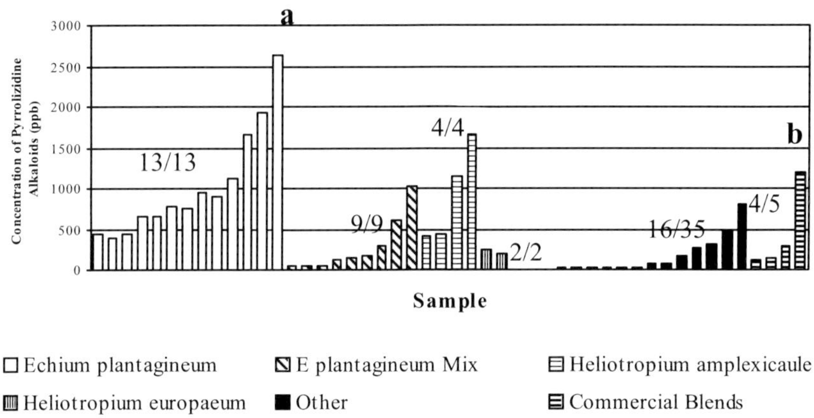

Figure 2. Pyrrolizidine alkaloid-positive samples relative to negative samples (p/n) determined using HPLC-APCI-MS analysis of 67 Australian-sourced honeys and one honey from New Zealand. The floral sources are as attributed by the apiarist and represent the major floral source in the collection area. The concentration of pyrrolizidine alkaloids is expressed as microgram equivalents of echimidine per kilogram of honey (ppb). **a** is the highest level found whilst **b** is the 'Alpine Borage Honey' referred to in the text.

The total pyrrolizidine alkaloid contents were summed and are shown in Figure 2. The highest level, in excess of 2500 μg equivalents of echimidine kg^{-1} honey (ppb), was observed with an *E. plantagineum*-derived sample (column **a** in Figure 2). The pyrrolizidine alkaloids identified (HPLC-APCI-MS) in these 68 samples reflected the floral sources indicated by the apiarist or clearly indicated the floral source where none was nominated. Thus honeys attributed to *E. plantagineum*, *Heliotropium europaeum* and *H. amplexicaule* were determined by the presence of echimidine, heliotrine and indicine respectively as the major alkaloid observed. As expected, those samples attributed to an *E. plantagineum* mix also contained echimidine as the major pyrrolizidine alkaloid. The pyrrolizidine alkaloid-positive samples of those attributed to non-pyrrolizidine alkaloid-producing plants e.g. *Eucalyptus* spp., mainly reflected *E. plantagineum* as the contaminating floral source. However the foraging or co-foraging of *H. europaeum* and

Senecio spp. was indicated by the identification of heliotrine or senecionine respectively in some of the samples. Analysis of three of the commercial blends with unknown plant source indicated the presence of *E. plantagineum* whilst a fourth (column **b** in Figure 2), attributed to *E. vulgare*, showed the presence of echimidine (*m/z* 398) plus an unidentified alkaloid with *m/z* 480 (Figure 1A).

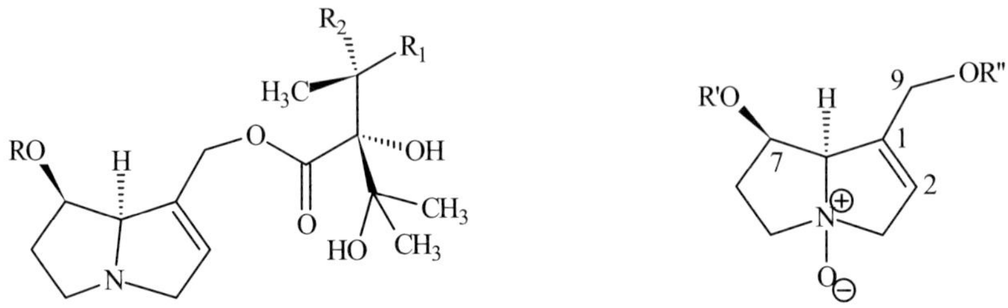

Pyrrolizidine-*N*-oxides

Structure	**R**	$\mathbf{R_1}$	$\mathbf{R_2}$	
1	H	OH	H	**leptanthine**
2	H	H	OH	**echimiplatine**
3	Ac	OH	H	**uplandicine**
4	H	undefined C9 ester		**echiuplatine**
5	$C_5H_7O_2$	OH	H	**echimidine**
6	H	$C_5H_7O_2$	H	**vulgarine**
7	$C_5H_7O_2$	OAc	H	**acetylechimidine**
8	$C_5H_7O_2$	$C_5H_7O_2$	H	**echivulgarine**

Where : Ac = acetyl
and $C_5H_7O_2$ = angelyl : R_1=R_3=CH_3, R_2=H
tiglyl : R_2=R_1=CH_3, R_3=H
seneciyl : R_3=R_2=CH_3, R_1=H

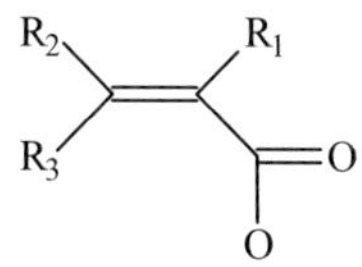

Figure 3. Structures and tentative structures for the pyrrolizidine alkaloids/*N*-oxides identified in *Echium vulgare* honey and pollen; the structure numbers refer to Figure 1.

The subsequent use of the silica-based SCX SPE columns for the analysis of nine honeys attributed to *E. vulgare* (Figure 1B), pure pollen collected from *E. vulgare* (Figure 1C), and bee-collected pollen from *E. vulgare* (Figure 1D) allowed the tentative identification of a new suite of pyrrolizidine alkaloids/*N*-oxides characteristic of *E. vulgare* (Betteridge *et al.* 2005; Boppré *et al.* 2005) (Figure 3) in addition to those also reported for *E. plantagineum* (Colegate *et al.* 2005). As with the extract of *E. plantagineum*, the pyrrolizidine alkaloids naturally occurred mainly as the *N*-oxides. Confirmation of *N*-oxide character was obtained by mixing a methanolic solution of the extracts with an indigocarmine-based redox resin for about 2–4 hours at 37ºC and then reanalyzing the solution using HPLC-ESI-MS (Figure 1E). The formation of new peaks, usually eluting slightly earlier than the putative *N*- oxides, with molecular ion adducts 16 mu less than the *N*-oxides and no evidence of dimeric molecular ion adducts, strongly supported the pyrrolizidine alkaloid/*N*-oxide relationship. For quantitation purposes, the pyrrolizidine alkaloids identified in the *E. vulgare* honeys and pollens were expressed as equivalents of lasiocarpine whilst the *N*-oxides were expressed as equivalents of lasiocarpine-*N*-oxide. The highest level determined was approximately 5000 ppb. Four samples returned levels of between 1200 and 2850 ppb and the final four between 550 and 950 ppb. The level of pyrrolizidine alkaloids/*N*-oxides in the pure pollen on the other hand was considerably higher at about 11,000 ± 3000 ppm (mg kg^{-1}) whereas levels in the 'pollen baskets' harvested from honeybees ranged from 925 to 4000 ppm with a mean (from four samples) of 2125 ppm.

Conclusions

The combination of silica-based SCX SPE and HPLC-ESI-MS analysis confirmed the natural predominance of *N*-oxides over the parent, tertiary-base pyrrolizidine alkaloids in the honey and pollen samples. The very high levels of pyrrolizidine alkaloids/*N*-oxides found in pure pollen compared to the honey indicated that pollen could be a major source of the pyrrolizidine alkaloids/*N*-oxides in honey. Analysis of the pollen collected by bees returned lower levels of the alkaloids, compared to pure pollen, consistent with the pollen baskets being made up of about 30% pollen in nectar-regurgitate 'glue'. This latter result supports the hypothesis that nectar contains little or no pyrrolizidine alkaloid/*N*-oxides, as might be expected of a substance that constitutes an attractant for pollinating insects. On the other hand, toxic or repellent secondary metabolites would be expected in the pollen to help protect the genetic material from predation.

The potential threat to human health that is posed by the pyrrolizidine alkaloids and their *N*-oxides has, for example, been considered by the German Government, which has instigated regulations for the presence of pyrrolizidine alkaloids in herbal preparations. Pregnant and lactating women are specifically advised to avoid exposure to the pyrrolizidine alkaloids. Recent, neonatal deaths ascribed to pyrrolizidine alkaloid-induced liver failure (Rasenack *et al.* 2003), when assessed in conjunction with levels of the pyrrolizidine alkaloids found in honey and the dietary intake figures for honey, support consideration of regular monitoring of honey and other bee products for the presence of pyrrolizidine alkaloids.

References

Beales KA, Betteridge K, Colegate SM, and Edgar JA (2004a). Solid phase extraction and LCMS analysis of pyrrolizidine alkaloids in honeys. *Journal of Agriculture and Food Chemistry* 52:6664–6672.

Beales KA, Colegate SM, and Edgar JA (2004b). Experience with the quantitative trace analysis of pyrrolizidine alkaloids using GCMS and LCMS. In *Poisonous Plants and Related Toxins* (T Acamovic, CS Stewart, and TW Pennycott, eds), pp. 453–458. CAB International, Wallingford, Oxon.

Betteridge K, Cao Y, and Colegate SM (2005). Improved method of extraction and LC-MS analysis of pyrrolizidine alkaloids and their *N*-oxides in honey: Application to *Echium vulgare* honeys. *Journal of Agriculture and Food Chemistry* 53:1894–1902.

Boppré M, Colegate SM, and Edgar JA (2005). Pyrrolizidine alkaloids of *Echium vulgare* honey found in pure pollen. *Journal of Agriculture and Food Chemistry* 53:594–600.

Colegate SM, Edgar JA, Knill AM, and Lee ST (2005). Solid phase extraction and LCMS profiling of pyrrolizidine alkaloids and their N-oxides: A case study of *Echium plantagineum*. *Phytochemical Analysis* 16:108–119.

Crews C, Startin JR, and Clarke PA (1997). Determination of pyrrolizidine alkaloids in honey from selected sites by solid-phase extraction and HPLC-MS. *Food Additives and Contaminants* 14:419–428.

Culvenor CCJ, Edgar JA, and Smith LW (1981). Pyrrolizidine alkaloids in honey from *Echium plantagineum* L. *Journal of Agriculture and Food Chemistry* 29:958–960.

Deinzer ML, Thompson PA, Burgett DM, and Isaacson DL (1977). Pyrrolizidine alkaloids: Their occurrence in honey from tansy ragwort (*Senecio jacobaea* L.). *Science* 195:497–499.

Edgar JA, Roeder E, and Molyneux RJ (2001). Honey from plants containing pyrrolizidine alkaloids: A potential threat to health. *Journal of Agriculture and Food Chemistry* 50:2719–2730.

Rasenack R, Müller C, Kleinschmidt M, Rasenack J, and Wiedenfeld H (2003). Veno-occlusive disease in a fetus caused by pyrrolizidine alkaloids of food origin. *Fetal Diagnosis and Therapy* 18:223–225.

Chapter 17

An Overview of Pyrrolizidine Alkaloid Toxicity in Livestock: Microbial and Metabolic Perspectives

A.M. Craig, J.M. Duringer, and L.L. Blythe

College of Veterinary Medicine, Oregon State University, Corvallis, OR 97331, USA

Introduction

Pyrrolizidine alkaloids (PAs) are among the most prolific groups of plant toxins affecting livestock, wildlife and human health throughout the world. More than 350 PAs have been isolated from over 6000 plant species (Stegelmeier *et al.* 1999). It is estimated that 3% of the world's flowering plants contain toxic PAs (Smith and Culvenor 1981). Pyrrolizidine alkaloids are largely hepatotoxins and are capable of causing acute and chronic toxicoses (Mattocks 1986). Acute toxicosis rarely occurs under natural conditions since it usually requires large doses of PAs, resulting in severe hemorrhagic necrosis of the liver and rapid death. Chronic toxicosis following oral ingestion of PAs usually results in development of veno-occlusive disease of the liver, which occludes the small branches of hepatic veins. This eventually causes ascites, edema and reduced urinary output, and can lead to cirrhosis and death in extreme cases. PAs can also affect other organs, such as the lungs and heart, in addition to initiating cancer (Mattocks 1986).

Pyrrolizidine-containing plants are a major poisonous plant concern for livestock producers in the northwest United States, as they are estimated to cost $20 million in lost production each year (Craig *et al.* 1991). The plants most commonly involved in livestock poisoning belong to the *Senecio* and *Crotalaria* genera, with tansy ragwort (*Senecio jacobaea*) being one of the more intensively studied species. Susceptibility to PA toxicity is influenced by many factors including species, age and sex as well as other temporary factors such as the mode of PA exposure and the biochemical, physiological and nutritional status of the individual (Stegelmeier *et al.* 1999). Cattle are susceptible to PA toxicosis and develop terminal cirrhotic liver disease after ingestion of tansy ragwort plants equal to 5–10% of their body weight, whereas sheep are more resistant and can consume up to 200% of their body weight without adverse effects (Cheeke 1984; Craig *et al.* 1991). A long term controversy has evolved regarding the mechanism of resistance to PA toxicity in sheep versus cattle. Both ruminal microorganism detoxification and hepatic metabolism differences between sheep and cattle have been examined and are reviewed here.

Ruminal Microbe Metabolism of Pyrrolizidine Alkaloids

Substantial evidence supports ruminal microbe metabolism of PA toxins as a major mechanism accounting for the sheep's resistance to PA-induced toxicosis (Craig *et al.* 1986, 1992; Wachenheim *et al.* 1992; Hovermale and Craig, 2002; Lodge-Ivey *et al.* 2005). Attempts to isolate and define the protective mechanism in sheep were initiated by administering purified pyrrolizidine alkaloids to cattle and sheep both via catheter (using a chronic *in vivo* liver perfusion technique (Blythe and Craig 1986)) and orally (Craig *et al.* 1986). When PA metabolism of the liver was isolated from the rumen via direct infusion of alkaloid into the hepatic portal vein, sheep developed clinical, enzymatic and histopathological signs of PA toxicosis comparable to those seen in cattle. Control sheep, which were fed a PA-containing diet orally, showed no signs of toxicosis. This implied that the rumen microflora, absorption and/or metabolism of the alkaloid in the gastrointestinal tract may play a role in the sheep's resistance to PA toxicosis.

Next, an *in vitro* assay examined the biotransformation of PAs from *S. jacobaea* in sheep and cattle rumen fluid (Wachenheim *et al.* 1992). PAs were analyzed by high-performance liquid chromatography and the rates of alkaloid biotransformation were determined. Sheep metabolized the PAs at 19.2 $\mu g\ ml^{-1}\ h^{-1}$ versus cattle, which converted them at a rate of 2.9 $\mu g\ ml^{-1}\ h^{-1}$. Sheep also had a higher number of PA-transforming bacteria (3.0×10^7 bacteria/mL) than cattle (1.1×10^7 bacteria/mL) when counted using the most probable numbers technique. As the rates of PA disappearance were generally related to the most probable numbers of PA-degrading bacteria (Figure 1), it appeared that ruminal bacteria in animals with a larger number of PA-biotransforming bacteria are responsible for increased rates of PA degradation and protection from PA toxicosis.

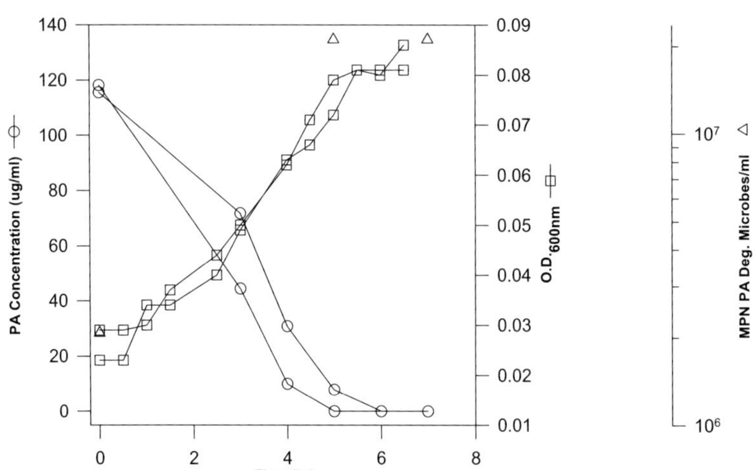

Figure 1. Characterization of growth and degradation of pyrrolizidine alkaloids from *Senecio jacobaea* by a PA-degrading enrichment culture. (○) PA concentration ($\mu g\ ml^{-1}$), (□) optical density of culture measured at 600nm, and (△) most probable number of cells mL^{-1}. Lines represent duplicate measurements.

Metabolites from degradation of PAs by a bacterial consortium isolated from the rumen of sheep (L4M2) were then analyzed by GC/MS and TLC (Hovermale and Craig 2002). L4M2 was found to degrade the PAs heliotrine and lasiocarpine to the 1-methylene metabolites 7α-hydroxy-1-methylene-8α-pyrrolizidine and 7α-angelyl-1-methylene-8α-pyrrolizidine/7α-tiglyl-1-methyl-ene-8α-pyrrolizidine, respectively. Monocrotaline was metabolized to 7β-hydroxy-1-methylene-8α-pyrrolizidine and PAs from an *S. jacobaea* extract were converted to a small amount of the same 1-methylene metabolite.

L4M2 was then characterized using molecular techniques (Lodge-Ivey *et al.* 2005). Length-heterogeneity PCR (LH-PCR) and restriction fragment length polymorphism (RFLP) analysis as well as cloning and sequencing were used to identify members of the consortium. Phylogenetic analysis of sequences from both methods was conducted. First, serial dilutions of L4M2 were made and plated onto agar plates of media E. Colonies were chosen based on size and morphology for representation of total culturable bacteria. Representative colonies were streaked on agar E for isolation. Isolates were characterized by Gram reaction, cellular morphology and size. Five morphotypes were observed.

Second, two different molecular techniques were used to reveal the identities of the bacteria in the L4M2 consortia. For RFLP, DNA was obtained from isolate colonies using a freeze–thaw method before PCR amplification of the 16S rRNA gene. PCR products were subjected to restriction enzyme digestion, and restriction fragments were separated on an agarose gel. Five different ribotypes corresponding to five different isolates were observed. PCR products from these isolates were sequenced and were used in the phylogenetic analysis of L4M2. DNA was also extracted from overnight cultures using a bead-beating, phenol-chloroform method with ethanol precipitation. Amplification of chromosomal 16S rRNA (rDNA) by PCR was performed. These products were cloned into a plasmid vector (pCR 2.1, Invitrogen) to create two libraries. One library consisted of clones with long-fragment inserts while the other was made up of short-fragment-insert clones. LH-PCR was performed on the short-insert clones. These analyses resulted in an electropherogram displaying seven independent peaks representing seven fragments differing by 2 to 27 base-pairs. Clones of representative peaks were sequenced and used in the phylogenetic analysis of L4M2. Phylogenetic analysis of both methods identified six organisms (Figure 2), <97.0% homologous with known bacteria, indicating that this consortium contains novel organisms.

Hepatic Metabolism of Pyrrolizidine Alkaloids in Livestock

Hepatic metabolism of senecionine (a PA found in *S. jacobaea*) was studied in liver microsomes and cytosol from sheep and cattle (Duringer *et al.* 2004). Senecionine, senecionine *N*-oxide (*N*-oxide, non-toxic metabolite) and 6,7-dihydro-7-hydroxy-1-hydroxymethyl-5H-pyrrolizine (DHP, toxic metabolite) were measured in microsomal incubations. In addition, the turnover number (k_{cat}) was determined for DHP and *N*-oxide formation. Chemical and immunochemical inhibitors assessed the role of cytochrome P450s (CYPs), flavin-containing monooxygenases (FMOs) and carboxylesterases in senecionine metabolism. CYP3A, CYP2B and FMO concentrations and activities were also determined.

DHP concentration did not differ between species (Table 1). Sheep formed more *N*-oxide, had a higher *N*-oxide k_{cat} and metabolized senecionine faster than cattle. CYP levels and isoforms had a larger influence on DHP formation in cattle than in sheep. While CYP3A and CYP2B activities were similar, cattle had higher concentrations of CYP3A protein than sheep. FMOs had a large influence on formation of *N*-oxide, and their activity was greater in sheep than in cattle. These results suggest that more *N*-oxide production and the ability of the parent PA to be metabolized quickly are factors used by sheep to abate PA toxicity. However, it is likely that hepatic metabolism differences alone do not account for the nearly 20-fold variation in susceptibility seen between sheep and cattle.

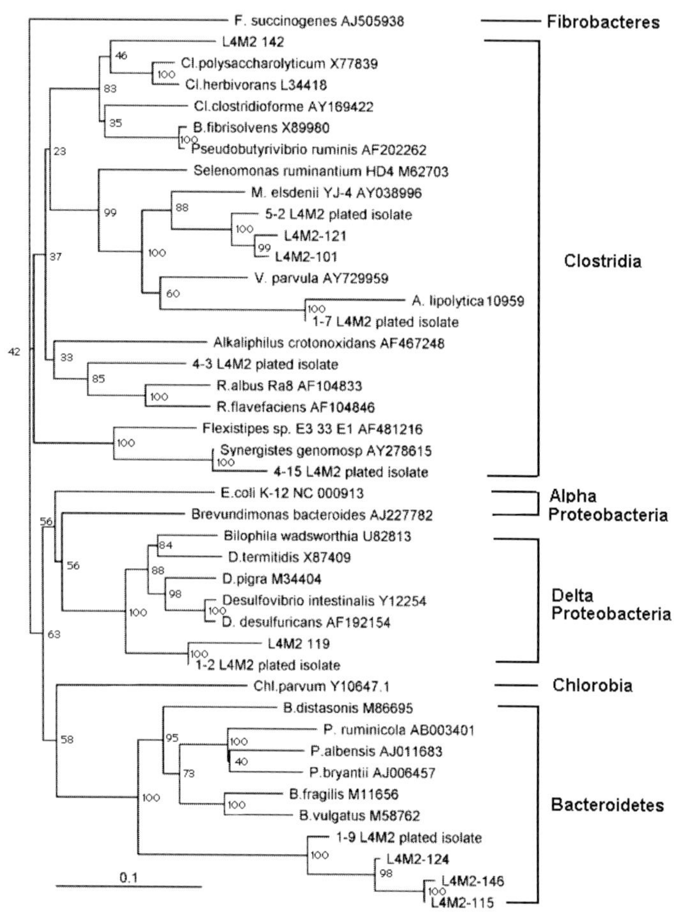

Figure 2. Phylogenetic tree showing the relationships of 16S rDNA sequences from a consortium of PA-degrading bacteria analyzed by LH-PCR. The scale bar represents genetic distance (10 substitutions per 100 nucleotides). The tree was constructed using the neighbor-joining analysis of a distance matrix obtained from a multiple-sequence alignment. Bootstrap values (expressed as percentages of the value for 100 replications) are shown at the branch points; values of 97% or more were considered significant.

Table 1. *In vitro* metabolism of senecionine by sheep and cattle liver microsomes to 6,7-dihydro-7-hydroxy-1-hydroxymethyl-5H-pyrrolizine (DHP) and senecionine *N*-oxide (mean ± SEM). SN_B, SN_A = amount of senecionine before (B) and after (A) incubation.

Species	Sex	DHP (nmol min^{-1} mg protein^{-1})	*N*-oxide (nmol min^{-1} mg protein^{-1})	Ratio of DHP/ *N*-oxide	SN_A/SN_B (%)
Cattle					
	Female	0.42±0.13	0.33±0.08	1.24±0.19	88.6±2.1
	Male	0.39±0.08	0.19±0.04	2.11±0.18	87.0±2.0
	Both	0.41±0.01	0.26±0.07	1.68±0.29	87.8±1.9
Sheep					
	Female	0.28±0.09	1.11±0.38	0.29±0.05	76.4±2.1
	Male	0.50±0.10	1.08±0.06	0.46±0.09	73.5±4.6
	Both	0.39±0.11	1.10±0.26	0.37±0.08	74.9±3.4

Conclusion

Increased ruminal metabolism of PAs in sheep appears to be an important protective mechanism in exerting their resistance to PA toxicosis, while hepatic enzymes provide a secondary means to degrade any PAs absorbed from the rumen. Further understanding of the metabolic pathways used by the L4M2 consortium to degrade PAs could lead to use of the consortium as a probiotic therapy for livestock afflicted with tansy ragwort toxicosis.

References

Blythe L and Craig A (1986). A chronic *in vivo* liver perfusion technique in ruminants. *Veterinary and Human Toxicology* 28(3):201–204.

Cheeke PR (1984). Comparative toxicity and metabolism of pyrrolizidine alkaloids in ruminants and nonruminant herbivores. *Canadian Journal of Animal Science* 64:201–202.

Craig AM, Blythe LL, Lassen ED, and Slizeski ML (1986). Resistance of sheep to pyrrolizidine alkaloids. *Israel Journal of Veterinary Medicine* 42(4):376–384.

Craig AM, Latham CJ, Blythe LL, Schmotzer WB, and O'Connor OA (1992). Metabolism of toxic pyrrolizidine alkaloids from tansy ragwort (*Senecio jacobaea*) in ovine ruminal fluid under anaerobic conditions. *Applied and Environmental Microbiology* 58(9):2730–2736.

Craig AM, Pearson EG, Meyer C, and Schmitz JA (1991). Serum liver enzyme and histopathologic changes in calves with chronic and chronic-delayed *Senecio jacobaea* toxicosis. *American Journal of Veterinary Research* 52(12):1969–1978.

Duringer JM, Buhler DR, and Craig AM (2004). Comparison of hepatic *in vitro* metabolism of the pyrrolizidine alkaloid senecionine in sheep and cattle. *American Journal of Veterinary Research* 65(11):1563–1572.

Hovermale J and Craig A (2002). Metabolism of pyrrolizidine alkaloids by *Peptostreptococcus heliotrinreducens* and a mixed culture derived from ovine ruminal fluid. *Biophysical Chemistry* 101-102:387–399.

Lodge-Ivey SL, Rappe MS, Johnston WH, Bohlken RE, and Craig AM (2005). Molecular analysis of a consortium of ruminal microbes that detoxify pyrrolizidine alkaloids. *Canadian Journal of Microbiology* 51(6):455–465.

Mattocks AR (1986). *Chemistry and Toxicology of Pyrrolizidine Alkaloids*. Plenum Press, New York.

Smith LS and Culvenor CCJ (1981). Plant sources of hepatotoxic PAs. *Journal of Natural Products* 44:129–152.

Stegelmeier BL, Edgar JA, Colegate SM, Gardner DR, Schoch TK, Coulombe RA, and Molyneux RJ (1999). Pyrrolizidine alkaloid plants, metabolism and toxicity. *Journal of Natural Toxins* 8(1):95–116.

Wachenheim DE, Blythe LL, and Craig AM (1992). Characterization of rumen bacterial pyrrolizidine alkaloid biotransformation in ruminants of various species. *Veterinary and Human Toxicology* 34(6):513–517.

Chapter 18

Acute Bovine Liver Disease: Hepatocytotoxicity of *Cynosurus echinatus*-related *Drechslera biseptata*

M.R. Aslani[2], I. Pascoe[3], M. Kowalski[1], A. Michalewicz[1], M.A.S. Retallick[1], and S.M. Colegate[1]

[1]CSIRO Livestock Industries, Australian Animal Health Laboratory, Private Bag 24, Geelong, Victoria, 3220, Australia; [2]Department of Clinical Sciences, School of Veterinary Medicine, Ferdowsi University of Mashhad, PO Box 91775-1793, Mashhad, Iran; [3]Department of Primary Industries and Environment, Knoxfield, Victoria, 3180, Australia

Introduction

Acute Bovine Liver Disease (ABLD) is an acute hepatopathy of beef and dairy cattle in southern parts of Australia (Gunn and Clarke 2003). Clinical signs include reduced milk production, acute photosensitization, profound morbidity and even sudden death, with some mortality rates in excess of 50% of affected animals. Deaths can occur within 12 hours of access to a toxic paddock at the autumn break of the season. The causative factor(s) of ABLD are undefined but outbreaks have been consistently associated with the presence of the annual grass *Cynosurus echinatus*. Stands of *C. echinatus* within a pasture seemed to be indicative of a less well-managed pasture that also contained other weeds amongst the predominant pasture grasses.

Initial suspected involvement of the sporidesmin-producing *Pithomyces chartarum* or the microcystins produced by blue-green algae were dismissed on histological grounds. The pathology of the intoxication is quite unique, involving hepatic periportal necrosis (Gunn and Clarke 2003).

To address the possibility of mycotoxin involvement in the aetiology of the disease, a mycological assessment of toxic pastures was conducted (I Pascoe, unpublished). Samples of *C. echinatus* collected from the site and at the time of ABLD outbreaks have yielded a known phytopathogenic fungus, *Drechslera biseptata*, which has a worldwide occurrence and has been suspected of producing mycotoxins (Leach and Tulloch 1972). Infertility in mice and guinea pigs has been associated with *Drechslera* spp. on *Romulea rosea* (Fisher and Finnie 1967) and other mammalian toxicity has been associated with *Drechslera campanulata* (syn. *Pyrenophora semeniperda*) (Schneider *et al.* 1985; Collett *et al.* 1988).

A preliminary feeding trial in which single 9-month-old steers were fed either oats or *C. echinatus*, both of which had been inoculated and incubated with a *D. biseptata* culture, failed to elicit any liver damage or clinical signs (T Jubb, Victorian Department of Primary Industries, personal communication).

Materials and Methods

Hepatocyte culture

Clone 9 rat hepatocytes were purchased from the American Type Culture Collection (Manassas, Virginia, USA – ATCC accession number CRL-1439). This cell line was isolated from a normal male Sprague-Dawley rat liver in 1968 (Weinstein *et al.* 1975) and has been recently used for toxicology studies (Huang *et al.* 2001).

In preparation for the bioassay, the hepatocytes were seeded at a density of 1×10^4 cells cm^{-2} in F-12K media (1×10^5 cells mL^{-1}) into OptiCells™ (BioCrystal, Westerville, Ohio, USA) and maintained in an incubator at 37°C under a 7.5% CO_2/92.5% air atmosphere.

Culture of *Drechslera biseptata*

Cultures of *D. biseptata*, isolated from outbreaks of ABLD, on V8 agar were preserved by aseptically excising 20–30 small (approx. 4 mm) cubes of agar together with the fungal mycelium and storage at 4°C in a sterile, screw-capped bottle containing sterile, distilled water (10 mL).

A sample of the stored source culture was inoculated at a single central point onto V8 agar in 9 cm Petri plates. The inoculated plates were placed in a plastic bag and the fungus left to grow at room temperature exposed to daylight. The diurnal changes in light stimulated the sporulation of the fungus. After 7 days the mycelium formed a complete mat over the surface of the entire plate, providing an ongoing source of the fungal isolate. For bioassay of fungal metabolites in the hepatocyte culture, the fungal growth was allowed to develop for about 2 weeks before processing.

Generation of test samples

Batches of up to six Petri plates of V8 agar cultures of *D. biseptata* were each washed (3 x 2 mL) with hexane, ethyl acetate, methanol or water and the extracts were evaporated to dryness. Extracts that caused *in vitro* cytopathic effects in the hepatocyte bioassay were further fractionated on C18 Reverse Phase (RP) solid phase extraction (SPE) columns, which were subsequently washed (2 x 6 mL) with water, 80% water in methanol, 30% water in methanol and finally with methanol. The column eluates were evaporated to dryness using a centrifugal, vacuum evaporator and subsequently reconstituted in cell culture media (12 mL) for subsequent bioassay or in methanol (1 mL) in preparation for analysis using high pressure liquid chromatography (HPLC) coupled to ion trap electrospray ionization (ESI) mass spectrometry (MS).

Bioassay

Each of the test samples was mixed with sterile cell culture media (12 mL) and aseptically filtered (0.45 μm). The cell growth media was removed from OptiCells™ that contained confluent and adhered hepatocytes and replaced with the test sample solutions in growth media. The 'loaded' OptiCells™ were then incubated at 37°C under a 7.5% CO_2/ 92.5% air atmosphere. After 4, 24 and 48 h digital photomicrographs of the live, unstained cells within the OptiCells™ were captured by using a Leica DMLB Microscope attached to a DC300F digital camera (Leica Microsystems, Bannockburn, Illinois, USA). After 48 hours, excised OptiCell™ membranes to which the cells were adhered were stained with hematoxylin and eosin and mounted on microscope slides for assessment.

Plant collection and extraction

At the time of an outbreak of ABLD in a Friesian dairy herd in southwestern Victoria (May 2003), samples of the pasture grass components were collected. Pasture samples included green, fresh *Phalaris aquatica* and *Lolium perenne* (Perennial Ryegrass), and dried *Cynosurus echinatus* that was brown in colour and was the remnants of the previous season's growth. Within 4 hours of collection the pasture samples were immersed in methanol for 24 hours with gentle stirring at room temperature. The methanolic solutions were evaporated to dryness under vacuum and mild warmth (<40°C) and the residues re-extracted with sterile cell culture media in preparation for bioassay using the OptiCells™. An aliquot of the test solution was processed on the C18 SPE columns, as described for the cultured fungal extracts, for subsequent HPLC-ESI-MS analysis.

HPLC-ion trap ESI-MS

Aliquots (2 μL) of the analytical samples derived from the fungal cultures or the plant extracts were injected on to a C18 Reverse Phase (RP) 'Aqua' column (Phenomenex) and eluted with a decreasing gradient (93% to 30% over 15 min) of 0.1% aqueous formic acid into 0.1% formic acid in acetonitrile (flow rate 200 μL min^{-1}). The eluate from the column was monitored using an ion trap mass spectrometer (ThermoFinnigan LCQ) in the ESI mode. Additional structural information was acquired using MS-MS experiments in which the molecular ion adduct ($[M + H]^+$) was isolated in the first scan and further fragmented in the second scan of a two scan sequence.

Results and Discussion

The major benefit of the OptiCells™ system is that the cells can be observed in 'real time' and at various intervals after exposure without resorting to large numbers of test samples. The ethyl acetate extract of the *D. biseptata* mycelium and spores had no effect on the hepatocytes (Figure 1A) whilst both the hexane and water extracts elicited multinucleation (Figure 1B). The methanol extract of the spores and mycelium, and the methanolic extracts of the residues after initial hexane, ethyl acetate or water extraction, all resulted in severe degeneration (Figure 1D) of the hepatocytes within 2 hours of exposure. The same degenerative hepatocytotoxic effect was observed with culture media-solubles from the methanolic extract of the toxic pasture-derived *C. echinatus*.

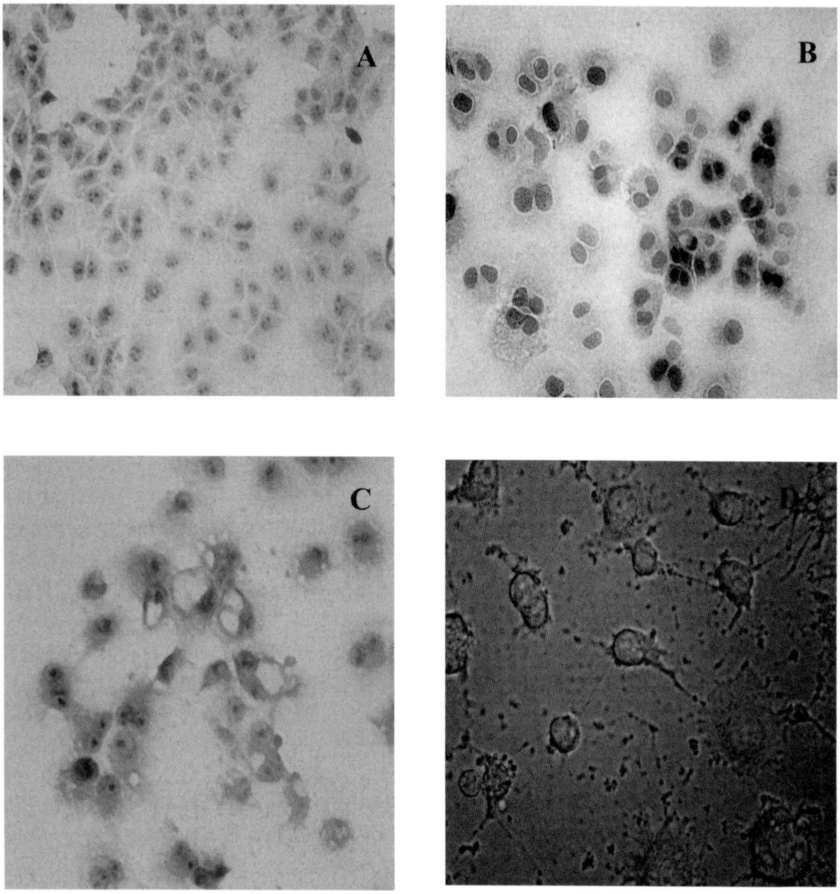

Figure 1. Cultured hepatocytes exposed to different extracts of *Drechslera biseptata*. **A.** ethyl acetate extract showing normal hepatocyte morphology (H&E, 400x); **B.** multinucleation associated with hexane or water extractables (H&E, 400x); **C.** cellular degeneration associated with exposure to the methanolic extractables after hexane extraction (H&E, 400x); and **D.** 'Real Time' photomicrograph (400x) of degenerate hepatocytes exposed to the methanolic extract of *D. biseptata* for 2 hours.

Bioassay-monitoring of further fractionation of the methanol extract of the *D. biseptata* cultures on the C18 RP SPE columns revealed that the hepatocytotoxicity was retained on the column and eluted with 70% methanol in water.

Previous work has identified cytochalasins from *D. campanulata* and *Drechslera wirreganensis* (Evidente *et al.* 2002; Capio *et al.* 2004). Binucleation of barley root tip cells and of mammalian cells have been reported following exposure to the cytochalasins (Carter 1967; Capio *et al.* 2004). HPLC-ESI-MS comparison of the hepatocytotoxic, methanol extract of the *D. biseptata* mycelium and spores with authentic samples of cytochalasins A and B indicated co-identity of one *D. biseptata* extract with cytochalasin B. However, the apparent discrepancy in the reported water insolubility of the cytochalasins A and B with the obvious water solubility of the hepatocytotoxic components from *D. biseptata* requires rationalization.

The HPLC-ESI-MS ion chromatogram of the 70% methanol eluate from the C18 SPE column displayed one main peak ($[M + H]^+$ *m/z* 480) that corresponded to the cytochalasin B peak in Figure 2B. The same peak, based on observed retention time and MS-MS spectra, was observed in the extracts of *C. echinatus*, and no other pasture species, collected from the toxic pastures. The later eluting peak ($[M + H]^+$ *m/z* 478/496, Figure 2A) was eluted from the SPE column with 20% methanol in water and did not correspond to cytochalasin A ($[M + H]^+$ *m/z* 478, Figure 2B).

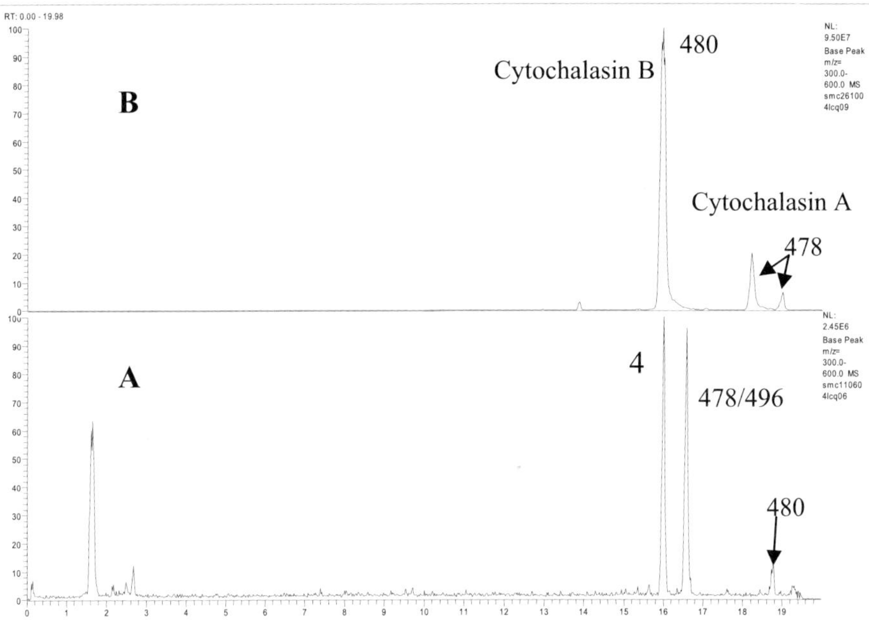

Figure 2. The HPLC-ESI-MS base ion (*m/z* 300-600) chromatograms for : **A.** a methanol extract of *Drechslera biseptata* culture (mycelium and spores), and : **B.** authenticated standards of cytochalasins B and A. The molecular ion adducts ($[M + H]^+$) are shown beside the peaks.

Conclusions

Extracts of agar cultures of *D. biseptata* isolated from outbreaks of acute bovine liver disease were shown to be cytotoxic to cultured rat hepatocytes in the same way that extracts of *C. echinatus*, collected from toxic pastures, were also hepatocytotoxic. The tentative identification of the major peak in the hepatocytotoxic fractions with cytochalasin B does not preclude the potential existence of another low abundance but highly potent hepatocytotoxic compound.

Future work is required to unambiguously identify the hepatocytotoxin(s) and ascertain whether they are indeed capable of causing ABLD.

Acknowledgement

Max Tate (School of Agriculture and Wine, University of Adelaide, South Australia) graciously supplied the authenticated cytochalasins A and B.

References

Capio ER, Tate ME, and Wallwork H (2004). Phytotoxic metabolites from *Drechslera wirreganensis* and *D. campanulata*. *Australian Plant Pathology* 33:23–28.

Carter SB (1967). Effects of cytochalasins on mammalian cells. *Nature* 213:261–264.

Collett MG, Fincham JE, Tustin RC, Joubert G, and Marasas WF (1988). The pathology of chronic *Drechslera campanulata* toxicosis in inbred rats. *Onderstepoort Journal of Veterinary Research* 55:51–57.

Evidente A, Andolfi A, Vurro M, Zonno MC, and Motta A (2002). Cytochalasins Z1, Z2 and Z3, three 24-oxa[14]cytochalasans produced by *Pyrenophora semeniperda*. *Phytochemistry* 60:45–53.

Fisher EE and Finnie EP (1967). Infertility in mice and guinea pigs induced by feeding with a fungal pathogen isolated from leaves of *Romulea rosea*. *Nature* 215:1276.

Gunn A and Clarke R (2003). Acute bovine liver disease (ABLD). *The Veterinarian* February:40–43.

Huang Q, Dunn II RT, Jayadev S, DiSorbo O, Pack FD, Farr SB, Stoll RE, and Blanchard KT (2001). Assessment of cisplatin-induced nephrotoxicity by microarray technology. *Toxicology Science* 63:196–207.

Leach CM and Tulloch M (1972). Worldwide occurrence of the suspected mycotoxin producing fungus *Drechslera biseptata* with grass seed. *Mycologia* 64(4):1357–1359.

Schneider DJ, Marasas WF, Collett MG, and van der Westhuizen GC (1985). An experimental mycotoxicosis in sheep and goats caused by *Drechslera campanulata*, a fungal pathogen of green oats. *Onderstepoort Journal of Veterinary Research* 52:93–100.

Weinstein IB, Orenstein JM, Gebert R, Kaighn ME, and Stadler UC (1975). Growth and structural properties of epithelial cell cultures established from normal rat liver and chemically induced hepatomas. *Cancer Research* 35(1):253–263.

Chapter 19

Switchgrass (*Panicum virgatum*) Toxicity in Rodents, Sheep, Goats and Horses

B.L. Stegelmeier[1], S.A. Elmore[2], S.T. Lee[1], L.F. James[1], D.R. Gardner[1], K.E. Panter[1], M.H. Ralphs[1], and J.A. Pfister[1]

[1]*USDA-ARS Poisonous Plant Research Laboratory, Logan, UT 84341, USA;* [2]*ILS Inc., Research Triangle Park, NC, 27709, USA*

Introduction

Switchgrass (*Panicum virgatum*) has been reported to cause a crystal-associated hepatic disease and photosensitivity similar to Kleingrass poisoning in livestock (Bridges *et al.* 1987; Cornick *et al.* 1988; Puoli *et al.* 1992). Widespread use of switchgrass in pastures, as a biofuel and as cover for crop reserve programs, has exposed many animals to this potentially toxic grass (McLaughlin *et al.* 2002; Kamm 2004). Though we have found saponins in switchgrass, switchgrass toxicity has been difficult to reproduce experimentally (Lee *et al.* 2001). Similar variation in response and crystal development in toxicity studies using *Tribulus terrestris, Narthecium ossifragum, Agave lecheguilla, Brachiaris decumbens* and various other *Panicum* species suggest that toxicity may not be entirely due to saponin-related crystals (Coetzer *et al.* 1983; Bridges *et al.* 1987; Button *et al.* 1987; Camp *et al.* 1988; Graydon *et al.* 1991; Holland *et al.* 1991; Kellerman *et al.* 1991; Flaoye *et al.* 1992). The relatively low incidence of poisoning and low saponin concentrations suggest that most switchgrass is not highly toxic to livestock (Lee *et al.* 2001). The purpose of this study is to establish a model for switchgrass toxicity, identify switchgrass toxins and determine their toxicity, document the clinical and histologic lesions of switchgrass poisoning, and predict when and under what conditions switchgrass poisons livestock.

Methods

Seeded switchgrass (*P. virgatum*) was cut, dried and baled near Lincoln, Nebraska. Ground switchgrass was incorporated into rodent pellets at 50%. Control rodent pellets contained nutritionally similar ground triticale hay. Six Syrian hamsters, six gerbils and six guinea pigs were randomly divided into two groups and fed the switchgrass pellets or control pellets for 45 days. After feeding, the animals were euthanized, necropsied and blood and tissues were collected for analysis. An additional four groups of three hamsters each were orally dosed with concentrated diosgenin and yamogenin from fenugreek seed to obtain doses of 0, 100, 300 and 900 mg saponins per day. After 10 days of treatment all animals were euthanized, necropsied and blood and tissues were collected for evaluation. Six horses, six sheep and six goats were also divided into two groups and fed the switchgrass hay for 90 days. Control animals were fed similar rations containing triticale hay. Both rations were supplemented with concentrates to form a maintenance ration. Blood and serum were collected weekly and a percutaneous liver sample was collected at the end of the study. An additional eight goats were fed fresh switchgrass for 30 days. Skin biopsies were collected and the photosensitized animals were euthanized and necropsied. The ground and fresh switchgrass was analyzed for glycosidic steroidal alkaloids using previously described techniques (Lee *et al.* 2001).

Results and Discussion

Diosgenin was identified as the major sapogenin with smaller amounts of yamogenin in both the fresh and dried switchgrass (Figure 1). Although quantification of saponin concentration was not possible, switchgrass appeared to have lower concentrations than Kleingrass. Switchgrass hay averaged 1.850 mg nitrogen with about 11.5% protein. It had poor digestibility, similar to that of wheat straw. Most animals fed switchgrass hay lost weight and body condition, but did not develop photosensitivity. None of the horses or sheep fed switchgrass as hay or any of the rodents fed switchgrass pellets developed serum biochemical changes suggestive of hepatic or renal disease or damage. Liver and skin lesions were not detected in any of the rodents fed switchgrass pellets nor in any of the hamsters gavaged with switchgrass. However, some goats fed green switchgrass hay and green switchgrass with green alfalfa developed skin lesions consistent with phylloerythrin-associated photosensitivity (Figure 2). The photosensitized goats had hepatocellular swelling, individual hepatocyte necrosis and characteristic perinuclear cytoplasmic clearing (Figure 3). There were minimal infiltrates of neutrophils and macrophages in the adjacent sinusoids. Ghosts of canilicular and bile duct crystalline inclusions were rare and no birefringent crystals were identified.

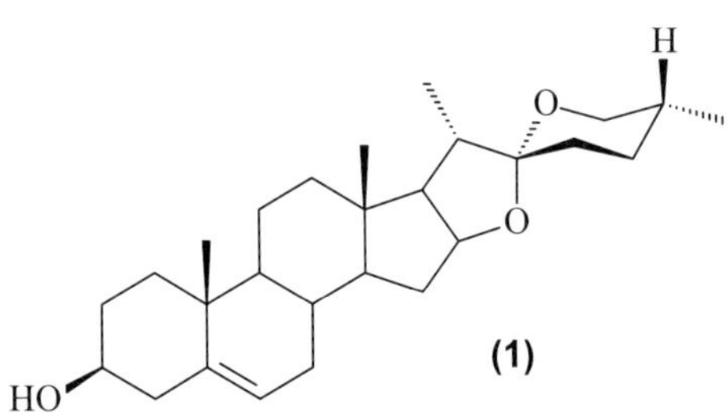

(1)

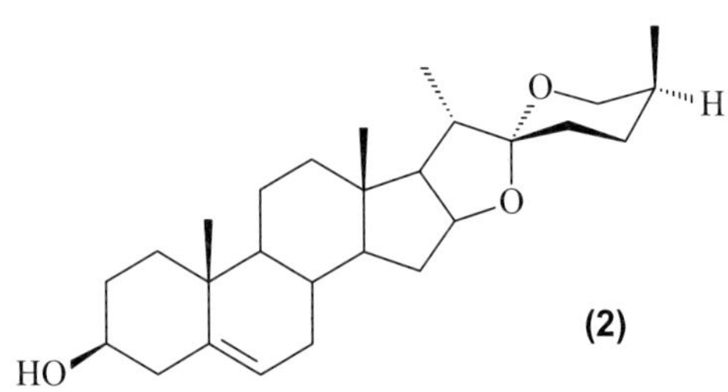

(2)

Figure 1. Diosgenin **(1)** was identified as the major sapogenin with smaller amounts of yamogenin **(2)**.

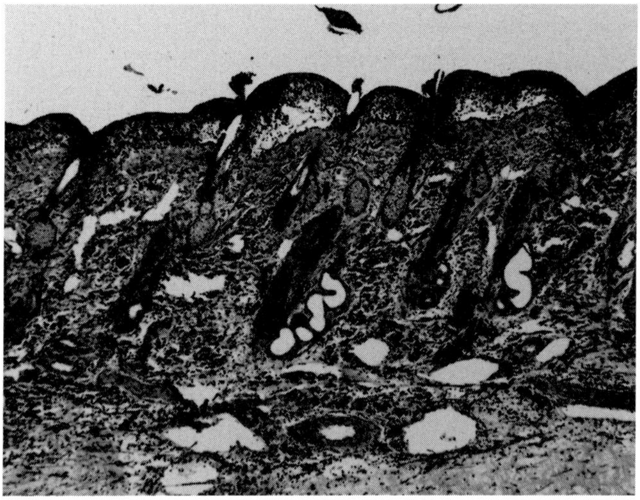

Figure 2. Photomicrograph of solar-induced dermatitis from the muzzle of a goat fed switchgrass. Notice the extensive dermal edema with neutrophilic infiltrates within the dermis and epidermis.

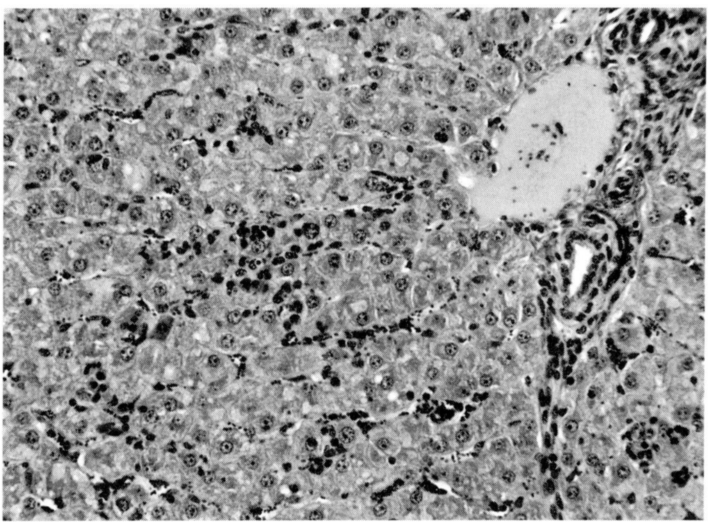

Figure 3. Photomicrograph of the liver of a goat fed switchgrass. Notice the hepatocellular swelling, focal hepatocellular necrosis and the characteristic perinuclear cytoplasmic swelling. There is mixed inflammation within the adjacent sinusoids. Crystalline inclusions were rare.

Others have also reported the sporadic nature of saponins-induced hepatogenous photosensitivity (Kellerman *et al.* 1991). Sporadic toxicity may be due to varying palatability, varying saponin concentrations, or the presence of other hepatotoxins. Hepatoxic sporodesmin was identified in many collections of *T. terrestris*, which was associated with photosensitization (Kellerman *et al.* 1980). Examination of the switchgrass hay from field outbreaks of photosensitivity and liver disease in horses found that it was extremely moldy. The cumulative effects of saponins and other hepatoxins may explain the sporadic photosensitivity associated with switchgrass.

The hepatic lesions seen in photosensitized goats were mild and crystalline inclusions were rare. These animals had serum phylloerythrin concentrations that were nearly double those of normal animals. Photosensitivity without extensive crystal production and cholestasis has been described in other studies (Kellerman *et al.* 1991). The multifocal patchy hepatocellular necrosis is a consistent finding. This suggests that the saponins are hepatoxic and alter phylloerythrin metabolism independently of morphologic biliary lesions. More work is needed to better identify the pathogenesis of saponin-induced hepatotoxicity and photosensitization and to better identify when saponin-containing plants are a potential danger to livestock.

Summary

These findings suggest that switchgrass can be toxic under certain conditions. Young goats may be most susceptible to poisoning. Green switchgrass fed with other chlorophyll-rich forage may be more likely to induce photosensitivity. Minimal crystalline hepatopathy is produced, but hepatocellular swelling and necrosis suggest toxicity is due to more than just crystalline disruption of the biliary system. Other hepatotoxins, such as aflatoxin, may exacerbate switchgrass toxicity.

References

Bridges CH, Camp BJ, Livingston CW, and Bailey EM (1987). Kleingrass (*Panicum coloratum* L.) poisoning in sheep. *Veterinary Pathology* 24:525–531.

Button C, Paynter DI, Sheil MJ, Colson AR, Paterson PJ, and Lyford RL (1987). Crystal-associated cholangiohepatopathy and photosensitization in lambs. *Australian Veterinary Journal* 64:176–180.

Camp BJ, Bridges CH, Hill DW, Patamalai B, and Wilson S (1988). Isolation of a steroidal sapogenin from the bile of a sheep fed *Agave lecheguilla*. *Veterinary and Human Toxicology* 30:533–535.

Coetzer JAW, Kellerman TS, Sadler W, and Bath GF (1983). Photosensitivity in South Africa. V. A comparative study of the pathology of the ovine hepatogenous photosensitivity diseases, facial eczema and geeldikkop (*Tribulosis ovis*) with special reference to their pathogenesis. *Onderstepoort Journal of Veterinary Research* 50:59–71.

Cornick JL, Carter GK, and Bridges CH (1988). Kleingrass-associated hepatotoxicosis in horses. *Journal of the American Veterinary Medical Association* 193:932–935.

Flaoye A, Borrebaek B, and Nordstoga K (1992). Glycogen accumulation and histological changes in the livers of lambs with alveld and experimental sporidesmin intoxication. *Veterinary Research Communications* 15:443–453.

Graydon RJ, Hamid H, Zahari P, and Gardiner CP (1991). Photosensitisation and crystal-associated cholangiohepatopathy in sheep grazing *Brachiaria decumbens*. *Australian Veterinary Journal* 68:234–236.

Holland PT, Miles CO, Mortimer PH, Wilkins AL, Hawkes AD, and Smith BL (1991). Isolation of the steroidal sapogenin epismilagenin from the bile of sheep affected by *Panicum dichotomiflorum* toxicosis. *Journal of Agricultural and Food Chemistry* 39:1963–1965.

Kamm J (2004). A new class of plants for a biofuel feedstock energy crop. *Applied Biochemistry and Biotechnology* 13:55–70.

Kellerman TS, Van der Westhuizen GCA, Coetzer JAW, Roux C, Marasas WF, Minne JA, Bath GF, and Basson PA (1980). Photosensitivity in South Africa. II. The experimental production of the ovine hepatogenous photosensitivity disease geeldikkop (*Tribulosis ovis*) by the simultaneous ingestion of *Tribulus terrestris* plants and cultures of *Pithomyces chartarum* containing the mycotoxin sporidesmin. *Onderstepoort Journal of Veterinary Research* 47:231–261.

Kellerman TS, Erasmus GL, Coetzer JAW, Brown JMM, and Maartens BP (1991). Photosensitivity in South Africa. VI. The experimental induction of geeldikkop in sheep with crude steroidal saponins from *Tribulus terrestris*. *Onderstepoort Journal of Veterinary Research* 58:47–53.

Lee ST, Stegelmeier BL, Gardner DR, and Vogel KP (2001). The isolation and identification of steroidal sapogenins in switchgrass. *Natural Toxins* 10:273–281.

McLaughlin SB, de la Torre Ugarte DG, Garten CTJ, Lynd LR, Sanderson MA, Tolbert VA, and Wolf DD (2002). High-value renewable energy from prairie grasses. *Environmental Science and Technology* 15:2122–2129.

Puoli JR, Reid RL, and Blelsky DP (1992). Photosensitization in lambs grazing switchgrass. *Agronomy Journal* 84:1077–1080.

Chapter 20

The Rise and Fall of Bighead: *Tetradymia* Poisoning of Sheep

D. J. Wagstaff

276 East 100 South, Pleasant Grove, UT 84062, USA

Introduction

Tetradymia glabrata Torr. & A. Gray (littleleaf horsebrush) and *T. canescens* DC. (spineless horsebrush) caused ovine hepatotoxicity and photosensitization in the Great Basin and adjacent regions of Western US. These plants in this report were identified by using the descriptions of Welsh *et al.* (1993), comparison to specimens at the Intermountain Herbarium, Logan, Utah and consultation with Alison Kelly, collections manager of that herbarium. They exemplify the 'tetra' part of the genus name because their flower parts are in sets of four: four involucral bracts, four yellow flowers in each composite head and four petals in each flower. A useful field mark to distinguish horsebrush, when it is not in flower, from associated shrubs is the knobby appearance of the twigs. Spine characteristics differentiate the Utah Great Basin horsebrush species. *Tetradymia canescens* is spineless compared to *T. glabrata*, which has short deciduous prickles. Two other *Tetradymia* not reported to be toxic are *Tetradymia nuttallii* Torr. & A. Gray (Nuttall's horsebrush), which has permanent stout straight spines, and *Tetradymia spinosa* Hook. & Arn. (cotton thorn), which has permanent stout spines recurved like cat's claws.

Tetradymia glabrata grows on well-drained volcanic and alluvial soils in semiarid basins that are dominated by either saltbrush (*Atriplex* spp.) or big sagebrush (*Artemisia tridentata* Nutt.). Typical habitat is benchland extending from the upper parts of the greasewood belt (*Sarcobatus vermiculatus* (Hook.) Torr.) into the lower edges of the pinyon-juniper belt (*Pinus edulis* Engelm.; *Juniperus osteosperma* (Torr.) Little). A typical patch of littleleaf horsebrush in an area dominated by big sagebrush is at the intersection of the Pony Express Trail and the Union Pacific Railroad near Faust, Utah at 40° 10.971' N, 112° 23.523' W.

Spring growth of this shrub coincides with the exodus of sheep from the desert. Leaves appear in March followed by flowers in late May to early June. By July, the leaves have turned yellow and dropped. A whitish bristle (pappus) replaces the flower. In dry seasons flowers are either absent or aborted. Contrary to popular opinion, littleleaf horsebrush is not one of the first desert plants to start growing. In fact, it is one of the last desert shrubs to flower in spring. Even among the desert horsebrushes, cotton thorn is the first to flower followed by Nuttall's horsebrush and finally littleleaf horsebrush.

Tetradymia canescens grows in big sagebrush areas of foothills and lower mountains. In contrast to the well-drained habitats of the desert horsebrushes, it is often found on the lower parts of slopes and in shallow swales. Flowers appear in late June and early July followed by pappus production. Leaves are shed in mid-summer or autumn. A typical stand of spineless horsebrush is at 40° 20.860' N, 111° 58.843 W at Eagle Mountain, Utah.

Historical Record

Tetradymia poisoning probably occurred in prehistoric times among the large native herbivores of the Great Basin. Weather extremes such as deep snow and drought would have forced them to eat distasteful and toxic plants. Petroglyphs indicate a strong interest of aboriginal people in the large herbivores, especially bighorn sheep, but any human role in plant poisonings is unknown.

The Great Basin was so remote that it was not explored until 1776 by the Spanish who never extensively utilized it. Later, American pioneers brought small herds of livestock into the area and probably encountered bighead. The sheep industry quickly boomed, and by 1900, there were several million sheep in the region. This was more than the ranges could support and sheep were forced to eat plants that were distasteful or toxic.

Overgrazing occurred because it was cost effective to concentrate sheep. One herder could manage 2000–3000 head if they were kept closely bunched. Sheep were driven to feed and central water holes then back each night to a perennial bedding ground. They were driven over crowded trails to lambing grounds, shearing sheds, summer ranges, and railroad shipping corrals. In addition to management practices, a number of natural factors played etiologic roles. Rain, misty weather and late season snowstorms preceded many outbreaks.

The first record of bighead I have found is an 1892 statement by a Utah sheep rancher that it caused his most serious losses (Carman *et al.* 1892). Bighead was first investigated in 1901 when the US Department of Agriculture (USDA) sent H.M. Rowe to Pocatello and Soda Springs, Idaho. He reported that there were hundreds of hotspots for the disease that sheepmen learned to avoid. One of note was a sheep trail near Pendleton, Oregon, called the Daley Road, along which losses were as high as 65%. Ranchers suspected a particular plant but Rowe did not stay in the West long enough to even identify the plant. This was unfortunate because the capability to investigate plant poisonings was nearby. In the prior year, 1900, USDA had sent V.K. Chesnut and E.V. Wilcox (1901) to Bozeman, Montana to investigate plant poisonings in that state. They had the cooperation of livestock owners, veterinarians, the experiment station, state officials, chemists and laboratory animal researchers. Moreover, they had the support of P.A. Rydberg of the New York Botanical Garden to verify plant identifications. However, Chesnut and Wilcox were from different bureaus of USDA than Rowe; there seems to have been no discussion among them regarding the matter. An opportunity was lost. Acute poisoning of sheep by *T. glabrata* was first researched in Nevada (Fleming 1918; Fleming *et al.* 1922) where the disease was recognized as a liver malady. Although swelling of the head and ears was seen in some sheep, it was not recognized as being bighead.

Recent Records

Many ranchers continued to believe the disease was caused by a plant. Finally in the 1930s, several events occurred that led to the discovery of the cause of bighead and started processes that brought about its eventual disappearance. In 1934, the Taylor Grazing Act was passed that controlled livestock grazing on public lands. The Bureau of Land Management (BLM) allotted a parcel of land for exclusive grazing by each permit holder. This allotment system not only enabled better monitoring of forage availability and restriction of grazing when necessary but it also eliminated the indiscriminate grazing that occurred when sheep were dogged in a race for the best forage. Although *Tetradymia* is not monitored in detail, the BLM allotment books contain data that could be used for evaluation of time trends of horsebrush populations. The economic depression and droughts of the 1930s reduced sheep numbers by bankrupting many ranchers. Overgrazing started to decline.

In that same year, 1934, the Desert Experimental Range was constructed in the West Utah Desert by the Civilian Conservation Corps and operated by the US Forest Service. Research conducted there and at other stations helped to balance grazing with availability of forage. Range studies have been augmented by study of prehistoric human activity, fire patterns and climate change to enhance our understanding of plant–animal interactions.

Bighead research was also started in the early 1930s. A.B. Clawson and Ward Huffman of the USDA investigated dozens of bighead outbreaks and theorized that the cause was *Tetradymia*. They confirmed this in experimental feedings of *T. canescens* and *T. glabrata*. It was further learned that *Tetradymia* alone was seldom responsible but that the disease occurred most often when sheep were preconditioned by feeding on black sagebrush (*Artemisia nova* A. Nelson) (Anonymous 1936). After Clawson's death in 1937 there was a large hiatus in research. Earl Johnson (1978) of the USDA-ARS Poisonous Plant Research Laboratory revived research by publishing previously unpublished data of Clawson and Huffman. Then he studied plant interactions and confirmed liver damage by measuring hepatic enzyme activity. He concluded that availability of chlorophyll was probably not the mechanism and eventually came to question whether bighead was a secondary photosensitization. Also during the 1970s, researchers at Montana State University isolated and structurally characterized toxic compounds in *T. canescens* (Jennings *et al.* 1978). One of these compounds, tetradymol, is metabolized by hepatic microsomal enzymes to a more toxic substance. This provides a plausible explanation for the marked differences in susceptibility of sheep prefed different plants.

By 1980, the bighead epizootic had virtually disappeared. Since then the disease is rarely reported and in very small numbers. In the approximate 100 years of bighead outbreaks, no accurate count of losses is available but it can be estimated that they exceeded 1 million sheep. It was one of the most devastating diseases.

Decline in Poisoning

Several factors are associated with this dramatic decline in poisoning. First, sheep populations declined as grazing on public lands shrunk, lamb consumption plummeted, and synthetic fibers replaced wool. Sheep allotments are being converted to cattle allotments.

Second, changes in management practices have decentralized sheep operations. Perennial bedding grounds have been abandoned. Water sources are dispersed through drilling of wells, piping, and use of tank trucks and portable troughs. Trailing is uncommon.

Trucks with portable corrals have replaced railroads and their associated shipping corrals. Shed lambing is replacing range lambing. Shearing sheds are replaced by portable shearing apparatus.

Third, perhaps littleleaf horsebrush populations have diminished and lost vigor. All of the plants I have examined have numerous dead branches. The causes are unknown but overgrazing is not among them; the occurrence of dead branches is similar inside and outside of exclosures.

And fourth, although invasion of annual weeds such as cheatgrass (*Bromus tectorum* L.) and redstem filaree (*Erodium cicutarium* (L.) L'Her.) has degraded ranges and fueled serious fires, these annuals provide spring feed and reduce pressure to graze horsebrush. In addition, grasses in general are not microsomal enzyme inducers. Therefore cheatgrass might reduce bighead risk even among sheep that consumed some horsebrush.

Conclusion

Lessons of history can be learned from this bygone intoxication. First, plant poisonings result from interactions of many natural and management factors. Therefore, poisonings should be viewed in a broad context. Second, cooperation of people in allied fields such as botany, range science, chemistry, and statistics should be obtained and acknowledged. Third, readers should critically evaluate reports and confirm them with personal observations whenever possible. And lastly, standard terms should be used in poisonous plant reports including voucher deposition documentation (Wagstaff *et al.* 1999), botanical names (Wiersema and Leon 1999), and geographic descriptors such as latitude and longitude.

References

Anonymous (1936). Sheep 'bighead' caused by two poisonous plants. *USDA Press Release* #1498-36.

Carman EA, Heath HA, and Minto J (1892). *Special Report on the History and Present Condition of the Sheep Industry of the United States*, p. 808. Government Printing Office, Washington, DC.

Chesnut VK and Wilcox EV (1901). *The Stock-poisoning Plants of Montana*. US Government Printing Office. Washington, DC.

Fleming CE (1918). Range plants poisonous to sheep and cattle in Nevada. *Nevada Agricultural Experiment Station Bulletin* #95, pp. 27–30.

Fleming CE, Miller MR, and Vawter LR (1922). The spring rabbit-brush (*Tetradymia glabrata*), a range plant poisonous to sheep. *Nevada Agricultural Experiment Station Bulletin* #104.

Jennings PW, Reeder SK, Hurley JC, Robbins JE, Holian SK, Holian A, Lee P, Pribanic JAS, and Hull M (1978). Toxic constituents and hepatotoxicity of the plant *Tetradymia glabrata* (Asteraceae). In *Effects of Poisonous Plants on Livestock* (RF Keeler, KR Van Kampen, and LF James, eds), pp. 217–228. Academic Press, New York.

Johnson AE (1978). Tetradymia toxicity - A new look at an old problem. In *Effects of Poisonous Plants on Livestock* (RF Keeler, KR Van Kampen, and LF James, eds), pp. 209–216. Academic Press, New York.

Wagstaff DJ, Wiersema JH, and Lellinger DB (1999). Poisonous plant vouchers. *Veterinary and Human Toxicology* 41:162–164.
Welsh SL, Atwood ND, Goodrich S, and Higgins LC (1993). *A Utah Flora*, 2nd ed. Rev. Brigham Young University Press, Provo, Utah.
Wiersema JH and Leon B (1999). *World Economic Plants - A Standard Reference*. CRC Press, Boca Raton, Florida.

REPRODUCTION

LUPINUS LEUCOPHYLLUS

Chapter 21

Antifertility Activity of *Carica papaya* in Male Rats

V.K. Gunawardana[1] and S.G. Tokhi[2]

[1]*Department of Basic Veterinary Sciences, Faculty of Veterinary Medicine and Animal Sciences, University of Peradeniya, Peradeniya, Sri Lanka;* [2]*Biochemistry Unit, School of Medical Sciences, Faculty of Life Sciences, RMIT University, Bundoora, Australia 3083*

Introduction

Traditional medical practice based on plants forms a substantial part of health care in many developing countries. This is also true of the veterinary practice. In Sri Lanka, rural farmers often seek the help of indigenous veterinary practitioners for the treatment of common ailments in livestock. A range of bioactivities has been reported from various plants; predominant among them are anthelmintic and antimicrobial activities. In recent times there has been much interest in scientifically testing the efficacy of plants so that they can be used as cheaper alternatives to standard drugs such as anthelmintics.

Carica papaya, native to tropical America, is a plant belonging to the family Caricaceae. It is cultivated in all parts of Sri Lanka and flowers throughout the year (Dassanayake 1996). Although the main reason for cultivation is its edible fruit, the plant is an important source of the enzyme papain. Ethnomedical reports indicate that the plant has been widely used for a variety of purposes and these include its use as an anthelmintic (Attygalle 1917).

The plant has been the subject of numerous investigations for its bioactivities. For example, the fruits are reported to contain antibacterial agents (Emeruwa 1982). Ripe and unripe fruits have been tested for their effects on reproduction in the female (Adebiyi *et al.* 2002). The seeds (Bose *et al.* 1961) and the latex (Satrija *et al.* 1994) have been investigated for anthelmintic activity. A compound, benzylisothiocyanate, which is known to be present in the seeds has tested positive for anthelmintic activity (Kermanshai *et al.* 2001).

Of all bioactivities reported to be present in the plant, perhaps the most important to livestock farmers is its anthelmintic activity. Seeds in particular have been used as an anthelmintic. However, some reports indicate that the seeds possess antifertility activity in the male (Lohiya *et al.* 1994, 1999a,b). If this is so, there is a distinct disadvantage in using the seeds for therapeutic purposes in livestock. This investigation was therefore undertaken to verify whether the seeds of *C. papaya* have adverse effects on male reproduction. Many previous studies on male antifertility effects of the seeds of *C. papaya* have used aqueous or solvent extracts. In the present study, bioassays for antifertility activity were carried out using finely ground seed since rural farmers using the seeds of *C. papaya* as an anthelmintic use it in this form.

Materials and Methods

Experimental animals

The animals used were locally bred, sexually mature Sprague-Dawley rats. Both males and females were of proven fertility. Two trials were conducted and at the commencement of each trial, male rats were weighed and assigned to a treatment or control group. The animals were distributed so that both groups had animals of comparable body weights. The rats were housed individually in wire cages and maintained under standard conditions of natural lighting (12 hours daylight) and the temperature was maintained at 21°C. Food and water was provided *ad lib* and the body weight was recorded daily. In each trial the treatment group was assigned ten animals. In Trial 1 the control group had ten animals while in Trial 2 the control group consisted of five animals.

Vaginal smears of female rats were checked each morning and only females showing two consecutive estrus cycles were selected.

Plant preparation and dosing

Seeds collected from ripe fruits of *C. papaya* were sun-dried and ground into a fine powder and suspended in Tween 80. In each trial the rats in the treatment group were administered the suspension orally using a feeding needle while control rats received an equal volume of Tween 80. In Trial 1, the treated rats received a dose of 0.67 g of the powder per kilogram body weight per day for a period of 20 days. In Trial 2, the rats were treated for 44 days with a higher dose of 1.34 g per kilogram body weight.

Five days before the end of dosing each male was paired with a proestrus female. Pairing was carried out in the evening and the females were separated from the male the following morning. The presence of sperm in the vaginal smear of the female was considered positive evidence of mating and date of mating was regarded as day one of pregnancy. If mating did not occur, a second opportunity was provided with a new proestrus female. Male rats were necropsied one day after the end of the dosing period and females were necropsied on day 16 of pregnancy.

Data recorded

In the male the left testes and epididymis were used for recording the weight. Thereafter the epididymis was homogenized in normal saline and the homogenate was stored at 4°C for estimating the epididymal sperm reserve (ESR). The maximum period of storage was 48 hours. The ESR was calculated based on sperm counts made using a Neubauer hemocytometer (Queiroz and Cardoso 1987).

To determine sperm motility and morphology, a small incision was made in the right cauda epididymis soon after sacrificing the animal. A drop of epididymal content which extruded from the incision was diluted with a drop of saline on a glass slide and examined immediately under 10x objective of the microscope and motility was scored.

A drop of the diluted suspension was mixed with a drop of Nigrosin–Eosin stain and smears prepared were examined under oil immersion for sperm abnormality. Two hundred spermatozoa were counted and the percentage of different types of abnormalities was calculated.

In the females, the number of implantation sites and normal fetuses were recorded at necropsy. In calculating the mean number of implantation sites and normal fetuses only those animals that were pregnant were taken into consideration. In this study a female with at least one implantation site was considered pregnant.

Results

In Trial 1, one treated animal died and one control animal was excluded from the trial. In Trial 2 four animals in the treated group died between day 9 and day 37 of dosing. These animals showed clinical signs of diarrhoea and emaciation prior to death.

Table 1 shows the body weight gain during the treatment period and organ weights at necropsy. In both trials the weight gain was very poor compared to the controls. The weight of the testis and epididymis expressed as a percentage of the body weight did not show a significant difference between treatment and control groups in both trials.

Table 1. Weight gain and organ weights

		Trial 1		Trial 2	
		Treated (n=9)	Control (n=9)	Treated (n=9)	Control (n=5)
Body weight	Weight gain (g)[a]	6.3	58.2	14.8	37.8
Testes	Absolute (g)[b]	1.17±0.11	1.34±0.14	1.23±0.21	1.37±0.10
	% body wt [a]	0.39	0.37	0.40	0.46
Epididymis	Absolute (g) [b]	0.34±0.03	0.48±0.08	0.41±0.03	0.48±0.02
	% body wt [a]	0.14	0.11	0.13	0.16

[a] values are means
[b] values are mean ± SD

All epididymal parameters were affected by treatment with *C. papaya*. However, as seen in Table 2, the most drastically affected parameter was the motility of cauda epididymal spermatozoa. In Trial 1, two treated animals had immotile spermatozoa and in the others motility was confined to slow vibration of the tail with no forward progression. In Trial 2, four treated animals had immotile spermatozoa while the other two had spermatozoa showing weak tail movements. A reduction in the ESR and an increase in sperm abnormality were also observed. Although the highest percentage abnormality was in the head region for all groups (Table 2), the tail abnormality in the treated group was 5.6 times higher than in the control in Trial 1 and it was four times higher in Trial 2.

Table 2. Epididymal parameters.

Parameter measured	Trial 1		Trial 2	
	Treated (n=9)	Control (n=9)	Treated (n=9)	Control (n=5)
ESR per g [a]	2.44×10^8	6.94×10^8	8.79×10^8	9.37×10^8
Sperm motility [b]	0 to +	+++	0 to +	+++ to ++++
% Sperm abnormality [a]	21.9	9.39	20.2	7.3
% Detached head	11.0	6.0	12.2	4.1
% Misshapen head	3.5	1.7	2.7	1.6
% Midpiece defect	5.2	1.3	1.7	0.8
% Retroverted tail	2.2	0.39	3.6	0.8

[a] values are means
[b] 0 = Immobile; + = Motile but no progressive motility; ++ = Poor progressive motility; +++ = Good progressive motility; ++++ = Vigorous progressive motility

The results of the fertility study are shown in Table 3. In the treated group the mean number of implantation sites was reduced by 35.43% compared to control in Trial 1. In this trial, of the seven animals that were pregnant two had all fetuses resorbed. As a result the number of normal fetuses is less than the number of implantation sites. At the higher dose only two animals were able to impregnate receptive females. Of the two pregnant animals one had only one implantation site and the other had three. In these implantation sites the fetuses were resorbed.

Table 3. Fertility of treated rats.

Parameter measured	Trial 1		Trial 2	
	Treated (n=9)	Control (n=9)	Treated (n=6)	Control (n=5)
Number mated	7	9	2	5
Implantation sites[a]	6.89±4.33	10.67±1.41	0.67±1.11	13.2±1.47
Normal fetuses[a]	6.67±4.27	10.67±1.41	0	13.2±1.47

[a] values are mean ± SD

Discussion

The end result of an antifertility agent is the prevention of production of a viable offspring. A male antifertility agent could achieve this in one of several ways. Disruption of sperm production, interference with epididymal function, interference with sperm transport and suppressing libido are possible ways in which male antifertility agents exert their actions. The results of this study indicate that the seeds of *C. papaya* contain antifertility agents. Even at the lower dose, the plant material was able to disrupt pregnancy in several females mated with treated males. The available data suggest that the action of the antifertility agent is mainly on the epididymis. The epididymis provides the milieu where spermatozoa acquire the potential for motility and undergo maturation and become fertile. From the present results it is clear that the immediate cause of infertility is the effect on sperm motility. In this investigation cauda epididymal spermatozoa were either immotile or showed only sluggish movement. Lohiya *et al.* (1994) also observed reduced sperm motility with aqueous extracts. Progressive motility of spermatozoa is an important factor for successful fertilization and in the present study the very poor fertility seen in the treated group particularly in Trial 2 may have been due to lack of progressive motility. This is further exemplified by the fact that even the pregnant animals had very few implantation sites. Acquisition of motility by mammalian spermatozoa is thought to depend on secretions of the epididymis and it has been linked to the production of certain proteins (Pholpramool *et al.* 1983). In this connection it is of interest to note that Verma and Chinoy (2001) found a decrease in the total protein content in epididymal fluid after a 5-day period of intramuscular treatment with an aqueous extract of the seeds of *C. papaya.*

In this study some of the treated animals that became pregnant were unable to carry the pregnancy to day 16. Resorption of unhealthy fetuses is well recognized in polytocus species such as the rat. It is also recognized that abnormal spermatozoa produce unhealthy fetuses. Thus it would appear that fetal resorption may be related to the increase in the percentage of sperm abnormality observed in the treated rats. This merits further investigation. Histological examination of the testes in animals from Trial 2 did not show any appreciable changes in testicular structure suggesting that spermatogenesis was unimpaired. However, some of the sperm abnormalities seen were primary abnormalities which originate in the testes.

Lohiya *et al.* (1999a) reported severe oligospermia in rabbits treated with a chloroform extract of the seed for a period of 150 days. The benzene fraction of this extract produced azoospermia (Lohiya *et al.* 1999b). Our results on ESR are not compatible with this observation. The difference may be due to the longer period of dosing in the previous study. In the present work, although the ESR in the treated group was lower than in controls, it remained at reasonable levels. This substantiates further the claim that reduction or absence of sperm motility rather than reduction in numbers of spermatozoa is responsible for the observed antifertility activity.

In conclusion it may be stated that the seeds of *C. papaya* have antifertility activity when administered orally to rats. This factor needs to be taken into account if this material is to be used as an anthelmintic in livestock. Further studies are needed to determine if effective anthelmintic doses of *C. papaya* could produce antifertility effects in livestock.

References

Adebiyi A, Ganesan Adaikan P, and Prasad RNV (2002). Papaya (*Carica papaya*) consumption is unsafe in pregnancy: fact or fable. Scientific evaluation of a common belief in some parts of Asia using a rat model. *British Journal of Nutrition* 88:199–203.

Attygalle J (1917). *Sinhalese Materia Medica,* Colombo Apothecaries Co Ltd, Colombo.

Bose BC, Saifi AQ, Vijayavargiya R, and Bhagwat AW (1961). Pharmacological study of *Carica papaya* seeds with special reference to its anthelmintic action: preliminary report. *Indian Journal of Medical Science* 15:888.

Dassanayake MD (1996). Caricaeae. In *A Revised Handbook to the Flora of Ceylon, Volume X* (MD Dassanayake and WD Clayton, eds), pp. 51–53. Oxford & IBH Publishing Co. Pvt. Ltd., New Delhi.

Emeruwa AC (1982). Antibacterial substance from *Carica papaya* fruit extract. *Journal of Natural Products* 45:123–127.

Kermanshai R, McCarry BE, Rosenfeld J, Summers PS, Weretilnyk EA, and Sorger GJ (2001). Benzylisothiocyanate is the chief or sole anthelmintic in papaya seed extracts. *Phytochemistry* 57:427–435.

Lohiya NK, Goyal RB, Jayaprakash D, Ansar AS, and Sharma S (1994). Antifertility effects of aqueous extract of *Carica papaya* seeds in male rats. *Planta Medica* 60:400–404.

Lohiya NK, Pathak M, Mishra PK, and Manivannan B (1999a). Reversible contraception with chloroform extract of *Carica papaya* Linn. seeds in male rabbits. *Reproductive Toxicology* 13:59–66.

Lohiya NK, Mishra PK, Pathak M, Manivannan B, and Jain SC (1999b). Reversible azoospermia by oral administration of the benzene chromatographic fraction of the chloroform extract of the seeds of *Carica papaya* in rabbits. *Advances in Contraception* 15:141–161.

Pholpramool C, Lea OA, Burrow PW, Dott HM, and Setchell BP (1983). The effects of acidic epididymal glycoprotein (AEG) and some other proteins on the motility of rat epididymal spermatozoa. *International Journal of Andrology* 6:240–248.

Queiroz GF and Cardoso FM (1987). Testicular and epididymal sperm reserves of Brazilian hairy rams. *Journal of Veterinary Medicine Series A* 34:657–660

Satrija F, Nansen P, Bjorn H, Murtini S, and He S (1994). Effect of papaya latex against *Ascaris suum* in naturally infected pigs. *Journal of Helminthology* 68:343–346.

Verma RJ and Chinoy NJ (2001). Effect of papaya seed extract on microenvironment of cauda epididymis. *Asian Journal of Andrology* 3:143–146.

Chapter 22

Fetotoxic Effects of Locoweed (*Astragalus lentiginosus*) in Pregnant Goats

S. Furlan[1], K.E. Panter[2], J.A. Pfister[2], and B.L. Stegelmeier[2]

[1]*College of Veterinary Medicine, University of São Paulo, São Paulo, Brazil;* [2]*USDA-ARS Poisonous Plant Research Laboratory, Logan, UT 84341, USA*

Introduction

Locoweed species (*Astragalus* and *Oxytropis* spp.) are widely distributed throughout the western US, and are a serious toxic plant problem for grazing livestock. The toxin in locoweed is the indolizidine alkaloid swainsonine, which inhibits lysosomal α-mannosidase and mannosidase II (Dorling *et al.* 1980). Inhibition of these enzymes results in oligosaccharide accumulation in lysosomes with characteristic vacuolated cells (Stegelmeier *et al.* 1995). Clinically, locoweeds cause depression, proprioceptive deficits, intention tremors, nervousness, emaciation, and death (James *et al.* 1970). Poisonings with similar etiology have been reported in Brazil and Mozambique from *Ipomoea carnea* (Damir *et al.* 1987; De Balogh *et al.* 1999) and in Australia from Darling Peas (*Swainsona* spp.; Hartley 1978). Swainsonine is found in these species also and is believed to be responsible for the reported toxicoses. Additionally, locoweeds cause abortions, reproductive dysfunction and occasional birth defects when ingested. There has been a substantial amount of research in sheep given locoweed from 60 to 130 days of gestation (James 1971; Panter *et al.* 1999). Even so, there has been no research on fetotoxic effects of locoweed during gestation days 30–60, and little has been done to describe the toxic effect of poisoning in pregnant goats. Thus the purpose of this study was to describe the clinical and ultrasonographic effects of locoweed poisoning in pregnant goats.

Materials and Methods

Nine Spanish goats were hand-mated to Spanish bucks and divided into two groups. Five pregnant does were dosed twice daily with finely ground *Astragalus lentiginosus* via gavage beginning on day 30 of gestation. The plant dosage was calculated to deliver 8 mg swainsonine kg^{-1} body weight per day (140–230 g dry ground *Astragalus lentiginosus*). Four similarly bred goats were dosed with equivalent amounts of ground alfalfa hay as negative controls. Blood samples were collected via the jugular vein on days 1–7, and every 5 days thereafter or until fetal death was confirmed. Serum was collected and frozen at –20°C and later analyzed for swainsonine and progesterone. Fetal movement was measured using ultrasound every 5 days beginning on day 30 of gestation. An Aloka Model 900 ultrasound with a 5 mhz abdominal transducer was used for ultrasonography observations. Briefly, an echogenic gel was applied on either side of the udder and 2–5 cm anterior. Once the fetus was observed a 5 min scan was performed and each voluntary movement of the fetus was recorded to videotape for future analysis. The fetal heartbeat was noted to monitor fetal viability but no attempt was made to determine the fetal heart rate.

Swainsonine analysis

Swainsonine concentrations were determined using a method previously published with some modification (Stegelmeier *et al.* 1995). Briefly, 0.6 mL of serum was combined with 0.3 mL sodium acetate (0.25 M, pH 4.0), vortexed, and boiled for 10 min. The supernatant was removed and 75 μL was put into a multiwell plate in duplicate with 15 μL of 0.0008 U/well Jack Bean α-mannosidase (Sigma Chemical Co., St. Louis, MO) and 10 μL of 10 mM p-nitrophenyl-α-D-mannosidase (Sigma Chemical Co, St Louis, MO). The mixture was incubated for 30 min at 37°C, after which 100 μL glycine (2.5 M, pH 10.3) was added to stop the reaction. The amount of swainsonine (α-mannosidase inhibition) in the test samples was then determined photometrically at 405 nm using a microplate reader and microplate manager software (Bio-Rad 3550 UV, Bio-Rad laboratories, Melville, NY). Analysis of swainsonine standards with concentrations ranging from 1000 to 1.2 ng mL^{-1} was performed in a similar manner using blood serum from normal, unexposed animals spiked with known amounts of swainsonine. The control swainsonine concentrations were verified by capillary gas chromatography using previously described methods (Molyneux *et al.* 1989). Samples higher than the linear range of the standard curve (>800 ng mL^{-1}) were diluted with sodium acetate (0.25 M, pH 4.0). The standards for these assays were similarly modified by diluting the normal serum.

Progesterone analysis

Serum progesterone levels were determined by radioimmunoassay using a no-extraction, solid phase I-125 radioimmunoassay kit (Diagnostic Products Corporation, Los Angeles, CA). The Coat-A-Count progesterone kit is highly specific with low cross reactivity to other steroids. Specific cross reactivities are listed in the procedure manual (Diagnostic Products Corporation). Inter- and intra-assay coefficients of variation were 6.15% and 4.25%, respectively. Progesterone kits were validated at 0.5, 2.5, 5, 10, 20 and 40 ng mL^{-1}.

Results

Locoweed–poisoned goats developed mild proprioceptive deficits within 9 to 10 days of treatment. These clinical signs became more severe, with poisoned animals becoming lethargic, hesitant to move, prominent intention tremors and proprioceptive deficits. Several animals developed severe rear limb weakness and partial paresis. These animals were euthanized and necropsied early.

Serum swainsonine was rapidly elevated in treated goats, reaching concentrations of > 400 ng mL^{-1} after the first day of treatment and peaking above 600 ng mL^{-1} (Figure 1). No swainsonine was detected in sera from control goats.

Serum progesterone significantly ($P<0.05$) declined in locoweed-fed goats after 5 days on locoweed, beginning on gestation day 35 and continuing throughout the treatment period (Figure 1). This was indicative of the negative effect of locoweed on the fetus, placenta, ovary and other structures important in the maintenance of pregnancy in goats.

All pregnant goats, including treatment and controls, had viable fetuses on day 30 when the treatments began, and embryonic vesicles appeared normal between days 30 and 32. Fetal movement was detected in most animals between days 36 and 38, which is typical in sheep and goats (Panter *et al.* 1990). Fetal movement was similar between control and locoweed-fed goats until after day 40 of gestation ($P>0.05$; Figure 1). Fetal death was observed by ultrasound in one goat on day 40 of gestation (10 days after treatment began). Fetal movement in the surviving fetuses was reduced ($P<0.05$) compared to controls and fetal death in two more locoweed-treated goats was observed on day 45 and the fourth on day 50 (Figure 1). Fetal movement remained normal in control goats throughout pregnancy and parturition, and goat kids were all normal.

All fetuses were negatively impacted by locoweed dosing, and fetal death was observed in all locoweed-treated goats between 10 and 20 days after the beginning of locoweed treatment (Figure 1). By treatment day 5 (day 35 of gestation) one fetus had no visible fetal heartbeat and it was presumed dead. This goat was dropped from the study on day 35 and was not necropsied. Fetal death was observed in the other four goats on treatment days 10, 15 (two animals), and 20. No abortions occurred before the goats were euthanized. Histologically the fetuses were autolytic, but no cellular vacuolations indicative of locoweed poisoning were found. However, all of the poisoned does had neurovisceral vacuolation characteristic of locoweed poisoning.

Discussion

Pregnant goats and their fetuses were very sensitive to high doses of locoweed during early pregnancy. This is important as many animals are exposed to locoweed during early gestation and field reports often indicate that many females are not pregnant at the end of the breeding season. Pregnant goats were greatly affected by locoweed, showing ataxia, loss of proprioception, and rear limb paralysis. These clinical signs had a relatively rapid onset, indicating the severity of the intoxication, and the rear limb paralysis was typically seen when animals were excited or stressed. Rear limb paralysis has not been a commonly reported clinical effect from intoxication by swainsonine-containing locoweeds, but has been reported for nitro-bearing species (Mathews 1940; James *et al.* 1980). The detailed histopathology on these goats will be reported elsewhere.

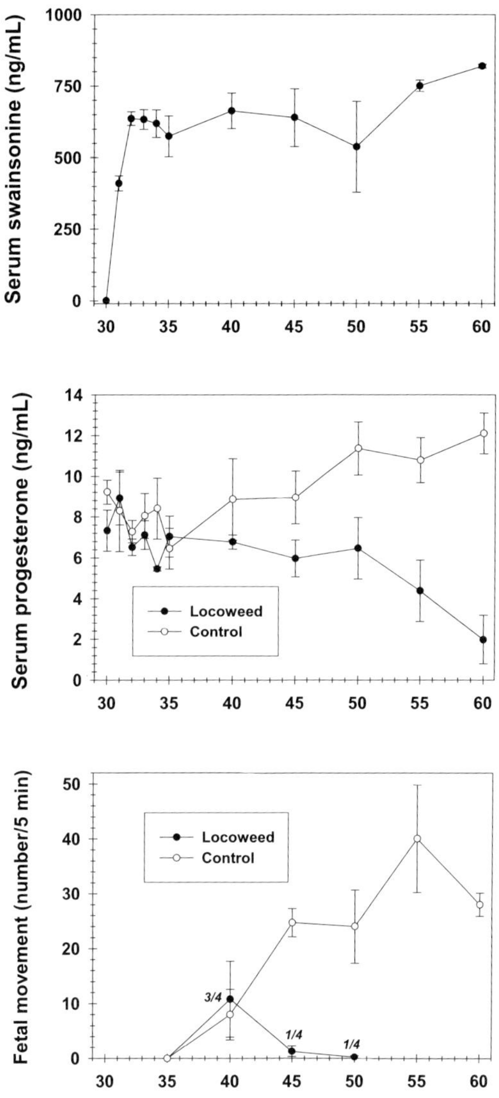

Figure 1. Average serum swainsonine and progesterone concentrations (ng/mL), and fetal movement (number/5 min), in pregnant goats dosed with 8 mg swainsonine/kg bw per day as ground locoweed (*Astragalus lentiginosus*). No swainsonine was detected in control sera. For fetal movement, the number above the data point represents live fetuses/total detected fetuses.

Locoweed exposure *in utero* also had severe effects on fetal goats. Fetal movement was depressed, which is indicative of fetotoxicty. Fetal death occurred as early as 5 days after treatment began and all fetuses were dead by 20 days. The fetuses were autolytic but the lack of fetal lesions, the ultrasound observations and the progesterone levels suggest that fetal death is possibly due to the maternal effects of locoweed poisoning. Locoweed given to pregnant sheep has been shown to affect fetal and placental development (Van Kampen and James 1971; Panter *et al.* 1987). Gross histological findings in sheep include fluid accumulation in the placenta (hydrops allantois, hydrops amnii), altered cotyledonary development, with fetal death followed by abortions. Panter *et al.* (1987) also showed that fetal heart rate was reduced and fetal heart contractions were irregular and weak. Fetal cardiac insufficiency and right heart failure may contribute to the fluid accumulation in the fetus and placenta (Panter *et al.* 1999) and cause fetal death and abortion (Panter *et al.* 1987). James (1971, 1976) fed ewes locoweed during specific gestational time periods and studied lamb fetal development. Sheep fetuses exposed to locoweed *in utero* before 100 days of gestation did not have any lesions associated with locoweed intoxication (James 1971, 1976). Cytoplasmic vacuolization occurred in lambs whose dams were fed locoweed from 100 to 120 days of gestation (James 1971).

Conclusions

This research suggests that Spanish goats are very sensitive to the negative effects of locoweed on reproduction. The early reduction in fetal movement and fetal death is indicative of this sensitivity. Also, the clinical signs of poisoning were pronounced relatively early but slightly different than what has been reported, *i.e.*, the rear end paresis. Additional studies of the hormonal, placental and early fetal lesions in poisoning are needed to better understand the mechanism of locoweed-induced fetal death. This work demonstrates that dosage of swainsonine ingested is an important aspect in predicting risk of poisoning. We know that low doses of swainsonine will also induce reproductive dysfunction; however, ingestion must occur over a longer period of time as expected. This threshold dosage is understood in many species but has not been fully described in goats and will require further research.

Acknowledgements

We thank Terrie Wierenga, Kermit Price, Seth Lundquist, Rex Probst, Danny Hansen, and Al Maciulis for valuable assistance with the study and for animal care.

References

Damir AH, Adam SEI, and Tartour G (1987). The effects of *Ipomoea carnea* on goats and sheep. *Veterinary and Human Toxicology* 29(4):316–319.

De Balogh KIM, Dimande AP, Van Der Lugt JJ, Molyneux RJ, Naude TW, and Welman WGA (1999). Lysosomal storage disease induced by *Ipomoea carnea* in goats in Mozambique. *Journal of Veterinary Diagnostic Investigation* 11: 266–273.

Dorling PR, Huxtable CR, and Colegate SM (1980). Inhibition of lysosomal a-mannosidase by swainsonine, an indolizidine alkaloid isolated from *Swainsona canescens*. *The Biochemical Journal* 191:649–665.

Hartley WJ (1978). A comparative study of Darling pea (*Swainsona* spp.) poisoning in Australia with locoweed (*Astragalus* and *Oxytropis* spp.) poisoning in North America. In *Effects of Poisonous Plants on Livestock* (RF Keeler, KR Van Kampen and LF James, eds), pp. 363–369. Academic Press, New York.

James LF (1971). Lesions in neonatal lambs resulting from maternal ingestion of locoweed. *The Cornell Veterinarian* 61:667–670.

James LF (1976). Effect of locoweed (*Astragalus lentiginosus*) feeding on fetal lamb development. *Canadian Journal of Comparative Medicine* 40:380–384.

James LF, Van Kampen KR, and Johnson AE (1970). Physiopathologic changes in locoweed poisoning in livestock. *American Journal of Veterinary Research* 31:663–672.

James LF, Hartley WJ, Williams MC, and Van Kampen KR (1980). Field and experimental studies in cattle and sheep poisoned by nitro-bearing *Astragalus* or their toxins. *American Journal of Veterinary Research* 41:377–382.

Mathews FP (1940). The toxicity of red-stemmed peavine (*Astragalus emoryanus*) for cattle, sheep, and goats. *Journal of the American Veterinary Medical Association* 47:125–134.

Molyneux RJ, James LF, Panter KE, and Ralphs MH (1989). The occurrence and detection of swainsonine in locoweeds. In *Swainsonine and Related Glycosidase Inhibitors* (LF James, A Elbein, RJ Molyneux, and CD Warren, eds), pp. 100–117. Iowa State University Press, Ames.

Panter KE, Bunch TD, James LF, and Sisson DV (1987). Ultrasonographic imaging to monitor fetal and placental developments in ewes fed locoweed (*Astragalus lentiginosus*). *American Journal of Veterinary Research* 48:686–690.

Panter KE, Bunch TD, Keeler, RF, Sisson DV, and Callan, RJ (1990). Multiple congenital contractures (MCC) and cleft palate induced in goats by ingestion of piperidine alkaloid-containing plants: Reduction in fetal movement as the probable cause. *Clinical Toxicology* 28(1):69–83.

Panter KE, James LF, Stegelmeier BL, Ralphs MH, and Pfister JA (1999). Locoweeds: effects on reproduction in livestock. *Journal of Natural Toxins* 8:53–62.

Stegelmeier BL, James LF, Panter KE, and Molyneux RJ (1995). Serum swainsonine concentration and alpha-mannosidase activity in cattle and sheep ingesting *Oxytropis sericea* and *Astragalus lentiginosus* (locoweeds). *American Journal of Veterinary Research* 56:149–154.

Van Kampen KR and James LF (1971). Ovarian and placental lesions in sheep from ingesting locoweed (*Astragalus lentiginosus*). *Veterinary Pathology* 8:193–199.

Chapter 23

Cross-fostering as a Means to Study Transplacental Toxicosis of *Ipomoea carnea* in Rats

I.M. Hueza, J.L. Guerra, and S.L. Górniak

Research Center for Veterinary Toxicology (CEPTOX), Department of Pathology, Faculty of Veterinary Medicine and Zootechny, University of São Paulo, CEP 05508-900, Brazil

Introduction

Ipomoea carnea Jacq. spp. *fistulosa* Choisy (Convolvulaceae) is a toxic plant widely distributed in Brazil (Tokarnia *et al.* 2000). Two toxins have been isolated from the plant, the nortropane alkaloids calystegines B_1, B_2, B_3 and C_1, and predominately the indolizidine alkaloid swainsonine (De Balogh *et al.* 1999). The latter alkaloid has a known mechanism of action as a potent inhibitor of two distinct intracellular enzymes, the acidic or lysosomal α-mannosidase and the Golgi mannosidase II. Conversely, there are few published reports about the possible biological effects of calystegines.

The inhibition of α-mannosidase results in lysosomal accumulation of incompletely processed oligosaccharides rich in α-mannosyl and β-*N*-acetyl glucosamine moieties inside vacuoles, which progress to loss of cellular function and ultimately cell death (Tulsiani *et al.* 1988).

Swainsonine inhibition of Golgi mannosidase II enzyme causes alteration of the N-linked glycoprotein process (Elbein 1989), producing increased numbers of high-mannose, hybrid or complex types of oligosaccharide structures which participate in different cellular molecules which can alter hormonal and endocrine function (Stegelmeier *et al.* 1995) and cause abnormal gastrointestinal function (Pan *et al.* 1993), immunological function (Karasuno *et al.* 1992), reproductive and embryonic function (Nelson *et al.* 1980).

Anecdotal information suggests that animals chronically intoxicated with *I. carnea* and other swainsonine-containing plants show increased incidence of abortion, stillbirth and teratological effects. Thus, the objective of this study was to perform a perinatal study to verify the possible toxic effects of *I. carnea* on the offspring from dams treated during the gestational period, using cross-fostering (CF) as a tool to evaluate the possible transplacental toxicosis of *I. carnea* in rats.

Materials and Methods

Ipomoea carnea leaves were collected in May, 2004 from plants cultivated at the Research Center for Veterinary Toxicology (CEPTOX), University of São Paulo (USP), Pirassununga, Brazil. The aqueous fraction (AF) resulting from the extraction of dry *I. carnea* leaves was obtained according to previously described methods (Hueza *et al.* 2003).

Male and female Wistar rats, 10 weeks of age, were obtained from our own colony. For breeding, males were housed overnight with females and females were checked for sperm-positive vaginal smears the next morning, which was designated as gestation day 0. All animals received food and water *ad libitum* and were maintained under controlled conditions of temperature (22–25°C), humidity (50–65%) and lighting (12/12 light/dark cycle). The experiments were carried out in accordance with the ethical principles for animal research adopted by the Bioethics Committee of the Faculty of Veterinary Medicine and Zootechny, University of São Paulo.

Pregnant rats from the experimental group were treated, by gavage, with aqueous extract equivalent to 15.0 g kg^{-1} body weight of dry *I. carnea* leaves from day 6 to 20 of gestation. Pregnant rats from the control group received only tap water. Total body weight gain and water and food consumption were measured in mothers during the experimental period. One hour after parturition, litters from treated and untreated dams were weighed and the subgroups for the CF procedure were designated as follows: Cont-Cont group – litters from untreated females fostered by other untreated females; Cont-AF group – litters from untreated females fostered by treated females; AF-AF group – litters from treated females fostered by other treated females; AF-Cont group – litters from treated females fostered by other untreated females. All the fostered pups were tattooed on one paw with China ink. Offspring body weight gain was measured every other day until postnatal day (PND) 7. On the 3rd and 7th PND, three pups of each subgroup were chosen at random, euthanized and necropsied to obtain representative tissue samples for histopathology evaluation. The renal tissue was fixed as an indication of *I. carnea* toxicosis.

Student's *t* test was used to compare two groups, and a significant F test ($P<0.05$) in the analysis of variance (ANOVA) was followed by Dunnett's *post hoc* test to compare controls with ≥ two treatment groups. Data are expressed as mean ± S.D.

Results and Discussion

The present study showed that the body weight of pups from dams treated with the *I. carnea* AF was diminished immediately after birth, when compared with those of pups from untreated mothers (Table 2). The result could be due to the direct toxic effect of swainsonine; alternatively, this finding may be a result of decreased food consumption by treated dams during pregnancy (Table 1). It is well known that the nutritional status of mothers affects embryonic and fetal development in rats (Khera 1985).

In a previous study performed in our lab (Hueza *et al.* 2003), the histological evaluation of pups born to swainsonine-treated dams, and necropsied immediately after birth, indicated that lesions may not be visible using the HE staining technique. Pup cells are relatively undifferentiated at birth, and this may result in an incorrect diagnosis of cellular vacuolation. However, when pups from treated mothers were evaluated on post-gestational day 7, the histological study revealed the presence of vacuolar lesions in different organs. These litters were necropsied during the suckling period, and therefore the observed lesions may have resulted from swainsonine excreted in the milk and not from *in utero* exposure to swainsonine. This possibility is supported by the work of James and Hartley (1977), who found that sufficient swainsonine was excreted in milk to intoxicate suckling offspring.

Table 1. Treatment effect of 15.0 g kg^{-1} of *Ipomoea carnea* aqueous fraction administered for dams, by gavage, from day 6 to day 21 of gestation, on pups body weight on parturition day (day 0) and on body weight gain from day 0 to postnatal day 3 and from day 3 to postnatal day 7, and total body weight gain during the 7 postnatal days of different subgroups of fostered pups after the parturition.

	Groups					
	Body weight (g)		Body weight gain (g)			
Period (days)	Untreated[a] (n=10)[c]	Treated (n=8)	Cont[b] (n=10)	AF (n=5)	Cont-AF (n=6)	AF-Cont (n=5)
0	6.0 ± 0.3	5.2 ± 0.5†	-	-	-	-
0–3	-	-	3.3 ± 0.4	2.5 ± 1.2	2.9 ± 0.9	2.6 ± 0.7
3–7	-	-	4.6 ± 1.0	2.6 ± 1.7*	5.0 ± 1.3	4.3 ± 1.8
Total	-	-	7.9 ± 1.0	5.1 ± 1.2*	7.9 ± 1.7	6.9 ± 2.6

† Significantly different from untreated group ($P<0.05$); * Significantly different from Cont-group ($P<0.05$).

[a] Group of litters from dams treated or untreated with *Ipomoea carnea* aqueous fraction before cross-fostering procedure.

[b] Group of litters from dams treated or untreated with *Ipomoea carnea* aqueous fraction after cross-fostering procedure.

[c] Number of litters per group.

Table 2. Total water and food consumption, and body weight gain during the gestational period from dams treated with 15.0 g kg^{-1} of dry leaves of *Ipomoea carnea* aqueous fraction, by gavage, from gestational day 6 to gestational day 21.

Groups	Water consumption (mL)	Food consumption (g)	Body weight gain (g)
Untreated females (n=10)	831.7 ± 237.3	313.7 ± 47.9	103.6 ± 11.6
Treated females (n=8)	825.2 ± 220.5	238.6 ± 15.5 *	75.4 ± 20.1 *

* Significantly different from untreated group ($P<0.05$).

Thus, the purpose of this study was to use the CF procedure to determine if pups were intoxicated by swainsonine during the lactation period. We found that only pups from the AF-AF group had a decrease in body weight gain during the postnatal period, even when compared with animals from AF-Cont group (Table 2). In addition, histological evaluation of renal tissue of AF-AF pups revealed marked vacuolar degeneration (Figure 1). Conversely, pups from the Cont-AF group showed no decrease in body weight gain. Moreover, pups from this group did not exhibit vacuolar lesions in renal epithelium tissue on post-gestational days 3 and 7.

These data indicate that the fetus is exposed *in utero* to swainsonine, and that females treated with the plant during lactation will continue to excrete swainsonine into the milk. This study suggests that swainsonine excreted in the milk during this short period is not enough to induce toxic effects *per se*, but it could enhance the swainsonine-induced intoxication when the plant is ingested during gestation.

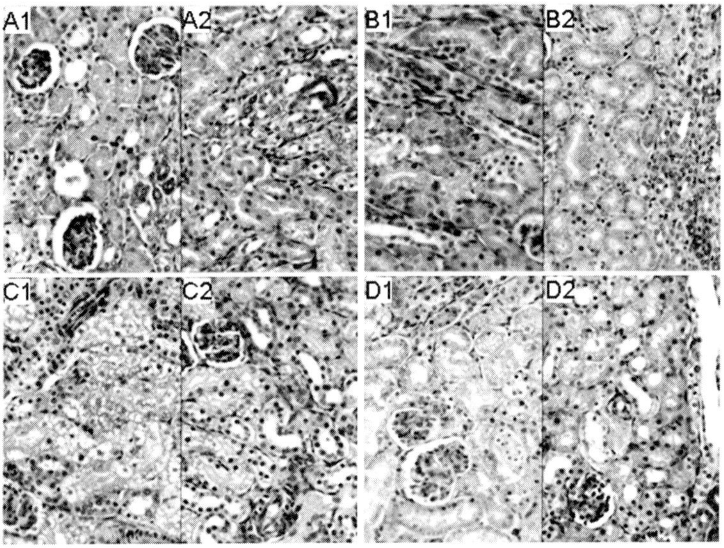

Figure 1. Tissue section of fostered pups from untreated and treated female rats with 15.0 g kg^{-1} of aqueous fraction of *I. carnea* dry leaves, by gavage, during the gestation. Kidney section of pups from Cont group on postnatal day 3 (A1) and postnatal day 7 (A2). Kidney section of pups from Cont-AF group on postnatal day 3 (B1) and postnatal day 7 (B2), no vacuolar lesion was observed in both periods. Kidney section of pups from AF group on postnatal day 3 (C1) and postnatal day 7 (C2), note the marked vacuolar lesions on convoluted epithelium cells in both periods. Kidney section of pups from AF-Cont group on postnatal day 3 (D1) and postnatal day 7 (D2), note the vacuolar lesions on convoluted epithelium cells that were less severe on the last day of evaluation. H&E.

Acknowledgements

This study was supported by grants from Fundação de Amparo à Pesquisa do Estado de SãoPaulo – FAPESP, Brazil (Proc. No. 01/11444-0).

References

De Balogh KIM, Dimande AP, van der Lugt JJ, Molyneux RJ, Naudé TW, and Welman WG (1999). A lysosomal storage disease induced by *Ipomoea carnea* in goats in Mozambique. *Journal of Veterinary Diagnostic Investigation* 11:266–273.

Elbein AD (1989). The effects of plant indolizidine alkaloids and related compounds in glycoprotein processing. In *Swainsonine and Related Glycosidase Inhibitors* (LF James, AD Elbein, RJ Molyneux and CD Warren, eds), pp. 87–155. Iowa State University Press, Ames.

Hueza IM, Dagli MLZ, Górniak SL, and Paulino CA (2003). Toxic effects of prenatal *Ipomoea carnea* administration to rats. *Veterinary and Human Toxicology* 45:298–302.

James LF and Hartley WJ (1977). Effects of milk from animals fed locoweed on kittens, calves, and lambs. *American Journal of Veterinary Research* 38:1263–1265.

Karasuno T, Kanayama Y, Nishiura T, Nakao H, Yonezawa T, and Tarui S (1992). Glycosidase inhibitors (castanospermine and swainsonine) and neuraminidase inhibit pokeweed mitogen-induced B-cell maturation. *European Journal of Immunology* 22:2003–2008.

Khera KS (1985). Maternal toxicity: a possible etiological factor in embryo-fetal deaths and fetal malformations of rodent-rabbit species. *Teratology* 31:129–153.

Nelson BK, James LF, Sharma RP, and Cheney CD (1980). Locoweed embryotoxicity in rats. *Clinical Toxicology* 16:149–166.

Pan YT, Ghidoni J, and Elbein AD (1993). The effects of castanospermine and swainsonine on the activity and synthesis of intestinal sucrase. *Archives of Biochemistry and Biophysics* 303:134–144.

Stegelmeier BL, Molyneux RJ, Elbein AD, and James LF (1995). The lesions of locoweed (*Astragalus mollissimus*), swainsonine, and castanospermine in rats. *Veterinary Pathology* 32:289–298.

Tokarnia CH, Döbereiner J, and Peixoto PV (2000). In *Plantas Tóxicas do Brasil*, pp. 123–125. Editora Helianthus, Rio de Janeiro.

Tulsiani DR, Broquist HP, James LF, and Touster O (1988). Production of hybrid glycoproteins and accumulation of oligosaccharides in the brain of sheep and pigs administered swainsonine or locoweed. *Archives of Biochemistry and Biophysics* 264:607–617.

Chapter 24

Effect of *Solanum malacoxylon* in Goats: a Prenatal Study

S.L. Górniak, I.M. Hueza, P.C.F. Raspantini, B. Shumaher-Henrique, M. Barbosa-Ferreira, and P.C. Maiorka

Research Center for Veterinary Toxicology (CEPTOX), Department of Pathology, Faculty of Veterinary Medicine and Zootechny, University of São Paulo, CEP 05508-900, Brazil

Introduction

Since the end of the 19th century, a disease of cattle and sheep called 'espichamento' has been known in Brazil. 'Espichamento' is a disease involving calcinosis, *i.e.*, pathologic deposition of calcium phosphate in soft tissue, and occurs in animals grazing leaves of the shrub *Solanum malacoxylon*, which contains a glycoside conjugated to Vitamin D3 ($1,25[OH]_2D_3$) (Wasserman *et al.* 1976). *Solanum malacoxylon* produces hypercalcemia, hyperphosphatemia, and extensive mineralization of the soft tissue (Döbereiner *et al.* 1975; Done *et al.* 1976).

Many experimental studies on the toxicity of the plant were conducted in different animal species as follows: cattle, sheep, rabbits, rats and chickens (Mello 2003). In all species, pathological features are similar to those of animals given massive doses of vitamin D. There is, however, little information about the effects of *S. malacoxylon* during pregnancy. Previous studies conducted in our laboratory showed that administration of *S. malacoxylon* to rabbits (Górniak *et al.* 1999) and rats (Górniak *et al.* 2003) during pregnancy produces litters with alterations of soft tissue and increased phosphorus and calcium serum levels. However, no work was found in literature aimed at understanding the effects of this plant on pregnancy in ruminants.

Thus, the objective of the present work was to evaluate the possible toxic effects on both dams and kids with maternal exposure of goats to *S. malacoxylon* during pregnancy. A previous study showed that this species is an excellent animal model for trials on the toxicity of *S. malacoxylon* (unpublished data).

Material and Methods

Plant material

Leaves of *S. malacoxylon* were collected from the Pantanal region, Mato Grosso do Sul, Brazil, and air dried, ground and incorporated into a ration for pregnant goats.

Experimental design

Twenty mixed-breed female goats, 9–14 months old, were bred to one Alpine buck. Estrus was synchronized in the female using D (+) cloprostenol (Veteglan, Serono). The day of breeding was defined as day 1 of gestation and pregnancies were confirmed with ultrasound (US) (Scanner 100, Vet-Pie Medical linear transducer, 5–7.5 MHZ) on day 27. Pregnant goats were allocated into four equal groups and dosed from day 27 through day 91 of gestation (DG) with 0.0 (control), 12.5 (Sm12.5), 25.0 (Sm25) and 50.0 (Sm50) mg kg^{-1} day of *S. malacoxylon* leaves incorporated into the diet. All goats were fed 100 g concentrate and sugar cane (*Sacharum officinarum*) *ad libitum*.

Each week, the animals were clinically evaluated and their body weights were recorded. Before the beginning of the trial, and then every other week, blood samples were obtained from the jugular vein in order to determine calcium (Ca), phosphorus (P) and alkaline phosphatase (AP) serum levels until the 18th week of the gestation. Moreover, on the 27th, 35th, 48th and 62nd DG ultrasound image analysis was done to evaluate fetal death and/or the following parameters: crown-rump length (cm), fetal movements and heartbeat (min) The animals were permitted to give birth naturally except for one female of each group which was delivered via caesarian section on the 140th DG; tissue samples were collected from both mother and fetus for histological study.

The body weight of each neonate was determined at birth, and the neonate examined carefully for gross abnormalities, as described by Szabo (1989). From birth to 2 months of age each female kid was weighed and blood samples were collected every other week as described above for their mothers in order to evaluate Ca, P and AP serum levels.

Student's *t* test was used to compare two groups, and analysis of variance (ANOVA) followed by Dunnett's *post hoc* test was used for more than two groups, with the level of significance set at $P<0.05$. Data are expressed as mean ± S.D.

Results

Clinically, all females from experimental groups showed, in a dose-response fashion, progressive physical deterioration such as emaciation, kyphosis and impaired gait. Two dams from Sm25 and two from Sm50 groups died between the 18th and 20th weeks of gestation; thus, body weight gain from these groups is not given. The body weight gain and length of gestation of goats in the Sm12.5 and control groups were not different. No aborted fetuses were observed in any group (Table 1). One female from Sm12.5 had fetal resorption detected by US at 35 DG. On the other hand, no alterations were detected by US in crown–rump length, fetal movements and heartbeat in all periods evaluated (data not shown). The two female survivors from groups Sm25 and Sm50 died between the 2nd and 4th weeks of lactation.

Table 1. Reproductive parameters from dams dosed with *S. malacoxylon* leaves from 24th to 91st gestational day[a].

Parameters	Sm (mg kg^{-1} in ration)			
	0	12.5	25	50
Body weight gain (kg ± SE)				
Pre-natal period	17.0 ± 2.6 (5)[a]	14.5 ± 1.5 (4)	5.8 ± 0.9 (2)	2.1 ± 4.3 (2)
Post-natal period	-1.2 ± 2.6 (4)	-5.5 ± 3.5 (3)	--[c]	--[c]
Length of gestation (days ± SE)	144.2 ± 2.2 (4)	146.5 ± 1.7 (3)	146.5 ± 0.7 (2)	151.5 ± 0.7 (2)
Fetal Death	0	1	0	0
Live kids	8	8	4	4
Males	2 (25.0)[b]	5 (62.5)	1 (25.0)	3 (75.0)
Females	6 (75.0)	3 (37.5)	3 (75.0)	1 (25.0)
Kids death	1 (12.5)	0	0	0
Twins	8 (100)	8 (100)	4(100)	2 (100)
Aborted fetuses	0	0	0	0

[a] five dams per group.
[b] data are expressed in percentages.
[c] Two goats died during lactation period.

Serum Ca, P and AP values from dams are shown in Figure 1. Treatment groups did not differ in baseline serum levels of Ca, P and AP. In contrast, from the 2nd week of dosing, the levels of Ca were higher in Sm50 and Sm25 mothers and from the 4th week for goats in the Sm12.5 group. No differences between control and experimental groups were detected in Ca levels in the 12th week of *S. malacoxylon* administration. In the 14th week there was a decrease in Ca levels in all experimental groups compared to controls. Phosphorus levels were increased in all experimental groups from the 2nd week of dosing; however, at the 10th week no differences between control and experimental groups were detected. No differences were detected in AP between the different groups during any periods evaluated.

For the neonates, no statistical analysis was performed for males because there were insufficient numbers of animals in the control group. No differences in weight gain and biochemical profile were observed in female kids between control and experimental groups throughout the study (data not shown).

Histopathological evaluation revealed alterations in all goats treated with *S. malacoxylon*. The histopathologic changes were dose-related; thus, the lesions were more conspicuous in those females treated with a larger amount of *S. malacoxylon*. The heart and lungs were the most affected organs, showing intense calcification of cardiac valves and cardiac muscle coronary (Figure 2), aorta and pulmonary artery calcification. Lungs displayed intense inflammatory infiltrates of mostly mononuclear cells around the bronchi, bronchioles and blood vessels. Kidneys were pale in the cortico-medullar area with severe calcification of glomerular and proximal convoluted tubules. Kids from Sm25 and Sm50 goats also showed the same lesions as described for their mothers; however, the lesions were much milder. No histopathologic observations were made in kids from the Sm12.5 and control groups.

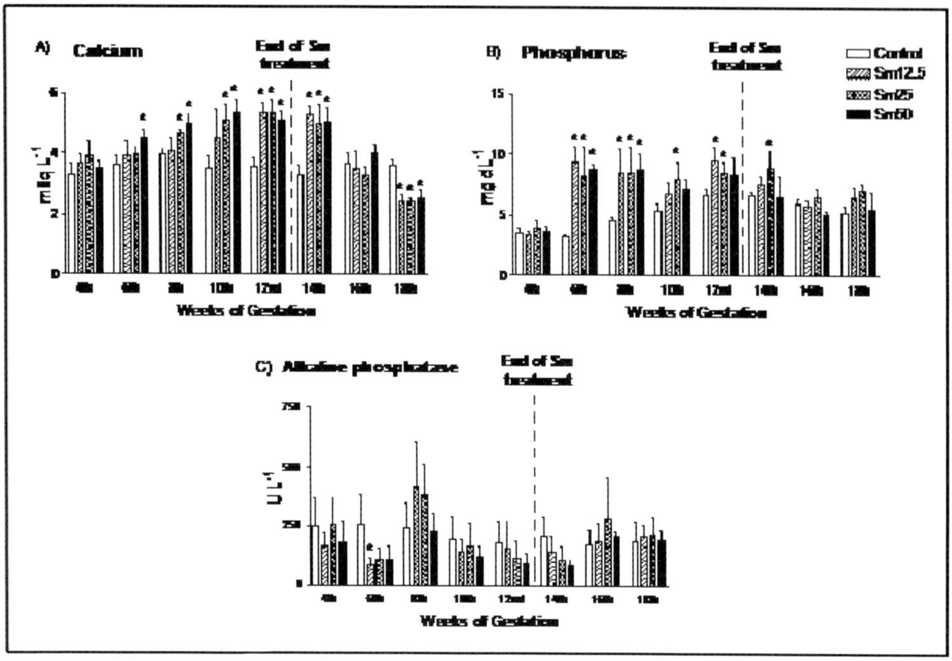

Figure 1. Serum levels of calcium (A), phosphorus (B) and alkaline phosphatase (C) from pregnant goats treated or not with *S. malacoxylon* leaves from 24th to 91st gestational day. Data are shown as mean ± SD. * $P<0.05$. Dunnett's test.

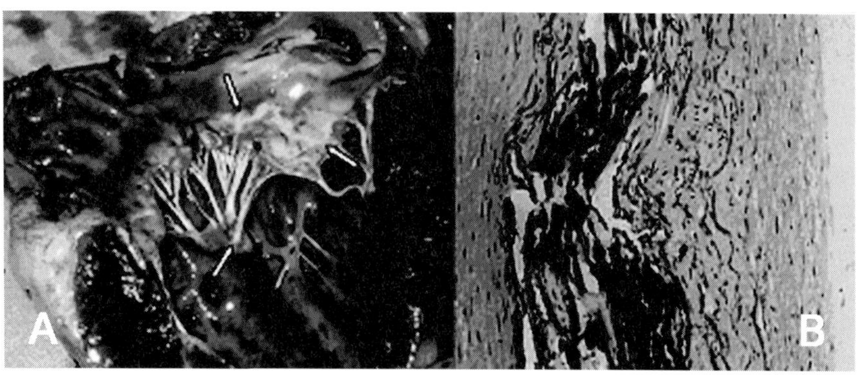

Figure 2. Heart of pregnant goat treated with 50 mg kg^{-1} of *Solanum malacoxylon* leaves from 24th to 91st gestational day and submitted to caesarean at the 140th day of gestation. Arrows show intense calcification of cardiac valves and cardiac muscle (A). Aorta section of the same pregnant goat cited above. Note the presence of marked calcium deposition within the elastic laminae of tunica intima (B). H&E, 20x.

Discussion

Enzootic calcinosis of domestic animals has been identified by many researchers as a disease that causes heavy economic losses to the cattle industry in Argentina and Brazil (Morris 1982; Boland 1988). A decrease of fertility has been related to *S. malacoxylon* ingestion and is a major factor in the losses (Morris 1982). Previous studies conducted in our laboratory showed that administration of *S. malacoxylon* to rabbits (Górniak *et al.* 1999) and rats (Górniak *et al.* 2003) during pregnancy produced litters with alterations of soft tissue. However, to date, no experimental studies on the effects of this plant on fetal development of ruminants have been published.

Anorexia and rapid loss of weight are initial and prominent symptoms of *S. malacoxylon* intoxication in rabbits, guinea pigs and cattle (Morris 1982). We also verified that male goats treated with *S. malacoxylon* showed a dramatic decrease in body weight from the first week of plant administration (unpublished data). In the present study, although food intake was not measured, we found a decrease in weight gain in those animals from the Sm groups. In addition, clinically these experimental goats had very poor body condition, and deaths were observed in pregnant goats from Sm25 and Sm50 groups. Therefore, this study revealed that pregnant goats are also affected by the anorexic effect produced by *S. malacoxylon*.

Hypercalcemia and/or hyperphosphatemia and soft-tissue mineralization are usually produced in naturally occurring calcinosis in cattle and experimentally in other domestic and laboratory animals exposed to *S. malacoxylon*. Goats fed with *S. malacoxylon* manifested increased serum Ca and P levels by the second week of dosing (*i.e.*, at the sixth week of gestation). Histopathological studies confirmed these biochemical changes as animals in both experimental groups showed marked mineralization, mainly in heart and lungs. It is interesting to note that, after *S. malacoxylon* had been withdrawn, there was a decrease in serum levels of Ca and P of the experimental mothers, and in the last week of evaluation (18th week of gestation) they were significantly lower than those found in goats from the control group. The reason for this result is unknown; thus, future experiments will be performed in order to measure the calcitonin and parathyroid hormone levels in late gestation to better understand the alterations observed here.

Hypercalcemia due to excessive vitamin D consumption is usually accompanied by hyperphosphatemia and normal AP serum levels because increased concentrations of blood Ca and P originate mainly from an increase in intestinal absorption rather than from resorption of bone (Capen and Rosol 1989). In fact, AP levels were not altered in the experimental goats.

The possibility that the calcinogenic effects observed in the fetus of treated mothers were induced by an excess of maternal serum Ca rather than by $1,25(OH)_2D_3$ is unlikely because maternal hypercalcemia during pregnancy tends to produce hypocalcemia in the offspring (Fairney and Weir 1970). In the present study fetuses from goats fed *S. malacoxylon* during gestation showed alterations in soft tissue similar to those of their mothers. Moreover, direct studies in sheep indicate placental transfer of both 25-hydroxyvitamin D_3 and $1,25(OH)_2D_3$ (Case *et al.* 1980).

In conclusion, the results of this study indicate that in a similar manner as in laboratory animals, the toxic principle of *S. malacoxylon* can affect fetuses from ruminants. Furthermore, this indirect form of intoxication may account for the heavy losses caused to the livestock industry by this plant.

Acknowledgement

This study was supported by grants from Conselho Nacional de Pesquisa (CNPq).

References

Boland RL (1988). *Solanum malacoxylon*: a toxic plant which affects animal calcium metabolism. *Biomedical and Environmental Sciences* 1:414–423.

Capen CC and Rosol TJ (1989). Calcium-regulating hormones and diseases of abnormal mineral (calcium, phosphorus, magnesium) metabolism. In *Clinical Biochemistry of Domestic Animals 4th ed* (JJ Kaneko, ed), pp. 678–752. Academic Press, San Diego.

Case AD, Ross R, Pikard DW, and Robinson JS (1980). The intra-uterine control of mineral metabolism. *Calcified Tissue International* 31(suppl): 29.

Döbereiner J, Done SH, and Beltran LE (1975). Experimental *Solanum malacoxylon* poisoning in calves. *British Veterinary Journal* 131:175–185.

Done SH, Döbereiner CH, and Tokarnia CH (1976). Systemic connective tissue calcification in cattle poisoned by *Solanum malacoxylon*: a histological study. *British Veterinary Journal* 132:28–38.

Fairney A and Weir AA (1970). The effect of abnormal maternal plasma calcium levels on the offspring of rats. *Journal of Endocrinology* 48:337–345.

Górniak SL, Dagli MLZ, Maiorka PC, Raspantini PC, and Jericó MM (1999). Evaluation in rabbits of the fetal effects of maternal ingestion of *Solanum malacoxylon. Veterinary Research Communications* 23:307–316.

Górniak SL, Mariorka PC, Raspantini PC, Hosomi R, Moraes AP, and Dagli MLZ (2003). Assessment of the perinatal effects of maternal ingestion of *Solanum malacoxylon* in rats. *Reproductive Toxicology* 17:67–72.

Mello JRB (2003). Calcinosis – calcinogenic plants. *Toxicon* 41:1–12.

Morris KML (1982). Plant-induced calcinosis. *Veterinary and Human Toxicology* 24:34–48.

Szabo KT (1989). *Congenital Malformations in Laboratory and Farm Animals*. Academic Press, San Diego.

Wasserman RH, Henion JD, Haussler MR, and McCain TA (1976). Calcinogenic factor in *Solanum malacoxylon* evidence that it is 1,25 dihidroxy-vitamin-D3-glicoside. *Science* 194:853–855.

Chapter 25

Evaluation of Ultrasonographic and Biochemical Profile of Fetal Development of Intoxicated Female Goats with *Senna occidentalis* Seeds

M. Barbosa-Ferreira and S.L. Górniak

Research Center for Veterinary Toxicology (CEPTOX), Department of Pathology, Faculty of Veterinary Medicine and Zootechny, University of São Paulo, CEP 05508-900, Brazil

Introduction

Senna occidentalis is a leguminous species found worldwide, often contaminating cultivated crops and pastures. *Senna occidentalis* has been shown to be toxic to several animal species. Skeletal muscle degeneration is the main lesion found in the majority of animals intoxicated by *S. occidentalis* (Barros *et al.* 1990); however, other alterations in myocardial muscle (Haraguchi *et al.* 1998), liver (Tasaka *et al.* 2000) and CNS (Barbosa-Ferreira *et al.* 2005) have also been described.

All parts of the plant are toxic but the seed is most toxic (Dollahite and Henson 1965). The toxic principle in *S. occidentalis* causes impairment of mitochondrial function (Calore *et al.* 1997), including swelling, loss of mitochondrial matrix, fragmented mitochondrial cristae and glycogen depletion. The toxins in *Senna* spp. are likely anthraquinones (Burrows and Tyrl 2001). Haraguchi *et al.* (1996) identified dianthrone, an anthranoid derivative in *S. occidentalis* seeds, and this substance might cause the characteristic mitochondrial myopathy (Calore *et al.* 1997).

Spontaneous cases of intoxication with *S. occidentalis* seeds have been reported in livestock when animals were fed with contaminated rations. Similarly, experimental studies conducted on different animal species have shown the occurrence of sickness and death with high levels of the seeds in the ration. However, very little is known about the possible effects of *S. occidentalis* exposure on reproductive toxicology (Tasaka *et al.* 2004). Thus, the purpose of the present study was to evaluate the impact of *S. occidentalis* seeds during gestation in goats. Previous work conducted in our laboratory (Barbosa-Ferreira 2003) clearly showed that *S. occidentalis* seeds cause toxicity in male goats, with prominent lesions in the liver and skeletal muscle.

Materials and Methods

Mature *S. occidentalis* (S.O) seeds were collected from stands growing at the Biological Institute of São Paulo State and at the Pirassununga Campus of the University of São Paulo. Twenty mixed-breed female goats, 9–14 months old, were bred to one Alpine buck. Estrus was synchronized in the females using D (+) cloprostenol (Veteglan, Serono). The day of breeding was defined as day 1 of gestation and pregnancies were confirmed by ultrasound (US) apparatus (Scanner 100 Vet, Pie Medical Linear probe, 5.0 / 7.0 MHz) at day 27. Pregnant goats were allocated into four treatment groups (n=5) given the following percentages of *S. occidentalis* seeds in the ration: 0% (control), 1% (S.O1), 2% (S.O2) and 4% (S.O4) from the 27th day of gestation to parturition.

The animals were clinically evaluated weekly and their body weights recorded. Before the beginning of the trial and then every other week blood samples were collected via jugular venipuncture to evaluate protein, albumin, aldolase A/B, glucose, aspartate amino transferase (AST), lactate dehydrogenase (LDH) and gamma glutamyltransferase (GGT). The ultrasonographic evaluation was performed on weeks 5, 9 and 13 to measure fetal movement, heartbeat, thoracic and abdominal diameter, and crown–rump length.

Data are expressed as the mean values and the standard error of the mean and were analyzed statistically by ANOVA followed by Duncan's test. Results were considered significant when $P<0.05$.

Results

Goats did not show any clinical signs of intoxication during the feeding period. Females from the S.O4 group gained 45% less weight ($P<0.05$) than did controls (Table 1), and two goats from this group resorbed their fetuses between the 27th and 35th day of pregnancy. Serum concentrations of aldolase, glucose, protein, albumin and LDH fluctuated during the trial, whereas AST and GGT concentrations were similar between control and all treatment groups. No changes in fetal development due to treatments were found during the ultrasound evaluation.

Table 1. Reproductive parameters from dams treated with *S. occidentalis* seeds from the 27th day of gestation to term.

Parameters	S.O seeds (%)			
	0	1	2	4
Pregnant / bred goats	5/5	4/5	4/5	5/5
Embryonic resorption	0	0	0	2
Weight gain (kg±SEM)	9.4 ± 1.2 (5)[a]	6.7 ± 2.07 (4)	9.5 ± 2.6 (4)	2.9 ± 0.4 (3)
Gestation (days±SEM)	144.4 ± 0.5(5)	147.8 ± 0.9 (4)	147.3 ± 0.9 (4)	148.0 ± 1.5 (4)
Live kids	7 (100)	5 (100)	5 (100)	3 (100)
Males	4 (57.1%)	3 (60.0%)	3 (60.0%)	1 (33.3%)
Females	3 (42.9%)	2 (40.0%)	2 (40.0%)	2 (66.7%)
Dead kids	0	0	0	1 (33.3%)
Twins	4 (57.1%)	4 (57.1%)	4 (57.1%)	0
Weak kids	0	2 (40.0%)	0	1 (33.3%)

[a] number of goats

Discussion

The aim of the present study was to evaluate the reproductive toxicity of *S. occidentalis* in goats and to improve the ruminant model for teratology studies that has been developed in our laboratory (Soto-Blanco and Górniak 2004). This model may be useful for the study of potential teratogenic and fetotoxic effects, not only of plant toxins, but also for a number of chemicals to which domestic ruminants are frequently exposed.

Anecdotal reports in the literature have associated *Senna* spp. consumption and infertility in rats and humans (Yadav and Jain 1999; Badami *et al.* 2003; Stuart 2003). Recently, Tasaka *et al.* (2004) verified that different concentrations of *S. occidentalis* seeds in rations of pregnant rabbits produced toxic effects in both mothers and fetuses.

In the present study two pregnant dams from the 4% *Senna* treatment resorbed their fetuses. These results indicate that, similar to rabbits, pregnant goats and their fetuses are affected by *S. occidentalis* when diets contain at least 4% seeds for extended periods of gestation. Abortion (or fetal resorption) is one of the first observed effects of prenatal toxicity of any substance to livestock (James *et al.* 1994). Affected animals were clinically evaluated, and we found no evidence that resorption was caused by infectious agents such as *Mycoplasma* sp., *Brucella* sp., *Toxoplasma* sp. and *Leptospira* sp.

Maternal toxicity can affect fetal development and/or death (Khera 1984). There are several possible explanations for our results. Although food consumption was not recorded, females in the 4% treatment failed to gain as much weight as controls. Thus, resorption could be related to maternal toxicity. On the other hand, considering that the putative toxin dianthrone disrupts oxidative phosphorylation in mitochondria (Cavaliere *et al.* 1997), and since most oxygen consumed in the placenta is used for oxidative phosphorylation (Carter 2000), resorption could be due to placental toxicity. Finally, dianthrone may pass through the placenta, causing direct toxic effects on the fetus. Future studies should be conducted to determine whether the toxic effect of *S. occidentalis* on goat fetuses was secondary to maternal or placental toxicity, or if it was a direct fetotoxic effect.

Acknowledgement

This work was supported by a grant from Coordenação de Aperfeiçoamento de Pessoal de Nível Superior (CAPES).

References

Badami S, Aneesh R, Sankar S, Sathishkumar MN, Suresh B, and Rajan S (2003). Antifertility activity of *Derris brevipes* variety coriacea. *Journal of Ethnopharmacology* 84(1):99–104.

Barbosa-Ferreira (2003). Estudo dos efeitos tóxicos produzidos pela administração prolongada de sementes de senna occidentalis. avaliação em ratos e caprinos. *Dissertação* (Mestrado)–Faculdade de Medicina Veterinária e Zootecnia, Universidade se São Paulo, Departamento de Patologia, São Paulo.

Barbosa-Ferreira M, Dagli MLZ, Maiorka PC, and Górniak SL (2005). Sub-acute intoxication by *Senna occidentalis* seeds in rats. *Food and Chemical Toxicology* 43(4):497–503.

Barros CSL, Pilati C, Andujar MB, Graça DL, Irigoyen LF, Lopes ST, and Santos CF (1990). Intoxicação por *Cassia occidentalis* (Leg. Caes.) em bovinos. *Pesquisa Veterinária Brasileira* 10(3–4):47–48.

Burrows GE and Tyrl RJ (2001). *Toxic Plants of North America*. Iowa State University Press, Ames, Iowa.

Calore EE, Cavalieri MJ, Haraguchi M, Górniak SL, Dagli MLZ, Raspatini PCF, and Calore NMP (1997). Experimental mitochondrial myopathy induced by chronic intoxication by *Senna occidentalis* seeds. *Journal of Neurological Sciences* 146:1–6.

Carter AM (2000). Placental oxygen consumption. Part I: in vivo studies – a review. *Placenta* 21 Suppl. A: S31–7.

Cavaliere MJ, Calore EE, Haraguchi M, Górniak SL, Dagli ML, Raspantini PC, Calore NM, and Weg R (1997). Mitochondrial myopathy in *Senna occidentalis*-seed-fed chicken. *Ecotoxicology and Environmental Safety* 37(2):181–185.

Dollahite JW and Henson JB (1965). Toxic plants as the etiologic agent of myopathies in animals. *American Journal of Veterinary Research* 26:749–752.

Haraguchi M, Górniak SL, Dagli MLZ, and Raspantini PCF (1996). Determinação dos constituintes químicos das frações tóxicas de fedegoso (*Senna occidentalis* (L.)). In *Annals of Reunião Anual da Sociedade Brasileira de Química*, Poços de Caldas, MG.

Haraguchi, M, Górniak, SL, Calore EE, Cavaliere MJ, Raspantini PCF, Calore NM, and Dagli MLZ (1998). Muscle degeneration in chicks caused by *Senna occidentalis* seeds. *Avian Pathology* 27:346–351.

James LF, Panter KE, Stegelmeier BL, and Molyneux RJ (1994). Effect of natural toxins on reproduction. *The Veterinary Clinics of North America Food Animal Practice* 10(3):587–603.

Khera KS (1984). Maternal toxicity - A possible factor in fetal malformations in mice. *Teratology* 29(3):411–416.

Soto-Blanco Band Górniak SL (2004). Prenatal toxicity of cyanide in goats – a model for teratological studies in ruminants. *Theriogenology* 62(6):1012–1026.

Stuart AG (2003). Herbs to avoid during pregnancy and lactation – Plantas a evitar durante embarazo y lactancia. University of Texas El Paso/University of Texas Austin Cooperative Pharmacy Program. In http://www.herbalsafety.utep.edu/populations/documents/Pregnancy%20Plants.pdf.

Tasaka AC, Weg R, Calore EE, Sinhorini IL, Dagli MLZ, Haraguchi M, and Górniak SL (2000). Toxicity of *Senna occidentalis* seed in rabbits. *Veterinary Research Communications* 24:573–582.

Tasaka AC, Sinhorini IL, Dagli MLZ, Haraguchi M, and Górniak SL (2004). Perinatal study of *Senna occidentalis*. Intoxication in rabbits. In *Poisonous Plants and Related Toxins* (T Acamovic, CS Stewart, and TW Pennycott, eds), pp. 459–464. CABI Publishing, Wallingford, UK.

Yadav R and Jain GC (1999). Antifertility effect of aqueous extract of seeds of *Cassia fistula* in female rats. *Advances in Contraception* 15(4):293–301.

Chapter 26

Lupine-induced Crooked Calf Disease: Plasma Disposition of Lupine Alkaloids Following Repeated Challenge in Cattle That Have Given Birth to Affected Calves or Those That Have Borne Normal Calves

C.C. Gay[1], A. Tibary[1], K.E. Panter[2], K.L. Mealey[1], J.M. Gay[1], S.W. Hjartarson[1], E. Motteram[1], T. Wierenga[2], and L.F. James[2]

[1]*Field Disease Investigation Unit, College of Veterinary Medicine, Washington State University, Pullman WA 99164-6610*, USA; [2]*USDA-ARS Poisonous Plant Research Laboratory, Logan UT 84341, USA*

Introduction

Lupine-induced arthrogryposis (crooked calf disease) is manifest with a pattern of essentially lethal congenital anomalies characterized by the presence of arthrogryposis of the forelimbs, typically involving both the elbow and the carpal joints, and by the variable presence of hind limb contractures, torticollis, scoliosis, kyphosis and cleft palate. It occurs in the offspring of range cattle that graze species of lupine that contain the teratogenic alkaloids anagyrine or ammodendrine in early pregnancy. The syndrome is prevalent on ranches in the Cow Creek drainage in the Channeled Scablands of Adams County in east-central Washington state. The at-risk area covers approximately 800 square miles. On ranches in this area this syndrome occurs at a relatively low incidence (1 to 5% of calves born) every year but periodically there are years with very high incidence and severity of deformities.

Almost all scabland range pastures in Adams County have *Lupinus leucophyllus*, which is the lupine in this area that contains the teratogenic alkaloid anagyrine. These plants emerge in March, start to flower in mid-May, and are in seed pod stage in June and July. Bulls are introduced in early spring with the main calving periods in February and March. Very few ranches in the at-risk area are lupine-free and consequently pregnant cattle are exposed to teratogenic lupines throughout the susceptible period of pregnancy.

Despite the constant presence of lupines on these pastures there are variable patterns in the occurrence of crooked calf, which, if understood, might allow the formulation of intervention strategies to mitigate the occurrence or severity of the disease. In some years there is a high incidence and severity of disease and other years a low occurrence, despite the presence of *L. leucophyllus* in all years.

There are range areas in the Cow Creek drainage that have an historic high incidence and severity of disease in the offspring of pregnant cattle that graze them and high risk for occurrence in most years that they are grazed, and there are pastures that have low risk. Both risk-type pastures have *L. leucophyllus*.

Within a herd the cows that are bred to calve in the spring have exposure to teratogenic alkaloids for much of the period of fetal susceptibility. Despite this, in the same management group only a proportion of cows will give birth to crooked calves, even though all cows in the management group have had the same lupine-exposure risk during pregnancy.

This study was established to address reasons for difference in birth outcome of cows that have the same exposure risk. It is possible that this difference in birth outcome reflects a difference between the dams in the adaptive response to anagyrine, or in its metabolism, with repeated exposure to anagyrine. In this study we hypothesized that the pharmacokinetic disposition of anagyrine following repeated challenge, and the response of the fetus, is different in cows that give birth to calves with lupine-induced arthrogryposis than in those that give birth to clinically normal calves.

Materials and Methods

Animals and challenge

Twelve mature cows were obtained from ranches on which calves with lupine-induced arthrogryposis had been born. From each ranch a cow that had given birth to an affected calf (group A) was purchased along with a cow that had been kept in the same management group, grazed the same pasture the previous spring and summer, and given birth to a clinically normal calf (group B). Lupine was collected from an historically high risk pasture for crooked calf disease. The cows were challenged with 2 g of dried lupine kg^{-1} body weight between 46 and 51 days gestational age for the first challenge, and 2.5g kg^{-1} body weight as the second and third challenges when fetal gestational age was between 60 to 63 days and 68 to 75 days, respectively. The concentration of anagyrine in the dried challenge lupine was 3.3 mg kg^{-1} of dry weight. Blood samples were collected at 0, 2, 4, 8, 12, 24, 36 and 48 hours after challenge.

Ultrasound

Fetal activity was determined by transrectal ultrasonography before and after lupine challenge. Ultrasound examinations were performed for a period of 20 min at 10 hours and 4 hours before lupine challenge and again at 4, 12, 24, 36 and 48 hours after challenge. Examinations were done transrectally with 5 MhZ linear transducer for pregnancies less than 70 days and with a 3.5 MhZ sector transducer for 120-day pregnancies. All examinations were videotaped and viewed at later dates by a treatment-blinded person. Fetal activity was determined by counting the number of two types of activity (slow limb *vs.* whole body movement) for each examination period. Repeated measures analysis of variance was performed on number of fetal movements (slow, whole body and total) in a model which included cow, time, day of pregnancy and cow x days of pregnancy interaction as factors.

Alkaloid and pharmacokinetic analysis

Analysis of alkaloids in lupine and plasma was by gas chromatography (Gardner and Panter 1993). Plasma alkaloid concentration versus time curves were generated. The AUC (area under the plasma concentration versus time curve) was calculated by use of the trapezoidal method; C_{max} (maximum plasma concentration), T_{max} (time to reach C_{max}), and MRT (mean residence time) of each alkaloid were determined by use of a nonlinear curve-fitting program (PK Analyst; MicroMath, Salt Lake City, Utah).

Means and standard deviations of alkaloid content were calculated using a commercial spreadsheet. Because the pharmacokinetic data were not normally distributed, the data were summarized by use of median and interquartile range, and statistical comparisons were performed by use of the Mann-Whitney U test. A commercial statistical software package was used. Values of $P \leq 0.05$ were considered significant.

Results

Plasma alkaloid concentrations are shown for the two groups of cattle. For the three outcome variables, the differences between the two groups over the three time periods are not statistically significant (AUC, $\underline{P} = 0.567323$; MRT, $\underline{P} = 0.537554$; C_{peak}, $\underline{P} = 0.683645$). A significant ($\underline{P} < 0.05$) decrease in fetal activity was observed in all cows 4 hours and 12 hours after challenge both at 60 days and 70 days of pregnancy, and there was no difference between the two groups. Cow and cow x days of pregnancy were also highly significant, showing that the response to the challenge may be different amongst cows and may depend on the stage of pregnancy.

Discussion

The AUC is an indicator of total exposure to an administered compound and is directly proportional to dose and bioavailability of the compound and inversely proportional to clearance. We expected that if the repeated exposure of a pregnant cow to anagyrine influenced the likelihood of giving birth to an affected calf, a significant difference would be detected in the AUCs between groups A and B; however, this was not observed. Suppression of movement in the fetuses occurred in both groups. This suppression suggests that the dose of anagyrine was a sufficient challenge and this dose was selected to be equivalent to, or greater than, that used in previous challenge studies that experimentally induced arthrogryposis in calves (Shupe *et al.* 1967; Keeler 1976; Keeler *et al.* 1977; Panter *et al.* 1994). There is no evidence in this study to support differences in metabolism and disposition of anagyrine as a reason for differing birth outcomes in cows under the same lupine-exposure risk. It is probable that this difference between cows reflects differences in amount and timing of lupine consumption.

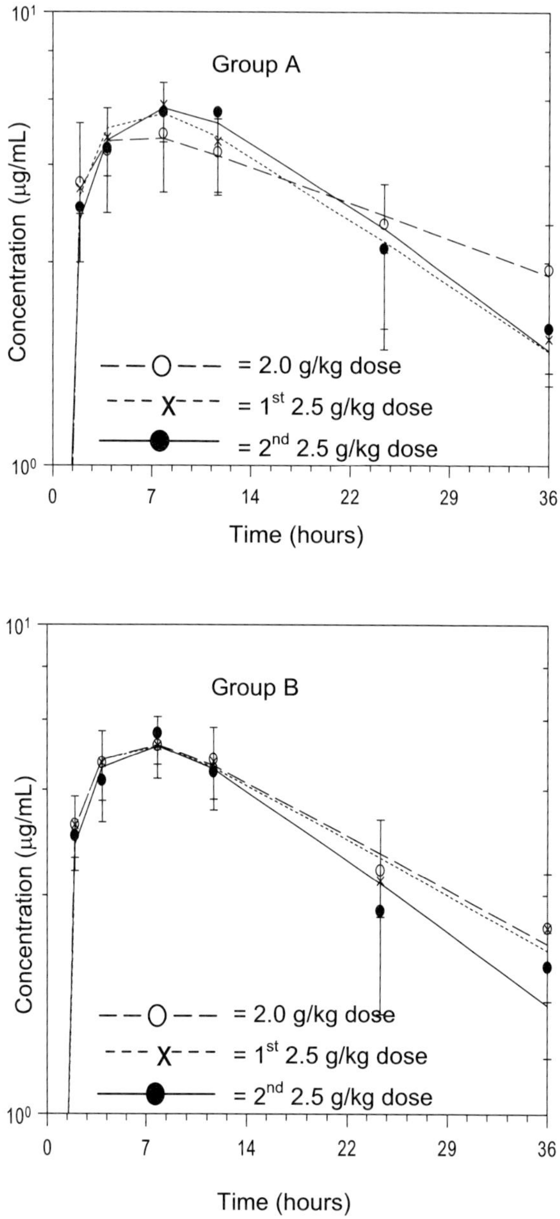

Figure 1. Mean ± SD plasma concentrations of anagyrine before (time = 0 hr) and at various time points after administration of lupine to Groups A and B cows. No significant ($P \leq 0.05$) differences were found between groups.

References

Gardner DR and Panter KE (1993). Comparison of blood plasma alkaloid levels in cattle, sheep, and goats fed *Lupinus caudatus*. *Journal of Natural Toxins* 2:1–11

Keeler RF (1976). Lupine alkaloids from teratogenic and nonteratogenic lupines. III. Identification of anagyrine as the probable teratogen by feeding trials. *Journal of Toxicology and Environmental* Health 1:887–898.

Keeler RF, James LF, Shupe JL, and Van Kampen KR (1977). Lupine-induced crooked calf disease and a management method to reduce incidence. *Journal of Range Management* 30:97–102.

Panter KE, Garner DR, and Molyneux RJ (1994). Comparison of toxic and teratogenic effects of *Lupinus formosus, L arbustus,* and *L caudatus* in goats. *Journal of Natural Toxins* 3:83–93.

Shupe JL, Binns W, James LF, and Keeler RF (1967). Lupine, a cause of crooked calf disease. *Journal of the American Veterinary Medical Association* 151:198–203.

Chapter 27

Risk Factors for Lupine-induced Crooked Calf Disorder in East-central Washington State

C.C. Gay[1], K.E. Panter[2], E. Motteram[1], J.M. Gay[1], H. Hantz[1], T. Wierenga[2], and T. Platt[3]

[1]Field Disease Investigation Unit, College of Veterinary Medicine, Washington State University, Pullman, WA 99164-6610, USA; [2]USDA-ARS Poisonous Plant Research Laboratory, Logan, UT 84341, USA; [3]Department of Animal Science, Cooperative Extension, College of Agriculture and Home Economics, Washington State University, Pullman, WA 99164-6310, USA

Introduction

Crooked calf disease is the colloquial name given to a pattern of congenital anomalies that occurs in the offspring of range cows that graze certain species of lupines in early pregnancy. The anomalies are characterized by the presence of arthrogryposis of the forelimbs typically involving both the elbow and the carpal joints, and by the variable presence of hind limb contractures, torticollis, scoliosis, kyphosis, and cleft palate. It is recorded in several areas in the western continental United States, British Columbia in Canada, and Kodiak Island in Alaska. In Washington state it occurs sporadically in several areas east of the Cascades but it is particularly prevalent and recurrent in parts of the Channeled Scablands in the central part of the state. This is an area of low rainfall and shallow soils with basalt outcrops created by huge, catastrophic flooding associated with the draining of glacial Lake Missoula. The severity of the problem in all at-risk regions varies from year to year, both with respect to the number of affected calves born and the severity of the deformities in affected calves. Calves with severe deformities cannot stand to suckle or follow the dam and are destroyed for humane reasons. Less severely affected calves (bent limbs) can stand and walk but are often not suited to range conditions.

Epidemiological and feeding studies have incriminated the quinolizidine alkaloid anagyrine as a putative teratogen (Keeler 1973, 1976; Panter *et al.* 1997) with the period of fetal susceptibility between 40 and 100 days gestation (Shupe *et al.* 1967; Keeler 1976; Panter *et al.* 1997).

The pattern of the occurrence of lupine-induced crooked calf disease and its severity on at-risk ranches suggests that factors other than simple exposure influence its occurrence. Ranches at risk for crooked calf experience years with a low incidence and years with high incidence without a change in stocking density or an apparent change in lupine presence in pastures. They commonly have a small percentage of calves born each year with minor forelimb deformities but periodically there are years in which a large proportion of calves born are affected with severe deformities. In February of 1997 it was apparent early in the calving period that this was going to be a year of high incidence and severe deformity. This epidemiology study was established to examine risk factors for the occurrence of severe disease.

Materials and Methods

Survey and prevalence data

A meeting of ranchers in the affected area was called in early March of 1997 and attending ranchers were requested to compile weekly records of birth dates, the presence or absence of deformity, and the severity of the deformity (mild or severe) for the calving period. Seventeen ranches in the problem area were subsequently visited to examine calves and to collect current and historical incidence data, demographic data, and data on cattle pasture locations during the 1996 spring and summer grazing periods. Incidence was analyzed with respect to species of lupine present, lupine density, lupine alkaloid concentration and profile, and stocking rates on each ranch. Time limitations on obtaining field measurements during the lupine growing period restricted this analysis to ten ranches with incidence rates of 0, 0, 7, 17, 20, 28, 32, 53, 64, and 100%, respectively.

Lupine density and alkaloid analysis

Following visual inspection over the pasture, transects were established at sites considered to be representative of the distribution of plant species on that pasture. Lupine species identification, measurements of lupine density, and collections for alkaloid analysis were made at each transect site. A 1.0 m^2 quadrat frame was randomly tossed and from this central frame 100 m transects established to the north, east, south and west. A 1.0 m^2 quadrat was placed at 33, 66, and 100 m on each transect line. Lupine plants within, or partly within, the boundaries of each quadrat were counted and speciated. Lupines were also counted on each transect line for a ten-pace distance from each quadrat. Lupine counts are presented as total count of lupine plants at each transect site. A minimum of four transect sites were measured on each ranch except one with two measured sites. Site locations were recorded using geographic positioning technology (GPS model 12XL, GARMIN International, Olathe, KS). Collections of lupines for alkaloid analysis were made from field identified species at individual transects between 9 a.m. and 1 p.m. and dried and stored at room temperature until analysis. Total alkaloid level and percent of individual alkaloids were measured for the aerial parts of various lupines by gas chromatography (GC) using a slightly modified procedure previously reported (Gardner and Panter 1993, 1994; Wink *et al.* 1995).

Stocking rate

Stocking rate was determined from ranch records with the numerator being the total acres grazed by the breeding cow group during the period from spring turn out of cows and calves until August 1, 1996 and the denominator the number of cows grazing this acreage. Cows with no calves were culled at branding on all ranches, and all cows had nursing calves at the time of turn out. The majority of ranches did not keep records of cattle movements on pasture during the grazing season. There was no supplemental feeding on any ranch during the spring/summer grazing period. All ranches offered free or block salt containing added selenium but no records were kept of consumption. None of the ranch pastures were irrigated.

Results

The rancher's recording of a crooked calf was accepted as a case, regardless of the severity of the deformity. Thirteen of the 17 herds visited in the study experienced crooked calves in the 1997 calving season. On the 13 ranches 628 of 2210 calves were born deformed. The birth-week-specific incidence of crooked calf disease on 12 ranches that kept sufficient records for analysis is shown in Figure 1. The date that bulls were introduced for breeding in the individual herds varied (early April to May 1) as did the length of time of exposure to the bulls but there was no observable influence of breeding strategy. On the majority of ranches, calves with deformity were born throughout the calving period, but on two ranches affected calves were born clustered in a 2-week period early in the calving period and on two others they were clustered in the latter part of the calving period. The nature of the deformities observed was similar to that described previously in calves from a ranch in this region (Abbott *et al.* 1986). The severity of the deformities varied both between calves on a ranch and between groups of calves on different ranches. Approximately 44% of calves born with the disorder were classified as severely affected by the ranchers and approximately 45% of dams with crooked calves required assistance at birth.

One ranch owner inspected the mouth and recorded the presence or absence of cleft palate in all calves born. Of the 428 calves born, 187 had skeletal deformity and a normal palate, 24 had skeletal deformity and cleft palate, and in 16 cleft palate was the sole deformity. On this ranch, calves with cleft palate were born throughout the calving period. Maternal age-specific rates of crooked calf were available from the records of this ranch. Cow age ranged from 2 to 14 years. Analysis for maternal age-specific incidence showed that although the risk of producing an affected calf was not independent of the dam's age category, (X^2 = 22.93, 12 df, p = 0.028), the change in risk did not appear to be linearly associated, either positively or negatively, with advancing age (Cochrane-Armitage trend X^2 = 17.941, 11 df, p = 0.083).

Lupine species and alkaloid profiles

Lupinus leucophyllus was the sole lupine species identified on eight (crooked calf incidence range 0% to 100%) of the ten intensive-study ranches. *Lupinus leucophyllus* was the dominant species on the other 2 ranches (crooked calf incidence 0% and 29%) where *Lupinus sericeus* was also present and approached 50% of the lupine population.

The two lupine species had distinct alkaloid profiles (Table 1). Those alkaloids labeled as unknown were not chemically identified for this study. None of the collected *L. sericeus* contained anagyrine or ammodendrine.

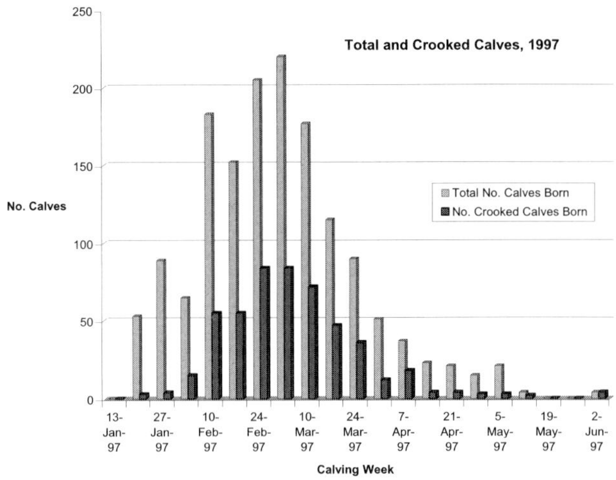

Figure 1. Birth-week-specific incidence of normal and crooked calves. Data from 12 ranches.

Table 1. Alkaloid profiles and concentrations (mg g^{-1} ± SD) for lupine species collected from study ranches.

Alkaloid	*L. leucophyllus* (n=81)		*L. sericeus* (n=14)	
Total alkaloid	12.8 ± 5.3		36.9 ± 11.0	
Anagyrine	1.9 ± 1.5	(n=81)		
Unknown F	1.6 ± 1.0	(n=81)		
5,6 dehydrolupanine	2.5 ± 1.4	(n=81)		
Lupanine	0.6 ± 0.5	(n=79)		
Unknown N	3.2 ± 1.9	(n=69)		
Unknown T	1.3 ± 1.8	(n=81)		
Ammodendrine	0.4	(n=2)		
Wink's #9			11.8 ± 5.0	(n=6)
Unknown U			17.5 ± 8.7	(n=14)
Unknown V			17.1 ± 6.5	(n=8)

Incidence, lupine density and stocking rate

Lupine density on pastures varied between ranches and was lower on the two ranches with no occurrence of crooked calf disease than the majority of those that had crooked calves. However, with the exception of the ranch on which all calves born were deformed, there was only minor difference in lupine density between the ranches that experienced crooked calf disease. There was a minor negative correlation (r^2 = -0.1909) between stocking rate and ranch incidence of crooked calf disease (data not shown).

Lupine alkaloids and risk

There was no apparent relation between total alkaloid concentration or anagyrine concentration (Figure 2) in lupines on the individual ranches and ranch occurrence of crooked calf. Mean anagyrine concentrations in lupines on the two ranches with no incidence of disease were equivalent to, or higher than, those on four of the ranches that had crooked calves, including the ranch with 100% incidence. Similarly there was no association of 5,6-dehyrolupanine or unknown T concentration with occurrence of crooked calf (data not shown) but there was a trend to a lower incidence with higher concentrations of lupanine (Figure 3).

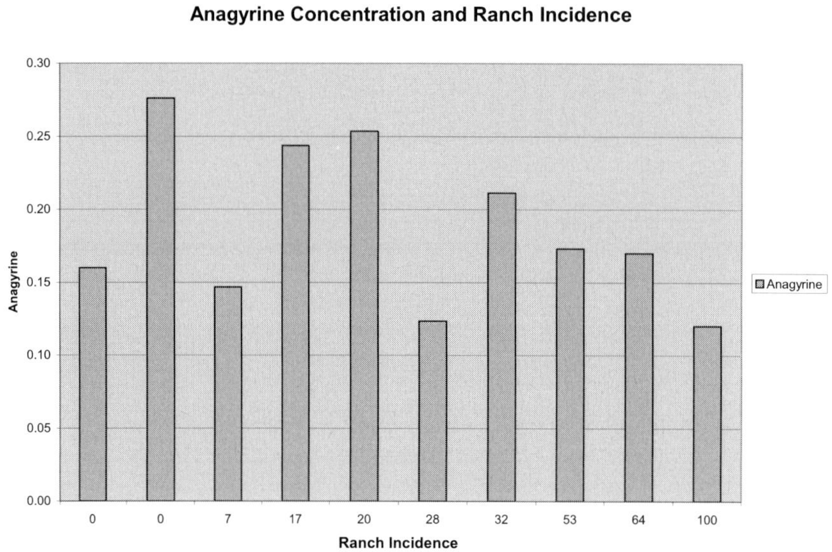

Figure 2. Anagyrine concentration in *Lupinus leucophyllus* and ranch incidence of crooked calf.

Discussion

Lupinus leucophyllus was the only lupine species present on all ranches that experienced crooked calf disease and it contained the alkaloid anagyrine, which has been previously identified as a teratogen for crooked calf. However, differences in the concentration of anagyrine in lupines on individual ranches were not related to differences in the ranch incidence of crooked calf disease. Ranches that had no occurrence of the disease had lupines with similar alkaloid concentrations and profiles to ranches with a high incidence of disease. The anagyrine concentration and the total alkaloid content in *L. leucophyllus* in this study are above the level considered necessary to produce crooked calf disease (Keeler 1976; Davis 1982; Davis and Stout 1986) and are similar to those found by Keeler (1973) in *L. sericeus* and *Lupinus caudatus* collections from ranches with significant crooked calf incidence at the Crawford and Antelope sites in Wyoming.

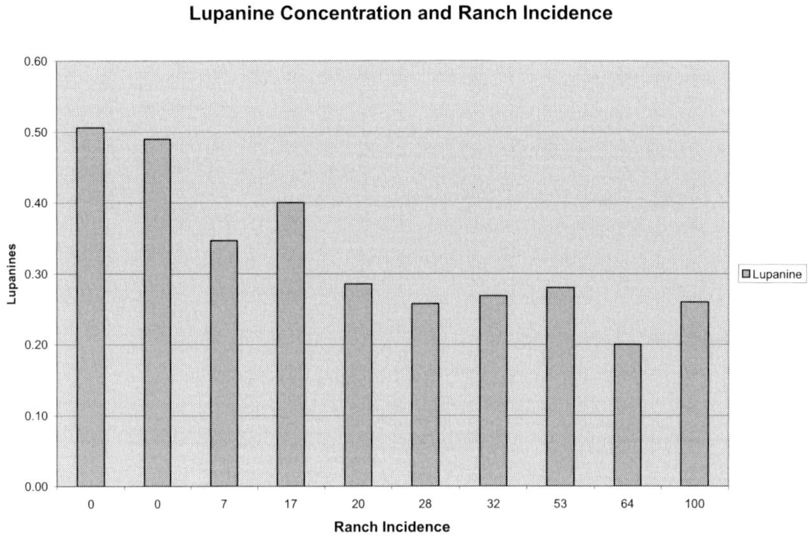

Figure 3. Lupanine concentration in *Lupinus leucophyllus* and ranch incidence of crooked calf.

In view of this, the lack of a relationship between ranch lupine anagyrine concentration and ranch incidence of disease was surprising. It indicates that the presence of lupines with a high concentration of anagyrine is not the sole determinant of risk for crooked calf disease and suggests that the lupine-eating behavior of pregnant cattle is an important contributing determinant. Factors determining grazing behavior and plant selection by cattle are complex and include the relative palatability of individual plant species on the pasture, the stage of plant growth, the relative proportions of individual plant species on the pasture and their community associations, stocking rates, grazing pressure, topographic features of the pasture, and the learned eating behavior of the cattle. We examined some of these.

Western range lupine species are not highly palatable to cattle (Keeler 1973, 1976) and it is reasonable to expect that lupine density on pastures could influence lupine consumption of cattle by reducing the relative proportion of more palatable and non-teratogenic plant species. We found between-ranch differences in lupine density but found no relationship between lupine density and ranch incidence of disease. Whereas extremes of lupine density, logically, must be a determinant of risk, the range of densities found in this study did not have an effect.

Bitterness of lupine could also influence lupine palatability but there was no association between the incidence of disease and ranch-specific total alkaloid, or individual alkaloid, concentrations in *L. leucophyllus*, although there was a trend to a lower ranch incidence of crooked calf with higher ranch lupine lupanine and 5,6-dehyrolupanine concentrations. Similarly, high stocking rates could force the consumption of less palatable plants, such as lupine, once palatable plants have been consumed, but we found no relationship between stocking rate and ranch incidence of disease. A caveat is that our measure was simply the number of acres available for grazing per cow during the spring and summer grazing period. However, we observed that the proportion of basalt outcrops, cliffs, and bodies of surface water varied between pastures. The topography of the area is not digitized so the actual area available for grazing could not be considered in this analysis. Of the two ranches with the highest incidence, one rancher did not move the cattle through this period and the other had the highest stated number of pasture movements.

Descriptions of some natural outbreaks of crooked calf disease report a high incidence in cattle calving for the first time (Shupe *et al.* 1967; Abbott *et al.* 1986), which could suggest a learned eating behavior, with consumption early in life leading to aversion, a consequent lower consumption with increasing age and a resultant lower risk for crooked calves in older cattle. However, first-calf heifers are commonly run as a separate group from older cows and these reported differences may simply reflect a different pasture risk. On one ranch where mixed parities were run together and where maternal age-specific rates of crooked calf could be determined, there was no age-specific trend in birth incidence of crooked calves.

Alkaloid concentrations are high in lupine seeds and, for this reason, it has been postulated that grazing lupines in the seed pod stage of growth presents the major risk for the occurrence of crooked calf disease (Keeler *et al.* 1977). There is, however, little information on lupine grazing behavior of cattle and of the risk for crooked calf disease incurred from grazing lupines at any stage of growth. Some inferences can be drawn from this study. The time of introduction of bulls and the varying bull-exposure time in the herds resulted in a calving period that extended from mid-January to June with the majority of calves born in February through April (Figure 2). These calves, as fetuses, entered the highly susceptible period for lupine-induced teratogenesis (40 to 70 days fetal age, gestation period 284 days) from approximately May 18, 1996 for the early-born calves to September 3, 1996 for calves born May 5, 1997. On the Scabland ranges, lupines emerge in early March; some start to flower by mid-May and the majority flower and set seed by late July. There was little difference in the incidence of crooked calves born in February and March, which indicates that cows graze lupines prior to the seed pod stage and that this is associated with risk for the development of crooked calves. This inference is supported by the occurrence of cleft palate throughout the calving period in one herd. Cleft palate produced by teratogenic alkaloids results from suppression of oral movement and mechanical interference imposed by the tongue between the palatine shelves during the critical closure period in early gestation (Panter *et al.* 1990; Panter and Keeler 1993). It only occurs in fetuses exposed during a narrow time period in early gestation (Panter *et al.* 1994, 1998), which in cattle includes gestation days 40 to 50. The occurrence of calves with cleft palate early in the calving period in this herd indicates that these cattle grazed lupine during the early growth and flower stages.

A further inference concerning lupine grazing behavior can be drawn from the pattern of abnormalities recorded in this herd. Fetal skeletal malformations result from exposure to teratogenic alkaloids during a later gestational age than that associated with the production of cleft palate (Panter *et al.* 1990, 1994, 1998). The variation in deformities in this herd, with the occurrence of cleft palate as a single defect in some calves, in combination with skeletal malformation in others and the occurrence of skeletal deformity without the presence of cleft palate in yet others, indicates that lupine consumption by individual cows grazing the same pasture is not constant during its period of availability or that its teratogenicity varies.

It is apparent that risk determinants for crooked calf disease are more complex than expected determinants of lupine consumption, such as stocking rates and lupine density on pastures. Further studies of risk will need to incorporate a design that measures lupine eating by individual cattle and the determinants of this lupine eating behavior.

References

Abbott LC, Finnell RH, Chernoff GF, Parish SF, and Gay CC (1986). Crooked calf disease: a histological and histochemical examination of eight affected calves. *Veterinary Pathology* 23:734–740.

Davis AM (1982). The occurrence of anagyrine in a collection of western American lupines. *Journal of Range Management* 35:81–84.

Davis AM and Stout DM (1986). Anagyrine in Western American lupines. *Journal of Range Management* 39:29–30.

Gardner DR and Panter KE (1993). Comparison of blood plasma alkaloid levels in cattle, sheep, and goats fed *Lupinus caudatus*. *Journal of Natural Toxins* 2:1–11.

Gardner DR and Panter KE (1994). Ammodendrine and related piperidine alkaloid levels in the blood plasma of cattle, sheep and goats fed *Lupinus formosus*. *Journal of Natural Toxins* 3:107–116.

Keeler RF (1973). Lupine alkaloids from teratogenic and nonteratogenic lupines. I. Correlation of crooked calf disease incidence with alkaloid distribution determined by gas chromatography. *Teratology* 7:23–30.

Keeler RF (1976). Lupine alkaloids from teratogenic and nonteratogenic lupines. III. Identification of anagyrine as the probable teratogen by feeding trials. *Journal of Toxicology Environmental Health* 1:887–898.

Keeler RF, James LF, Shupe JL, and Van Kampen KR (1977). Lupine-induced crooked calf disease and a management method to reduce incidence. *Journal of Range Management* 30:97–102.

Panter KE and Keeler RF (1993). Quinolizidine and piperidine alkaloid teratogen from poisonous plants and their mechanism of action in animals. *Veterinary Clinics of North America: Food Animal Practice* 9(1):33–40.

Panter KE, Bunch TD, and Keeler RF (1990). Multiple congenital contractures (MCC) and cleft palate induced in goats by ingestion of piperidine alkaloid-containing plants. Reduction in fetal movement as the probable cause. *Clinical Toxicology* 28:69–84.

Panter KE, Garner DR, and Molyneux RJ (1994). Comparison of toxic and teratogenic effects of *Lupinus formosus, L. arbustus* and *L. caudatus* in goats. *Journal of Natural Toxins* 3:83–93.

Panter KE, Gardner DR, Gay CC, James LF, Mills R, Gay JM, and Baldwin TJ (1997). Observations of *Lupinus sulphureus*-induced crooked calf disease. *Journal of Range Management* 50:587–592.

Panter KE, Gardner DR, and Molyneux RJ (1998). Teratogenic and fetotoxic effects of two piperidine alkaloid-containing lupines (*L. formosus* and *L. arbustus*) in cows. *Journal of Natural Toxins* 7(2):131–140.

Shupe JL, Binns W, James LF, and Keeler RF (1967). Lupine, a cause of crooked calf disease. *Journal of the American Veterinary Medical Association* 151:198–203.

Wink ML, Meibner C, and Witte L (1995). Patterns of quinolizidine alkaloids in 56 species of the genus *Lupinus*. *Phytochemistry* 38:139–153.

Chapter 28

Pregnancy Loss in Mares Associated with Exposure to Caterpillars in Kentucky and Australia

N.R.Perkins[1], M.M. Sebastian[2], K.H. Todhunter[3], R.M. Wylie[4], A.P. Begg[5], G. Muscatello[6], J.R. Gilkerson[6], D.J. Racklyeft[7], C. Chicken[8], M.C. Wilson[9], A.J. Cawdell-Smith[10], and W.L. Bryden[10]

[1]AusVet Animal Health Services, South Brisbane, Qld, Australia; [2]College of Veterinary Medicine, University of Georgia, Athens GA, USA; [3] Tails and Scales Veterinary Services, The Junction, NSW, Australia; [4]Gundy Veterinary Services, Scone, NSW, Australia; [5]MayneHealth Vetnostics, Kotara, NSW, Australia; [6]Faculty of Veterinary Science, University of Melbourne, Vic, Australia; [7]Satur Veterinary Clinic, Scone, NSW, Australia; [8]Scone Veterinary Hospital, Scone, NSW, Australia; [9]EquiVet Australia, Southbrook, Qld, Australia; [10]School of Animal Studies, University of Queensland, Gatton, Qld, Australia

Introduction

Two syndromes of pregnancy loss in mares that occurred in midwestern USA (2001–2002) and Australia (2004–2005) are described in this chapter. The syndromes were associated with similar clinical and pathological characteristics which appeared to have a toxicological basis, and where caterpillars have been implicated as a causal factor.

Mare Reproductive Loss Syndrome (MRLS)

Mare reproductive loss syndrome (MRLS) was observed in Ohio and central Kentucky during 2001 and 2002 (Dywer 2003; Riddle 2003). The clinical syndromes associated with MRLS were abortion, fetal death, opthalmitis and pericarditis.

Late-term abortion and delivery of poorly viable foals

Abortions occurred in late gestation (Brown 2003; Dywer 2003), and neonatal care units in Kentucky reported increased case load during MRLS outbreaks (Byars and Seahorn 2003). Fetuses were of normal size and weight and generally fresh. Pathologic findings included funisitis (inflammation of the umbilical cord) of the amniotic portion of the umbilical cord, amnionitis, fetal pneumonia, bacterial infection and hemorrhages. The most common bacterial isolations from fetal tissue, stomach contents or placental samples included non-beta hemolytic *Streptococci* and *Actinobacilli* (Donahue *et al*. 2003; Williams *et al*. 2004).

Early fetal loss

Most early fetal loss occurred between days 40 and 80 of gestation (range 32 to 140 days). Common findings were observation of a dead fetus on ultrasound examination with echogenic allantoic and amniotic fluid. Mares less than 30 days pregnant in 2001 were apparently not affected since there was no increase in losses between days 15 and 30 of pregnancy (Riddle 2003).

Opthalmitis

The typical appearance was an acute onset, unilateral, exudative endopthalmitis (inflammation of the internal structures of the tissues in the eyeball), with consistent changes including corneal edema and exudates in the anterior and posterior chambers (Latimer 2003).

Pericarditis

During the course of the 2001 and 2002 MRLS outbreaks, larger numbers of fibrinous pericarditis were observed in central Kentucky than expected (Seahorn *et al*. 2003). Clinical signs, progression, outcome and bacteria involved were all consistent with previous reports of equine pericarditis. Cases were temporally and spatially clustered and appeared to be associated with the MRLS outbreak (Seahorn *et al*. 2003).

Causal Hypotheses for MRLS

Known causes of abortion in horses were ruled out through field and laboratory investigation. Monitoring of pasture and soil factors and climatic indices on both MRLS-affected and unaffected Kentucky farms during 2002 did not identify any association between measured parameters (plant cyanogens, phytoestrogens, ergot alkaloids and other mycotoxins) and MRLS (see Powell *et al.* 2003). Retrospective epidemiological studies identified factors that warranted further study, in particular exposure to eastern tent caterpillars (ETC) (Cohen *et al.* 2003a,b; Dywer *et al.* 2003). Experimental studies involving exposure of mares to ETC have demonstrated increased abortion compared to control animals. Abortions have been induced by feeding ETC on pasture, mixed with sweet feed and by direct administration of ETC homogenate in water (Sebastian *et al.* 2003a,b, 2004; Bernard *et al.* 2004; Webb *et al.* 2004). Feeding exoskeleton caused abortions whereas feeding gut and frass did not (Bernard *et al.* 2004). When intact ETC were autoclaved and then fed, mares did not abort, indicating that the abortive agent could be removed by autoclaving (Webb *et al.* 2004). Administration of irradiated ETC also caused abortion, supporting a hypothesis that MRLS may be associated with an unidentified toxin carried and/or deposited by the ETC and that is either toxic to the placenta or the fetus (Sebastian *et al.* 2004).

Equine Amnionitis and Fetal Loss (EAFL)

During the winter of 2004 a number of abortions were reported from thoroughbred farms in Australia that involved an unusual and consistent clinical and pathological appearance (Perkins 2005). Abortions were detected from mid-pregnancy to term. Inflammatory changes appeared to initiate within the amnion and amniotic portion of the umbilical cord and extend to involve the allantois (particular the extra-embryonic coelomic space or allantoic cavity). In chronic cases the inflammatory changes extended through the allantois to involve portions of the chorion around the site of the attachment of the umbilical cord. Fetuses were typically fresh and showed variable non-specific inflammatory changes. A range of unusual and normally non-pathogenic bacteria were recovered on culture of samples from uncontaminated fetal stomach or lung or amnion/umbilical cord including *Cellulomonas* spp., *Microbacterium aggregans*, *Arthrobacter* spp., *Curtobacterium* spp., non-beta hemolytic slow-growing *Streptococci*, other coryneform bacteria, and some Gram-negative non-fermenting rods.

A positive diagnosis of EAFL required the following conditions to be met (Perkins 2005):

- examination of fetus, chorioallantois and amnion incorporating gross examination, histopathology and microbiology (culture or virus isolation);
- absence of any evidence of other known causes of equine abortion on gross pathology, histology or culture/virus isolation;
- typical gross and/or histopathological changes in amnion, amniotic portion of umbilical cord and allantois;
- absence of gross and histological evidence of significant chorionitis particularly cervical pole and body chorionitis; in chronic cases there may be focal severe chorionitis usually in the body of the chorion, but this should be associated with more extensive, severe chronic changes in the amnion and amniotic umbilical cord and allantois; and
- growth of environmental bacteria from culture of fetal stomach contents/lung or from uncontaminated amnion. This requirement was not essential to a diagnosis of EAFL because of difficulties in obtaining positive culture results from some abortion cases.

Possible Causal Factors for EAFL

There is not yet any firm evidence linking possible causal factors with EAFL risk. Investigations of environment, soil and pasture have not identified any abnormalities. Similarities between the clinical and pathological features of EAFL and MRLS led to consideration of caterpillars as a possible risk factor in Australia. Although ETC do not occur in Australia, a number of other caterpillar species occur in EAFL-affected regions including processionary caterpillars (*Ochragaster lunifer*), white cedar moth caterpillars (*Leptocneria reducta*), and the mistletoe brown tail moth caterpillar (*Euproctis edwardsi*).

Australian native pennyroyal (*Mentha satureioides*) was also reported in large amounts on one affected property and this plant has been described by Everist (1981) as a possible cause of abortion in domestic livestock.

Conclusion

Both EAFL and MRLS show sufficient similarities to warrant consideration of commonalities in pathogenesis. The interrelationships between potential causal factors are not clearly understood nor is it clear whether causal factors may be the same in the two regions. The two conditions offer an opportunity for international, multidisciplinary, collaborative studies to investigate aspects of the pathogenesis of both MRLS and EAFL and achieve an outcome that will benefit both equine breeding industries.

References

Bernard WV, LeBlanc MM, Webb BA, and Stromberg AJ (2004). Evaluation of early fetal loss induced by gavage with eastern tent caterpillars in pregnant mares. *Journal of the American Veterinary Medical Association* 225:717–721.

Brown S (2003). Field and clinical observations related to late fetal loss in mares affected with mare reproductive loss syndrome. In *Proceedings of a Workshop on the Equine Placenta* (DG Powell, D Furry, and G Hale, eds), pp. 14–15. Kentucky Agricultural Experiment Station, Lexington, KY.

Byars TD and Seahorn TL (2003). Clinical observations of mare reproductive loss syndrome in critical care mares and foals. In *Proceedings of a Workshop on the Equine Placenta* (DG Powell, D Furry, and G Hale, eds), pp. 15–16. Kentucky Agricultural Experiment Station, Lexington, KY.

Cohen ND, Carey VJ, Donahue JG, Seahorn JL, Donahue JK, Williams DM, and Harrison LR (2003a). Case-control study of late term abortions associated with mare reproductive loss syndrome in central Kentucky. *Journal of the American Veterinary Medical Association* 222:199–209.

Cohen ND, Donahue JG, Carey VJ, Seahorn JL, Piercy D, Donahue JK, Williams DM, Brown SE, and Riddle TW (2003b). Case-control study of early term abortions (early fetal loss) associated with mare reproductive loss syndrome in central Kentucky. *Journal of the American Veterinary Medical Association* 222:210–217.

Donahue JM, Sells S, Giles R, Harrison L, Hong C, Poonacha K, Roberts J, Sebastian M, Smith R, Swerczek T, Tramontin R, Vickers M, and Williams N (2003). Bacteria associated with mare reproductive loss syndrome: late fetal losses. In *Proceedings of a Workshop on the Equine Placenta* (DG Powell, D Furry and G Hale, eds), pp. 27–29. Kentucky Agricultural Experiment Station, Lexington, KY.

Dwyer RM (2003). Fitting the pieces of the puzzle: mare reproductive loss syndrome update. *Equine Veterinary Education* 15:45–47.

Dwyer RM, Garber LP, Traub-Dagartz JL, Meade BJ, Powell D, Pavlick MP, and Kane AJ (2003). Case-control study of factors associated with excessive proportions of early fetal loss associated with mare reproductive loss syndrome in central Kentucky during 2001. *Journal of the American Veterinary Medical Association* 222:613–619.

Everist SL (1981). *Poisonous Plants of Australia*, 2nd Edition. Angus & Robertson, Sydney.

Latimer C (2003). Endopthalmitis: A syndrome associated with mare reproductive loss syndrome? In *Proceedings of a Workshop on the Equine Placenta* (DG Powell, D Furry and G Hale, eds), pp. 17–18. Kentucky Agricultural Experiment Station, Lexington, KY.

Perkins NR (2005). Equine amnionitis and foetal loss (EAFL). pp. 19–48. Hunter Valley Equine Research Centre, Scone, NSW.

Powell DG, Furry D and Hale G (eds) (2003). *Proceedings of a Workshop on the Equine Placenta*. Kentucky Agricultural Experiment Station, Lexington, KY.

Riddle WT (2003). Clinical observations associated with early fetal loss in mare reproductive loss syndrome during the 2001 and 2002 breeding seasons. In *Proceedings of a Workshop on the Equine Placenta* (DG Powell, D Furry, and G Hale, eds), pp. 12–14. Kentucky Agricultural Experiment Station, Lexington, KY.

Seahorn JL, Slovis NM, Reimer JM, Carey HJ, Donahue JG, Cohen ND (2003). Case-control study of factors associated with fibrinous pericarditis among horses in central Kentucky during spring 2001. *Journal of the American Veterinary Medical Association* 223:832–838.

Sebastian M, Gantz M, Tobin T, Harkins JD, Bosken J, Hughes C, Harrison L, Bernard B, RichterD, and Fitzgerald TD (2003a). The mare reproductive loss syndrome and the eastern tent caterpillar: a toxicokinetic/statistical analysis with clinical, epidemiological and mechanistic implications. *Veterinary Therapeutics* 4:324–339.

Sebastian M, Williams D, Harrison L, Donahue J, Seahorn T, Slovis N, Richter D, Fuller T, Trail C, Douglas R, and Tobin T (2003b). Experimentally induced mare reproductive loss syndrome late fetal losses with eastern tent caterpillar. In *Proceedings of a Workshop on the Equine Placenta* (DG Powell, D Furry, and G Hale, eds), pp. 80–81. Kentucky Agricultural Experiment Station, Lexington, KY.

Sebastian M, Bernard W, Harrison L, Donahue J, Brown S, Tobin T, Newman K, Lynn B, and Fitzgerald T (2004). Experimental induction of mare reproductive loss syndrome with irradiated eastern tent caterpillars to assess whether the primary pathogen is a toxic molecule or a microorganism. In *Proceedings of a Workshop on the Equine Placenta* (DG Powell, D Furry, and G Hale, eds), pp. 103–104. Kentucky Agricultural Experiment Station, Lexington, KY.

Webb BA, Barney WE, Dahlman DL, DeBorde SN, Weer C, Williams NM, Donahue JM, and McDowell KJ (2004). Eastern tent caterpillars (*Malacosoma americanum*) cause mare reproductive loss syndrome. *Journal of Insect Physiology* 50:185–193.

Williams NM, Donahue JM, Bolin DC, Giles RC, Harrison L, Hong CB, Jackson CB, Poonacha KB, and Vickers ML (2004). Equine placental pathology: Kentucky perspective. In *Proceedings of a Workshop on the Equine Placenta* (DG Powell, D Furry, and G Hale, eds), pp. 88–91. Kentucky Agricultural Experiment Station, Lexington, KY.

TREATMENT

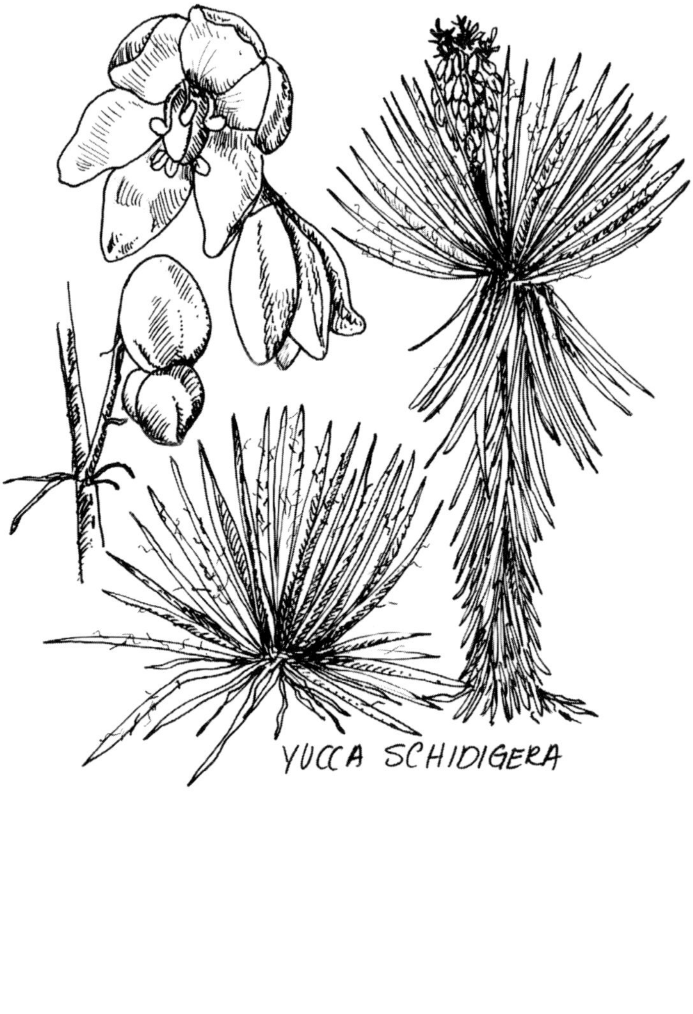

Chapter 29

Clinical Signs and Treatment of Cattle Grazing High Molybdenum Forage

W. Majak and D.Steinke

Kamloops Range Research Unit, Agriculture and Agri-Food Canada, Kamloops, British Columbia V2B 8A9

Introduction

The Highland Valley Copper mine (HVC), near Kamloops, British Columbia, Canada is in the process of restoring rangeland disturbed by over 20 years of mining. Of primary concern are the tailings ponds that can contain elevated levels of molybdenum (Mo), a residue from the copper (Cu) and Mo extraction process. With a milling recovery of 50% Mo, the mineral is a prominent residue in some tailings areas. Environmental reclamation of tailings areas consists primarily of establishing agronomic vegetation with the aid of fertilizers. The proposed end land use plan for these areas includes livestock grazing, thus restoring the terrain to sustainable agricultural use. Studies have shown that residual Mo in the tailings is imbibed by vegetation and can accumulate to extremely high levels depending on plant species and site (Gardner *et al.* 2003; Majak *et al.* 2004). Animal health guidelines recommend a maximum level of 5 ppm Mo in feed for beef cattle (NRC 1996). Accordingly, grazing studies were initiated at HVC to determine the feasibility of utilizing these reclaimed sites for cattle production.

Elevated levels of Mo in forage can induce molybdenosis, a form of Mo toxicity in ruminants. The clinical signs of molybdenosis can include lameness with a characteristic stiff gait, severe diarrhea, weight loss, depigmentation of hair coat, reduced conception rates and death (Swan *et al.* 1998). Younger cows, less than 3 years old, appear to be most susceptible to molybdenum toxicosis (Radostits *et al.* 2000). Molybdenosis is also known as secondary Cu deficiency, as ingestion of excess Mo can result in the reduction of biologically available Cu.

Clinical Evaluations at Highmont (1999–2001)

In 1999, grazing studies with cow/calf pairs were initiated at the Highmont tailings area where Mo levels in forages were among the highest in the area. The major forages, alfalfa (*Medicago sativa*) and orchardgrass (*Dactylis glomerata*), contained 95–460 ppm Mo (Majak *et al.* 2004). In addition, drinking water contained on average 7 ppm Mo, which is a hundredfold greater than the recommended limit of 0.06 ppm Mo for livestock (Puls 1994). The dietary contribution of Mo from water was estimated as 20% of total Mo intake (Majak *et al.* 2004). The trial was conducted to determine if grazing animals would be affected by the high levels of Mo in forage and water and whether cattle grazing was a viable option for the reclamation plan at Highmont.

The average stocking rate was maintained at 0.63 ha AUM^{-1}. The herd consisted of 3- to 10-year-old Hereford and Red Baldy cows with nursing Charolais-sired calves. Twenty cow/calf pairs were used in 1999 and thirty in 2000 and 2001. Cattle grazed the Highmont site for 11 weeks during July to October (1999–2001). In 1999, clinical abnormalities were observed in four cows after 4–5 weeks of exposure and the signs persisted for 10–20 days. Cattle with abnormalities displayed a stiff, shuffling gait and favored their tiptoes when traveling. This lameness was the major clinical sign. All other functions including temperature, pulse, respiration and body condition appeared to be normal (Majak *et al.* 2004). The lameness did not persist and each animal had fully recovered by the end of the trial without treatment. In 2000, pasture expansion allowed for a larger herd size. Lameness was observed within a week of arrival to Highmont with two-thirds of the cows eventually showing clinical signs. All affected cattle had again fully recovered by the end of the trial without treatment.

In 2001, the focus of the project was shifted from strictly clinical observation to therapy that might alleviate the disorders. A study was initiated to test the efficacy of a Cu bolus that might counteract Mo toxicity. Of the 30 cows on test, 15 received two Cu boluses (ALLTRACE, Agrimin Ltd., UK) and the other 15 served as controls (Majak *et al.* 2004). Clinical disorders were again evident, with four bolus and five control cows exhibiting lameness 3 weeks after arrival to the site. All animals had fully recovered by the end of the trial. The boluses increased hepatic Cu but there was no evidence that the bolus was therapeutic. The Cu dosage provided by the bolus may not have been sufficient.

During the 3–year clinical evaluation study, the proportion of cows that displayed signs of lameness was 20%, 70% and 30% in 1999, 2000 and 2001, respectively. The proportions of affected cattle were similar in 1999 and 2001 but were significantly higher in 2000. The onset and frequency of clinical signs was linked to prevailing moisture conditions, which may have enhanced plant growth and the forage source of Mo in 2000 (Majak *et al.* 2004). Average daily gains were adequate during 1999 to 2001 as were conception rates. Hepatic Cu levels for the cows were in the marginal to deficient range prior to grazing each year, suggesting that the herd may have been predisposed to the effects of elevated Mo levels in forage. Except for cows with Cu boluses, liver biopsies showed marginal to deficient Cu; all cows showed toxic levels of hepatic Mo as a result of grazing at Highmont each year. Clinically affected animals did not necessarily show highest hepatic Mo or lowest hepatic Cu, indicating that liver Mo and Cu status did not directly correlate to clinical signs.

The 3–year clinical evaluation study at Highmont demonstrated that cows grazing the high Mo forage were susceptible to some degree of molybdenosis and that some form of supplementation other than the Cu boluses would be required if cattle grazing was to be included in the land use plan for the Highmont site. A supplement such as copper sulfate in salt has been recommended for forage rations or pastures with excessive amounts of Mo (Merck 1967; Radostits *et al.* 2000).

Supplement Trials at Highmont (2002–2004)

A second 3–year study was initiated to test the efficacy of a Cu supplement to prevent the onset of clinical signs. The supplement trials at Highmont were conducted from August to October each year. The period of the trial was reduced to permit increased stocking rates. The average stocking rate was 0.89 ha AUM^{-1} and 53, 46, and 51 cow/calf pairs were used in 2002, 2003, and 2004, respectively. The herd was randomly divided into two groups and turned out to separate pastures. Copper sulfate ($CuSO_4 \cdot 5H_2O$), 2.5% in loose salt, was provided free choice to one group, while the other group, which served as a control, was given only loose salt. In addition to lameness, secondary clinical signs including diarrhea and hair coat condition were monitored during the supplement trials (Table 1).

Clinical signs were prominent in 2002 and lameness was observed in one-third of the control group within a week of arrival to Highmont (Table 1). There was a marked difference in the general condition of the control group compared to the supplement group. Only one supplement cow exhibited mild clinical signs of lameness, while acute signs were clearly evident in the control group. All affected cattle had fully recovered by the end of the trial without treatment, except for fading on red portions of the hair coat, which was resolved by the following spring.

Only five animals showed signs of lameness in 2003 and the incidence of diarrhea and hair coat problems was also substantially reduced (Table 1). The inclement weather may have been a factor. Precipitation was reduced significantly and temperatures were extremely elevated. The site received half the normal rainfall for the growing season at Highmont (Highland Valley Copper 2004). Less moisture yielded less regrowth, less available forage for grazing cattle and possibly the intake of Mo was also reduced. The lower incidence of lameness in 2003 could not be attributed to a decrease in levels of forage Mo since forage content of Mo was not significantly different between 2002 and 2003 (Majak *et al.* 2006). In 2004, eight control cows and one supplement cow suffered from varying degrees of lameness (Table 1) and the control group accounted for 89% of the documented signs of lameness, 83% of the cases of diarrhea and 88% of hair coat problems in the last year of the trial.

Table 1. Clinical signs observed in cattle during grazing trials at Highmont.

Year	Treatment	n	Lameness	Diarrhea	Hair coat
2002	Control	27	10	18	18
	Supplement	26	1	10	11
2003	Control	23	4	2	1
	Supplement	23	1	2	0
2004	Control	26	8	10	7
	Supplement	25	1	2	1
Totals	Control	76	22	30	26
	Supplement	74	3	14	12

Both lameness and diarrhea were reduced in animals that had previous exposure to the pasture during 2002–2004 (Majak *et al.* 2006). These results suggested that animals could develop a tolerance to grazing pastures with high levels of Mo. As a result, the benefits of supplementation were mainly in reducing the incidence of lameness in animals with no previous exposure. Supplementation had little effect on animals with previous exposure. Average daily gains were adequate during 2002–2004 (Majak *et al.* 2006). Liver biopsies showed, on average, marginal to deficient Cu levels in control cows but adequate levels of Cu in supplement cows. Hepatic levels of Mo were in the toxic range in both groups.

In summary, the observational trials at Highmont (1999–2001) confirmed that grazing cows could suffer ill effects at the site. Lameness, the primary sign, was resolved in all animals by the end of each trial without treatment. Results from the grazing trials during 2002–2004 clearly demonstrated that with the aid of a 2.5% $CuSO_4{\bullet}5H_2O$ supplement in loose salt, cattle could graze the site without major problems. Clinical signs were not completely eliminated in the supplement group, probably because the supplement was free-choice. The overall results showed that 29% of control cows exhibited primary signs of lameness whereas only 4% of supplement cows were affected (Table 1), usually with attenuated signs. Secondary signs of diarrhea and hair coat depigmentation were also reduced in the supplement group. Calves were much less susceptible as only seven of the 230 calves showed clinical signs of lameness during the 6-year period. Cattle grazing might be included in the land use plan for Highmont if a $CuSO_4{\bullet}5H_2O$ supplement is provided or if previously exposed animals are used that have developed a tolerance to Mo.

Forages at Highmont greatly exceed NRC guidelines for acceptable levels of Mo. At first, the inorganic molybdate anion (MoO_4^{2-}) was characterized as the configuration of Mo in alfalfa (Surridge *et al.* 2001) and the anion has been detected in drinking water at Highmont (Majak *et al.* 2004). More recent spectroscopic studies indicate that the Mo in forage is bound to a ligand. This bound form of Mo may well explain the lack of widespread acute toxicity in grazing cattle at Highmont.

Acknowledgements

This project was partially funded by Highland Valley Copper. We thank the Rey Creek Ranch for providing livestock.

References

Gardner WC, Broersma K, Popp JD, Mir Z, Mir PS, and Buckley WT (2003). Copper and health status of cattle grazing high molybdenum forage from a reclaimed mine tailings site. *Canadian Journal of Animal Science* 83:479–485.

Highland Valley Copper (2004). *Summary of Growing Season Weather Conditions 1967 – 2004.* HVC, Logan Lake, British Columbia.

Majak W, Steinke D, McGillivray J, and Lysyk T (2004). Clinical signs in cattle grazing high molybdenum forage. *Journal of Range Management* 57:269–274.

Majak W, Steinke D, Lysyk T, Ogilvie K, and McGillivray J (2006). Efficacy of copper supplementation in the prevention of molybdenosis in cattle. *Rangeland Ecology and Management* 59:285–292.

Merck. (1967). *The Merck Veterinary Manual,* p. 1012. Merck and Co., Inc., Rahway, New Jersey.

National Research Council (NRC) (1996). *Nutrient Requirements of Beef Cattle*, 7th revised ed., pp. 62–68. National Academy Press, Washington, D.C.

Puls R (1994). *Mineral Levels in Animal Health: Diagnosis Data*, 2nd ed. Sherpa International, Clearbrook, Canada.

Radostits OM, Gay CC, Blood DC, and Hinchcliff KW (2000). *Veterinary Medicine: A Textbook of the Diseases of Cattle, Sheep, Pigs, Goats and Horses*, 9th ed., pp. 1598–1599. W.B. Saunders Company Ltd., San Francisco, California.

Surridge B, Sorensen TS, and Majak W (2001). Purification and characterization of molybdenum species from *Medicago sativa* (alfalfa) grown on mine tailings. *Journal of Agricultural and Food Chemistry* 49:3853–3858.

Swan DA, Creeper JH, White CL, Ridings M, Smith GM, and Costa ND (1998). Molybdenum poisoning in feedlot cattle. *Australian Veterinary Journal* 76:345–349.

Chapter 30

New Applications of Yucca and Quillaja Products in Animal Nutrition

P.R. Cheeke

Professor Emeritus, Department of Animal Sciences, Oregon State University, Corvallis 97331, and Scientific Advisor, Desert King International, 7024 Manya Circle, San Diego, CA 92154, USA

Yucca and quillaja products are produced from *Yucca schidigera* and *Quillaja saponaria*. *Yucca schidigera* is an herbaceous plant of the lily family, native to the Mexican state of Baja California. *Quillaja saponaria* is a hardwood tree native to Chile. Both plants inhabit arid areas, and while much different taxonomically and in appearance, share some similar characteristics. For example, these two plants are the major commercial sources of saponins.

Phytochemicals in Yucca and Quillaja

Traditionally, yucca and quillaja have been used as sources of saponins. Saponins are natural detergents, or surfactants, because they contain both water-soluble and fat-soluble entities. They consist of a fat-soluble nucleus, having either a steroid or a triterpenoid structure, with one or more side chains of water-soluble carbohydrates. Saponins are excellent foaming agents; yucca and quillaja extracts are used in certain beverages (e.g. Slurpees and other crushed ice drinks, and root beer), where a stable foam is desired.

In animal nutrition, saponins have a number of applications and implications. These include anti-protozoal and anti-nematode activity, modification of intestinal mucosa, reduction of egg cholesterol content, modification of rumen fermentation, reduction of stillbirths in swine, reduced carcass fat, and possibly other effects. In addition, quillaja saponins have a long history of use as adjuvants in veterinary vaccines.

Recently, it has been recognized that other phytochemicals besides saponins are important constituents of yucca and quillaja. Of major interest are certain polyphenolic compounds. Oleszek *et al.* (2001a) identified a number of unique phenolic compounds in yucca powder. These included two stilbenes (resveratrol and methoxystilbene) and three novel compounds, yuccaols A, B and C. These compounds occur only in the yucca bark (Table 1). Hence they are present in the whole yucca powder products, but not in yucca extracts.

Table 1. Concentrations of phenolics (mg g^{-1}) in different fractions of yucca (from Oleszek *et al.* 2001b).

Item	Resveratrol	Yuccaols
Yucca bark	21.7	72.6
Yucca whole plant powder	3.2	10.0
Yucca extract	n.p.*	n.p.

*n.p. = not present

Yucca/Quillaja Products and Ammonia Suppression

Yucca products have been used for many years for the control of atmospheric ammonia and odors in livestock and poultry confinement facilities. Because this application is well established, it will be discussed only briefly here.

Inclusion of dietary levels of yucca products at 100–150 ppm in nonruminant (swine, poultry, rabbit) diets consistently produces a reduction in the ammonia concentrations in the animal facility. The effects of yucca on ammonia are probably not attributable to saponins (Killeen *et al.* 1998a). These authors determined that the effects on ammonia and nitrogen metabolism are caused by the non-butanol extractable fraction; saponins are removed in the butanol extract. Oligosaccharides (glycocomponents) have been suggested as the active fraction. Polyphenolics are also involved. Phenolics are well known for their ability to bind to amino groups of amino acids, so they likely bind ammonia as well.

Physiological and Biochemical Effects of Saponins

Saponins have diverse biological activities. Saponins form micelles with sterols such as cholesterol and bile acids. The hydrophobic portion of the saponin (the aglycone or sapogenin) associates (lipophilic bonding) with the hydrophobic sterol nucleus in a stacked micellar aggregation (Oakenfull and Sidhu 1989). Interactions of saponins with cholesterol and other sterols account for many of their biological effects, particularly those involving membrane activity. It was demonstrated over 45 years ago that dietary saponin reduces blood cholesterol levels in chickens (Griminger and Fisher 1958). This effect is likely a result of saponins binding to the cholesterol excreted in the bile, thus inhibiting entero-hepatic recycling of cholesterol. Yucca and quillaja extracts lower total and LDL cholesterol levels in hypercholesterolemic humans (Kim *et al.* 2003).

Saponins affect the permeability of intestinal cells by forming complexes with cholesterol in mucosal cell membranes (Johnson *et al.* 1986). These authors found that saponins increase the permeability of intestinal mucosal cells, inhibit active nutrient transport, and may facilitate the uptake of substances to which the gut would normally be impermeable. By this means, saponins may act as adjuvants in orally administered vaccines.

Saponins have anti-protozoal activity, by complexing with cholesterol in protozoal cell membranes. This has been well demonstrated with rumen protozoa *in vivo* (Lu and Jorgensen 1987; Wallace *et al.* 1994; Klita *et al.* 1996) and *in vitro* (Makkar *et al.* 1998;

Wang *et al.* 1998). The anti-protozoal activity requires the intact saponin structure with both the nucleus and side chain(s) present (Newbold *et al.* 1997).

Activity against ruminal protozoa raises the question as to whether saponins would be effective against protozoal diseases which afflict humans, livestock and poultry. Those protozoal diseases in which part of the life cycle occurs in the gastrointestinal tract would be expected to be responsive to anti-protozoal activity of saponins. An example is the disease giardiasis, caused by the protozoan *Giardia lamblia* (also known as *Giardia duodenalis*). It is one of the most common intestinal pathogens in humans and animals throughout the world. Yucca saponins are as effective as the drug metronidazole in killing giardia tropozoites in the intestine (McAllister *et al.* 2001).

The effect of saponins on other common livestock protozoal diseases such as coccidiosis is an area which should be investigated. Another example is equine protozoal myeloencephalitis (EPM), a disease of horses reported throughout North America. Horses ingest the protozoal sporocysts in contaminated feed and pasture. The sporocysts sporulate in the intestine, producing sporozoites which enter the intestinal epithelial cells, where they undergo asexual reproduction to produce merozoites. These invade CNS tissue, causing fatal neurologic disease. An intriguing possibility is that daily supplementation of horses with yucca may kill the sporozoites before they enter the intestinal epithelium.

Quillaja saponins are toxic to plant parasitic nematodes (San Martin 2004). A crude quillaja extract was more effective than a purified saponin faction, suggesting that both saponins and polyphenols are involved in nematocidal activity. Nematodes are also internal parasites of animals (e.g. *Strongyles* or stomach worms). Forages containing polyphenols suppress internal parasites in livestock (Butter *et al.* 2001). Yucca and quillaja products, and especially whole plant powders, contain saponins and polyphenols, both of which have anti-nematode activity. Thus their use as feed additives may assist in control of internal parasites.

Anti-arthritic and Anti-inflammatory Effects of Yucca Products

Yucca products have been used for many years for reputed anti-arthritic effects. The only direct studies of anti-arthritic effects are those of Bingham *et al.* (Bingham 1976; Bingham *et al.* 1978), who reported that symptoms of pain and swelling in arthritic human patients were relieved by consumption of yucca tablets. These reports led to the widespread use of yucca products for treatment and prevention of arthritis in horses and companion animals. Bingham (1976) proposed that yucca saponins have anti-protozoal activity, which suppresses protozoal infestation of the intestine. Bingham (1976) reported that a protozoal organism is involved in causation of rheumatoid arthritis. Subsequent studies have well established the ant-protozoal effects of saponins, as discussed above. There are well-known interactions between rheumatoid arthritis, chronic inflammatory disease, and food and nutrition (Parke *et al.* 1996; Martin 1998). In a recent review, Cordain *et al.* (2000) stated, 'Despite the almost universal clinical observation that inflammation of the gut is frequently associated with inflammation of the joints and vice-versa, the nature of the relationship remains elusive.'

Cordain *et al.* (2000) reported that arthritis is associated with intestinal bacterial overgrowth of *Escherichia coli* and *Lactobacillus lactis*. Yucca saponins have anti-bacterial properties (Katsunuma *et al.* 2000; Wang *et al.* 2000b). Thus a beneficial effect of yucca on arthritis could involve both the anti-protozoal and anti-bacterial properties. Of interest is that the drug metronidazole attenuates gastrointestinal inflammation and can prevent activation of arthritis in animal models (Cordain *et al.* 2000). As mentioned earlier, metronidazole is anti-protozoal, and is used to treat giardia. Yucca saponins are as effective as metronidazole in treating giardia infections (McAllister *et al.* 2001), supporting an anti-protozoal effect of yucca saponins in arthritis prevention.

Recent research suggests another possible mode of action of yucca in preventing arthritis, by anti-inflammatory activity. Yucca contains a number of phenolic compounds with antioxidant activity. These include resveratrol and yuccaols A, B and C (Oleszek *et al.* 2001a) and yuccaols D and E (Piacente *et al.* 2004). Resveratrol is well known for its anti-inflammatory activity (Bertelli *et al.* 2002; Bhat and Pezzuto 2002). Yucca bark is rich in resveratrol (Table 1), while yucca extract is devoid of activity. Thus whole plant yucca powder is a good source of resveratrol and yuccaols, while yucca extracts are poor sources.

Marzocco *et al.* (2004) demonstrated that yucca phenolics inhibit inducible nitric oxide synthase (iNOS) expression. Nitric oxide is an inflammatory agent, and its content increases in tissues during inflammatory responses. The expression of iNOS is controlled by NF kappa B (NFkB), a transcription factor which regulates gene expression. Resveratrol and yucca phenolics strongly inhibit NFkB (Marzocco *et al.* 2004). Yuccaol C is particularly effective. Thus whole plant yucca powder has powerful anti-inflammatory activity, mediated via inhibition of NFkB activation.

Platelet aggregation is characteristic of inflammation. Yucca phenolics have inhibitory activity against platelet aggregation (Olas *et al.* 2002, 2003). Yucca phenolics also have antioxidant activity (Piacente *et al.* 2004), and free-radical scavenging effects (Oleszek *et al.* 2001b). Thus there are a number of well-documented mechanisms by which yucca phenolics, present only in whole yucca plant powder, have anti-inflammatory activity, supporting a role for yucca powder in the treatment and prevention of arthritis.

The inflammatory response and its cytokine mediators cause alterations of nutritional requirements, reducing animal performance (Klasing 1988, Lundeen 2004). The reduction in feed intake that accompanies an inflammatory response is responsible for about 70% of the decrease in animal growth, while the remaining 30% is due to inefficiencies in nutrient absorption and metabolism (Klasing 1988, Lundeen 2004). Protein metabolism is altered, with a decrease in use of amino acids for muscle growth. Reactive oxygen species are produced in an immune response, increasing the antioxidant requirements. Thus the anti-inflammatory (Marzocco *et al.* 2004) and antioxidant (Piacente *et al.* 2004) effects of yucca phenolics may account for the improvement in animal performance observed with the feeding of yucca powder.

Effects of Yucca and Quillaja on Rumen Fermentation

A consistent finding when yucca products are administered to ruminants is a reduction in rumen ammonia concentrations (Wallace *et al.* 1994). A major source of rumen ammonia is from the proteolysis of bacterial protein, occurring as a result of ingestion of rumen bacteria by protozoa. Saponins have pronounced anti-protozoal activity, especially against rumen protozoa. Saponins form irreversible complexes with cholesterol in protozoal cell walls, causing cell lysis and death. The anti-protozoal activity requires the intact saponin structure with both the nucleus and side chain(s) present. A complicating factor is that saponins are hydrolyzed by rumen bacteria which remove the carbohydrate side chains

(Makkar and Becker 1997; Wang *et al.* 1998). Primarily as a result of suppression of rumen protozoa, dietary saponins increase the outflow of bacterial protein from the rumen (Wallace *et al.* 1994; Makkar and Becker 1996). Makkar and Becker (1997) observed that quillaja saponins were stable in the rumen for up to 6 hours after administration. It is possible that this time period may be adequate for the saponins to have anti-protozoal activity. Thus, the fact that saponins are rapidly degraded in the rumen may not necessarily eliminate their capacity to suppress rumen protozoa. The practicality of using yucca to suppress rumen protozoa has been questioned by Killeen *et al.* (1998b) because effective concentrations (1000 to 10,000 mg L^{-1}) are much higher than those commonly utilized in livestock feeds (60–250 mg kg^{-1}).

Although the most obvious effect of saponins on ruminal microbes is the suppression of protozoa, there are effects on ruminal bacteria (Wallace *et al.* 1994). Using pure cultures of ruminal bacteria, Wallace *et al.* (1994) observed that yucca extract stimulated growth of *Prevotella ruminicola*, whereas the growth of *Streptococcus bovis* was suppressed. The anti-bacterial properties were most pronounced against Gram-positive bacteria, similar to the action of ionophores.

It would be interesting to determine if there is an interaction between saponins and ionophores in ruminal fermentation. Both saponins and ionophores suppress Gram-positive bacteria and protozoa, so synergistic effects would not be surprising. In the anti-protozoal activity, they act via different mechanisms: saponins cause cell lysis by interacting with cholesterol in the protozoal cell membrane, while ionophores disrupt ion transport. Ruminal protozoa are unable to adapt to or detoxify saponins (Newbold *et al.* 1997).

In several studies, Wang *et al.* (1998, 2000a,b) noted that yucca saponins tended to promote ruminal amylolytic activity and depress cellulolytic activity. Effects on amylolytic bacteria were species dependent (Wang *et al.* 2000b). Gram-positive bacteria were inhibited by yucca saponin, whereas Gram-negative species were either stimulated or unaffected. The effects were similar to those of ionophores. Rumen fungi were extremely sensitive to yucca saponins (Wang *et al.* 2000b), as they are to ionophores. Wang and coworkers suggest that yucca supplementation would be most likely to be of benefit in high-grain diets for ruminants.

The mode of action of anti-bacterial effects of saponins seems to involve membranolytic properties, rather than simply altering the surface tension of the extracellular medium (Killeen *et al.* 1998b). Thus, their inhibitory activity is associated with adsorption to microbes, and is, therefore, influenced by microbial population density. Sen *et al.* (1998) observed a concentration-dependent growth response of *E. coli* K-12 to quillaja and yucca saponins, with growth-promoting activity at low concentrations and inhibition at higher levels. Thus the impact on a mixed bacterial population such as in the rumen is difficult to predict.

In virtually all studies on the effects of yucca, quillaja or other saponin sources on rumen fermentation, there is an increase in the relative proportion of propionate. Yucca saponins stimulate growth of *Selenomonas ruminantium* (Wang *et al.* 2000a). This species is very important in the utilization of high concentrate diets and the production of propionate. By stimulating this organism, yucca and quillaja may promote the utilization of grain-based diets for dairy and feedlot cattle. Ryan *et al.* (2003) suggested that surfactant activity of yucca saponins may encourage microbial adhesion and consortia formation, resulting in more rapid substrate degradation.

All of the studies on effects of yucca and quillaja on rumen fermentation have emphasized the roles of saponins. There has been no research on the effects of resveratrol or the yucca phenols in the rumen. This is an area of obvious research need.

Yucca saponins have been used to facilitate the tempering of grains for feedlot diets (Wang *et al.* 2003). The surfactant properties improve the interaction of water with the waxy surface of corn grain and increase rate of water absorption during the tempering of barley (Wang *et al.* 2003). In addition, yucca extract inhibits aflatoxin production in grains (Sanchez-Garcia *et al.* 2000).

Stillbirth Reduction in Swine

As with other monogastric species, yucca and quillaja products are used for ammonia and odor control in swine. Dietary levels of 100–125 ppm for yucca powder and 150–200 ppm for quillaja powder are typically used.

Considerable research has been conducted recently on the effects of yucca and quillaja products on stillbirths in swine. About 9% of piglets born are stillbirths (Ilsley *et al.* 2003). This represents a significant economic loss to swine producers. Numerous studies have shown that yucca and quillaja products reduce stillbirths when fed to late-gestation sows.

The first report of an effect of yucca extract on stillbirths in pigs was that of Cline *et al.* (1996). These researchers at the University of Illinois fed a corn-soy lactation diet with either 0 or 125 ppm yucca extract to sows upon entering the farrowing room (about 5 days pre-farrowing) until weaning at 21 days. A total of 81 multiparous cross-bred sows were allotted to the treatment groups based on parity. Blood oxygen values at the time of birth were obtained for 483 piglets.

Feeding 125 ppm yucca extract resulted in a 41% reduction in incidence of stillbirths (Table 2), while pre-weaning mortality was 26.2% lower with yucca extract (Table 2). Sows fed yucca extract tended to consume slightly more lactation diet (300 g/day) and lose slightly less weight (740 g/sow). The greatest effect of yucca was seen with first litter gilts and older (>5 parity) sows. Blood oxygen levels of piglets at birth were greater ($\underline{P}<0.006$) for first litter gilts fed yucca extract. The greatest incidence of stillbirths in control sows occurred in the last third of the birth order. Sows fed yucca extract had reduced stillbirths in the last third of the birth order. Cline *et al.* (1996) concluded that feeding yucca extract to sows prior to farrowing increased the blood oxygen supply to the fetus during birth, thus lowering stillbirth incidence and pre-weaning mortality. The mode of action of yucca in improving blood oxygenation is unclear. Cline *et al.* (1996) suggested that by reducing intestinal ammonia, yucca reduces intestinal mucosal turnover rate and tissue oxygen demand, thus allowing greater oxygen transfer to the fetuses.

Table 2. Effect of dietary yucca extract on reproduction and stillbirths in swine (Cline *et al.* 1996).

Item	Dietary Treatment	
	Control	Yucca extract
Number of sows	41	40
Stillbirths per litter	0.85	0.50 (−41.2%)
Pre-weaning mortalty (%)	18.1	13.4 (−26.2%)

Following the publication by Cline *et al.* (1996), Herpin *et al.* (2004) of INRA, France, conducted a trial on stillbirths, using yucca powder. Dried whole plant yucca powder (DK Sarsaponin 40) was fed to sows at 120 ppm from day 107 of gestation to day 21 of lactation. Feed intake was adjusted to 2.6 kg day^{-1} during gestation and increased by 0.5 kg day^{-1} after farrowing. Parturition was induced on day 113. There were nine sows on the control diet and ten on the yucca powder. The effects on sow and piglet performance and reproductive data are summarized in Table 3. The number and percentage of stillbirths per litter averaged 0.78 and 7.2%, respectively, in the control group, and 0.5 and 4.7% for the yucca powder group. Most of the stillbirths (71.4%) occurred in the last third of the birth order in control sows. This percentage was reduced to 40% in the yucca group. There was a trend towards reduced mortality in the pigs from the yucca-fed sows. The number of litters with no mortality was only 11% in the control group, and 40% in the yucca group. In the study of Herpin *et al.* (2004), there were no differences between the two groups in sow or piglet blood oxygen levels. Thus this study confirmed the stillbirth reduction reported by Cline *et al.* (1996) but did not support the hypothesis that the effect is due to improved blood oxygenation. The piglets from yucca-fed sows had significantly higher rectal temperatures in the first 24 hours of life, suggesting an improvement of the thermoregulatory ability of the baby pigs.

Saponins have immuno-stimulatory properties; favorable effects on the immune system might be a factor in stillbirth reduction. Yucca powder contains phenolic antioxidants such as resveratrol and yuccaols (Oleszek *et al.* 2001a), which might also have a role in improving neonatal survival because of their anti-inflammatory activity.

Ilsley *et al.* (2003) evaluated the effects of feeding extracts of yucca and quillaja on stillbirths. Dietary levels were 250 ppm quillaja and 200 ppm yucca. There were 20 sows per treatment. As in the studies of Cline *et al.* (1996) and Herpin *et al.* (2004), both saponin-containing extracts (yucca and quillaja) reduced stillbirths and baby pig mortality (Table 4). In a subsequent study, purified quillaja saponin when fed to sows reduced stillbirths, suggesting that saponins are the stillbirth-reducing factors in yucca and quillaja (Ilsley and Miller, 2005). In a field trial in Iowa with 2400 sows, conducted by the Conti-Agri Group (formerly Continental Grain), feeding DK sarsaponin 40 resulted in a 33.3% reduction in stillbirths. Although the mode of action is still unclear, there is abundant published evidence that dietary yucca and quillaja reduce stillbirths in swine.

Garcia *et al.* (2004) investigated the effects of orally administered quillaja extract on the passive immunization of neonatal pigs. It was hypothesized that the effects of saponins on enhancing the permeability of intestinal mucosal cells would enhance the uptake of immunoglobulins. Although quillaja saponin supplementation did not appear to affect intestinal uptake in well-nourished neonates, improved passive immunity and increased growth rates were noted in low-weight, under-nourished piglets. This effect, combined with the improved thermoregulatory abilities for low weight piglets reported by Herpin *et al.* (2004), suggests that saponin administration improves viability of neonatal pigs. Isley *et al.* (2005) found that administration of quillaja saponin to weaned pigs increased IgG and IgA levels and enhanced C-reactive protein (CRP). They concluded that dietary quillaja saponins enhance the immune function of the weanling pig in normal commercial rearing situations even in the absence of any pathogen challenge.

Table 3. Effect of DK Yucca Powder on piglet body temperature and pre-weaning mortality (Herpin *et al.*).

Treatment	Control	DK Yucca Powder
Number of litters	9	10
Number of pigs/litter	10.8 ± 0.7	10.6 ± 0.7
Stillbirths		
Per litter	0.78 ± 0.32	0.50 ± 0.22
% of total pigs born	7.2% (7/97)	4.7% (5/106)
In the last third of the birth order, %	71.4% (5/7)	40% (2/5)
Postnatal mortality %	8.2% (8/97)	6.6% (7/106)
Total mortality %	15.5% (15/97)	11.3% (12/106)
Number of litters with no mortality	1	4
Rectal temperature at 1d, °C	38.8 ± 0.05	39.0 ± 0.03**

Means ± SE. Rectal temperature at 1d was recorded from 70 control and 89 DK Powder piglets.
Statistical significance: **$\underline{P}$<0.01.

Table 4. Stillbirth and baby pig mortality as influenced by feeding yucca and quillaja extracts (Ilsley *et al.* 2003).

	Control	Yucca	Quillaja
% Stillbirths	8.91	4.20	3.83
% Piglet mortality	10.04	11.16	8.41
No. born alive	10.3	10.6	11.8

Conclusions

Yucca and quillaja have traditionally been regarded primarily as sources of saponins. Saponins have diverse biological effects, some of which are mediated by their ability to form complexes with cholesterol in cell membranes. Recent research indicates that the membrane effects of saponins result in anti-protozoal and anti-nematode activity, and modifications in rumen microbe populations.

Yucca is also a good source of polyphenolic compounds, including resveratrol and the yuccaols. These compounds function as anti-inflammatory agents and as antioxidants. Anti-inflammatory activity may account for some of the favorable effects of yucca products on animal performance. Finally, yucca and quillaja supplements fed to sows in late gestation reduce the incidence of stillbirths in swine and may enhance the immune function of the weanling pig.

References

Bertelli AA, Migliori M, Panichi V, Origlia N, Filippi C, Das DK, and Giovannini L (2002). Resveratrol, a component of wine and grapes, in the prevention of kidney disease. *Annals of the New York Academy of Sciences* 957:230–238.

Bhat KP and Pezzuto JM (2002). Cancer chemopreventive activity of resveratrol. *Annals of the New York Academy of Sciences* 957:210–229.
Bingham R (1976). New and effective approaches to the prevention and treatment of arthritis. *Journal of Applied Nutrition* 28:38–47.
Bingham R, Bellow BA and Bellow JG (1975). Yucca plant saponin in the management of arthritis. *Journal of Applied Nutrition* 27:45–51.
Bingham R, Harris DH, and Laga T (1978). Yucca plant saponin in the treatment of hypertension and hypercholesterolemia. *Journal of Applied Nutrition* 30:127–136.
Butter NL, Dawson JM, Wakelin D, and Buttery PJ (2001). Effect of dietary condensed tannins on gastrointestinal nematodes. *Journal of Agricultural Sciences* 137:461–469.
Cline JL, Fisher BA, Trottier NL, Walker RD, and Easter RA (1996). Effect of feeding MICRO-AID on stillbirths, pre-weaning mortality, blood oxygen values of piglets and blood urea nitrogen in sows. *Journal of Animal Science* 74(Suppl. 1):189 (Abstract).
Cordain L, Toohey L, Smith MJ, and Hickey MS (2000). Modulation of immune function by dietary lectins in rheumatoid arthritis. *British Journal of Nutrition* 83:207–217.
Garcia MR, Lopez P, Williams RH, Lukefahr SD, and Laurenz JC (2004). Effect of *Quillaja saponaria* extract on passive immunization in a pig model. *Journal of Animal and Veterinary Advances* 3(8):535–541.
Griminger P and Fisher H (1958). Dietary saponin and plasma cholesterol in the chicken. *Proceedings of the Society for Experimental Biology and Medicine* 99:424–426.
Herpin P, Vincent A, and Cheeke PR (2004). Effect of feeding *Yucca schidigera* (DK powder) to the sow on piglet blood oxygenation and survival. *Proceedings of the Western Section of the American Society of Animal Science* 55:145–150.
Ilsley SE and Miller HM (2005). Effect of dietary supplementation of sows with quillaja saponins during gestation on colostrum composition and performance of piglets suckled. *Animals Science* 80:179–184.
Ilsley SE, Miller HM, Greathead HMR, and Kamel C (2003). Plant extracts as supplements for lactating sows: effects on piglet performance, sow food intake and diet digestibility. *Animal Science* 77:247–254.
Ilsley SE, Miller HM, and Kamel C (2005). Effects of dietary quillaja saponin and curcumin on the performance and immune status of weaned piglets. *Journal of Animal Science* 83:82–88.
Johnson IT, Gee JM, Price KR, Curl C, and Fenwick GR (1986). Influence of saponins on gut permeability and active nutrient transport *in vitro*. *Journal of Nutrition* 116:2270–2277.
Katsunuma Y, Nakamura Y, Toyoda A, and Minato H (2000). Effect of *Yucca schidigera* extract and saponins on growth of bacteria isolated from animal intestinal tract. *Animal Science Journal* 71(2):164–170.
Killeen GF, Connolly CR, Walsh GA, Duffy CF, Headon DR, and Power RF (1998a). The effects of dietary supplementation with *Yucca schidigera* extract or fractions thereof on nitrogen metabolism and gastrointestinal fermentation processes in the rat. *Journal of the Science of Food and Agriculture* 76:91–99.
Killeen GF, Madigan CA, Connolly CR, Walsh GA, Clark C, Hynes MJ, Timmins BF, James P, Headon DR, and Power RF (1998b). Antimicrobial saponins of *Yucca schidigera* and the implications of their *in vitro* properties for their *in vivo* impact. *Journal of Agriculture and Food Chemistry* 46:3178–3186.

Kim S-W, Park S-K, Kang S-I, Kang H.-C, Oh H-J, Bae C-Y, and Bae, D-H (2003). Hypocholesterolemic property of *Yucca schidigera* and *Quillaja saponaria* extracts in the human body. *Archives of Pharmaceutical Research* 26:1042–1046.

Klasing KC (1988). Nutritional aspects of leukocytic cytokines. *Journal of Nutrition* 118:1436–1446.

Klita PT, Mathison GW, Fenton TW, and Hardin RT (1996). Effects of alfalfa root saponins on digestive function in sheep. *Journal of Animal Science* 74:1144–1156.

Lu CD and Jorgensen NA (1987). Alfalfa saponins affect site and extent of nutrient digestion in ruminants. *Journal of Nutrition* 117:919–927.

Lundeen T. (2004). Inflammatory responses may alter nutritional requirements. *Feedstuffs* 76(10), October 4.

Makkar HPS and Becker K (1996). Effect of quillaja saponins on *in vitro* rumen fermentation. *Advances in Experimental Medicine and Biology* 405:387–394.

Makkar HPS and Becker K (1997). Degradation of quillaja saponins by mixed culture of rumen microbes. *Letters in Applied Microbiology* 25:243–245.

Makkar HPS, Sen S, Blummel M, and Becker K (1998). Effects of fractions containing saponins from *Yucca schidigera*, *Quillaja saponaria*, and *Acacia auriculoformis* on rumen fermentation. *Journal of Agriculture and Food Chemistry* 46:4324–4328.

Martin RH (1998). The role of nutrition and diet in rheumatoid arthritis. *Proceedings of the Nutrition Society* 57:231–234.

Marzocco S, Piacente S, Pizza C, Oleszek W, Stochmal A, Pinto A, Sorrentino R, and Autore G (2004). Inhibition of inducible nitric oxide synthase expression by yuccaol C from *Yucca schidigera* Roezl. *Life Sciences* 75:1491–1501.

McAllister TA, Annett CB, Cockwill CL, Olson ME, Yang Y, and Cheeke PR (2001). Studies on the use of *Yucca schidigera* to control giardiasis. *Veterinary Parasitology* 97:85–99.

Newbold CJ, El Hassan SM, Wang J, Ortega ME, and Wallace RJ (1997). Influence of foliage from African multipurpose trees on activity of rumen protozoa and bacteria. *British Journal of Nutrition* 78:237–249.

Oakenfull D and Sidhu GS (1989). Saponins. In *Toxicants of Plant Origin, Vol. 2,* (PR Cheeke, ed.), pp. 97–141. CRC Press, Boca Raton, Florida.

Olas B, Wachowicz B, Stochmal A, and Oelszek W (2002). Anti-platelet effects of different phenolic compounds from *Yucca schidigera* roezl. bark. *Platelets* 13:167–173.

Olas B, Wachowicz B, Stochmal A, and Oleszek W (2003). Inhibition of oxidative stress in blood platelets by different phenolics from *Yucca schidigera* roezl. bark. *Nutrition* 19: 633–640.

Oleszek W, Sitek M, Stochmal A, Piacente S, Pizza C, and Cheeke P (2001a). Resveratrol and other phenolics from the bark of *Yucca schidigera* Roezl. *Journal of Agriculture and Food Chemistry* 49:747–752.

Oleszek W, Sitek M, Stochmal A, and Cheeke P (2001b). Antioxidant properties of *Yucca schidigera* products. In *Biologically-active Phytochemicals in Food. Analysis, Metabolism, Bioavailability and Function* (W Pfannhauser, GR Fenwick, and S Khokhar, eds), pp. 303–306. Royal Society of Chemistry.

Parke AL, Parke DV, and Jones FA (1996). Diet and nutrition in rheumatoid arthritis and other chronic inflammatory diseases. *Journal of Clinical Biochemistry and Nutrition* 20:1–26.

Piacente, S, Montoro, P, Oleszek, W, and Pizza, C (2004). *Yucca schidigera* bark: Phenolic constituents and antioxidant activity. *Journal of Natural Products* 67:882–885.

Ryan JP, Quinn T, Mullally J, and Leek BF (2003). Effects of *Yucca schidigera* plant extract (*De-Odorase*) on the fermentation of hay, straw and concentrates by sheep's ruminal fluid *in vitro* and *in vivo*. *Irish Veterinary Journal* 56:209–213.

Sanchez-Garcia CA, Heredia NL, and Garcia JS (2000). Extracts of *Yucca sp*, and *Larrea tridentate* inhibit aflatoxin production in corn kernels. *American Society of Microbiology (ASM) Abstract Database*.

San Martin R (2004). Use of Quillaja saponins to control nematodes. Development of a commercial product: QL AGRI.® In *International Conference on Saponins* (Abstracts), p. 6. Pulawy, Poland. Sept. 8–10, 2004.

Sen S, Makkar HPS, Muetzel S, and Becker K (1998). Effect of *Quillaja saponaria* saponins and *Yucca schidigera* plant extract on growth of *Escherichia coli*. *Letters on Applied Microbiology* 27:35–38.

Wallace RJ, Arthaud L, and Newbold CJ (1994). Influence of *Yucca schidigera* extract on ruminal ammonia concentrations and ruminal microorganisms. *Applied Environmental Microbiology* 60:1762–1767.

Wang Y, McAllister TA, Newbold CJ, Rode LM, Cheeke PR, and Cheng K-J (1998). Effects of *Yucca schidigera* extract on fermentation and degradation of steroidal saponins in the rumen simulation technique (RUSITEC). *Animal Feed Science and Technology* 74:143–153.

Wang Y, McAllister TA, Yanke LJ, Xhong JX, and Cheeke PR (2000a). *In vitro* effects of steroidal saponins from *Yucca schidigera* extract on rumen microbial protein synthesis and ruminal fermentation. *Journal of the Science of Food and Agriculture* 80:2114–2122.

Wang Y, McAllister TA, Yanke LJ, and Cheeke PR (2000b). Effect of steroidal saponin from *Yucca schidigera* extract on ruminal microbes. *Journal of Applied Microbiology* 88:887–896.

Wang Y, Greer D, and McAllister TA (2003). Effects of moisture, roller setting, and saponin-based surfactant on barley processing, ruminal degradation of barley, and growth performance by feedlot steers. *Journal of Animal Science* 81:2145–2154.

Chapter 31

Cutting for Silage May Not Prevent Annual Ryegrass Toxicity

J.G. Allen, A.M. Masters, and G. Doncon

Animal Health Laboratories, Department of Agriculture, Locked Bag No. 4, Bentley Delivery Centre WA 6983, Australia

Introduction

Annual ryegrass toxicity (ARGT) is a major disease of livestock in Western Australia. It is caused when livestock eat ryegrass infected with the toxigenic bacterium *Rathayibacter toxicus*. The bacterium is introduced into the ryegrass by the nematode *Anguina funesta* that reproduces in galls formed within the seedhead of the ryegrass. If the bacterium is introduced into the nematode gall it may multiply to kill the nematodes and form a toxic bacterial gall. The toxins produced by *R. toxicus* are called corynetoxins (Allen 2004).

Rathayibacter toxicus only commences to produce the corynetoxins when the ryegrass begins to flower, with most toxicity developing during the period that the ryegrass forms its seeds. Recommendations for the control of ARGT revolve around stopping the ryegrass from flowering (Brown *et al.* 1989). One of these recommendations is to cut crops or pasture for silage, as this is anticipated to be done well before the ryegrass has flowered (Riley *et al.* 1995). There is also the possibility that within the acid environment of the silage any *R. toxicus* or corynetoxins present might be killed or denatured/detoxified, respectively. There has been only one case of ARGT resulting from the feeding of silage in Western Australia, and it is reported here.

History

In September 2001 a farmer in the Cunderdin area of the wheatbelt in Western Australia decided that ryegrass control within a crop of lupins was so poor that he would cut the crop for silage. This was the 4th year that the paddock had been in a wheat:lupin crop rotation. The crop was cut on 24 September, which in an average year would have been about when ryegrass would start flowering in this area. There was a sudden dry spell and the crop was left on the ground for approximately a week. At this time it received 1 mm of rain and was then immediately baled into plastic silage bags. During February 2002 the silage was fed to sheep, cattle and horses, resulting in several hundred sheep, two bulls, four cows and one horse dying from ARGT. Samples were taken from six silage bales in late February 2002 and tested with an ELISA for *R. toxicus* (based on a monoclonal antibody to a specific polysaccharide antigen of *R. toxicus* (A.M. Masters *et al.*, unpublished data) and standardized to provide results in bacterial gall equivalents). Extraction for the ELISA was done with water.

In May 2004 it was determined that the farmer had not destroyed the silage. The farm was again visited at this time and 20 silage bales sampled and tested with the ELISA for *R. toxicus* and an ELISA for the corynetoxins (Than *et al.* 2004). Extraction was done with 1% cyclodextrin so both ELISAs could be performed on the same extract (Anderton *et al.* 2004). None of the bales sampled in 2004 had any markings or damage to indicate that they had been previously sampled, so it is believed that they were different bales to those sampled in 2002.

Results

In 2002 bacterial galls were detected in all nine bales, and eight had concentrations of bacterial galls consistent with the ARGT that was experienced (Table 1). The Animal Health Laboratories regard feeds containing more than 1000 bacterial galls/kg as presenting a very high risk of causing ARGT. In contrast, in 2004, bacterial galls were detected in only 12 of the 20 bales, and only eight had concentrations consistent with presenting a very high risk (Table 1). However, corynetoxins were detected in all 20 bales (Table 1).

Conclusion

Between 2002 and 2004 the detection rate for *R. toxicus* in the silage bales fell from 100% to 60%, and it appeared that when the bacterium was detected in 2004 it was in lower concentrations than in 2002. On the other hand corynetoxins were detected in all the samples in 2004, and the three greatest concentrations were present in bales in which no *R. toxicus* could be detected. These findings indicate that within the acidic (pH 4.0–4.5) environment of the silage the antigen measured to detect *R. toxicus* is less stable than the corynetoxins over time. It is possible that the polysaccharide antigen of *R. toxicus* is used as an energy source by the fermentation bacteria in the silage.

A similar finding of less stability of the *R. toxicus* antigen was reported when old seed galls in *Festuca nigrescens* were tested for the presence of *R. toxicus* and corynetoxins (Riley *et al.* 2003). Antigens of the bacterium could be weakly detected in galls 22 years old, but not at all in galls 58 years old. On the other hand the corynetoxins were found in all the galls, although the 58-year-old galls contained approximately 50% less than the 22-year-old galls.

Why the silage was toxic is uncertain. If it had been cut before the ryegrass had flowered little or no toxin production would have been expected. Toxicity in the silage may have resulted from the crop being cut too late (that is, the ryegrass had already flowered and seed set had commenced), or the bacterial galls may have continued to develop toxicity after the ryegrass had been cut. The latter would not normally be expected to occur (Riley *et al.* 1995).

Table 1. Concentrations of *R. toxicus* bacterial galls in 2002, and of *R. toxicus* bacterial galls and corynetoxins in 2004, in silage bales that had caused ARGT when fed to livestock in 2002

2002		2004		
Silage bale*	*R. toxicus* galls/kg	Silage bale*	*R. toxicus* galls/kg	Corynetoxins (µg/kg)
1	2574	1	ND	27000
2	3478	2	2291	850
3	3610	3	ND	683
4	3428	4	ND	2045
5	3719	5	ND	455
6	1930	6	>2667	5852
7	3608	7	1687	1035
8	2292	8	1513	6306
9	198	9	ND	7355
		10	42	50
		11	38	165
		12	ND	25450
		13	ND	6830
		14	102	71
		15	1021	568
		16	1529	758
		17	17	134
		18	1021	1180
		19	ND	635
		20	>2667	5436

* Different silage bales were sampled in 2002 and 2004.
ND = not detected

Acknowledgements

The CSIRO Plant Toxins Research Group provided the reagents for the corynetoxins ELISA.

References

Allen JG (2004). Annual ryegrass toxicity. In *Clinical Veterinary Toxicology* (KH Plumlee, ed), pp. 422–424. Mosby, St. Louis, Missouri.

Anderton N, Cao Y, Colegate SM, Edgar JA, Knill AM, Michalewicz A, Olsen V, and Than KA (2004). Improved methods for the extraction and purification of corynetoxins for preparative and analytical purposes using water solubilizing additives. In *Poisonous Plants and Related Toxins* (T Acamovic, CS Stewart and TW Pennycott, eds), pp. 70–75. CABI Publishing, Wallingford, Oxon, UK.

Brown A, Madin R, Burdass J, Maling I, and Stanton J (1989). Annual ryegrass toxicity. *Bulletin No. 4156.* Western Australian Department of Agriculture, South Perth, Western Australia.

Riley I, Mayberry C, Kingdon B, Arkell P, and Smart B (1995). Avoiding ARGT in high rainfall areas. *Farmnote 65/1995.* Western Australian Department of Agriculture, South Perth, Western Australia.

Riley IT, Gregory AR, Allen JG, and Edgar JA (2003). Poisoning of livestock in Oregon in the 1940s to 1960s attributed to corynetoxins produced by *Rathayibacter* in nematode galls in Chewings fescue (*Festuca nigrescens*). *Veterinary and Human Toxicology* 45:160–162.

Than KA, Cao Y, Michalewicz A, Olsen V, Anderton N, Cockrum P, Colegate S, and Edgar JA (2004). Analysis of corynetoxins: a comparative study of an indirect competitive ELISA and HPLC. In *Poisonous Plants and Related Toxins* (T Acamovic, CS Stewart and TW Pennycott, eds), pp. 402–407. CABI Publishing, Wallingford, Oxon, UK.

Chapter 32

The Pelleting Process or Treatment with Sodium Hydroxide Do Not Detoxify the Toxins of Annual Ryegrass Toxicity

J.G. Allen[1], M. Hoxey[2], C. Thompson[3], A.M. Masters[1], J.T.B. Milton[3], and P.E. Vercoe[3]

[1]*Animal Health Laboratories, Department of Agriculture, Locked Bag No. 4, Bentley Delivery Centre WA 6983, Australia;* [2]*17 Seahaven Street, Safety Bay WA 6169, Australia;* [3]*Faculty of Natural and Agricultural Science, The University of Western Australia, Crawley WA 6009, Australia*

Introduction

Annual ryegrass toxicity (ARGT) is a major disease of livestock in Western Australia. It is caused when livestock eat ryegrass infected with the toxigenic bacterium *Rathayibacter toxicus*. The bacterium is introduced into the ryegrass by the nematode *Anguina funesta* that reproduces in galls formed within the seedhead of the ryegrass. If the bacterium is introduced into the nematode gall it may multiply to kill the nematodes and form a toxic bacterial gall. The toxins produced by *R. toxicus* are called corynetoxins (Allen 2004).

Toxic ryegrass may contaminate hay and crop fines, which may cause ARGT when fed to livestock (Roberts and Devenish 2001; McKay *et al.* 2002). Large quantities of cereal hay are rejected for export from Western Australia each year because of the presence of *R. toxicus* (30% of samples rejected during 2001 hay season, Animal Health Laboratories records). If a practical means could be found to detoxify this contaminated hay the need to destroy it would be averted and financial losses reduced. The process of pelleting feeds involves heat and steam and these physical treatments may detoxify the corynetoxins. Alkali treatment is known to reduce the toxicity of lupin stubbles (Allen *et al.* 1986), and has the potential advantage of increasing the digestibility of straws and crop residues (Jackson 1977; Chaudhry and Miller 1996). The potential of these procedures to detoxify the corynetoxins is evaluated in the studies reported here.

Materials and Methods

Toxic ryegrass

Several hundred kg of toxic ryegrass was harvested with a small crop harvester in 1999 from a paddock in which ARGT had caused the death of approximately 600 sheep. This material was stored in wool bags in a shed.

Assay for bacterial gall equivalents

The number of the bacterial gall equivalents in a sample was determined using an ELISA based on a monoclonal antibody to a specific polysaccharide antigen of *R. toxicus* (A.M. Masters *et al.*, unpublished data). Aqueous cyclodextrin (1%), added at five times the volume of the sample, was used to prepare the extract for testing with the ELISA.

Quantitative assay for corynetoxins

The concentration of corynetoxins in a sample was determined using a quantitative ELISA for corynetoxin-like substances (Than *et al.* 2004). As for the bacterial ELISA, extracts were prepared using 1% cyclodextrin, so both ELISAs could be performed on a single extract.

Bacterial bioassay for assessing the toxicity of bacterial galls

A bacterial bioassay was used to assess the toxicity of bacterial galls in the second alkali experiment in case the corynetoxins ELISA was detecting non-toxic breakdown products of the corynetoxins. This bioassay was based on the inhibition of the growth of *Bacillus cereus* by the corynetoxins as demonstrated by Vogel *et al.* (1982).

Bacterial galls that had received treatment with alkali were macerated in Teflon canisters in a ball mill at slow speed for 4 min. The canisters were cleaned with methanol between samples. The macerated sample was placed in a 2 mL Eppendorf tube and 1 mL of tris-HCl buffer solution (pH 8.5) was added. The sample was mixed in the buffer thoroughly and then left for an hour to allow sedimentation. A 100 μL aliquot of the liquid phase was removed and diluted to 250 μL with additional tris-HCl buffer. This was the test extract. A tunicamycin standard (tunicamycin is a chemically related compound with similar bioactivity to the corynetoxins (Vogel *et al.* 1982; Jago *et al.* 1983)) containing 25 μg/mL was prepared using tris-HCl buffer.

A stock culture of *B. cereus* from the Western Australian Animal Health Laboratories was used. It was grown overnight at 28ºC in Luria-Bertani broth, then 100 μL of the culture was spread onto 523M agar (adjusted to pH 8 with concentrated HCl) plates to seed a bacterial lawn. Sterile pipette tips and forceps were used to create three or four wells in each agar plate. Fifty μL of the tunicamycin standard was placed in one of the wells and 50 μL of individual test extracts placed into the other wells. The plates were then incubated at 28ºC for 24 hours when zones of inhibition were measured using a ruler and compared against the tunicamycin standard. Tris-HCl buffer and a 1:2 solution of 7.5% NaOH in tris-HCl buffer were tested separately to establish that neither of these solutions caused inhibition of the *B. cereus*.

Pelleting experiment

Commercial livestock pellets containing 50% finely chopped oaten chaff, 15.51% wheat, 10% lupin seed meal, 10% millmix, 8.9% canola meal, 2% gypsum, 2% molasses, 1.09 % salt and 0.1% mineral mix were prepared using a Lister Junior (UK) pelleting machine, which was suitable for pelleting the small mixes involved in this experiment.

Three 25 kg mixes of these constituents were prepared, to which were added 368, 736 or 2941 g of the toxic material. This was expected to provide materials for pelleting containing approximately 500, 1000 and 4000 μg corynetoxins kg^{-1} and referred to as mixes M1, M2 and M3, respectively. The feed mixes were then mixed thoroughly and, as this was being done, 5 x 100 g samples were taken from each, placed in plastic bags, sealed and labeled. Each feed mix was then put through the pelleting machine, with the machine being cleaned thoroughly between mixes. The peak temperature was recorded while each mix was being pelleted. The pellets from each mix were stored in separate bags for 24 hours to allow them to cool. The pellets in each bag were then placed in a large plastic container and mixed thoroughly and, during this process, 5 x 100 g samples were taken from each, placed in plastic bags, sealed and labeled.

All samples were extracted using cyclodextrin (1%) and tested for the presence of bacterial and corynetoxin antigens at the same time. The feed mixes were extracted as received, whereas the pellets were milled prior to extraction.

Alkali experiments

Two experiments were conducted with alkali treatments using NaOH. In the first four aqueous alkali treatments were applied, namely 0 (control), 2.5, 5.0 and 7.5% NaOH (w/w), with four replicates of each. Before commencing, impurities like cape weed seeds and clover burr were removed by hand and discarded (approximately 20% of the original material) from the toxic ryegrass material to be used. Sixteen 10 g samples of the cleaned ryegrass were each placed in individual aluminium trays (20 x 10 cm), and spread out in a thin layer. The NaOH treatments were applied in 2.5 mL, with the controls receiving water only. An initial 1.5 mL was sprayed onto the ryegrass sample, which was then mixed for 30 seconds with a plastic spatula before the remaining volume was sprayed on and the sample mixed again for another 60 seconds. The trays were left on the bench for 1 week to allow the NaOH to react, before the samples were tested for corynetoxin concentration using the ELISA.

In the second alkali experiment the same four treatments were used, but with three replicates of each. In this experiment bacterial galls were isolated and treated to increase the concentration of toxin to within the range of sensitivity of the *B. cereus* bioassay. The toxic ryegrass material was cleaned using a mechanical sieve and then the seed examined on a light box to identify and separate out the bacterial galls. Twelve 125 mg samples of galls were prepared, and each was placed into a separate 2 mL Eppendorf tube. The NaOH treatments were applied in 0.5 mL solutions that were added to the galls in the tubes, which were shaken vigorously to ensure even exposure. The samples were left in the tubes with the lids on for 4 days to allow reaction with the NaOH. The samples were then placed into separate Petri dishes without lids to dry for 24 hours at 39°C before being tested for toxicity using the *B. cereus* bioassay.

Results

Pelleting did not result in any reduction in the concentrations of bacterial galls or the corynetoxins detected by the respective ELISAs (Table 1). In fact, the tendency was for concentrations to increase (significant increases for bacterial galls M3 and corynetoxins M2 and M3, in Table 1).

In the first alkali experiment the concentration of corynetoxins increased as the strength of the alkali increased (adjusted $r^2 = 0.46$, $P < 0.002$) (Table 2). The mean concentration in the 7.5% NaOH treatment was significantly greater than the means for the other treatments.

In the second alkali experiment the toxicity of the bacterial galls did not change with NaOH treatment. The zones of inhibition were the same for all treatments, and not different from the controls (Table 3).

Conclusion

Neither the pelleting process nor treatment with a strong alkali (NaOH) reduced the toxicity of toxic ryegrass or toxic bacterial galls. Future work in this area should evaluate oxidizing agents as it has been shown that periodate renders both corynetoxins and tunicamycin non-toxic to nursling rats (Stynes and Vogel 1983).

Table 1. The effect of pelleting of three feed mixes (M1, M2 and M3) containing three different concentrations of corynetoxin/kg on their apparent mean bacterial gall (galls kg^{-1}) and mean corynetoxins (µg kg^{-}) concentrations, and the temperatures (°C) of the pelleting process.

Feed mix	Pelleting temperature	Bacterial galls		Corynetoxins	
		In feed mix	In pellets	In feed mix	In pellets
M1	86	627^{aA}	588^{a}	206^{aA}	228^{a}
M2	82.5	685^{aAB}	991^{a}	362^{aA}	484^{b}
M3	79.7	1312^{aB}	4473^{b}	1449^{aB}	1877^{b}

Within rows, when comparing concentrations in the feed mix and the pellets, means with lower case superscript that differ are significantly different (P < 0.05).
Within columns, when comparing the concentrations in the feed mixes before pelleting, means with upper case superscripts that differ are significantly different (P < 0.5).

Table 2. The mean corynetoxin concentrations in toxic ryegrass after treatment with three concentrations of NaOH.

Alkali treatment	Corynetoxins (µg/kg)
0	7581^{a}
2.5% NaOH	9358^{a}
5.0% NaOH	9501^{a}
7.5% NaOH	12369^{b}

Mean concentrations with different superscripts are significantly different (P < 0.05).

Table 3. The mean radii of zones of inhibition in lawns of *B. cereus* caused by extracts of bacterial galls after they had been treated with three concentrations of NaOH.

Alkali treatment	Radius of zone of inhibition (mm)[a]
0	3
2.5% NaOH	2
5.0% NaOH	2
7.5% NaOH	3

None of the mean radii are significantly different ($P < 0.05$).

In the pelleting experiment the apparent increase in the concentrations of bacterial galls and corynetoxins after pelleting was probably due to improved extraction of the antigens from the milled pellets. The feed mixes were not milled.

In the first alkali experiment the apparent increase in corynetoxin concentration with increasing concentration of alkali may have resulted from the alkali actually cleaving the corynetoxins into smaller molecules that still had sites to which the antibodies in the ELISA could attach. The increased number of molecules would give the impression of an increased concentration of corynetoxins. Another possibility is that the alkali treatment just increased the efficiency of extraction of the corynetoxins. Than *et al.* (2002) reported that treatment with NaOH did not increase the extraction efficiency of the corynetoxins, but they only used concentrations up to 0.2% (w/w).

Acknowledgements

The CSIRO Plant Toxins Research Group provided the reagents for the corynetoxins ELISA. Mr. L. den Hollander assisted with the *B. cereus* bioassay.

References

Allen JG (2004). Annual ryegrass toxicity. In *Clinical Veterinary Toxicology* (KH Plumlee, ed), pp. 422–424. Mosby, St. Louis, Missouri.

Allen JG, Croker KP, Hancock GR, and Southey IN (1986). Control of ovine lupinosis: The treatment of lupin stubble materials with alkali. *Australian Veterinary Journal* 63:350–352.

Chaudhry AS and Miller EL (1996). The effect of sodium hydroxide and alkaline hydrogen peroxide on chemical composition of wheat straw and voluntary intake, growth and digesta kinetics in store lambs. *Animal Feed Science and Technology* 60:69–86.

Jackson MG (1977). Review article: The alkali treatment of straws. *Animal Feed Science and Technology* 2:105–130.

Jago MV, Payne AL, Peterson JE, and Bagust TJ (1983). Inhibition of glycosylation by corynetoxin, the causative agent of annual ryegrass toxicity: A comparison with tunicamycin. *Chemico-Biological Interactions* 45:223–234.

McKay A, Allen J, and Davis J (2002). ARGT and hay exports – R&D and industry working together to solve a crucial issue. *RIRDC Export Hay R&D Update*, Issue 2, August 2002. Rural Industries Research and Development Corporation, Barton, Australian Capital Territory.

Roberts D and Devenish K (2001). Chaff heaps – a useful source of feed for sheep and cattle. *Farmnote 55/2001*. Western Australian Department of Agriculture, South Perth, Western Australia.

Stynes BA and Vogel P (1983). A bacterial inhibition assay for corynetoxins from parasitized annual ryegrass. *Australian Journal of Agricultural Research* 34:483–489.

Than K, Colegate S, and Edgar J (2002). Development of a quantitative ELISA for screening fodder for corynetoxins. *RIRDC Report 02/118*. Rural Industries Research and Development Corporation, Barton, Australian Capital Territory.

Than KA, Cao Y, Michalewicz A, Olsen V, Anderton N, Cockrum P, Colegate S, and Edgar JA (2004). Analysis of corynetoxins: a comparative study of an indirect competitive ELISA and HPLC. In *Poisonous Plants and Related Toxins* (T Acamovic, CS Stewart and TW Pennycott, eds), pp. 402–407. CABI Publishing, Wallingford, Oxon, UK.

Vogel P, Stynes BA, Coakley W, Yeoh GT, and Petterson DS (1982). Glycolipid toxins from parasitized annual ryegrass: a comparison with tunicamycin. *Biochemical and Biophysical Research Communications* 105:835–840.

Chapter 33

Butylated Hydroxytoluene Reduces Aflatoxin B_1 Bioavailability and Hepatic Adduct Formation in Turkeys

J.A. Guarisco, J.O. Hall, and R.A. Coulombe, Jr.

Interdepartmental Graduate Toxicology Program and Department of Veterinary Sciences, Utah State University, Logan, UT 84322-4620, USA

Introduction

Aflatoxin B_1 (AFB_1), a mycotoxin produced by *Aspergillus flavus* and *A. parasiticus*, is a potent hepatotoxin in many animals, especially poultry. Turkeys are extremely sensitive to the effects of this toxin (Dalvi 1986). AFB_1 is a nearly universal contaminant of poultry feeds, and consumption of AFB_1-contaminated feed results in a variety of toxic effects, such as poor feed conversion, muscular weakness, and immunotoxicity (Newberne and Butler 1969). In turkeys, AFB_1 is metabolically activated to the reactive electrophilic AFB_1-8,9-epoxide (AFBO), a reaction apparently catalyzed primarily by a cytochrome P-450 (CYP) 1A2 homolog, and to a minor extent, 3A4 (Klein *et al.* 2000). Toxicity is reputedly mediated by reaction between AFBO and critical cellular macromolecules such as DNA, resulting in the formation of AFB_1-DNA adducts among other products (McLean and Dutton 1995).

Previous research in our laboratory revealed that this extreme sensitivity is due to a combination of efficient hepatic CYP-mediated activation and deficient glutathione S-transferase-mediated detoxification of AFB_1 (Klein *et al.* 2000). We also found that dietary food-grade antioxidant butylated hydroxytoluene (BHT) is a powerful chemoprotectant against many of the toxic effects of AFB_1 in turkeys (Klein *et al.* 2002). The apparent mode of action is via competitive inhibition of CYP-mediated conversion of AFB_1 to AFBO *in vitro* (Klein *et al.* 2003). Given these observations, we hypothesize that dietary BHT would reduce AFB_1 bioavailability and formation of AFB_1-DNA adducts in turkeys *in vivo*. Here, we confirm this hypothesis, showing that dietary BHT reduces AFB_1 bioavailability as well as hepatic AFB_1-DNA adducts in turkeys.

Materials and Methods

Animals and treatments

One-day-old male white turkeys were maintained on a corn-based commercial diet for 10 days, then one-half of the animals were placed on a diet containing 4000 ppm BHT (treatment group) while the remaining birds were maintained on commercial diet (control group). At 21 days of age, both groups of animals were administered a single oral dose of ^{3}H-AFB$_1$ (0.5 μg kg^{-1}; 10 μCi/animal). The birds were euthanized by CO_2 asphyxiation at 2, 4, 8, 12, and 24 hours post ^{3}H-AFB$_1$ administration. Blood was collected by cardiac puncture; liver and muscle samples were rapidly removed. Livers were perfused with cold phosphate-buffered saline (PBS; pH 7.4), and both liver and muscle samples were frozen on dry ice and stored at –80°C until analysis.

AFB$_1$ activation assay

Due to its short half-life, formation of the AFBO was determined as the stable, trapped AFB$_1$GSH conjugate, which provides an accurate and quantitative measurement of AFBO formation (Ramsdell and Eaton 1990), using conditions previously published (Klein *et al.* 2000).

Tissue analysis

The tissues and serum were digested with peroxide and perchloric acid and counted in a Beckman LS 6500 LSC counter (Coulombe and Sharma 1985). Liver DNA adducts were measured by isolating the genomic DNA with the Wizard® Genomic DNA Purification kit following the manufacturer's instructions. The samples were quantified with a Genesis 6 UV spectrophotometer (ThermoSpectronic, Rochester, NY) and counted in a Beckman LS 6500 LSC counter (Beckman Instrument, Inc., Fullerton, CA).

Data analysis

Data were analyzed using SigmaStat32, (SPSS Inc., Chicago, IL). The values obtained were analyzed statistically using the one-way ANOVA and *t*-test. If a significant F was determined, means were separated using the Tukey test. A level of $\underline{P}<0.05$ was chosen as a significant difference between control and treated means.

Results

The addition of BHT to liver microsomes decreased activation of AFB$_1$ to the active AFBO formation in a dose-dependent manner (Figure 1). This inhibitory effect was statistically significant at the two highest BHT concentrations.

In birds receiving dietary BHT, serum concentrations of AFB$_1$ were significantly decreased at all time points after oral administration of this mycotoxin (Figure 2). The inhibition was dose-dependent, as increasing BHT concentrations resulted in concomitant reductions in metabolic conversion of AFB$_1$ to AFBO. Likewise, tissue concentrations of

AFB_1 were significantly lower in birds receiving dietary BHT compared to controls, but at 2, 12 and 24 hours post AFB_1 administration (Figure 3).

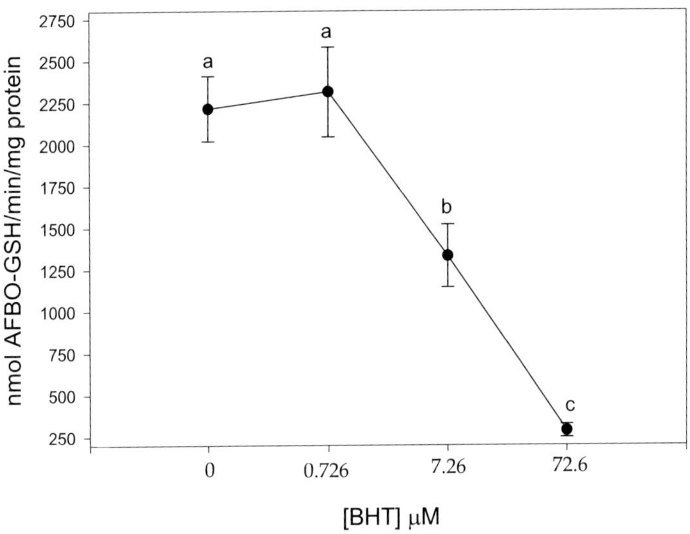

Figure 1. The inhibitory effect of BHT on liver microsomal activation of AFB_1. Because of its inherent instability, AFBO was measured as the stable AFB_1GSH conjugate by HPLC. Different letters denote a statistically significant difference ($\underline{P}<0.05$). Values are the mean ± SD (n = 3).

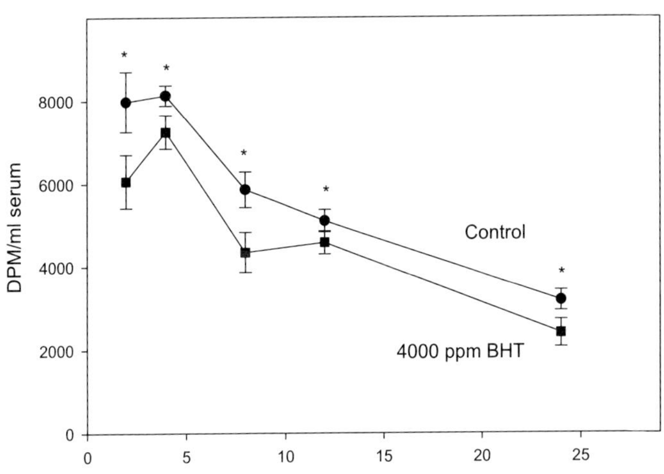

Figure 2. Addition of 4000 ppm dietary BHT decreases the amount of ^3H-AFB_1 levels in 21-day-old turkey poult serum. Values are in mean ± SD, (n=4). A * indicates a statistically significant difference between treatments $\underline{P}< 0.05$.

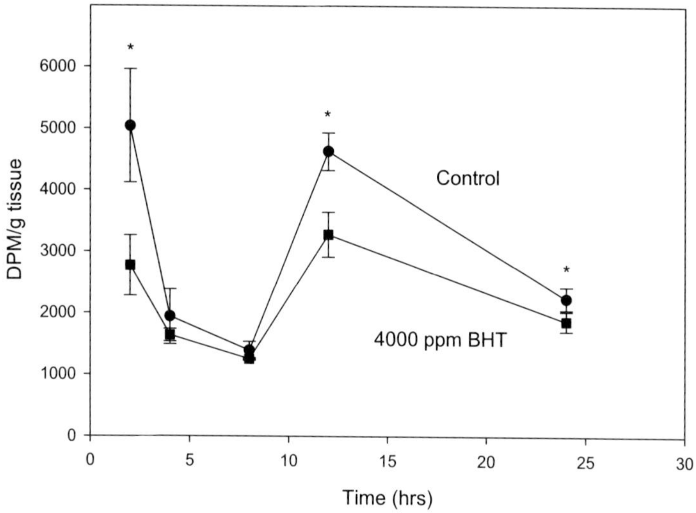

Figure 3. Addition of 4000 ppm dietary BHT decreases the amount of ^{3}H-AFB_1 levels in 21-day-old turkey poult breast muscle. Values are in mean ± SD (n=4). A * indicates a statistically significant difference between treatments $\underline{P} < 0.05$.

In the same way, dietary BHT significantly reduced the formation of hepatic AFB_1-DNA adducts compared to control birds 12 and 24 hours post AFB_1 administration (Figure 4). At these time points, adduct formation in the BHT group was decreased 48 and 61%, respectively, compared to control birds.

Conclusions

Like thousands of natural and synthetic toxins, AFB_1 must first be bioactivated to exert its toxicity in animals and people. In the case of AFB_1, this conversion is mediated primarily by CYP1A homolog (Klein *et al.* 2000), which catalyzes the formation of AFBO, which can bind to DNA, forming AFB_1-DNA and other adducts. These and other events have been postulated to underlie both short- and long-term toxicity of AFB_1 (Coulombe 1993).

Chemoprevention is one strategy employed in human and veterinary medicine to reduce the impact of toxins through prior treatment with natural or synthetic chemicals that alter metabolism and bioavailability of toxic chemicals. Indeed, inhibition of bioactivation is one of many chemopreventive strategies to reduce or eliminate chemical toxicity and/or carcinogenesis. We have previously shown that dietary BHT reduces liver microsomal AFB_1 bioactivation *in vitro* (Klein *et al.* 2003). That BHT added to liver microsomal preparations also inhibits AFB_1 bioactivation confirms the earlier observation.

Here, we show that dietary BHT also causes significant reductions in AFB_1 bioavailability and AFB_1-DNA adduct formation in the whole animal, events both reasonable to assume are a result of inhibition of AFB_1 bioactivation. Thus, in addition to reducing the clinical signs of aflatoxicosis (Klein *et al.* 2000), dietary BHT may have the added benefit of reducing AFB_1 residues in poultry, thereby resulting in a safer food for consumers.

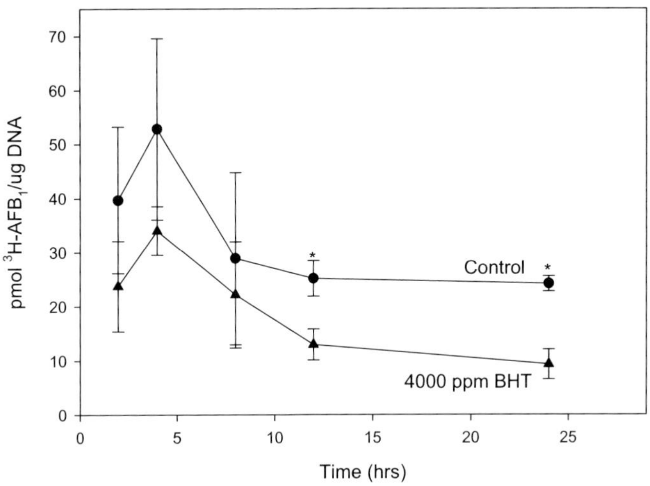

Figure 4. Addition of 4000 ppm dietary BHT decreases the amount of ^3H-AFB_1-DNA adducts in 21-day-old turkey poult liver. Values are in mean ± SD, (n=4).
A * indicates a statistically significant difference between treatments $\underline{P} < 0.05$.

Acknowledgements

This work was supported by USDA-NRI competitive grant 2002-35204-12294, and by the Utah Agricultural Experiment Station, where this paper is published as number 7718.

References

Coulombe RA, Jr (1993). Biological action of mycotoxins. *Journal of Dairy Science* 76(3):880–891.

Coulombe RA, Jr and Sharma RP (1985). Clearance and excretion of intratracheally and orally administered aflatoxin B1 in the rat. *Food and Chemical Toxicology* 23(9):827–830.

Dalvi RR (1986). An overview of aflatoxicosis of poultry: Its characteristics, prevention, and reduction. *Veterinary Research Communications* 10:429–443.

Klein PJ, Buckner R, Kelly J, and Coulombe RA Jr (2000). Biochemical basis for the extreme sensitivity of turkeys to aflatoxin B1. *Toxicology and Applied Pharmacology* 165:45–52.

Klein PJ, Van Vleet TR, Hall JO, and Coulombe RA, Jr (2002). Dietary butylated hydroxytoluene protects against aflatoxicosis in turkeys. *Toxicology and Applied Pharmacology* 182:11–19.

Klein PJ, Van Vleet TR, Hall JO, and Coulombe RA, Jr (2003). Effects of dietary butylated hydroxytoluene on aflatoxin B1-relevant metabolic enzymes in turkeys. *Food and Chemical Toxicology* 41:671–678.

McLean M and Dutton MF (1995). Cellular interactions and metabolism of aflatoxin: An update. *Pharmacology and Therapeutics* 65:163–192.

Newberne PW and Butler WH (1969). Acute and chronic effects of aflatoxin on the liver of domestic and laboratory animals: A review. *Cancer Research* 29:236–250.

Ramsdell HS and Eaton DL (1990). Species susceptibility to aflatoxin B1 carcinogenesis: Comparative kinetics of microbial biotransformation. *Cancer Research* 50(3):615–620.

CNS

Chapter 34

Poisonous Plants Affecting the Central Nervous System (CNS) of Cattle in Brazil

R.R. Rech[1], A. Rodrigues[1], D.R. Rissi[1], F. Riet-Correa[2], and C.S.L. Barros[1]

[1]*Laboratório de Patologia Veterinária, Universidade Federal de Santa Maria, Santa Maria, RS, 97105-900, Brazil;* [2]*Centro de Saúde e Tecnologia Rural, Universidade Federal de Campina Grande, Patos, PB, 58700-000, Brazil*

Introduction

With the growing importance of surveillance for BSE (bovine spongiform encephalopathy), pursuing and accurately diagnosing CNS (central nervous system) disease in cattle has increased substantially. In Brazil, there are many plants that will induce CNS disease in cattle, including those plants causing lysosomal storage disease (*Ipomoea carnea* sub. *fistulosa*, *Phalaris angusta*, *Sida carpinifolia* and *Solanum fastigiatum* var. *fastigiatum*), primary neuronal degeneration (*Ipomoea asarifolia*, *Prosopis juliflora*) and plants causing spongy degeneration of the CNS white matter (*Ateleia glazioviana*, *Tetrapterys* spp. and *Senecio* spp.). In addition, recently *Marsdenia* sp. was described causing CNS disease in cattle but no morphological changes could be associated with the disease.

Ipomoea carnea sub. *fistulosa*

Poisoning by *Ipomoea fistulosa* (Convolvulaceae) causes lysosomal storage of glycoprotein-derived oligosaccharides (mannosidosis) due to swainsonine inhibition of α-mannosidase in neurons and visceral cells of cattle, sheep and goats from the Amazon, Pantanal and northeast regions of the country. Cattle need to consume large amounts of the plant for several weeks to develop the disease, which runs a protracted course. Clinical signs are nonspecific and include lethargy, slow movements and rough hair coat (Tokarnia *et al.* 1960). No gross lesions are found; microscopically there is cytoplasmic vacuolation of neurons (mainly Purkinje cells) and several other cells, (e.g. tubular epithelium, hepatocytes and thyroid cells). Toxic substances in *Ipomoea fistulosa* include the indolizidine alkaloid swainsonine, 2-epi-lentiginosine, calystegines and N-metyl-trans-4-hydroxi-1-proline (Haraguchi *et al.* 2003).

Phalaris angusta

Phalaris spp. (Gramineae) poisoning in ruminants occurs in three forms: a 'sudden death' syndrome; acute Phalaris poisoning associated with transient nervous signs; and a chronic form referred to as Phalaris staggers, characterized by gradual development of neurological signs and characteristic lesions in the CNS. Outbreaks of Phalaris staggers, a neuronal lipofuscinosis, have been described in cattle in southern Brazil associated with the consumption of *P. angusta* (Gava *et al.* 1999). Morbidity and mortality rates are 5% and 50% respectively. Cattle poisoned by *P. angusta* present constant movement of the ears, apprehension, hypersensitivity to external stimuli such as touch and noise, spasticity, convulsive seizures and lateral recumbency with paddling (Gava *et al.* 1999). In the chronic form there are diffuse brown-green bilateral symmetrical areas of discoloration of the thalamus, brain stem and cerebellar gray matter. Microscopically pigmented neurons, usually involving specific nuclei, are evident. The pigment is loosely dispersed in the perykarion as golden or dark brown pigment granules of lipofuscin (Sousa and Irigoyen 1999) which are strongly positive using the PAS stain but negative for hemosiderin (Prussian blue stain). Tryptamine and β-carboline are active principles extracted from *Phalaris* spp. but the toxic principle and pathogenesis remain to be fully clarified. While the observation of the pigment in neurons is diagnostically useful in suspected cases of toxicity, especially in young animals, the accumulation of lipofuscin seems insufficient to explain the clinical signs and mortality. Some have proposed that central nervous system signs result from serotonergic effects of toxins on upper motor neurons. The principal differential diagnoses for neuronal pigmentation include old age lipofuscinosis and poisoning by other plants such as *Trachyandra laxa* and *T. divaricata*, which is not reported in Brazil.

Sida carpinifolia

The consumption of *S. carpinifolia* (Malvaceae) induces a chronic degenerative disease in goats (Driemeier *et al.* 2000), horses (Loretti *et al.* 2003), sheep (Seitz *et al.* 2005) and cattle (Gava 2003, personal communication) in southern Brazil. The plant is palatable to livestock but morbidity and mortality rates are low; the toxic principle of *S. carpinifolia* (swainsonine) inhibits the degradation of oligosaccharides. Cattle seem less susceptible than goats, but there was one as yet unpublished outbreak where five out of 20 cattle were affected. Clinical signs include tremors and ataxia which deteriorate if the animal is disturbed. There are no gross changes. Microscopically there is fine vacuolation of the cytoplasm of neurons of the CNS and peripheral ganglia and of several other cells of other organs. These lesions are reversible if the ingestion of the plant is discontinued. Neuronal lesions caused by the ingestion of *S. carpinifolia* were observed as incidental lesions in two cows that died of rabies (Driemeier 2003, personal communication).

Solanum fastigiatum var. *fastigiatum*

The ingestion of large amounts of *S. fastigiatum* (Solanaceae) for long periods causes a neurological disease in cattle of southern Brazil characterized by transient convulsive seizures with low to moderate morbidity (1–20%) and very low mortality (Riet-Correa *et al.* 1983). The consumption of the plant occurs mostly due to dearth of forage, and cattle of all breeds and both sexes over 1 year of age can be affected. Clinical signs are those of cerebellar ataxia and include transient seizures with loss of balance, hypermetria, nystagmus, opisthonous and falling. The frequency of crises is variable but they usually occur when cattle are disturbed. The disease is chronic and the clinical signs may persist for several years after cattle are withdrawn from infested pastures. Death is rare but may result from acquired trauma or drowning during the seizures. There are no specific lesions at necropsy but those related to trauma can be found. Histologically there is fine vacuolation of the cytoplasm of the Purkinje cells with progressive loss of these cells and depopulation of Purkinje layer and reactive Bergman glia proliferation; spheroids are found within the granule cell layer and in the white matter of the cerebellum (Riet-Correa *et al.* 1983; Barros *et al.* 1987). In severely affected cows there is thinning of the molecular layer of the cerebellum (unpublished observations). The vacuolation of the cytoplasm of the Purkinje cell neurons decreases if the ingestion of plant is discontinued (Zambrano *et al.* 1985). The active principle is unknown but ultrastructural (Barros *et al.* 1987) and lectin histochemical studies (Paulovich *et al.* 2002) indicate that *S. fastigiatum* induces glycolipid lysosomal deposits in the cytoplasm of Purkinje cells. The reason that the lesions are restricted to Purkinje neurons is unclear.

Ipomoea asarifolia

Ipomoea asarifolia (Convolvulaceae) is a tremorgenic plant with an unknown active principle. The intoxication occurs in the northeast (Döbereiner *et al.* 1960) and affects mainly young sheep, but goats and cattle are also affected (Medeiros *et al.* 2003). Ingestion occurs during the dry season due to lack of more suitable forage. Clinical signs include muscle tremors, hypermetria and hypersensitivity to noise. When the bovine is left alone the clinical signs subside but a wide base stance is maintained. If the ingestion of the plant is discontinued, clinical signs gradually abate. Histologically there are degeneration and loss of Purkinje cells, axonal spheroids in the granular layer and foci of demyelination in the white matter of the cerebellum.

Prosopis juliflora

The intoxication by the pods of *P. juliflora* is a chronic neurological disease frequent in cattle in the semiarid region of northeastern Brazil and is characterized by atrophy of the masseter muscle, protusion and involuntary movements of the tongue, decreased tonus of mandibular muscles, head tilt during rumination, abundant drooling, yawning, dysphagia and continuing mastigatory movements (Tabosa *et al.* 2000a). The pods are used as feed for cattle in the dry season and when ingested for long periods cause degeneration of neurons of motor (trigeminal) nuclei in the brainstem, which results in atrophy and paralysis of masticatory muscles, leading to starvation due to difficulty in apprehension and maintenance of food in the mouth. Piperidinic alkaloids were isolated from the plant (Tabosa *et al.* 2000b) but their role in the intoxication is uncertain. Necropsy findings include changes related to starvation and atrophy of masticatory muscles. Microscopically there is chromatolysis and fine perikaryal vacuolation in neurons of the trigeminal motor nuclei with eventual loss of these cells. Ultrastructurally the neuronal lesion consists of mitochondrial swelling with disorientation and fragmentation of mitochondrial cristae.

Ateleia glazioviana, Tetrapterys spp. and *Senecio* spp.

These plants induce a condition in cattle which was referred to as spongiform myelinopathies (Van der Lugt 2002) 'in which the dominant pathological feature is vacuolation of myelin in the absence of a significant degree of myelin breakdown or phagocytosis.' All of the plants listed are causes of prominent vacuolation in the white matter of the CNS by splitting of myelin lamellae (Raffi *et al.* 2005), a condition known as spongy degeneration. Spongy degeneration associated with *A. glazioviana* and *Tetrapterys* spp. is caused by the primary action of an unknown toxic principle(s) in the CNS (Tokarnia *et al.* 2000; Gava and Barros 2001) whereas the same lesion found in *Senecio* spp. affected cattle is secondary to liver failure due to cirrhosis (hepatic encephalopathy). *Senecio* spp. and *A. glazioviana* are important causes of death in cattle in southern Brazil, and *Tetrapterys* spp. affects cattle in the southeastern region.

The ingestion of the roots of *Marsdenia* sp. was recently described as causing neurological disease in cattle from northeastern Brazil (Riet-Correa *et al.* 2004). Within a few hours after ingestion, cattle were recumbent. Clinical signs included muscular tremors, muscle fasciculations, hyperesthesia, marked salivation, continuous masticatory movements, opisthotonus, clonic convulsions, and dyspnea. No gross or microscopic lesions were found and the toxic active principle in the plant is unknown.

References

Barros SS, Riet-Correa F, Andujar MB, Barros CSL, Méndez MC, and Schild AL (1987). *Solanum fastigiatum* and *Solanum* spp. poisoning in cattle ultrastructural changes in the cerebellum. *Pesquisa Veterinária Brasileira* 7:1–5.

Döbereiner J, Tokarnia CH, and Canella CFC (1960). Intoxicação experimental pela 'salsa' (*Ipomoea asarifolia*) em ruminantes. *Arquivos do Instituto Biológico* 3:39–57.

Driemeier D, Colodel EM, Gimeno EJ, and Barros SS (2000). Lysosomal storage disease caused by *Sida carpinifolia* poisoning in goats. *Veterinary Pathology* 37:153–159.

Gava A and Barros CSL (2001). Field observations of *Ateleia glazioviana* poisoning in cattle in Southern Brazil. *Veterinary and Human Toxicology* 43:37–41.

Gava A, Sousa RS, de Deus MS, Pilati C, Cristani J, Mori AM, and Neves DS (1999). *Phalaris angusta* (Gramineae) como causa de enfermidade neurológica em bovinos no Estado de Santa Catarina. *Pesquisa Veterinária Brasileira* 19:35–38.

Haraguchi M, Gorniak SL, Ikeda K, Minami Y, Kato A, Watson AA, Nash RJ, Molyneux RJ, and Asano N (2003). Alkaloidal components in the poisonous plant *Ipomoea carnea* (Convolvulaceae). *Journal of Agriculture and Food Chemistry* 51:4995–5000.

Loretti AP, Colodel EM, Gimeno EJ, and Driemeier D (2003). Lysosomal storage disease in *Sida carpinifolia* toxicosis: An induced mannosidosis in horses. *Equine Veterinary Journal* 35:434–438.

Medeiros RMT, Barbosa RC, Riet-Correa F, Lima EF, Tabosa IM, Barros SS, Gardner DR, and Molyneux RJ (2003). Tremorgenic syndrome in goats caused by *Ipomoea asarifolia*. *Toxicon* 41:933–935.

Paulovich FB, Portiansky EL, Gimeno EJ, Schild AL, Méndez MC, and Riet-Correa F (2002). Lectin histochemical study of lipopigments present in the cerebellum of *Solanum fastigiatum* var. *fastigiatum* intoxicated cattle. *American Journal of Veterinary Medicine* 49:473–477.

Raffi MB, Rech RR, Sallis ESV, Oliveira FN, Barros SS, and Barros CSL (2005). Chronic cardiomyopathy and cerebral spongy changes in sheep experimentally fed with *Ateleia glazioviana*. *Journal of Comparative Pathology* (submitted)

Riet-Correa F, Méndez MC, Schild AL, Summers BA, and Oliveira JA (1983). Intoxication by *Solanum fastigiatum* var. *fastigiatum* as a cause of cerebellar degeneration of cattle. *The Cornell Veterinarian*. 73:240–256.

Riet-Correa F, Silva DM, Oliveira OF, Simões SVD, Medeiros RMT, and Nóbrega RS (2004). Intoxicação por *Marsdenia* spp. (mata-calado) em ruminantes. *Pesquisa Veterinária Brasileira* (Suppl.): 50–51.

Seitz AL, Colodel EM, Schmitz M, Gimeno EJ, and Driemeier D (2005). Use of lectin histochemistry to diagnose *Sida carpinifolia* (Malvaceae) poisoning in sheep. *The Veterinary Record* 156:386–388.

Sousa RS and Irigoyen LF (1999). Intoxicação experimental por *Phalaris angusta*. *Pesquisa Veterinária Brasileira* 19:116–122.

Tabosa IM, Pamplona FV, and Quintans Júnior A (2000a). Isolamento e identificação de juliprosopina, um dos constituintes tóxicos da *Prosopis juliflora*. *Revista Brasileira de Farmácia* 9:11–22.

Tabosa IM, Souza JCA, Barbosa-Filho JM, Almeida RN, Graça DL, and Riet-Correa F. (2000b). Neuronal vacuolation of the trigeminal nuclei in goats caused by the ingestion of *Prosopis juliflora* pods (mesquite beans). *Veterinary and Human Toxicology* 42:155–158.

Tokarnia CH, Döbereiner J, and Peixoto PV (2000). *Plantas Tóxicas do Brasil*. Helianthus, Rio de Janeiro.

Tokarnia CH, Döbereiner J, and Canella CFC (1960). Estudo experimental sobre a toxidez do 'canudo' (*Ipomoea fistulosa*) em ruminantes. *Arquivos do Instituto de Biologia Animal* 3:57–71.

Van der Lugt JJ (2002). Myelinopathies. In *The Clinicopathology and Pathology of Selective Toxicoses and Storage Diseases of the Nervous System of Ruminants in Southern Africa*. (JJ Van der Lugt, ed.), p. 67. Doctoral Thesis, University of Utrecht.

Zambrano MS, Riet-Correa F, Schild AL, and Méndez MC (1985). Intoxicação por *Solanum fastigiatum* var. *fastigiatum*: evolução e reversibilidade das lesões em bovinos e suscetibilidade de ovinos, coelhos, cobaios e ratos. *Pesquisa Veterinária Brasileira* 5: 133–141.

Chapter 35

Comparative Study of Bovine Cerebellar Cortical Degenerations Caused by Different South American *Solanum* spp.

J. M. Verdes[1,2], A. Moraña[1], F. Gutiérrez[1], F. Guerrero[2], A. Goicoa[2], L.E. Fidalgo[2], and F. Riet-Correa[3]

[1]*Facultad de Veterinaria, Universidad de la República, Av. A. Lasplaces 1550, CP 11600, Montevideo, Uruguay;* [2]*Facultade de Veterinaria de Lugo, Universidade de Santiago de Compostela, Campus Universitario s/n, 27002 Lugo, Spain;* [3]*Centro de Saúde e Tecnologia Rural. Federal University of Campina Grande, Patos, Paraíba, Brazil 58700-000*

Introduction

The Solanaceae family includes 90 genera and more than 2300 species, 75% belonging to the *Solanum* genus, with worldwide distribution. This genus is particularly abundant in tropical and subtropical regions of Central and South America (Ruiz-Díaz *et al.* 2004), and encompasses many toxic plant species which cause multiple disease conditions and death by different mechanisms. Many clinical syndromes including hypervitaminosis D and acquired cerebellar degeneration have been reported and attributed to ingestion of various *Solanum* species (Baker *et al.* 1989).

Irreversible cerebellar disorders in ruminants due to spontaneous or experimental poisoning with different *Solanum* species have been described in South Africa (Pienaar *et al.* 1976; van der Lugt *et al.* 2002), USA (Menzies *et al.* 1979; Porter *et al.* 2003), Australia (Bourke 1997), and South America (Riet-Correa *et al.* 1983; Barros *et al.* 1987; Medeiros *et al.* 2004; Verdes *et al.* 2006). *Solanum fastigiatum* Willd. var. *fastigiatum* (Jurubeba) in Brazil and *S. bonariense* L. (Naranjillo) in Uruguay cause spontaneous disease, and the intoxication has been experimentally reproduced in cattle (Riet-Correa *et al.* 1983; Verdes *et al.* 2006). *Solanum paniculatum* L. (also known as Jurubeba), a shrub commonly used as an infusion in Brazilian folk medicine for the treatment of liver and gastrointestinal disorders in humans with a potent inhibitory effect in gastric acid secretion in mice (Mesia-Vela *et al.* 2002), also causes cerebellar degeneration when administrated experimentally in cattle (Medeiros *et al.* 2004). The disease was also induced by an unknown species (Barros *et al.* 1987), later identified as *S. paniculatum* (Riet-Correa, 2000, unpublished data). As in other cases of *Solanum* intoxication (Pienaar *et al.* 1976; Menzies *et al.* 1979; Bourke 1997; van der Lugt *et al.* 2002; Porter *et al.* 2003), the main pathophysiological features in South

American species are the vacuolation of the perikaria in Purkinje cells followed by loss of these neurons, and the presence of axonal torpedoes (or spheroids) as well as microcavitations in the cerebellar granular layer and/or white matter (Riet-Correa *et al.* 1983; Barros *et al.* 1987; Medeiros *et al.* 2004; Verdes *et al.* 2006). Different hypotheses about the pathogenic mechanism of this intoxication have been proposed. Riet-Correa *et al.* (1983) suggested that these plants cause an induced lysosomal storage disease, probably gangliosidosis. The main feature of the intoxication is the selective damage of Purkinje cells. Barros *et al.* (1987) suggested that the disease is not a typical storage disease, proposing a physico-chemical interaction between the active principle of the plants and lipids, with concomitant formation of a complex which would be less susceptible to enzymatic degradation than normal lipids. Ultrastructural studies in Purkinje cell perikaria also showed alteration of the endoplasmic reticulum, a decrease in the number of neurofilaments and microtubules (Barros *et al.* 1987), and accumulation of vesicles (Verdes *et al.* 2006). Similar alterations of the cytoskeleton and accumulation of vesicles are observed in axonal spheroids of granular layer and sub-cortical white matter (Barros *et al.* 1987; Verdes *et al.* 2006). These details taken together suggest the hypothesis that the occurrence of axonal spheroids in granular layer and white matter is caused by an axonal transport alteration (Verdes *et al.* 2006). The toxic compound causing neurologic dysfunction in ruminants has not been definitively determined (Baker *et al.* 1989), although in the case of *crazy cow syndrome* induced by *Solanum dimidiatum* in USA, calystegin B_2 has been proposed as a potential causative agent (Molyneux *et al.* 1994). To date, attempts to isolate active fractions responsible for the disease in *Solanum* spp. have been unsuccessful, and no susceptible laboratory animals have been found (Riet-Correa *et al.* 1983; Zambrano *et al.* 1985; Verdes *et al.* 2000; Ruiz-Díaz *et al.* 2004). The identification of the active principle, its molecular mechanism, and the development of bioassays could be an important step to define human and animal health concerns with these *Solanum* species.

The objective of this paper was to review the epidemiology, clinical signs, lesions, ultrastructural changes, lectin- and immuno-histochemical patterns of the spontaneous and experimental disease in cattle. The identification of other susceptible species and the active fractions in *Solanum* species are also reviewed.

Epidemiology and Clinical Signs of Spontaneous Disease in Cattle

In Brazil and Uruguay the spontaneous disease affects cattle older than 8 months of various breeds raised, but dairy cattle and crossbreeds are most affected. Morbidity varies from 1% to 20%. Mortality is always less than 1% and the deaths are associated with progressive weakening or drowning accidents in some affected animals (Riet-Correa *et al.* 1983; Verdes *et al.* 2006). A major feature in affected cattle is the occurrence of periodic seizures without loss of consciousness, lasting up to one minute. Animals appear to be clinically normal between episodes. Nervous signs appear spontaneously or are induced when affected animals are excited or intentionally stressed by the Head Raising Test (HRT) (Pienaar *et al.* 1976). Clinical observations include ataxia, hypermetria, hyperesthesia, staggering gait, muscle tremors, head and thoracic limbs extension, opisthotonus, nystagmus, and in most affected animals falling to the side or backward. In a few animals with permanent neurological signs, dissymmetric gait or 'star gazing' attitude are observed (Riet-Correa *et al.* 1983; Verdes *et al.* 2006).

Bovine Experimental Cases and Other Animal Models

Experiments with *S. fastigiatum* var. *fastigiatum*

Cerebellar signs in cattle were first observed by HRT after administration of 0.4 kg day^{-1} of dried leaves via rumen cannula for 155 days (total dose 0.18 kg of dried leaves/kg bw), or after administration of commercial ration containing 10% of dry plant for 76 days (total dose 0.14 kg of dried leaves kg^{-1} bw) (Riet-Correa *et al.* 1983). In sheep, similar cerebellar lesions were observed after administration of a commercial ration containing 20% of dry plant for 202 to 370 days (total dose 0.43 kg of dried leaves kg^{-1} bw). Neither guinea pigs, rabbits or rats are susceptible after being given an oral dose via commercial ration containing 10% of dry plant for 120 days (Zambrano *et al.* 1985).

Experiments with *S. paniculatum* L.

Cerebellar signs in cattle were first observed by HRT after administration of green leaves *ad libitum* for 639 days (total dose 0.94 kg of green leaves kg^{-1} bw) in one animal ingesting a commercial ration (Barros *et al.* 1987), or after administration of 5 g kg^{-1} day^{-1} of dried leaves via rumen cannula from 93 to 142 days (total dose 0.17 kg of dried leaves kg^{-1} bw) (Barros *et al.* 1987; Medeiros *et al.* 2004). No toxic signs were observed in mice following administration of different aqueous extracts up to 2 g kg^{-1} bw intraduodenally, but these doses promote an antiulcer activity (Mesia-Vela *et al.* 2002).

Experiments with *Solanum bonariense* L.

Cerebellar signs in cattle (induced by HRT) were observed after the daily administration of green leaves, in amounts equivalent to 1% bw, via rumen cannula for 128 to 136 days (total dose 1.024 kg of green leaves kg^{-1} bw) (Verdes *et al.* 2006). Microhistological analysis of fecal samples obtained from experimental calves showed *S. bonariense* epidermis in 93% of slides studied (García y Santos *et al.* 2004). An equivalent dose of 1.024 kg of green leaves kg^{-1} bw of a lyophilized hydro-alcoholic extract solubilized in 1 mL of sterile water injected intraperitoneally produced acute neurological signs without lesions in mice and rats. Thin layer chromatography of hydro-alcoholic extract showed specific polar alkaloids different from solanine, solanidine, chaconine or foliumin B (Verdes *et al.* 2000; Ruiz-Díaz *et al.* 2004).

Histopathology, Electron Microscopy, Lectin- and Immuno-histochemistry

There were no significant gross lesions in all cattle studied. The microscopic lesions were specifically localized in the cerebellum. Vacuolation of the perikaria was present in Purkinje neurons. Eosinophilic, round, homogeneous, occasionally fine granular axonal spheroids (or torpedoes) and microcavitations were found in the granular layer (Riet-Correa *et al.* 1983; Barros *et al.* 1987) and cerebellar white matter (Riet-Correa *et al.* 1983; Barros *et al.* 1987; Verdes *et al.* 2006). Ultrastructural study of degenerative Purkinje cells revealed perikaria filled with electron-dense residual bodies, which were sometimes accompanied by lamellar arrays of endoplasmic reticulum without ribosomes, and membranous organelles constituted by polar lipids and membrane debris. Axonal spheroids are the result of myelinated axon enlargement, with many heterogeneous membranous organelles filled with residual bodies and mitochondria (Riet-Correa *et al.* 1983; Barros *et al.* 1987; Verdes *et al.* 2004b, 2006). Lectin binding pattern of the stored material present in Purkinje cells in *S. fastigiatum* var. *fastigiatum* poisoning is compatible with a glycolipid storage disease (Paulovich *et al.* 2002).

Immunohistochemical study of the cytoskeleton of Purkinje cells in *S. bonariense* L. poisoning (Verdes *et al.* 2004a) shows that it resembles the ultrastructural cytoskeletal alteration reported in S. *fastigiatum* var. *fastigiatum* and *S. paniculatum* L. poisonings (Barros *et al.* 1987), thus reinforcing the hypothesis of a cytoskeletal alteration with a concomitant axonal transport deficit as one of the pathogenic mechanisms of these neurodegenerative diseases.

Acknowledgements

Financial support was provided by Comisión Sectorial de Investigación Científica (CSIC, UdelaR, Uruguay), and Universidade de Santiago de Compostela (Spain).

References

Baker DL, Keeler RF, and Gaffield W (1989). Pathology in hamsters administered *Solanum* plant species that contain steroidal alkaloids. *Toxicon* 27(12):1331–1337.

Barros SS, Riet-Correa F, Andujar MB, Lombardo de Barros CS, Méndez MC, and Schild AL (1987). *Solanum fastigiatum* var. *fastigiatum* and *Solanum* sp. poisoning in cattle: ultrastructural changes in the cerebellum. *Pesquisa Veterinária Brasileira* 7(1):1–5.

Bourke CA (1997). Cerebellar degeneration in goats grazing *Solanum cinereum* (Narrawa burr). *Australian Veterinary Journal* 75(5):363–365.

García y Santos MC, Verdes JM, Moraña A, Gutiérrez F, Battes D, Capella A, and Bartolomé J (2004). Análisis microhistológico de epidermis de *Solanum bonariense* como método diagnóstico durante la intoxicación experimental en terneros. *Pesquisa Veterinária Brasileira* 24(Suppl.):24–25.

Medeiros RMT, Guilherme RF, Riet-Correa F, Barbosa RC, and Lima EF (2004). Experimental poisoning by *Solanum paniculatum* (jurubeba) in cattle. *Pesquisa Veterinária Brasileira* 24(Suppl.):41.

Menzies JS, Bridges CH, and Bailey EM (1979). A neurological disease of cattle associated with *Solanum dimidiatum*. *The South Western Veterinarian* 32(1):45–49.

Mesia-Vela S, Santos MT, Souccar C, Lima-Landman MT, and Lapa AJ (2002). *Solanum paniculatum* L. (Jurubeba): potent inhibitor of gastric acid secretion in mice. *Phytomedicine* 9(6):508–514.

Molyneux RJ, James LF, and Nash RJ (1994). Identification of calystegine B_2 in *Solanum dimidiatum*: a potential causative agent of crazy cow disease. In *International Research Congress on Natural Products*, p. 26. Halifax, Canada.

Paulovich FB, Portiansky EL, Gimeno EJ, Schild AL, Méndez MC, and Riet-Correa F (2002). Lectin histochemical study of lipopigments present in the cerebellum of *Solanum fastigiatum* var. *fastigiatum* intoxicated cattle. *Journal of Veterinary Medicine. Serie A, Physiology, Pathology, Clinical Medicine* 49(9):473–477.

Pienaar JG, Kellerman TS, Basson PA, Jenkins WL, and Vahrmeijer J (1976). Maldronksiekte in cattle: a neuronopathy caused by *Solanum kwebense* N. E. Br. *Onderstepoort Journal of Veterinary Research* 43(2):67–74.

Porter MB, Mac Kay RJ, Uhl E, Platt SR, and de Lahunta A (2003). Neurologic diseases putatively associated with ingestion of *Solanum viarium* in goats. *Journal of American Veterinary Medical Association* 223(4):501–504.

Riet-Correa F, Mendez MC, Schild AL, Summers BA, and Oliveira JA (1983). Intoxication by *Solanum fastigiatum var. fastigiatum* as a cause of cerebellar degeneration in cattle. *Cornell Veterinarian* 73(3):240–256.

Ruiz-Díaz A, Verdes JM, Moraña A, Gutiérrez F, Battes D, and Vázquez Á (2004). Caracterización y actividad biológica de la fracción hidroalcohólica de *Solanum bonariense* L. en ratones. *Pesquisa Veterinária Brasileira* 24(Suppl.):57–58.

Van der Lugt JJ, Bastianello SS, van Ederen AM, and van Wilpe E (2002). Cerebellar cortical degeneration in cattle caused by *Solanum kwebense*. In *The Clinicopathology and Pathology of Selective Toxicoses and Storage Diseases of the Nervous System on Ruminants in Southern Africa* (JJ Van der Lugt, PhD thesis), pp. 49–65. Faculty of Veterinary Medicine, Utrecht University, Utrecht, The Netherlands.

Verdes JM, Moraña A, and Gutiérrez F (2000). Toxicity of fractions obtained from *Solanum bonariensis*. *World Buiatrics Congress* (21):10381–10384.

Verdes JM, Battes D, Ruiz P, Gutiérrez F, Moraña A, Guerrero F, Fidalgo LE, Goicoa A, and Kun A (2004a). Estudio inmunohistoquímico de las alteraciones de citoesqueleto en células de Purkinje del cerebelo de bovinos intoxicados con *Solanum bonariense* (Linnaeus, 1753). *Pesquisa Veterinária Brasileira* 24(Suppl.):75–76.

Verdes JM, Moraña A, Battes D, Gutiérrez F, Goicoa A, Fidalgo LE, and Guerrero F (2004b). Modificaciones ultraestructurales en la degeneración cerebelosa inducida experimentalmente por *Solanum bonariense* (Linnaeus, 1753). *Pesquisa Veterinária Brasileira* 24(Supl.):76–77.

Verdes JM, Moraña A, Gutiérrez F, Battes D, Fidalgo LE, and Guerrero F (2006). Cerebellar degeneration in cattle grazing *Solanum bonariense* ('Naranjillo') in Western Uruguay. *Journal of Veterinary Diagnostic Investigation* 18(3):299–303.

Zambrano MS, Riet-Correa F, Schild AL, and Méndez MC (1985). Intoxicacao por *Solanum fastigiatum* var. *fastigiatum*: evolucao e reversibilidade das lesoes em bovinos, e susceptibilidade de ovinos, coelhos, cobaios e ratos. *Pesquisa Veterinária Brasileira* 5(4):133–141.

Chapter 36

Poisoning by Cyanogenic Plants in Ruminants in the State of Paraíba, Northeastern Brazil

S.L. de Amorim[1], R.M.T. Medeiros[1], C.H. Tokarnia[2], and F. Riet-Correa[1]

[1]Centro de Saúde e Tecnologia Rural, Universidade Federal de Campina Grande, Patos, PB 58700-00, Brazil; [2]Instituto de Zootecnia, Universidade Federal Rural do Rio de Janeiro, Seropédica, RJ 23835-000, Brazil

Introduction

Cyanogenic toxic plants in Brazil are *Manihot* spp. (Euphorbiaceae), including *Manihot esculenta* Crantz. (cassava) and many wild species of *Manihot*; *Anadenanthera macrocarpa* (= *Piptadenia macrocarpa*) and *Piptadenia viridiflora*, trees from the Leguminosae Mimosoideae sub-family (Tokarnia *et al.* 1994, 1999); *Sorghum* spp. (Tokarnia *et al.* 2000); and *Prunus sellowii* (= *P. sphaerocarpa)*, a tree from the Rosaceae family (Saad and Camargo 1967; Gava *et al.* 1992). Few outbreaks of poisoning by *Cynodon dactylon* (Tifton) have been reported (Gava *et al.* 1998). *Holocalyx glaziovii*, a tree of the Leguminosae Mimosoideae sub-family, is also cyanogenic, but the spontaneous poisoning has not been reported (Armien *et al.* 1995). The object of this paper is to review the cyanogenic plants of northeastern Brazil, with emphasis on those occurring in the state of Paraíba.

Cyanogenic Plants of Northeastern Brazil

Poisoning by *Manihot esculenta*

Poisoning by *M. esculenta* (cassava; mandioca in Portuguese) is a well-known cause of livestock death in Brazil. There are different varieties of cassava, containing linamarin and lotaustralin in variable concentrations. The less toxic varieties, known as 'mansas' (mild), 'aipim' or 'macaxeira,' are used for human food. Other varieties, known as 'bravas' (fierce), containing higher concentrations of cyanide, are used for flour, fine flour or latex production. During the production of flour, cyanogenic glycosides are destroyed by heating. The roots of the more toxic (fierce) cassava varieties are also used for animal food. Before being given to ruminants, the roots must be chopped and left in the open air for some hours.

Because of the variation in the concentration of HCN in different varieties of cassava, this process requires variable amounts of time. Some farmers recommend caution feeding naive animals cassava because they appear to be more susceptible than those that have been routinely ingesting the roots. Poisoning occurs when ruminants consume the roots shortly after collection, without previous detoxification treatment, or when animals invade fields of the plant, ingesting their leaves. Poisoning also occurs when animals ingest the liquid residue (called manipuera) from the compression of ground cassava for flour production (Tokarnia *et al.* 2000).

Poisoning by *Manihot* spp.

In Brazil, there are many species of wild *Manihot* shrubs or small trees. Ruminants ingest this plant in natural pastures or as hay or silage. Some *Manihot* species are used for animal food because of their resistance to drought, good palatability, and nutritional value, containing about 12% protein. In the beginning of the 20th century some of these species, including *Manihot glaziovii*, *M. dichotoma*, and *M. cearulescens* were used commercially for latex production. Another five species are important in the semiarid region: *Manihot diamantinensis*, *M. jacobinensis*, *M. janiphoides*, *M. maracasensis*, and *M.* sp. (mandioca tapuio). There is also a hybrid between wild species and cassava (prinunça, pornuncia, mandioca-de-sete-anos or maniçoba-de-jardim), which is used as an ornamental, but it was also used for flour production (Araújo and Cavalcanti 2002). The concentration of HCN in different species of *Manihot* is variable. Araújo and Cavalcante (2002) report the concentration of 1289 mg of cyanide per kg dry matter in *Manihot dichotoma* var. *dichotoma*, 1104 mg kg^{-1} in *M. carthaginensis,* 1002 mg kg^{-1} in *M. epruinosa*, 304 and 529 mg kg^{-1} in two samples of *M. glaziovii* from different places, and 387 mg kg^{-1} in *M. cearulescens.* Many species of *Manihot,* including *M. glaziovii, M. pseudoglaziovii, M. carthaginensis,* and *Manihot* spp. (mandioca tapuio) have been used for forage, mainly for hay or silage production (Araújo and Cavalcanti 2002).

Poisoning by *Manihot* spp. occurs when cattle or goats have access to fields of the plant or when the animals are fed recently collected fresh plants. Farmers report that field poisoning in goats and cattle is more frequent after the first rains, when *Manihot* spp. are sprouting, sometimes earlier than the other vegetation. It is reported that the poisoning also occurs when the first rains are followed by a drought period; under this condition of forage shortage the animals ingest the green or the wilting plants of *Manihot* in larger amounts than other forages. In experiments in cattle with *Manihot piauhyensis*, poisoning was produced with doses from 2.5g kg^{-1} bw and higher (Tokarnia *et al.* 2000), and in experiments with *M. glaziovii* the plant caused death at a dose of 9.3 g kg^{-1} bw (Tokarnia *et al.* 1994, 1999).

Poisoning by *Anadenanthera macrocarpa* and *Piptadenia viridiflora*

Poisoning by these two species occurs when ruminants ingest leaves from fallen trees or branches after storms, or after the trees are cut down. *Anadenanthera macrocarpa,* known as angico preto, is found in the entire northeastern region. Experiments with the leaves of *A. macrocarpa* from different states of the northeastern region demonstrated that there are toxic and non-toxic plants (Tokarnia *et al.* 1994, 2000). The cyanogenic plants cause death when given above 9.3 g kg^{-1} bw (Tokarnia *et al.* 1999; Medeiros *et al.* 2004);

the leaves from sprouting plants in October were more toxic than mature leaves collected in March (Tokarnia *et al.* 1999).

Piptadenia viridiflora, known as espinheiro or surucucu, is reported to be toxic from the state of Bahia. Experimental poisoning was produced with doses of 10 and 14.1 g kg^{-1} bw (Tokarnia *et al.* 1999). Farmers in the northeastern region report that wilted leaves of *A. macrocarpa* and *P. viridiflora* are more toxic than the fresh leaves; but in experiments with wilted and fresh plants there was no difference in their toxicity (Tokarnia *et al.* 1994, 1999). The probable reason for this wrong interpretation by the farmers is that wilted plants are more palatable than fresh ones, and with fallen trees, animals have access to large amounts of leaves, which are mostly already wilted. Another important feature of these species in relation to other cyanogenic plants is that after being dried, the leaves maintain their toxicity for long periods (Tokarnia *et al.* 1999). Experiments in rabbits demonstrated that the dry leaves of *A. macrocarpa* and *P. viridiflora* maintain their toxicity for 6 and 14 months, respectively (Brito *et al.* 2000). Leaves from these two species also react slower to the picrate test for cyanide (10–30 min) than those of *M. glaziovii* (less than 5 min). Also, poisonings induced by *A. macrocarpa* and *P. viridiflora* have a more prolonged clinical course than the poisoning by *M. glaziovii*. These differences are probably due to the presence of more stable cyanide glycosides in *A. macrocarpa* and *P. viridiflora* than those found in *M. glaziovii* (Tokarnia *et al.* 1999).

Experiments with Cyanogenic Plants in Paraíba

Experiments with *Manihot glaziovii* in goats

These experiments were performed to determine the toxicity of *M. glaziovii* in goats, the conditions of poisoning, the toxicity of the plant in different phonological stages of growth, and the reduction of toxicity after harvest. Samples of fresh, wilting and dry leaves of *M. glaziovii* were administered orally to 104 Moxotó goats in single doses of up to 12 g kg^{-1} bw. The cyanide content of the plant samples was determined by the picrate paper test. The plant was collected from January to June 2004. When the goats with clinical signs were in lateral recumbence, they were treated intravenously with 50 mL per 100 kg bw of a 20% aqueous solution of sodium thiosulfate.

Three experiments were performed. In Experiment 1, the plant was given immediately after harvest to six goats; two ingested the ground plant and four ingested the intact plant. In Experiment 2, the intact (*i.e.*, unground) plant samples were maintained in the shade in the open air or inside plastic bags. The bagged plant was given to 18 goats at 4, 8, 12, 16, 20, 24, 48, 72, 96 and 120 hours after harvest. The unbagged plant was given to 13 goats at 4, 24, 48, 72 and 96 hours and 9, 10, 23 and 30 days after collection. In Experiment 3, the ground and bagged or unbagged plant was administered 4, 8, 12, 16, 20, 24, 48, 72 and 96 hours after collection. Seventeen goats received the bagged ground plant and 16 received the unbagged ground plant.

In Experiments 2 and 3, two or three goats were used for each period after plant collection, and the plant was given until the loss of its toxicity. Forty goats were used as controls for comparison of cardiac and respiratory functions.

In Experiment 1, the ground and unground plant had similar toxicity. In Experiment 2, the unground plant maintained in the open air maintained its toxicity during the whole experiment (30 days), and the plant maintained in plastic bags was toxic until 96 hours after harvest. In Experiment 3, the ground plant, maintained in plastic bags or in the open air, was toxic for 72 hours after harvest. In all experiments, clinical signs were characteristic of cyanide poisoning. All poisoned goats were treated successfully and survived (Amorim *et al.* 2005).

This experiment demonstrated that *M. glaziovii* has similar toxicity from January to June, in all phenological stages of growth, and during the entire rainy period. Occurrence of poisoning is dependent on the consumption of large amounts of the plant within a short time rather than only on the toxicity of the plant. It was also seen that unground *M. glaziovii* maintains its toxicity for at least 30 days and the ground plant for up to 72 hours. It is suggested that hay or silage not be prepared with the whole plant, and it should not be fed to animals for at least 3 days after preparation. The results also demonstrated that the plant cannot be used for feeding in the fresh stage under any circumstances (Amorim *et al.* 2005).

Experiments with *Manihot glaziovii* and *Anadenanthera macrocarpa* in cattle

Six collections of *M. glaziovii* and six collections of *A. macrocarpa* from different regions were administered to cattle. The picrate paper test for cyanide was performed in all collections and only positive samples were administered to the cattle. Doses were 5, 7, 8 and 10 g kg^{-1} bw. Eight animals were fed *A. macrocarpa*, 12 fed *M. glaziovii*, and two were used as controls. Four out of six collections of *A. macrocarpa* and the six collections of *M. glaziovii* were positive in the picrate paper test for cyanide. Picrate reaction times were 5±2.7 min and 13±4.8 min for *M. glaziovii* and *A. macrocarpa*, respectively. *Manihot glaziovii* was 35.3±2.7% dry matter and *A. macrocarpa* was 62.15±1%. All collections tested were toxic, but animals which received 5 g kg^{-1} bw had slight signs of cyanide poisoning and did not need treatment. All animals which received 6–10 g kg^{-1} bw of both plants showed severe signs of poisoning and were treated successfully with 50mL per100 kg bw of 20% sodium thiosulfate. Animals poisoned with *M. glaziovii* showed clinical signs within 20 min, and animals poisoned with *A. macrocarpa* had symptoms 41 min after administration.

The results of these experiments confirm the information previously reported (Tokarnia *et al.* 1999) regarding the variability of the toxicity of different samples of *A. macrocarpa*. Two collections were negative in the picrate paper test for cyanide. The other four demonstrate a longer picrate reaction time compared to *M. glaziovii*, which explains the delayed toxicosis for the latter.

Acknowledgements

This work was financially supported by the Programa de Apoio a Núcleos de Excelência (PRONEX-Grant N° 001/04, CNPq, FAPESQ, MCT). We thank Graça Marinho, PhD, University of Campina Grande, Campus of Patos, Patos, PB, for the identification of the plants.

References

Amorim SL, Medeiros RMT, Riet-Correa F, Barbosa RC, Lima EF, Oliveira ACP, and Araújo JAS (2005). Intoxicação experimental por *Manihot glaziovii* (Euphorbiaceae) em caprinos na Paraíba. *Pesquisa Veterinária Brasileira* 25(3):179–187.

Araújo GGL and Calvancante J (2002). Potencial de utilização da maniçoba. *Proceedings of the 3rd Simpósio Paraibano de Zootecnia, Areia, PB*, on CD-ROM.

Armién AG, Peixoto PV, Döbereiner J, and Tokarnia C (1995). Intoxicação experimental por *Holocalyx glaziovii* (Leg.Mimosoideae) em bovinos. *Pesquisa Veterinária Brasileira* 15:89–92.

Brito MF, França TN, Oliveira KD, and Cerqueira VD (2000). Estudos experimentais em coelhos com plantas cianogênicas. *Pesquisa Veterinária Brasileira* 20(20):65–70.

Gava A, Pilati C, Cristian J, Simões J, and Simões L (1998). *Intoxicação cianogênica em bovinos alimentados com Tifton (*Cynodon *sp).* Proceedings, VIII CAMEV, Lages, Santa Catarina.

Gava A, Stolft L, Neves DS, Stolft O, Varaschim MS, and Ferreira FMM (1992). Intoxicação experimental por *Prunus sellowi* (Rosaceae) em bovinos. *Pesquisa Veterinária Brasileira* 12(1/2):1–4.

Medeiros RMT, Riet-Correa F, Barbosa RC, Lima EF, Neto SAG, and Tabosa I (2004). Plants toxic for domestic animals in the state of Paraíba, Northeastern Brazil. In *Poisonous Plants and Related Toxins* (T Acamovic, CS Stewart, and TW Pennycott, eds), pp. 151–156. CABI Publishing, Wallingford, Oxon, UK.

Saad AD and Camargo WVA (1967). Intoxicação cianídrica em animais domésticos. *Biológico*, São Paulo, 33:211–220.

Tokarnia CH, Peixoto PV, and Döbereiner J (1994). Intoxicação Experimental por *Piptadenia macrocarpa* (Leg. Mimosoideae) em bovinos. *Pesquisa Veterinária Brasileira* 14(2/3):57–63.

Tokarnia CH, Peixoto PV, Brito MF, Duarte MD, and Brust LAC (1999). Estudo Experimental com plantas cianogênicas em bovinos. *Pesquisa Veterinária Brasileira* 19(2):84–90.

Tokarnia CH, Döbereiner J and Peixoto PV (2000). *Plantas Tóxicas do Brasil.* Editora Helianthus, Rio de Janeiro.

Chapter 37

Leukoencephalomalacia in Mules in Northern Brazil

G. Riet-Correa[1], M.D. Duarte[1], V.D. Cerqueira[1], C.M.C. Oliveira[1], and F. Riet-Correa[2]

[1]*Central de Diagnóstico Veterinário, Campus de Castanhal, Universidade Federal do Pará, Rua Maximino Porpino, 1000, Bairro Pirapora, Castanhal, Pará, 68740-080;*
[2]*Centro de Saúde e Tecnologia Rural, Universidade Federal de Campina Grande, 58700-000, Patos, PB*

Introduction

In Brazil, leukoencephalomalacia (LEM) is a common disease of horses. Outbreaks of LEM in Brazil from 1950 to 1996 have been reviewed (Riet-Correa *et al.* 1998)

Clinical signs of LEM in horses start abruptly and include anorexia, somnolence and depression or hyperexcitability, impaired food apprehension and mastication, ataxia, tremors, head pressing, circling, dullness, unilateral or bilateral blindness and recumbence. The clinical manifestation period varied between 2 and 72 hours, but most affected horses died in a period of 6–24 hours after the onset of signs. (Riet-Correa *et al.* 1998).

At necropsy one cerebral hemisphere is usually enlarged with flattened gyri. On the cut surface there are yellowish and hemorrhagic areas of malacia in the centrum semiovale and corona radiata of the cerebral hemispheres. Fluid-containing cavities are frequently observed within these areas. Internal capsule or thalamus are also frequently affected. Hemorrhagic areas are occasionally observed also in the colliculi, cerebellar peduncles, pons and medulla oblongata. Most lesions in the brain stem are probably due to compression secondary to brain edema. Lesions are usually unilateral, but on some occasions they are bilateral, being more marked in one side. After fixation the yellow areas are less evident and the hemorrhagic lesions appear brown to gray (Riet-Correa *et al.* 1998; Rabelo *et al.* 2004). Histologic lesions are malacic areas surrounded by edematous and hemorrhagic neuropil. Swollen astrocytes with eosinophilic cytoplasm and dark marginal nucleus (sometimes called clasmadendrocytes) are observed within the malacic areas. Hypertrophied and degenerative changes in the vascular endothelium, and perivascular edema, hemorrhages and eosinophilic globules are also observed. Some vessels have perivascular cuffs consisting of eosinophils, neutrophils or mononuclear cells (Riet-Correa *et al.* 1998).

The disease is well known in horses, but outbreaks in donkeys are also reported (Badiali *et al.* 1968; Rosiles *et al.* 1998), and the intoxication has been reproduced experimentally in these species (Badiali *et al.* 1968; Wilson and Maronpot 1971). In this paper an outbreak of LEM is reported in mules in the state of Pará, northern Brazil. The differences in clinical signs and location of the lesions in mules with those reported in horses and donkeys are discussed.

Materials and Methods

The farm where the disease occurred was visited to observe affected mules and to collect data on the epidemiology of the disease. Two mules were examined clinically. Eight mules were necropsied. Samples of different organs of the abdominal and thoracic cavities and the whole brains were fixed in 10% buffered formalin. After fixation the brains were sectioned in 4–5 mm thick transverse sections. Histologic evaluation of the CNS was performed in transverse sections taken from the cervical spinal cord, medulla oblongata, pons, cerebellar peduncles, cerebellum, rostral colliculi, thalamus, basal nuclei, cortex (frontal, occipital, temporal and parietal), and any area showing gross lesions. All sections were stained by hematoxylin and eosin.

A sample of corn collected on the farm at time of the necropsies was sent to Dr Janio M. Santúrio from the Federal University of Santa Maria, Rio Grande do Sul, Brazil, for the determination of fumonisins. The sample of corn collected was not from the same batch that caused the outbreak, but from another batch recently delivered to the farm from the same supplier.

Results

The outbreak was observed in mules in the state of Pará, northern Brazil, in June 2004, affecting 13 out of 80 mules. Until 1 week before the outbreak, the 80 mules were supplemented daily with 60 kg of ground whole corn (a mixture of corn ground together with straw and corncobs) and 30 kg of African palm oil meal. One week before the outbreak, the amount of food was increased to 150 kg of the ground whole corn, 60 kg of African palm oil meal, and 120 kg of poultry litter.

Clinical signs observed in two mules were ataxia, muscular tremors, aggressiveness, absence of palpebral reflex, decreased tongue and upper and lower lip tone, decreased sensibility of the face and palate, and mandibular slackening. The farmer reported that one mule had compulsive walking, blindness, and head pressing, and others had lip and mandibular paralysis with food remaining in the mouth for long periods. All animals died after a clinical manifestation period of 10–24 hours.

No evidence of asymmetry between cerebral hemispheres was observed at necropsies. On transverse sections of the brain after fixation, cavitations, hemorrhages and yellow discoloration were observed mainly in the thalamus (Figure 1) and mesensephalus. Only two horses had lesions in the centrum semiovale and corona radiata. Lesions in those locations and also in the internal and external capsules were always mild. Mild hemorrhagic lesions were also observed in the pons. The localization and degree of the gross lesions in the eight mules are presented in Table 1. Histologically, areas of malacia in the white matter were characterized by a vacuolated neuropil with edema and hemorrhages. Swollen astrocytes with eosinophilic cytoplasm and dark emarginated nuclei (clasmadendrocytes) were also observed. Cavitations were sometimes observed in the center of the lesion. Hypertrophied and degenerative changes in the vascular endothelium and perivascular edema, hemorrhages and eosinophilic globules were also observed. Some vessels have perivascular cuffs consisting of eosinophils, neutrophils or mononuclear cells. Histologic lesions, mainly mild hemorrhages, were also observed in some areas without gross lesions (Table 1). No significant lesions were observed in other organs.

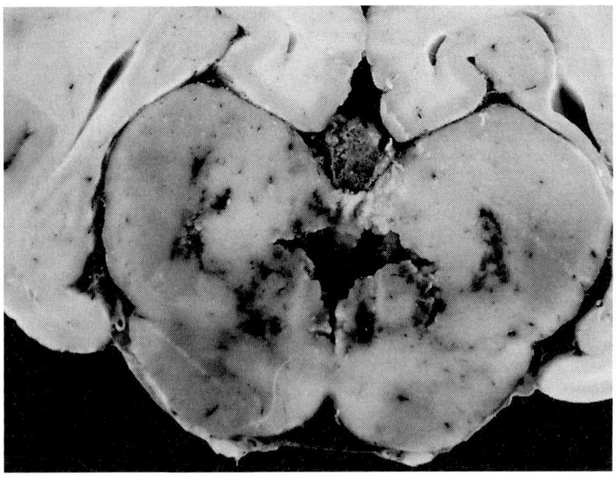

Figure 1. Thalamus of mule No. 1 affected with LEM. Severe cavitation and hemorrhages are observed.

The whole ground corn from another batch from the same supplier contained 3μg g^{-1} of fumonisin B_1.

Discussion

This is the first outbreak of LEM reported in northern Brazil and occurred in the tropical climate of the Amazon region; it is the first report in mules in Brazil. Previously, the disease was reported from other climatic regions in Brazil including temperate, subtropical, and tropical semiarid climates, demonstrating that the disease and the toxin occur in many climatic conditions. In the southern and southeastern regions the disease occurs more frequently during winter, and the only feature associated with the occurrence of outbreaks in those regions was the lower temperatures observed during the end of autumn, winter and early spring (Riet Correa *et al.* 1998). We also believe that the main reason for the occurrence of more outbreaks during winter is the necessity for additional grain supplementation during this time of the year because of the shortage of forage. There is little information about the season of occurrence of LEM in tropical regions. The outbreak reported in this paper and the outbreak observed in Paraíba (Riet-Correa *et al.*

2003) both occurred during the rainy season, although one outbreak reported in the state of Mato Grosso occurred during the dry season (Dantas *et al.* 2001).

Table 1. Distribution of the lesions in the central nervous system in 8 mules with LEM.

Mule no.	CR	CSO	BN	Thal	Mese	Pons	Obex	Cereb
1	Hist	Hist	X	XXX	XX	X		Hist
2	X	X	Hist	Hist	XX	X		
3			X	X	X			His
4				Hist	XX	X		
5			X	X		Hist		
6	X	X	X	X	X	X	Hist	
7	Hist	Hist	X	X				
8	Hist	Hist.	X	X	X	X		Hist

CR=corona radiata; CSO centrum semiovale; BN=basal nuclei (lesions in internal and/or external capsules); Thal= thalamus; Mese=mesencephalus; Cereb=cerebellar white matter; X= mild gross lesions; XX= moderate gross lesions; XXX= severe gross lesions Hist=only histologic lesions.

Because the sample of corn examined for fumonisin was from a batch other than that causing the disease, the concentration of 3 $\mu g\ g^{-1}$ of fumonisin found in the sample might not be representative of the food consumed by the mules. In Brazil, fumonisin B_1 ranging from 10 $\mu g\ g^{-1}$ to 500 $\mu g\ g^{-1}$ was detected in 24 (82.1%) out of 32 corn samples examined (Meireles 1993). In another 13 feed samples associated with LEM, fumonisin was detected in all samples, varying between 0.2 and 38.5 $\mu g\ g^{-1}$ for fumonisin B_1 and between 0.1 and 12 $\mu g\ g^{-1}$ for fumonisin B_2 with mean concentration of 11.9 $\mu g\ g^{-1}$ and 4 $\mu g\ g^{-1}$, respectively (Sydenham *et al.* 1992). In the USA, 13 samples of corn associated with LEM contained 1.3 to 27 $\mu g\ g^{-1}$ of fumonisin B_1 and 0.1 to 12.6 $\mu g\ g^{-1}$ of fumonisin B_2. In South Africa one corn sample associated with LEM had 8.85 $\mu g\ g^{-1}$ of fumonisin B_1 and 3.0 $\mu g\ g^{-1}$ of fumonisin B_2 (Kellerman *et al.* 1990).

In previous outbreaks affecting horses in the southern, southeastern, central-western, and northeastern regions of Brazil, the main gross lesions were in the centrum semiovale and corona radiata. In a recent paper the distribution of the lesions in the CNS were studied in 15 previously reported cases of LEM in the states of Rio Grande do Sul, Mato Grosso, and Paraíba; in all animals except one the main lesions were in the corona radiata and/or centrum semiovale. Lesions in the corpus callosum (1 horse), basal nuclei (5 horses), and thalamus (2 horses) were also observed. Hemorrhagic lesions in the brain stem interpreted as a consequence of cerebral edema were observed in one animal. Only one horse had lesions exclusively in the thalamus (Rabelo *et al.* 2004).

These results demonstrate that the main location for the lesions of LEM in horses is the white matter of the cerebral hemispheres, but lesions are also observed in other regions. Lesions in the brain stem can be due to the toxic effect or secondary to compression of the brain stem and cerebellum as a consequence of cerebral edema. In cases of LEM in horses reported from other countries, lesions were mainly in the cerebrum but other regions of the brain were also affected, such us the brain stem (Schwarte *et al.* 1937; Buck *et al.* 1979; Robertson-Smith *et al.* 1985), cerebellum (Buck *et al.* 1979) and spinal cord (Schwarte *et al.* 1937; Biester *et al.* 1940; Buck *et al.* 1979; Robertson-Smith *et al.* 1985). Also in an experimental case, the administration of fumonisin B_1 caused necrosis of the medulla oblongata (Marasas *et al.* 1988).

In the outbreak reported in this paper, the main lesions were in the thalamus and mesencephalus (Table 1). This difference in the location of the lesions with previously reported cases is probably due to the different species involved. In horses, clinical signs represent cerebral lesions, but there are also symptoms related to the brain stem. Also in donkeys, in spontaneous and experimental cases, the lesions are located in the white matter of the cerebral hemispheres, similar to those observed in horses (Badiali *et al.* 1968; Wilson and Maronpot 1971; Rosiles *et al.* 1998). In this outbreak in mules, lesions were mainly localized in the brain stem so the main clinical signs in the two mules we examined were impairment of the cranial nerves.

Acknowledgements

This work was financially supported by the Programa de Apoio a Núcleos de Excelência (PRONEX-Grant Nº 001/04, CNPq, FAPESQ, MCT).

References

Badiali L, Abou-Youssef MH, Radwan AI, Hamdy FM, and Hildebrandt PK (1968). Moldy corn as the major cause of an encephalomalacia syndrome in Egyptian equidae. *American Journal of Veterinary Research* 29:2029–2035.

Biester HE, Schwarte LH, and Reddy CH (1940). Further studies on moldy corn poisoning (leukoencephalomalacia) in horses. *Veterinary Medicine* 35:636–639.

Buck WB, Haliburton JC, Thilsted JP, Lock TF, and Vesonder RF (1979). Equine leukoencephalomalacia: comparative pathology of natural occurring and experimental cases. *Annual Proceedings of the American Association of Veterinary Laboratory Diagnosticians* 22:239–258.

Dantas AFM, Girardi N, Schein FB, Grecco FB, Colvara IG, and Ávila MO (2001). Leucoencefalomalacia em eqüinos no Estado de Mato Grosso. Proceedings. *Encontro Nacional de Patologia Veterinária*, 10, Pirassununga, SP, p. 155.

Kellerman TS, Marasas WFO, Thiel PG, Gelderblom WCA, Cawood M, and Coetzer AW (1990). Leukoencephalomalacia in two horses induced by oral dosing of fumonisin B_1. *Onderstepoort Journal of Veterinary Research* 57:269–275.

Marasas WFO, Kellerman TS, Gerdelblom WCA, Coetzer JAW, Thiel PG, and Van Der Lugt JJ (1988). Leukoencephalomalacia in a horse induced by fumonisin B_1 isolated from *Fusarium moniliforme. Onderstepoort Journal of Veterinary Research* 55:197–203.

Meireles MA (1993). *Leucoencefalomalacia eqüina no Brasil.* PhD thesis, University of São Paulo, São Paulo.

Rabelo JHO, Riet-Correa F, Barros CSL, Dantas AFM, Schild AL, and Tabosa IM (2004). Localização das lesões macroscópicas em 15 casos de leucoencefalomalácia dos eqüinos. *Pesquisa Veterinária Brasileira* 24:46–47.

Riet-Correa F, Meireles M, Barros C, and Gava A (1998). Equine leukoencephalomalacia in Brasil. In *Toxic Plants and Other Natural Toxicants* (T Garland and C Barr, eds), pp. 479–482. CABI Publishing, Wallingford, Oxon, UK.

Riet-Correa F, Tabosa IM, Azevedo EO, Medeiros RMT, Simões SVD, Dantas AA, Alves JC, Nobre VMT, Athaylde AC, Gomes AA, and Lima EF (2003). Doenças de ruminantes e eqüinos no semi-árido da Paraíba. *Semi-árido em foco*, Patos, 1:2–86.

Robertson-Smith RG, Jeffcott LB, Friend SCE, and Badcoe LM (1985). An unusual incidence of neurological disease affecting horses during a drought. *Australian Veterinary Journal* 62:6–12.

Rosiles MR, Bautista J, Fuentes VO, and Ross F (1998). An outbreak of equine leukoencephalomalacia at Oaxaca, Mexico, associated with fumonisin B1. *Zentralbl Veterinarmed A* 45:299–302.

Schwarte LH, Biester HE, and Murray C (1937). A disease of horses caused by feeding moldy corn. *Journal of the American Veterinary Medical Association* 909:76–85.

Sydenham EW, Marasas WFO, Shephard GS, Thiel PG and Hirooka EY (1992). Fumonisin concentration in Brazilian feeds associated with field outbreaks of animal mycotoxicoses. *Agricultural Food Chemistry* 40:994–997.

Wilson BG and Maronpot RR (1971). Causative fungus agent of leukoencephalomalacia in equine animals. *Veterinary Record* 88:484–486.

Chapter 38

ARGT-related Inhibition of GlcNAc-1-P Transferase Activity in Different Tissues of Sheep and Rats

P.L. Stewart, J. Greenwood, A. Michalewicz, and S.M. Colegate

CSIRO Livestock Industries, Australian Animal Health Laboratory, Geelong, Victoria 3220, Australia

Introduction

The inhibition of the enzyme *N*-acetylglucosamine-1-phosphate transferase (GPT) by corynetoxins (CTs) is a primary biochemical lesion associated with Annual Ryegrass Toxicity (ARGT) and its tunicamycins (TMs)-based experimental model. The determination of tissue microsomal GPT activity is a diagnostic tool for suspected tunicaminyluracil poisoning, an indicator of the long-term effects of poisoning, and a sensitive determinant of no effect levels for oral exposure to the toxins. Inhibition of GPT activity by tunicamycins has been demonstrated in liver microsomes from chickens (Mahoney and Duksin 1979), pig embryos (Kyosseva and Zhivkov 1983), rats (Jago *et al.* 1983; Chandra *et al.* 1991) and sheep (Stewart 1998). Similar inhibition has also been confirmed for corynetoxins (Jago *et al.* 1983).

The determination of microsomal GPT activity in different tissues provides an indirect way to assess the toxicodynamics of the TMs but can be subject to variation depending on the conditions of collection and storage prior to analysis (Stewart 1998). To enable assay of many samples the stability of the enzyme in sheep and rat liver tissue stored at −80 ºC was determined. In order to investigate whether systemic CTs/TMs can be distributed to tissues other than liver, and adversely affect the endogenous microsomal GPT, it was necessary to determine the native GPT activity for rough and total microsomal fractions from various tissues of sheep. It has previously been shown that the enzyme from rat liver rough microsomes is more sensitive to TMs than the GPT from any other tissue examined (Stewart *et al.* 2004).

Materials and Methods

Rats

Six-month-old male Hooded Wistar rats were used in the study of liver and microsome storage at −80°C. Rats were housed in pairs in polypropylene cages containing autoclaved wood shavings, and maintained in an air-conditioned room (23°C) with a 12-hour light/dark cycle. Rats had free access to deionized (reverse osmosis) water and pelleted feed. The protocol for preparation of rough microsomes and assay of GPT has been previously described (Stewart 1998).

To study the effect of storage on GPT activity, livers from four male rats were separately divided into 3 g portions. The GPT activity of the rough microsomal fraction derived from the fresh tissue portion was determined for each animal. The other portions were stored for up to 4 months at −80°C and at various times the rough microsomes were prepared and assayed.

Sheep

The *N*-acetylglucosamine-1-phosphate transferase activity assay conditions were essentially as previously described (Stewart 1998). The stability of GPT activity in liver and liver rough microsome samples stored at −80°C was investigated with liver tissue taken from three 18-month-old ewes.

To enable the determination of the effects of CTs/TMs on the GPT of different tissues, the specific GPT activities for the total, smooth and rough microsomal fractions isolated from liver, kidney, heart, lung, muscle and brain were assessed. The usual amount of tissue utilized in all assays was 3–4 g. The assays of those microsomal fractions (total, smooth or rough) that yielded the highest level of GPT activity for a particular tissue were further optimized relative to incubation time, membrane protein concentration and the requirements for the nucleotide pyrophosphatase inhibitors, adenosine monophosphate (AMP) and dimercaptopropanol (DMP).

Using the optimized assays, the *in vitro* sensitivity of microsomal GPT to TMs was determined for each tissue. The *in vivo* effect of TMs on GPT activity of liver, kidney, heart, lung, muscle and brain was assessed 72 hours after two groups of three 9-month-old merino wethers were injected subcutaneously with a single dose of TMs at one of two dose rates (5 or 20 μg kg^{-1} body weight).

Oral exposure of rats and sheep to tunicamycins

As part of a comprehensive, chronic toxicology study of the effects of low-level dietary TMs on rats and sheep (Colegate *et al.*, unpublished), the effects on liver GPT activity were monitored. Groups of 5 sheep (rams and ewes) were exposed to 40.5 or 4.5 μg TMs kg^{-1} bw day^{-1} in their diet over a 12-month period; groups of 20 rats were similarly exposed to 121.5 or 13.5 μg TMs kg^{-1} bw day^{-1} in their diet over a 9- and 12-month period respectively.

Results and Discussion

Rats

The hepatic rough microsomal GPT activity was stable when liver samples were stored for at least 4 months. However, the GPT activity in isolated liver rough microsomes, with a protein concentration of 2.2–2.3 mg mL^{-1} and stored at −80 ºC, decreased by 15% after 4 days and by 30% relative to the assay of freshly prepared microsomes after 5 months. The stability of stored liver and/or isolated microsomes over longer periods has not been investigated.

An examination of GPT assay data acquired (PL Stewart, unpublished data) over several years has highlighted three different sample preparation protocols. In the first, rough microsomes were prepared from fresh liver (n=14 rats) and analyzed without any freeze-storage. The second protocol involved freeze-storage (−80ºC) of rough microsomes prepared from fresh liver (n=21 rats) whilst the third (n=14 rats) involved freeze-storage of both liver and rough microsomes. The mean of the GPT activities of the rats in group 1 was set at 100% and the activities determined in the other two groups were compared relative to this. Thus, the activity of the stored microsomes (3–9 days at −80ºC) prepared from fresh liver was 95% of group 1 activity and the activity in rough microsomes prepared from frozen liver (2–9 weeks at −80ºC) and then refrozen until analysis (1 day–3 weeks at −80ºC) was 94% of the group 1 activity. Since the protein concentrations of the stored microsomes generally ranged from 3 to 6 mg mL^{-1}, in contrast to the more recent analyses described in the previous paragraph, these archival data indicate that GPT activity in stored (−80ºC) rat liver rough microsomes may be more stable at protein concentrations greater than about 3 mg mL^{-1}.

It has been shown that the liver rough microsomal GPT activity is the most sensitive indicator of exposure to the TMs (Stewart *et al.* 2004). Following 12 months of dietary exposure at 13.5 μg TMs kg^{-1} bw day^{-1} the residual liver GPT activity was reduced to about 60% of untreated controls. At the higher exposure rate of 121.5 μg TMs kg^{-1} bw day^{-1} 95% inhibition was already evident at the first observation, i.e., after 3 months continual dosing. This high level of inhibition continued until the end of the study at 9 months.

Sheep

The rough microsomal GPT activity was stable in sheep liver stored for at least 4 months at −80ºC. However, the activity in rough microsomes (protein concentration 3 mg mL^{-1}) isolated from fresh sheep liver decreased by 10% after 5 weeks and by 30% after 6 months storage at −80ºC. Stability of the GPT over longer periods of liver or rough microsomal fraction freeze-storage has not been investigated.

Of the three microsomal preparations (total, smooth and rough), specific GPT activity for various tissues was highest in the rough microsomes from liver, kidney and lung, and in the total microsomal fraction from brain and muscle. However, the rough microsomal yield from a 4 g sample of lung tissue was too low for efficient routine assay. Thus the total microsomal fraction from lung tissue was used in the assay along with the rough microsomal fraction from liver and kidney and the total microsomal fraction from brain and muscle. The times of incubation, the range of microsomal protein content, and the requirement for AMP or DMP to optimize GPT activity within a linear range for various tissues are shown in Table 1. Thus, the sheep microsomal GPT measurement in liver, brain, kidney, lung and muscle tissue required similar incubation times and protein concentration range. However, all tissues except muscle required the presence of a nucleotide pyrophosphatase inhibitor for optimal enzyme activity.

The rough or total microsomal GPT activities in all tissues were sensitive to TMs *in vitro*. The amounts of TMs resulting in a 50% inhibition of GPT activity (I_{50}) in different tissues are shown in Table 1.

Table 1. Optimal assay conditions for microsomal GPT activity in different sheep tissues. The I_{50} (50% inhibition) concentrations are listed as ng of tunicamycins per assay tube. AMP is adenosine monophosphate and DMP is dimercaptopropanol, both at 5 mM in the reaction mixture.

Tissue	Incubation time (min)	Protein (mg tube^{-1})	AMP or DMP	I_{50} (ng tube^{-1})
Liver	4	0.12–0.28	DMP	0.4
Kidney	4	0.1–0.3	DMP	0.3
Lung	4	0.2–0.36	AMP	0.6
Muscle	3	0.14–0.36		0.5
Brain	4	0.1–0.22	DMP	0.3

Application of the optimized assay conditions to tissues sampled from a sheep 72 hours after parenteral administration of 20 μg TMs kg^{-1} bw demonstrated inhibition of GPT activity in liver, kidney, lung and brain microsomes (Table 2). Liver microsomes provided the highest level of GPT activity and were affected to a greater extent by the TMs. At the lower dose rate there was clear inhibition only in liver microsomes. These results were similar to previous research on GPT activity from different tissues in rats (Stewart *et al.* 2004). At oral dose levels of 40.5 μg TMs kg^{-1} bw day^{-1}, the liver microsomal GPT activity in rams and ewes was inhibited by approximately 95% relative to untreated controls by the first observation time, *i.e.*, after 3 months of continual dosing, and remained at this level for the remaining 9 months of the study. At a dose level of 13.5 μg TMs kg^{-1} bw day^{-1} the activity of liver GPT dropped to about 60% that of controls after 6 months and persisted at about that level for the remaining 6 months of continual feeding in the 12 month study.

Table 2. Microsomal GPT activity in different tissues from sheep treated with tunicamycins as a percentage of the activity of non-treated controls. (R): rough microsomal fraction. (T): total microsomal fraction.

Tunicamycins Dose (μg kg^{-1} bw, *i.p.*)	GPT Activity (% of Controls)				
	Liver (R)	Kidney (R)	Lung (T)	Muscle (T)	Brain (T)
5	3	113	91	85	68
20	2	54	37	68	50

Conclusions

Liver samples can be successfully stored (−80°C) for at least 4 months without any adverse effect on the GPT activity. However the GPT stability of freeze-stored liver microsomes is not so reliable but does increase if the protein concentration is greater than 3 mg mL^{-1}, resulting in a loss of less than 10% of initial activity over a 4-week period. The effect of freeze-storage on other tissues is expected to mirror the experience with the liver but has not yet been formally determined.

Similarly to the *in vivo* effect on rat tissue GPT activity (Stewart *et al.* 2004), GPT activity in sheep liver microsomes was affected to a far greater extent *in vivo* by TMs than activity in other tissues and is therefore the tissue of choice as an indicator of tunicaminyluracil glycolipid exposure. However, the detection of microsomal GPT activity in sheep kidney, lung and brain indicates, as with the rats (Stewart *et al.* 2004), the translocation of the TMs to these tissues.

A dose-dependent effect on liver rough microsomal GPT activity was observed in rats and sheep of both genders when exposed to low levels of dietary TMs. There was no evidence of any compensating upregulation of GPT activity back to levels consistent with untreated animals in either study.

Acknowledgements

The dietary exposure of rats and sheep to tunicamycins was facilitated by Neil Anderton, Yu Cao, Chris Darcy, Adam Stein and Andrew Knill of CSIRO Livestock Industries.

References

Chandra NC, Doody MB, and Bretthauer RK (1991). Specific lipids enhance the activity of UDP-GlcNAc:dolichol phosphate GlcNAc-1-phosphate transferase in rat liver endoplasmic reticulum membrane vesicles. *Archives of Biochemistry and Biophysics* 290:345–354.

Jago MV, Payne AL, Peterson JE, and Bagust TJ (1983). Inhibition of glycosylation by corynetoxin, the causative agent of annual ryegrass toxicity: a comparison with tunicamycin. *Chemico-Biological Interactions* 45:223–234.

Kyosseva SV and Zhivkov VI (1983). Biosynthesis of lipid-linked oligosaccharides in embryonic liver. *International Journal of Biochemistry* 15:1051–1057.

Mahoney WC and Duksin D (1979). Biological activities of the two major components of tunicamycin. *Journal of Biological Chemistry* 25: 6572–6576.

Stewart PL (1998). Activity of *N*-acetylglucosamine-1-phosphate transferase in sheep liver microsomes: *in vivo* and *in vitro* inhibition by tunicamycin. *Research in Veterinary Science* 64:31–35.

Stewart PL, Hooper P, Lenghaus C, Raisbeck MF, and Colegate SM (2004) Effect of tunicamycins on GlcNAc-1-P transferase activity in rat tissues and toxic effects during pregnancy and lactation. In *Poisonous Plants and Related Toxins* (T Acamovic, CS Stewart and TW Pennycott, eds), pp. 448–452. CABI Publishing, Wallingford, Oxon, UK.

Chapter 39

A Toxicoproteomic Approach to Understanding Annual Ryegrass Toxicity

M.A.S. Retallick[1,2,3], S.M. Colegate[1], and W.P. Michalski[2]

[1]*Plant Toxins Research Group;* [2]*Protein Biochemistry and Proteomics Group, Australian Animal Health Laboratory, CSIRO Livestock Industries, Geelong, VIC 3220, Australia;* [3]*Plant Cell Biology Research Centre, School of Botany, The University of Melbourne, Parkville, VIC 3010, Australia*

Introduction

Tunicamycins (TMs) and corynetoxins (CTs) are tunicaminyluracil-based glycolipid toxins with antibiotic properties. They share a common *N*-acetylglucosamine-tunicaminyluracil core and differ in the length, terminal branching structure and hydroxylation states of their fatty-acid side chains (Cockrum and Edgar 1983). In the field, CTs are synthesized by *Rathayibacter toxicus* upon its co-colonization and domination of *Anguina funesta* nematode galls in the seed heads of annual ryegrass (*Lolium rigidum*) (Riley and Ophel 1992). Excessive consumption of CTs by grazing livestock results in a severe hepatocerebral disease known as annual ryegrass toxicity (ARGT) (Berry and Wise 1975). The principal biochemical effects of ARGT are associated with the toxins' inhibition of the enzyme *N*-acetylglucosamine-1-phosphate transferase (GPT) (Lehle and Tanner 1976). GPT is involved in the assembly of dolichol phosphate-linked glycans before their *N*-linked transfer to nascent proteins in the endoplasmic reticulum.

Previous *in vivo* studies have shown that acute exposure to TMs produces a syndrome that is indistinguishable from ARGT when compared on the basis of clinical and pathological observations (Jago *et al.* 1983; Finnie and Jago 1985; Jago and Culvenor 1987). To validate the experimental model at a molecular level, a toxicoproteomic approach, which involved a comparative study of the proteins secreted by an *in vitro* rat hepatocyte cell line in response to toxin-exposure, was used to evaluate the bioequivalence of CTs and TMs.

Materials and Methods

Preparation of tunicamycins- and corynetoxins-containing media

Cultured hepatocytes were exposed to the toxins via TMs- and CTs-spiked media preparations. Two types of media were utilized throughout the study: F-12K Nutrient

Mixture media (Invitrogen, Carlsbad, California, USA) and RPMI-1640 (MP Biomedicals, Irvine, California, USA). RPMI-1640 was supplemented with 10 μCi mL^{-1} of Trans^{35}S-Label (MP Biomedicals), containing >70% ^{35}S-methionine and >15% ^{35}S-cysteine, with a specific activity of 1175 Ci mmol^{-1}.

TMs (purchased from Sigma Aldrich, St. Louis, Missouri, USA) and CTs (purified from toxic ryegrass galls by Neil Anderton, CSIRO Livestock Industries, Plant Toxins Research Group) were solubilized and serially diluted in filter-sterilized 0.6 mM NaOH (AnalaR Grade, BDH Ltd., Merck, Darmstadt, Germany) then added to the media at 0.5% v/v such that the final concentrations of TMs and CTs in both F-12K and RPMI-1640 were 5 nM, 50 nM, 500 nM and 5000 nM. NaOH control samples, consisting of 0.5% v/v of toxin-free 0.6 mM NaOH in the two types of media, were also prepared in addition to solvent-free media controls, which were unmodified. This equated to four TMs doses, four equivalent doses of CTs and two controls.

Exposure of *in vitro* hepatocytes to tunicaminyluracil toxins

One hundred and fifty OptiCells (BioCrystal, Westerville, Ohio, USA) were seeded with Clone 9 rat hepatocytes (American Type Culture Collection, Manassas, Virginia, USA; cell line reference number #CRL-1439) at a density of 1 x 10^5 cells mL^{-1} in toxin-free F-12K media. At 48 hours post-seeding the OptiCells were equally distributed into five groups designated as 0–6 hours, 6–18 hours, 18–30 hours, 30–42 hours and 42–54 hours. Within each group, the 30 OptiCells were further separated into ten groups of three representing the eight TMs and CTs dose groups, the NaOH control and the solvent-free control (*i.e.*, three replicates per treatment per time span). The media was removed from all OptiCells belonging to the 6–18, 18–30, 30–42, and 42–54 hour groups and replaced with control F-12K media or with F-12K media containing the various amounts of toxins or NaOH. The media from the 0–6 hour group was also removed but replaced with the ^{35}S-met and ^{35}S-cys-fortified RPMI-1640 that was supplemented with the toxins, NaOH or was unmodified. All groups were incubated at 37°C in 95% humidified air + 5% CO_2. Following 6 hours of metabolic labeling, the media from the 0–6 hour group was harvested from the OptiCells for proteomic analysis, and the toxin-containing F-12K, NaOH F-12K or control F-12K was removed from the 6–18 hour cultures and replaced with toxin-containing RPMI-1640, NaOH RPMI-1640 or control RPMI-1640 accordingly. These latter cultures were then incubated in the presence of ^{35}S-met and ^{35}S-cys for 12 hours (i.e. from 6 until 18 hours post toxin-exposure) prior to harvesting. This pulse/harvest pattern continued for the 18–30, 30–42 and 42–54 hour cultures.

Proteomic analysis of hepatocyte secretomes

At each harvest point, the media was removed from the OptiCells using the specialized OptiCell tips and cleared of cellular material by centrifugation at 1000 x *g* for 5 min. The supernatants, containing the proteins that were secreted by the hepatocytes (the secretomes) including many of the components of serum, were then desalted and concentrated fivefold using Vivaspin 5000 kDa molecular weight cut-off centrifugal devices (Vivascience, Sartorius, Goettingen, Germany). Electrophoretic separation of the secretomes was performed according to Laemmli system of SDS-PAGE (Laemmli 1970). The resulting gels were dried and subjected to autoradiography using pre-flashed Hyperfilm MP (GE

Healthcare, Uppsala, Sweden). Once developed, the autoradiograms were densitometrically analyzed using Scion Imaging Software (Scion Corporation, Washington D.C., USA). Profile plots of individual samples, which schematically represented the densities of the various bands on the autoradiogram, were divided into 20 evenly spaced segments and the volumes of each segment were calculated. The variations in the volumes could then be compared across the toxins, dose-groups, and time spans by applying principal component analysis (PCA) (Wold 1987; Eriksson *et al.* 2004) using the statistical computing software package R (R Foundation for Statistical Computing, Vienna, Austria).

Results and Discussion

Exposure to TMs and CTs produced a number of changes in the secretomes of *in vitro* rat hepatocytes. Visual observation of the autoradiograms (Figure 1) identified two distinct profiles across all time periods: those of cells exposed to ≤ 5 nM TMs/CTs and those exposed to ≥ 50 nM TMs/CTs. However, PCA revealed differences between all dose groups and indeed between certain equivalent doses of TMs and CTs.

As indicated by the PCA scores plots (Figure 1), which graphically highlight the degree of variation between the samples based on the first (x-axis) and second (y-axis) principal components, the spatial distributions of the control and NaOH samples were very similar across all five time spans. Consequently, the effect of NaOH-exposure was considered to be minimal and, as such, any changes in the protein profiles of the toxin-exposed cultures could be directly attributed to the toxins and not the solvent they were dissolved in.

Across the time spans, the 5 nM TMs and CTs samples were consistently co-localized on the PCA scores plots. This indicated that at this dose, the effects of the two toxins were bioequivalent. Interestingly, the scores plots corresponding to the 0–6 and 6–18 hour time spans revealed that the 5 nM samples were initially displaced from the controls, but return to a ‘control-like’ profile from 18 hours onwards.

Exposure to 5000 nM doses of TMs and CTs also had an analogous impact upon the secretomes as indicated by the scores plots. At each time span the 5000 nM cultures showed very little variability and, accordingly, TMs and CTs were considered to be bioequivalent at this dose. The CTs and TMs profiles of the 500 nM and 50 nM dose groups, however, did not always indicate bioequivalence. This was particularly the case at the 42–54 hour time span, when the 500 nM CTs and TMs cultures showed a distinct degree of separation within the first principal component and the 50 nM cultures appeared to be different based on both the first and the second principal components. Once isolated, the identities of the secreted proteins that discriminate the CTs- and TMs-exposed cultured hepatocytes will be determined using mass spectrometry and N-terminal sequencing.

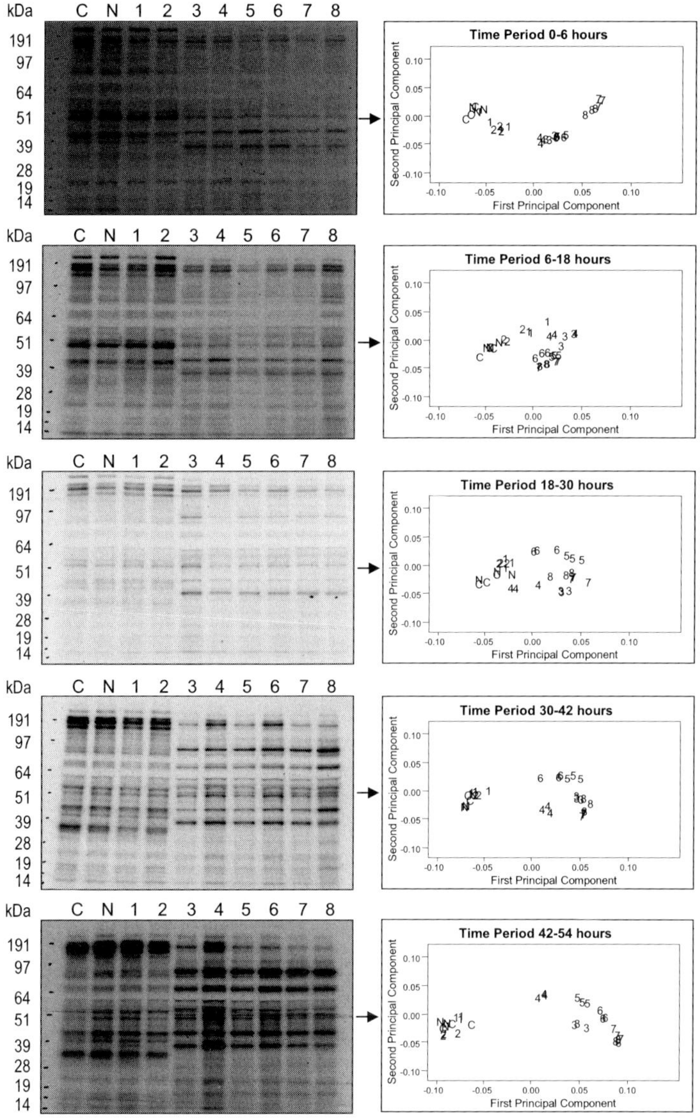

Figure 1. Autoradiograms and PCA scores plots of *in vitro* rat hepatocyte secretomes. Autoradiograms depict pools of triplicate cultures. PCA scores plots were generated from autoradiograms of individual cultures (three per dose group per time point – not shown). C = control media, N = NaOH-containing media, 1 = 5 nM TMs, 2 = 5 nM CTs, 3 = 50 nM TMs, 4 = 50 nM CTs, 5 = 500 nM TMs, 6 = 500 nM CTs, 7 = 5000 nM TMs, and 8 = 5000 nM CTs.

Conclusions

This study has indicated that exposure of cultured hepatocytes to doses of TMs and CTs > 5 and < 5000 nM resulted in proteomic changes that were not always identical between the two toxins. This finding highlights the importance of applying molecular techniques to comparative toxicology. A complementary investigation based on morphological and cell viability studies of *in vitro* hepatocytes exposed to the same doses of TMs and CTs as described here was unable to discriminate between the two toxins (see Allen *et al.*, Chapter 86, this publication).

Identifying molecular indicators of non-equivalence of tunicaminyluracil toxin-exposure *in vitro* has important implications for *in vivo* studies of ARGT that have been modelled with TMs rather than CTs, particularly in the field of biomarker discovery. Candidate protein biomarkers of clinical and sub-clinical disease may be investigated using TMs-exposed animals, but their capacities to act as diagnostic indicators of ARGT must be confirmed by experimental animal trials involving CTs-exposure.

References

Berry PH and Wise JL (1975). Wimmera ryegrass toxicity in Western Australia. *Australian Veterinary Journal* 51:525–530.

Cockrum PA and Edgar JA (1983). High-performance liquid chromatographic comparison of the tunicaminyluracil-based antibiotics corynetoxins, tunicamycin, streptovirudin and MM 19290. *Journal of Chromatography* 268:245–254.

Eriksson L, Antti H, Gottfries J, Holmes E, Johansson E, Lindgren F, Long I, Lundstedt T, Trygg J, and Wold S (2004). Using chemometrics for navigating in the large data sets of genomics, proteomics, and metabonomics (gpm). *Analytical and Bioanalytical Chemistry* 380:419–429.

Finnie JW and Jago MV (1985). Experimental production of annual ryegrass toxicity with tunicamycin. *Australian Veterinary Journal* 62(7):248–249.

Jago MV and Culvenor CC (1987). Tunicamycin and corynetoxin poisoning in sheep. *Australian Veterinary Journal* 64(8):232–235.

Jago MV, Payne AL, Peterson JE, and Bagust TJ (1983). Inhibition of glycosylation by corynetoxin, the causative agent of annual ryegrass toxicity: a comparison with tunicamycin. *Chemical and Biological Interactions* 45(2):223–234.

Laemmli UK (1970). Cleavage of structural proteins during the assembly of the head of bacteriophage T4. *Nature* 227:680–685.

Lehle L and Tanner W (1976). The specific site of tunicamycin inhibition in the formation of dolichol-bound N-acetylglucosamine derivatives. *FEBS Letters* 71:167–170.

Riley LT and Ophel KM (1992). *Clavibacter toxicus* sp.nov., the bacterium responsible for annual ryegrass toxicity in Australia. *International Journal of Systematic Bacteriology* 42:64–68.

Wold S (1987). Principal Component Analysis. *Chemometrics and Intelligent Laboratory Systems* 2:37–52.

Chapter 40

Effects on Bodyweight of Rats and Sheep due to Exposure to Tunicamycins: Implications for Annual Ryegrass Toxicity

N. Anderton, Y. Cao, C. Darcy, M. Ford, J. Greenwood, A. Knill, A. Michalewicz, P.L. Stewart, S. Wilson, and S.M. Colegate

CSIRO Livestock Industries, Australian Animal Health Laboratory, Geelong, VIC 3220, Australia

Introduction

Annual Ryegrass Toxicity (ARGT), caused by bacterially produced corynetoxins (CTs), has often been experimentally reproduced using tunicamycins (TMs) that are structurally very similar to the corynetoxins. Inappetence and resultant weight loss or failure to gain weight has been noted in previous instances of acute to sub-acute intoxication of rats and sheep with TMs or CTs (Berry *et al.* 1976, 1982; Peterson *et al.* 1996). However, the potential for adverse effects on body weight in cases of chronic, low-level exposure to the CTs is undefined. This latter potential has implications for on-farm productivity as well as food safety-related issues if CTs contaminate the human food supply.

Several studies have been conducted by the CSIRO Plant Toxins Research Group to help define the toxic effects of TMs and CTs including the effects on the transferase enzyme target of the toxins (Stewart *et al.* 1998), the testicular damage to male rats (Peterson *et al.* 1996) and a comprehensive toxicology study to determine a no observed effect level for the toxins in rats, sheep and pregnant rats (Colegate *et al.*, unpublished). Data collected in these studies included weight-related effects and feed intake levels for rats given TMs as single parenteral doses or exposed to different levels of dietary TMs for varying lengths of time. A single study was completed to assess the impact of chronic exposure of sheep to dietary TMs.

Materials and Methods

All animal experiments (Table 1) were conducted under the auspices of the Australian Animal Health Laboratory's Animal Ethics Committee. Same-gender pairs of Hooded

Wistar rats were housed in polypropylene cages containing autoclaved wood shavings, and maintained in an air-conditioned room (23 °C) with a 12-hour light/dark cycle. Rats had free access to reverse osmosis deionized tap water and pelleted feed. For the dietary exposure studies in rats a pelleted American Institute of Nutrition semi-purified rat diet (AIN 93) was spiked with tunicamycins and provided *ad libitum*. Growth diet was used for 6–13-week-old rats and maintenance diet was utilized from 13 weeks old onwards.

Table 1. Summary of the dosing experiments to determine effects on body weight.

Study	**Species** (Gender, Number of Animals, Age)	**Dose** (Route of Administration)	**Duration of Study**
1	Rat (males, 4, adults)	300 μg kg^{-1} bw (single sc)	72 hours
	Rat (males, 4, adults)	75 μg kg^{-1} bw (single sc)	72 hours
2	Rat (female, 28, adults)	200 μg kg^{-1} bw (single sc)	1 day–20 weeks
	Rat (female, 28, adults)	100 μg kg^{-1} bw (single sc)	1 day–20 weeks
3	Rat (male and female, 12, 6 weeks)	800 μg kg^{-1} bw day^{-1} (oral)	2 weeks
	Rat (male and female, 12, 6 weeks)	294 μg kg^{-1} bw day^{-1} (oral)	2 weeks
4	Rat (male, 6, 6 weeks)	800 μg kg^{-1} bw day^{-1} (oral)	1 week
	Rat (male, 6, 6 weeks)	800 μg kg^{-1} bw day^{-1} (oral)*	1 week
5	Rat (male and female, 12, 6 weeks)	108 μg kg^{-1} bw day^{-1} (oral)	6 months
6	Rat (male, 12, 6 weeks)	40 μg kg^{-1} bw day^{-1} (oral)	6 months
7	Rat (male and female, 30, 6 weeks)	121.5 μg kg^{-1} bw day^{-1} (oral)	9 months (males); 12 months (females)
	Rat (male and female, 30, 6 weeks)	40.5 μg kg^{-1} bw day^{-1} (oral)	12 months
8	Sheep (male and female, 5, 10 months)	40.5 μg kg^{-1} bw day^{-1} (oral)	12 months

* re-purified TMs to remove potential antifeedant impurities

Sheep from 10 months of age were housed in same-gender groups of five in open pens within an open-sided shed. The sheep groups had free access to feed and water. Tunicamycins were administered three times per week within special dosing diets presented to individual sheep in specifically designed dosing pens.

Results and Discussion

In Studies 1 and 2, involving single subcutaneous injections of TMs to adult rats, a reduction in body weight was observed. whereas in Studies 3 to 7, involving oral exposure of young, growing rats to TMs, a reduction in weight gain was observed.

Study 1

The effects of single subcutaneous injections of TMs on body weight, liver weight and the weight of the stomach and contents are summarized in Table 2. Under circumstances where untreated control rats gained 4 g or 1.2% of the initial body weight over the 72-hour study period, the groups that received 75 or 300 μg TMs kg^{-1} bw lost a mean of 8 g or 2.6% of the initial body weight and 25 g or 7.8% of the initial body weight respectively. The group mean liver weights as a percentage of final body weights were 3.9, 3.7 and 2.9% for the controls, low dose and high dose respectively (Table 2). The weights of the stomach and contents as a percentage of the final body weights for each group were 1.2, 1.0 and 6.3% for the controls, low dose and high dose groups respectively.

Table 2. Mean (n=4) body and organ weights 72 hours after subcutaneous injection of tunicamycins.

Dose	Body weight (g)		Liver	Stomach
(μg TMs kg^{-1} bw)	Initial	72 hours	weight (g)	weight (g)
0	322	326	12.71	3.86
75	311	303	11.27	3.10
300	322	297	8.62	18.7

The stomachs of the high dose rats were grossly enlarged with a solid firm mass of undigested food (Figure 1). There appears to be two mechanisms contributing to the weight loss. Whether these mechanisms are sequentially dependent on dose is undefined. At the low dose, about 20% of a lethal dose for male rats, there is weight loss but no significant impact upon the weights of the liver and the stomach and contents. However, at the higher dose, about 85% of a lethal dose for male rats, there was a dramatic impact on gastric motility and a resultant retention of food in the stomach. The combination of retained feed and a reduced liver weight with respect to the body weight in the high dose group is indicative of an inappetence mechanism of weight loss (Oishi *et al.* 1979). Similar effects on gastric motility also occur in suckling rats given TMs parenterally (Peterson and Jago 1977). Rats given single subcutaneous doses of 133, 200 or 300 μg TMs kg^{-1} BW have been shown to exhibit reduced food consumption and substantial, dose-related body weight loss in the first week following dosing (Peterson *et al.* 1996). Despite the substantial loss of body weight in the high dose group, rats have the capacity to regain the lost weight. For example, male rats given a single subcutaneous dose of 200 μg TMs kg^{-1} bw lost 20% of their initial body weight after 1 week, but regained the body weight of untreated controls within 8 weeks (Peterson *et al.* 1996).

Study 2

A study of the effect of a single subcutaneous dose of TMs on liver *N*-acetylglucosamine-1-phosphate transferase (a molecular target for TMs and CTs) (Stewart *et al.* 1998) also provided some body weight data as animals in the study were euthanased at various time points for enzyme activity assessment. Thus, female rats given a single subcutaneous dose of 200 μg TMs kg^{-1} bw lost 6% of their initial body weight after 1 week whereas the body weights of rats that received a single, subcutaneous dose of 100 μg TMs kg^{-1} bw were unaffected and similar to controls. Despite losing weight immediately after the injection of 200 μg TMs kg^{-1} bw the females regained the body weight of controls (and the group injected with the lower dose) by 4 weeks. Since female rats are about 50% more resistant to acute TM poisoning than males (Peterson *et al.* 1996) they might be expected to lose less weight and regain body weight more quickly than male rats that received a similar dose of TMs.

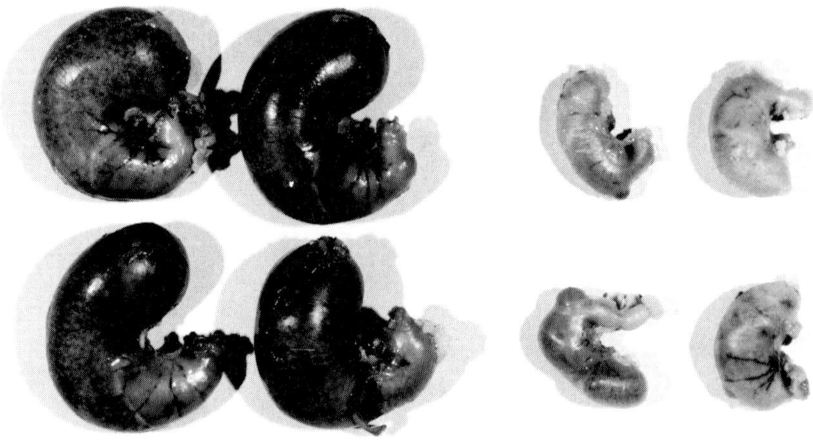

Figure 1. Stomachs from four male rats 72 hours after injection with 300 μg TMs kg^{-1} bw (left) or untreated controls (right).

Studies 3 and 4

After 2 weeks in Study 3, male and female rats receiving 800 or 294 μg TMs kg^{-1} bw day^{-1} in their semi-purified pelleted diet showed substantial reductions in weight gain compared to controls (Table 3). At the higher dose, males were 25% lighter and females 21% lighter than controls after 2 weeks dosing. At 294 μg TMs kg^{-1} bw day^{-1} males were 12% lighter and females 7% lighter than controls. Although the absolute feed intakes for the male high and low dose groups were 70 and 90% of control intake respectively and 75% and 100% of control group intake for the females, the weekly feed intake relative to final body weight was similar (50–56% for males and 55–63% for females) for all groups during the second week of the study.

Table 3. Mean (n=12) body weights after 1 week and 2 weeks of a feeding trial of semi-purified feed containing tunicamycins.

Dose (μg TMs kg^{-1} bw day^{-1})	Body weight (g) 0	Body weight (g) 1 week	% increase	Body weight (g) 2 weeks	% increase
Males					
0	155	209	35	254	64
294	155	202	30	224	45
800	154	176	14	190	23
Females					
0	129	158	22	184	43
294	128	157	23	171	34
800	127	145	14	146	15

To ensure that the TMs utilized in Study 3 were not contaminated with a potent antifeedant that led to the body weight loss rather than a direct effect of the TMs, the TMs were purified (Colegate *et al.*, unpublished) and reassessed against the original TMs used in Study 3. In the 1-week duration of Study 4, the rats orally exposed to the original or purified TMs gained 11 and 14% of initial body weight respectively whereas the untreated controls gained 32% of their initial body weight. The absolute food intakes by all groups were identical. Study 4 confirmed the body weight gains observed in Study 3 for the controls and high dose groups and supported a direct effect of the TMs rather than an impurity.

Based upon the effect on the liver *N*-acetylglucosamine-1-phosphate transferase, it has been suggested that a factor of 50 relates a subcutaneous dose to an oral dose of TMs in rats (Stewart *et al.*, unpublished). Therefore it can be estimated that the oral exposure of 800 μg TMs kg^{-1} bw day^{-1} used in Studies 3 and 4 is roughly equivalent to 16 μg TMs kg^{-1} bw day^{-1} administered subcutaneously or a total of 112 or 224 μg TMs kg^{-1} bw over 7 or 14 days respectively. This observation raises obvious questions about the cumulative nature of exposure to TMs insofar as body weight impacts are concerned. Contrary to the single subcutaneous doses, the oral administration in Studies 3 and 4 did not elicit weight loss, only a reduced weight gain relative to untreated controls.

Studies 5, 6 and 7

Extending the investigation into possible long-term accumulation of dietary TMs, body weight and feed intake data were collected from groups of male and female rats exposed to dietary TMs in semi-purified pelleted feed for 6, 9 and 12 months (Table 1). Exposure of female rats to 108 μg TMs kg^{-1} bw day^{-1} for 6 months had no impact upon body weight gain relative to untreated controls. However, males exposed to the same dose showed a reduction in body weight gain by week 15 and that progressed to 89% of the control group mean body weight at the end of Study 5. No adverse effect on body weight gain was observed for males exposed to 40 μg TMs kg^{-1} bw day^{-1} for 6 months (Study 6).

Exposed over 9 months to 121.5 μg TMs kg^{-1} bw day^{-1} and 40.5 μg TMs kg^{-1} bw day^{-1} for 12 months, male rats were still gaining weight slightly but the final group mean body weights were only 82% and 95% of the control group mean body weight respectively (Figure 2). After 12 months exposure to 121.5 μg TMs kg^{-1} bw day^{-1}, the group mean body weight of the females was 88% that of controls whereas exposure to 40.5 μg TMs kg^{-1} bw day^{-1} for 12 months had no adverse effect on body weight gain (Figure 2).

Study 8

Acute exposure to CTs has been associated with rumen stasis and anorexia (Berry *et al.* 1976, 1982) that would presumably lead to reductions in weight or weight gain. In a chronic, low-level exposure study, groups of five male or female sheep were exposed to 40.5 μg TMs kg^{-1} bw day^{-1} in their diet for 12 months. There was no difference in feed intake, body weight gains or final body weight in the treated sheep relative to the controls. This is despite the fact that the sheep were exposed to a cumulative dose approximately equivalent to ten times the estimated acute LD_{50} of 1 mg TMs kg^{-1} bw (Bourke and Carrigan 1993). This is a similar result to the lack of any adverse effect on live weight of

Merino ewes dosed with a total of 800 μg CTs kg^{-1} bw over a 77-day period prior to and during pregnancy (Davies *et al.* 1996).

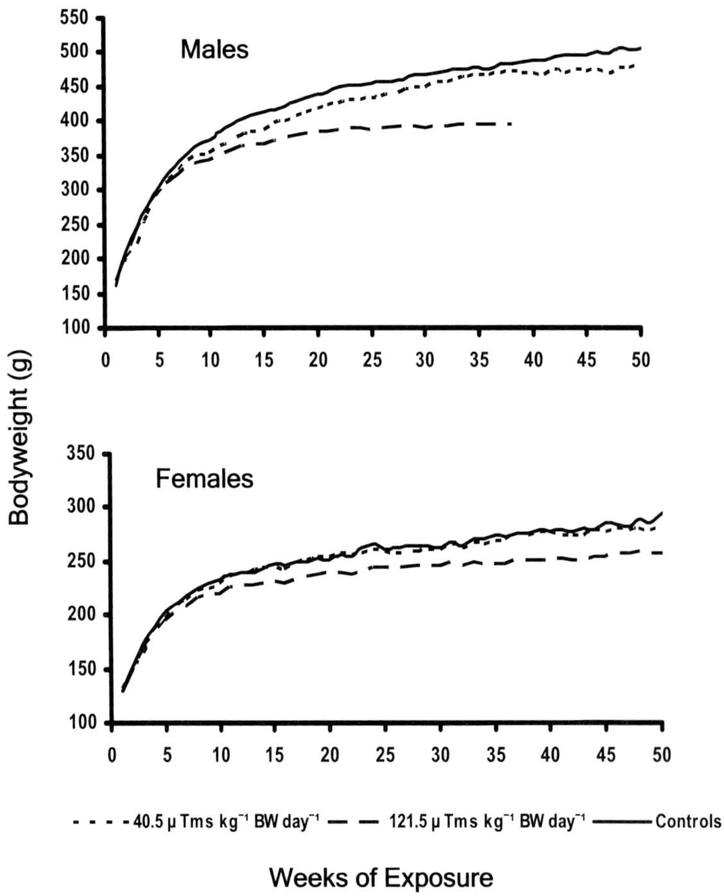

Figure 2. Mean body weight of male and female rats in a 9–12 month feeding trial of semi-purified feed containing tunicamycins (Study 7).

Conclusions

The studies with rats seem to indicate at least two mechanisms (related to feed intake and utilization) for adverse impact on body weight. It is not known whether these mechanisms are sequentially related via dose level. Actual loss of body weight, in contrast to a reduction in weight gain, only occurred in the immediate aftermath (about 1 week) of a single subcutaneous dose equivalent to about 50–70% of a subcutaneous LD_{50}. Significant reduction in weight gain was observed with rats receiving non-lethal amounts of dietary

TMs but concomitant adverse effects did dictate euthanasia after 9 months and would probably have caused death (Colegate *et al*., unpublished).

Assuming that a bacterial 'gall' on annual ryegrass contains about 1–2 μg of CTs and a sheep ingests about 1 kg feed day^{-1}, Study 8 implies that no adverse effect on body weight or feed intake is expected with sheep ingesting the equivalent of about 800–1620 bacterial 'galls' kg^{-1} feed day^{-1} for 12 months.

References

Berry PH, Cook RD, Howell JMcC, White RR, and Purcell DA (1976). Lesions in sheep and guinea pigs pen fed parasitised annual ryegrass (*Lolium rigidum*). *Australian Veterinary Journal* 52:540–541.

Berry PH, Richards RB, Howell JMcC, and Cook RD (1982). Hepatic damage in sheep fed annual ryegrass, *Lolium rigidum*, parasitised by *Anguina agrostis* and *Corynebacterium rathayi*. *Research in Veterinary Science* 32:148–156.

Bourke CA and Carrigan MJ (1993). Experimental tunicamycin toxicity in cattle, sheep and pigs. *Australian Veterinary Journal* 69:228–229.

Davies SC, White CL, Williams IH, Allen JG, and Croker KP (1996). Sublethal exposure to corynetoxins affects production of grazing sheep. *Australian Journal of Experimental Agriculture* 36:649–655.

Oishi S, Oishi H, and Hiraga K (1979). The effect of food restriction for 4 weeks on common toxicity parameters in male rats. *Toxicology and Applied Pharmacology* 47:15–22.

Peterson JE and Jago MV (1977). Brain damage by extracts of parasitized annual ryegrass (*Lolium rigidum*) in nursling rats. *Australian Journal of Experimental Biology and Medical Science* 55:233–244.

Peterson JE, Jago MV, and Stewart PL (1996). Permanent testicular damage induced in rats by a single dose of tunicamycin. *Reproductive Toxicology* 10:61–69.

Stewart PL, May C, and Jago MV (1998). Reduction and recovery of *N*-acetylglucosamine-1-phosphate transferase activity in the liver of sheep and rats after a single subcutaneous dose of tunicamycin. *Australian Veterinary Journal* 76:287–288.

Chapter 41

Dose Response of Tunicamycins in Sheep Following Intra-duodenal Administration

S.L. Hauptli[1,2], S.M. Colegate[3], S. Glover[2], and S.J. McClure[1]

[1] *F. D. McMaster Laboratory, CSIRO Livestock Industries, Locked Bag 1, Delivery Centre, Armidale, NSW 2350, Australia;* [2] *Chemistry, School of Biological, Biomedical and Molecular Sciences, University of New England, Armidale, NSW, 2351, Australia;* [3] *Australian Animal Health Laboratory, CSIRO Livestock Industries, Plant Toxins Research Group, Private Bag 24, Geelong, VIC, 3220, Australia*

Introduction

Annual ryegrass toxicity (ARGT) is a neurological disease of livestock that is caused by corynetoxins (CTs) and results in major economic losses to the Australian livestock industry annually. The commercially available tunicamycins (TMs) are structurally, biochemically and toxicologically very similar to the corynetoxins (Edgar *et al.* 1982; Vogel *et al.* 1982; Jago *et al.* 1983; Frahn *et al.* 1984) and have been used to model ARGT due to the intrinsic difficulties in isolating large amounts of pure corynetoxins. ARGT has been produced experimentally in sheep following oral dosing of infected ryegrass and by subcutaneous injection of tunicamycins (Berry *et al.* 1982; Finnie and Jago 1985).

In preparation for assessing the protective effect of mucosal vaccination for ARGT, it was necessary to determine the dose-response relationship, especially that of the liver, of tunicamycins surgically administered into the duodenum of sheep as a model for the natural route of exposure to the toxins. The aim of this study was to determine the minimum single, intra-duodenal dose of tunicamycins that would significantly alter serum biochemistry (liver enzymes, albumin, total protein and total bilirubin) but would not induce overt clinical signs of the disease. This dose would subsequently be used as a challenge dose in the mucosal vaccination experiments.

Materials and Methods

Animals and experimental design

All animal experiments were completed under the auspices of the CSIRO Armidale Ethics Committee consistent with the Australian Code of Practice for the care and use of Animals for

Scientific Purposes 1997. No animals died as a result of the intoxication. Those presenting with clinical signs of intoxication were humanely euthanased.

Six- to12-month-old fine wool Merino wethers, raised on pasture and with a body weight range of 22 to 41 kg, were penned individually in an animal shed and acclimatized for 3 weeks on standard pelleted feed prior to the start of the study.

Tunicamycins (Sigma Aldrich, Castle Hill, Australia) were dissolved in 0.05 M aqueous NaOH and administered to groups of sheep as a single dose of 12.5, 18.75, 25, 31.25, 37.5, 50, 62.5, 100, 150, 200, 250 or 1000 μg TMs kg^{-1} body weight (bw) (Table 1). At 16 days, or when clinical signs were observed, sheep were euthanased by captive bolt and exsanguination and liver samples were taken for histology. Blood from each animal was collected from the jugular vein into heparinized tubes at 0800–0900 hours each day and immediately prior to euthanasia.

Intra-duodenal dosing of tunicamycins

Fasted (24 hours), anaesthetized, and fully monitored sheep were positioned in left lateral recumbency in preparation for exteriorization of the duodenum, immediately distal to the pylorus, via a skin incision in the right paralumbar fossar, 3 cm posterior to the last rib and approximately 15 cm from midline. Doses of 1.6–5.5 mL tunicamycins of appropriate concentrations were injected into the exposed duodenum using a 22-gauge needle prior to replacement of the exposed section of duodenum and closure of the incision.

Post-operative sheep were fed pellets and hay *ad libitum* and were monitored twice-daily for body temperature, feed and water intake, feces and urine output, respiration and abdominal discomfort. Antibiotic and pain management treatment was maintained for 3 days following dosing.

Biochemical assays

The serum samples were analyzed for aspartate aminotransferase (AST), alkaline phosphatase (ALP), alanine aminotransferase (ALT), gamma glutamyltransferase (GGT), lactic dehydrogenase (LDH), total protein (TP), total bilirubin (TBIL) and albumin (ALB) using commercial kits on the Dimensions® Clinical Chemistry System (Dade Behring Diagnostics, Coorparoo, QLD, Australia). Results from these measurements were assessed, using an analysis of variance for repeated measures (Systat version 9 for Windows, SPSS inc), for normality in distribution, whether any effects were dose-related, whether there were any day-to-day changes in the effects within a dose group and whether any such day-to-day effects were consistent between dose groups.

Histology

Samples of liver were collected and stored in 10% phosphate-buffered formalin. Fixed sections, 7 μm in thickness, were cut and stained with hematoxylin and eosin and examined by Dr C. Bourke, NSW Agriculture, Orange, NSW, Australia.

Results and Discussion

The survival rate and incidence of neurological signs following the administration of the TMs is summarized in Table 1. Overt clinical signs, including episodic ataxia, tremor and lateral recumbency with convulsions, were precipitated in affected animals by mild stimulation involving movement of the sheep or making sudden noises. Sheep that received doses less than 150 μg TMs kg^{-1} bw showed no evidence of neurological signs. At 150 μg TMs kg^{-1} bw one sheep from the group showed episodes of moderate to severe tremor and ataxia after stimulation on Day 3 and was later euthanased. One case of mild neurological signs was seen following a dose of 200 μg TMs kg^{-1}bw; however, full recovery was seen once stimulation had ceased and no further periods of neurological signs were presented. A dose of 250 μg TMs kg^{-1} bw caused periods of lateral recumbency and convulsions in one sheep at Day 2.5 post-dosing and moderate stimulation-initiated tremor and ataxia in another at Day 4.5 post-dosing. All sheep given 1000 μg TMs kg^{-1} bw presented with spontaneous, repeated episodes of lateral recumbency, severe tremor and convulsions on Day 3. This latter effect is similar to the lethal effect on sheep within 4 days of receiving a dose of about 1000 μg TMs kg^{-1} bw (Bourke and Carrigan 1993).

Whilst there were statistically significant differences between levels of some serum liver enzymes and hematology parameters even between lower dose groups, the relevance of these differences is difficult to judge without further studies. Firstly, the levels remained within normal ranges expected for sheep and the apparent differences persisted throughout the 16 days. However, since the highest dose group (1000 μg TMs kg^{-1} bw) was equivalent to an oral LD_{50} (Bourke and Carrigan 1993) it was not unexpected that serum liver enzymes increased in concentration with the onset of clinical signs. This is represented by the effect on LDH from Day 0 to Day 3.5 when the animals were euthanased (Figure 1). The levels of TBIL spiked to abnormal levels 4 days following a dose of 250 μg TMs kg^{-1} bw (Figure 1).

Table 1. Toxicity of tunicamycins in sheep: single surgical intra-duodenal administration.

Dose (μg/kg)	Number Treated	Number of Survivors	Neurological Symptoms; Day of occurrence and number of sheep	Time to Euthanasia (Days)
12.5	5	5	-	16 (5)
18.75	5	5	-	16 (5)
25	5	5	-	16 (5)
31.25	5	5	-	16 (5)
37.5	5	5	-	16 (5)
50	5	5	-	16 (5)
62.5	5	5	-	16 (5)
100	4	4	-	16 (4)
150	4	3	3 (1)*	4 (1), 16 (3)
200	4	4	3 (1)	16 (4)
250	3	1	2.5 (1), 4.5 (1)	3 (1), 5 (1), 16 (1)
1000	3	0	3 (3)	3 (1), 3.5 (2)

* Numbers in parenthesis indicate the number of sheep with the same value

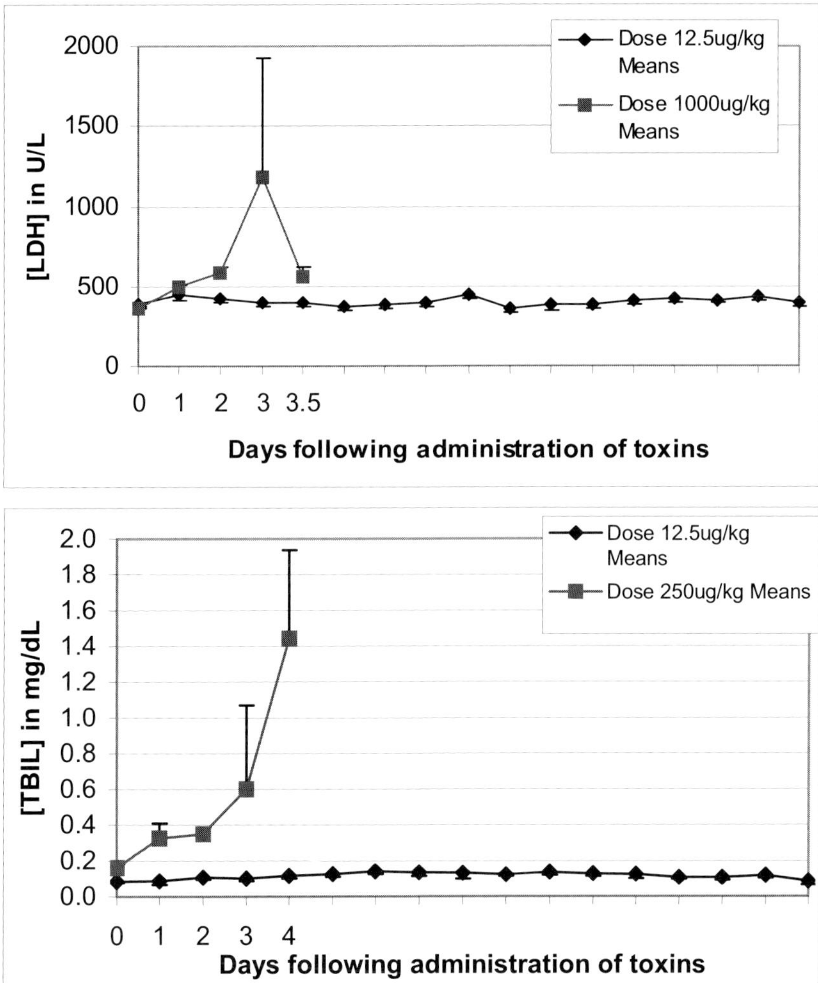

Figure 1. Dose 12.5 μg kg^{-1} means versus Dose 1000 μg kg^{-1} means for [LDH] in $U{\cdot}L^{-1}$ and Dose 12.5 μg kg^{-1} means versus Dose 250 μg kg^{-1} means for [TBIL] in mg dL^{-1}.

In a preliminary histological assessment, liver changes characteristic of tunicaminyluracil toxicity were present in sheep exposed to a single, intra-duodenal dose of 1000 μg TMs kg^{-1} bw. Thus, diffuse, mild lipidosis and hydropic degeneration of hepatocytes were observed along with occasional hepatocytes that were undergoing a coagulative necrosis that was accompanied by cytoplasmic shrinkage (Figure 2). There were also small numbers of neutrophils within some sinusoids.

Conclusions

The results of this experiment indicate that further studies need to be conducted to establish the relevance of lower dosage results, and to therefore establish whether serum liver enzymes or hematological parameters are of any value in detecting sub-clinical toxicity resulting from intra-duodenal administration of tunicamycins. However, whilst the data indicated that a study of the efficacy of mucosal vaccination is best assessed by challenging with doses of tunicamycins that cause overt clinical signs, regular assessment of LDH and TBIL during a challenge study may yet reveal a lessened impact at lower doses following vaccination.

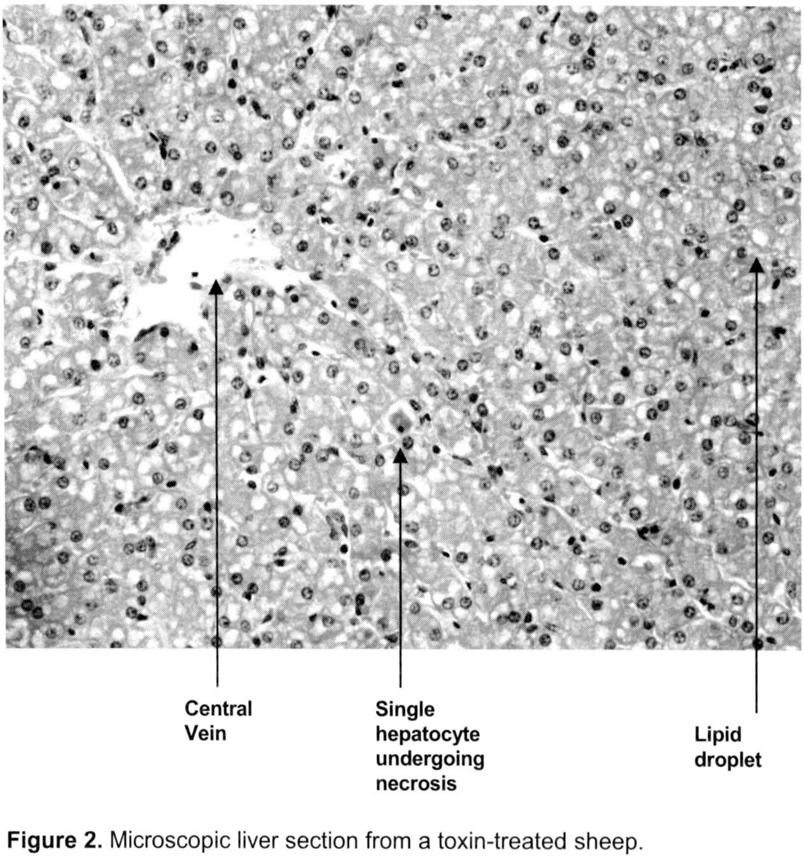

Figure 2. Microscopic liver section from a toxin-treated sheep.

References

Berry PH, Richards RB, Howell JMcC, and Cook RD (1982). Hepatic damage in sheep fed annual ryegrass, *Lolium rigidum*, parasitised by *Anguina agrostis* and *Corynebacterium rathayi*. *Research in Veterinary Science* 32:148–156.

Bourke CA and Carrigan MJ (1993). Experimental tunicamycin toxicity in cattle, sheep and pigs. *Australian Veterinary Journal* 69:228–229.

Edgar JA, Frahn JL, Cockrum PA, Anderton N, Jago MV, Culvenor CCJ, Jones AJ, Murray K, and Shaw KJ (1982). Corynetoxins, causative agents of annual ryegrass toxicity; their identification as tunicamycin group antibodies. *Journal of the Chemical Society, Chemical Communications*, 222–224.

Finnie JW and Jago MV (1985). Experimental production of annual ryegrass toxicity with tunicamycin. *Australian Veterinary Journal* 62:248–249.

Frahn JL, Edgar JA, Jones AJ, Cockrum PA, Anderton N, and Culvenor CCJ (1984). Structure of the corynetoxins, metabolites of *Corynebacterium rathayi* responsible for toxicity of annual ryegrass (*Lolium rigidum*) pastures. *Australian Journal of Chemistry* 37:165–182.

Jago MV and Culvenor CCJ (1987). Tunicamycin and corynetoxin poisoning in sheep. *Australian Veterinary Journal* 64:232–235.

Jago MV, Payne AL, Peterson JE, and Bagust TJ (1983). Inhibition of glycosylation by corynetoxin, the causative agent of annual ryegrass toxicity: a comparison with tunicamycin. *Chemico-Biological Interactions* 45:223–234.

Vogel P, Stynes BA, Coackley W, Yeoh GT, and Petterson DS (1982). Glycolipid toxins from parasitised annual ryegrass: a comparison with tunicamycin. *Biochemical and Biophysical Research Communications* 105:835–840.

Chapter 42

An Assessment of Spatial Learning and Memory in Sheep Chronically Exposed to Tunicamycins

C. Lee[1], A.D. Fisher[1], A. Michalewicz[2], and S.M. Colegate[2]

[1]*CSIRO Livestock Industries, Locked Bag 1, Armidale, NSW 2350, Australia;* [2]*CSIRO Livestock Industries, Australian Animal Health Laboratory, Private Bag 24, Geelong, Victoria 3220, Australia*

Introduction

Annual Ryegrass Toxicity (ARGT) is an often-fatal plant-associated intoxication of livestock that clinically manifests with neurological signs (Bourke 1994). It is caused by ingestion of corynetoxins, a group of potently toxic, tunicaminyluracil glycolipids that are produced by the bacterium *Rathayibacter toxicus* (Edgar *et al.* 1982). The bacterium is transported to the developing seedheads of *Lolium rigidum* (annual ryegrass) by the gall-forming nematode *Anguina funesta*. The bacterium outcompetes the nematode and 'takes over' the gall, producing the toxins as the host plant starts to senesce (Stynes and Bird 1983). Other plant species involving the same, or similar, bacteria have also produced toxic, bacterial 'galls' evidenced by livestock deaths (Bryden *et al.* 1994) or the demonstration of the presence of the toxins (Anderton *et al.* 2004).

The corynetoxins exert their impact by inhibiting *N*-acetylglucosamine-1-phosphate transferase, an enzyme involved in the first step of the biosynthesis of oligosaccharides destined for protein *N*-glycosylation (Jago *et al.* 1983). The disease has often been experimentally modelled using the structurally similar tunicamycins due to their more ready availability. Acute intoxication of sheep with tunicamycins causes clinical signs indistinguishable from ARGT. Ongoing concerns about the potential adverse effect of long-term, low-level exposure to environmental and dietary corynetoxins has prompted a No Observed Effect Level (NOEL) study investigating clinical, biochemical, pathological and behavioral parameters in rats and sheep (SM Colegate *et al.*, unpublished).

As part of a neurological assessment, the spatial learning ability and memory of male and female Merino sheep that had been exposed to various low levels of dietary tunicamycins for 42–48 weeks was investigated. For this purpose, a maze was specifically designed and modified to allow a relatively rapid assessment of the ability of individual sheep to learn and remember the best route through the maze. The motivating force was the desire for the individual to rejoin its flock mates visible to the individual at the other end of the maze. The capacity for the maze to provide a useful assessment of spatial learning and memory was validated by comparing the performance of control sheep with those that had been administered scopolamine, which is a CNS muscarinic receptor antagonist and a known cognition-impairing substance (Taffe *et al.* 1999).

Methods, Results and Discussion

Development of the maze

Previous studies of sheep behavior in a maze utilized the Hebb and Williams maze to assess 'mental ability' (Kilgour and Bruere 1970). However, this test requires 6 days of previous training to reach fixed, baseline performance criteria, which limits the number of animals that can be tested in a reasonable time. To measure cognitive performance and spatial learning in large numbers of sheep without the need for prior training, a maze that utilized conspecifics and the strong flocking instinct of sheep as motivation to move through the maze was developed (Lee *et al.* 2006).

An 18.4 x 8.2 m maze was constructed in the open with 1.2 m-high opaque external walls and a number of internal barriers (Figure 1). The inner walls were made of open-barred portable fence panels to enable animals to view conspecifics at the opposite end of the maze, thus providing motivation to traverse the maze and join their flock mates. To allow rapid assessment of large numbers of sheep, the maze configuration was adapted such that 5 min was the maximum time for most naive sheep to complete the maze.

Sheep (controls, scopolamine-treated and tunicamycins-treated) were assessed in the maze on each of 3 successive days (Days 1, 2 and 3) and this was repeated, for the tunicamycins groups, 6 weeks later (Days 42, 43 and 44). The performance of each sheep was a factor of the total time for the subject to complete the maze and the error time (sum of time spent in zones A and B; Figure 1).

Scopolamine study

Two groups of 15 2-year-old Merino wethers were injected (im) either with saline (1 mL) or scopolamine hydrobromide (100 μg kg^{-1} body weight) dissolved in saline (1 mL) 30 min prior to admission to the maze on each day of assessment. The scopolamine dose level was chosen based on research that described disturbed motor function in ewes that received scopolamine at 200 μg kg^{-1} body weight but not at 100 μg kg^{-1} body weight (Levy *et al.* 1997), and an observed delay in learning a visual discrimination task but without any accompanying motor disruption also at 100 μg kg^{-1} body weight (Ferreira *et al.* 2003).

The control sheep dramatically improved their performance in the maze over 3 consecutive days, reducing the time taken to negotiate the maze from 245 ± 59.3 sec on Day 1 to 38 ± 6.5 sec on Day 3. In contrast, by Day 3 sheep receiving scopolamine were significantly slower (198 ± 65.6 sec, $\underline{P}<0.05$) to complete the maze than control sheep and showed no improvement from Day 1. There were no differences in error times between control and scopolamine-treated sheep, with both groups showing a marked improvement from Day 1 (Controls, 84 ± 15 sec; Scopolamine-treated, 137 ± 49.1 sec) to Day 2 (Controls, 19 ± 5.5 sec; Scopolamine-treated, 33 ± 21 sec), which was maintained into Day 3.

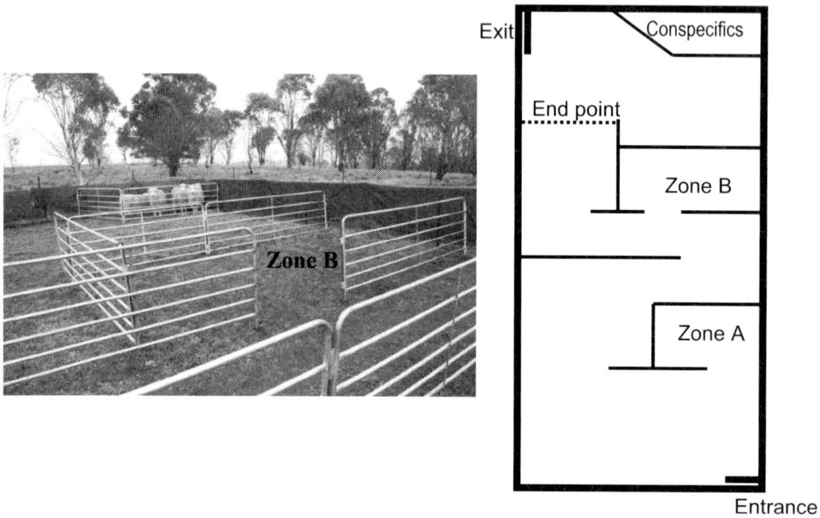

Figure 1. A plan and photograph of the maze. Note the conspecifics in the holding pen at the far end of the maze. For orientation, Zone B is also shown in the photograph.

Thus, whilst the error times of the scopolamine-treated sheep improved at the same rate as the control sheep, their transit times for the maze did not decrease, indicating that the maze can be used to differentiate performance. Although it is not possible to completely exclude motor function deficit as a contributing factor to the decreased transit time, the normal motor function at 100 μg scopolamine hydrobromide kg^{-1} body weight recorded by Levy *et al.* (1997) suggests that differences in cognition were the main contributors to the decreased performance of the scopolamine-treated animals. However, due to the complex nature of cognitive processes that have been measured by maze tests, it is not possible to identify the particular aspects of cognition that were measured in this study.

Tunicamycins study

Single gender groups of five 12-month-old Merino sheep (total N=60; 30 rams, 30 ewes) received 0, 0.5, 1.5, 4.5, 13.5 or 40.5 μg tunicamycins kg^{-1} body weight day^{-1} administered three times per week in their diet. The doses were selected based on previous research that demonstrated an adverse effect at sub-lethal doses. The highest dose was then refined in a 6-month chronic toxicity study (SM Colegate *et al.*, unpublished), made necessary by the reported bioaccumulation of the toxins, and lower doses selected as one-third decreases estimated to identify a No Observed Adverse Effect Level. After 42 weeks of dosing, all sheep were clinically normal and were tested individually in the maze on each of 3 successive days. When the testing was repeated 6 weeks later (Days 42, 43 and 44), sheep were still clinically normal on observation. Dosing with the tunicamycins continued during the maze test period.

There were no significant differences in the total times to negotiate the maze or in the error times for any of the groups. However, there was a significant improvement in maze performance (Total Transit Time and Error Time) by all Groups from Day 1 to Day 2 and Day 3, which was maintained when re-assessed at Days 42, 43 and 44 (Lee *et al.* 2006).

Conclusions

Exposure of male and female sheep to levels of 40.5, 13.5, 4.5, 1.5 and 0.5 μg tunicamycins kg^{-1} body weight day^{-1} of tunicamycins in their diet for 42–48 weeks did not have any discernible impact on the ability of the exposed sheep to negotiate and remember the maze. This is despite the top dose sheep cumulatively receiving approximately ten times the oral lethal dose of tunicamycins (approximately 1 mg kg^{-1} body weight; Bourke and Carrigan 1993). The lack of any adverse effect of the tunicamycins as assessed by the maze does not obviate the possibility of other adverse effects that may be used to define the NOEL.

This maze test, by virtue of measuring high-level biological function, may have applications in other areas, such as intensive management practices and including plant-associated intoxications other than ARGT, where the health and welfare of animals may be compromised.

References

Anderton N, Beales KA, Cao Y, Colegate SM, Edgar JA, Michalewicz A, Riley PL, Stewart PL, and Than KA (2004). The identification of corynetoxin-like tunicaminyluracil-glycolipids from nematode galls in *Festuca nigrescens* from North America and New Zealand. In *Poisonous Plants and Related Toxins* (T Acamovic, CS Stewart, and TW Pennycott, eds), pp. 204–209. CABI Publishing, Wallingford, Oxon, UK.

Bourke CA (1994). Tunicaminyluracil toxicity, an emerging problem in livestock fed grass or cereal products. In *Plant-associated Toxins: Agricultural, Phytochemical and Ecological Aspects* (SM Colegate, and PR Dorling, eds), pp. 399–404. CABI Publishing, Wallingford, Oxon, UK.

Bourke CA and Carrigan MJ (1993). Experimental tunicamycin toxicity in cattle, sheep and pigs. *Australian Veterinary Journal* 69:228–229.

Bryden WL, Trengove CL, Davis EO, Giesecke PR, and Curran GC (1994). Corynetoxicosis of livestock: A nematode-bacterium disease complex associated with different grasses. In *Plant-associated Toxins: Agricultural, Phytochemical and Ecological Aspects* (SM Colegate and PR Dorling, eds), pp. 410–415. CABI Publishing, Wallingford, Oxon, UK.

Edgar JA, Frahn JL, Cockrum PA, Anderton N, Jago MV, Culvenor CCJ, Jones AJ, Murray K, and Shaw KJ (1982). Corynetoxins, causative agents of annual ryegrass toxicity; their identification as tunicamycin group antibiotics. *Journal of the Chemical Society Chemical Communications*, 222–224.

Ferreira G, Poindron P, and Levy F (2003). Involvement of central muscarinic receptors in social and nonsocial learning in sheep. *Pharmacology, Biochemistry and Behavior* 74:969–975.

Jago MV, Payne AL, Peterson JE, and Bagust TJ (1983). Inhibition of glycosylation by corynetoxin, the causative agent of annual ryegrass toxicity: a comparison with tunicamycin. *Chemical and Biological Interactions* 45:223–234.

Kilgour R and Bruere AN (1970). Behaviour patterns in chromatin-positive Klinefelter's syndrome in sheep. *Nature* 225: 71–72.

Lee C, Colegate SM, and Fisher AD (2006). Development of a maze test and its application to assess spatial learning and memory in Merino sheep. *Applied Animal Behaviour Science* 96:43–51.

Levy F, Pichard P, Meurisse M, and Ravel N (1997). Scopolamine impairs the ability of parturient ewes to learn to recognize their lambs. *Psychopharmacology* 129:85–90.

Stynes BA and Bird AF (1983). Development of annual ryegrass toxicity. *Australian Journal of Agriculture* 34:653–660.

Taffe MA, Weed MR, and Gold LH (1999). Scopolamine alters rhesus monkey performance on a novel neuropsychological test battery. *Cognitive Brain Research* 8:203–212.

Chapter 43

Dose-related, Differential Gene Expression in Cultured Rat Hepatocytes Exposed to Tunicamycins: a Model for Annual Ryegrass Toxicity

M. Kowalski[1], T. Doran[1], D. Cahill[2], and S.M. Colegate[1]

[1] CSIRO Livestock Industries, Australian Animal Health Laboratory, Geelong, VIC 3220, Australia; [2] School of Life and Environmental Sciences, Deakin University, Geelong, VIC 3217, Australia

Introduction

Annual Ryegrass Toxicity (ARGT) is an often fatal hepatocerebral disease of livestock caused by ingestion of corynetoxins (CTs) produced by the bacterium *Rathayibacter toxicus* that colonize *Anguina festuca* nematode galls in the seedheads of *Lolium rigidum*. Diagnosis of non-clinical exposure of livestock to CTs is important when considering animal health, welfare and productivity issues. Currently, diagnosis of ARGT is limited to clinical observations of affected livestock and their clinical chemistry. Furthermore, corynetoxins are cumulative in nature so multiple sub-lethal exposures may potentially cause irreversible damage before the manifestation of overt clinical symptoms.

The commercially available tunicamycins (TMs) are a group of tunicaminyluracil glycolipids that are structurally very similar to CTs and have been routinely used as a substitute to experimentally induce ARGT. Several *in vivo* studies have shown TMs and CTs to exhibit similar antibiotic and biochemical activities, toxicities and clinical disease based on acute exposure (Jago *et al.* 1983; Finnie and Jago 1985; Jago and Culvenor 1987). The *in vitro* effects of acute exposure to CTs and TMs on primary rat liver hepatocytes have been reported separately (Yeoh *et al.* 1984; Finnie 2001).

To further investigate the effects of TMs at the molecular level, a toxicogenomic approach has been developed to examine differential gene expression between cultured rat hepatocytes exposed to TMs at low levels, compared to untreated cultured hepatocytes (control). Cultured rat hepatocytes have become a useful indicator of toxic exposure using current microarray technology (Huang *et al.* 2001).

Materials and Methods

Preparation of tunicamycin-containing media

Tunicamycins (av. MW 840; Sigma-Aldrich, St Louis, Missouri, USA) were solubilized and serially diluted in 0.6 mM AnalaR grade sodium hydroxide (NaOH) (BDH Ltd., Merck, Darmstadt, Germany). Tunicamycins were added to the media at 0.5% v/v such that the final concentrations in F-12K Nutrient Mixture Media (Invitrogen, Carlsbad, California, USA) were 0.05 μM and 50 μM. NaOH control samples consisting of 0.5% v/v toxin-free 0.6 mM NaOH in F-12K media were prepared in addition to the solvent-free F-12K media controls.

Exposure of *in vitro* hepatocytes to tunicamycin

Twelve Opticells™ (BioCrystal, Westerville, Ohio, USA) were seeded with Clone 9 rat hepatocytes (ATCC, Manassas, Virginia, USA; accession number CRL-1439) at a density of 1×10^5 cells mL^{-1} in toxin-free F-12K media. This provided four groups of triplicate samples representing the solvent-free control (F-12K media alone), NaOH control (F-12K media with toxin-free NaOH added), and 0.05 μM and 50 μM TMs in F-12K media. After 48 hours post-seeding incubation at 37 °C in a 5% CO_2 in air atmosphere, the media was removed from all Opticells™ and replaced with the various test media and incubation continued for a further 24 hours.

Cell treatment and RNA extraction

After incubation the test media were removed from each Opticell™ and the residual cells still adhering to the Opticell™ membrane were washed twice with phosphate-buffered saline (PBS). Cells were recovered from the Opticell™, following treatment with trypsin for 5 min at 37°C, pelleted by centrifugation and total RNA extracted (RNeasy® Mini Kit, Qiagen). The purity (from contaminating DNA) and the concentration of extracted RNA was assessed spectrophotometrically at 260/280 nm (Nanodrop Spectrophotometer, Nanodrop Technologies). Denaturing agarose gel electrophoresis was also carried out to assess the quality and quantity of mRNA in the total RNA sample.

Reverse transcription

Eight μg of total RNA was used in a reverse transcription reaction with indirect labeling (the incorporation of a biotin-dNTP mixture during cDNA synthesis) performed in triplicate for control and treated samples and then pooled to minimize variability according to the DualChip™ instruction manual. An internal standard mixture was used for quantification/normalization and estimation of experimental variation (de Longueville *et al.* 2002).

Microarray design and hybridization

The DualChip™ rat hepato (Eppendorf, Germany) has been designed for toxicogenomic analysis of rat hepatocytes and contains two arrays per slide with a range of genes involved in cellular processes such as drug metabolism, stress responses, cell proliferation, cell cycle activation, transcription, inflammation and apoptosis (de Longueville *et al.* 2002). To evaluate the reliability of the experimental data, several positive and negative hybridization and detection controls were included on the microarray. For normalization, three internal standard controls and eight housekeeping genes were arrayed on the slides.

Hybridization of the DualChip™ rat hepato with the biotin-labeled cDNA was completed in 16 hours at 60 °C and 1400 rpm using a Thermomixer comfort (Eppendorf, Germany). Hybridized products were detected using a Cy™3-conjugated IgG Fraction Monoclonal Mouse Anti-Biotin (Jackson Immuno Research Laboratories, USA).

Detection and data analysis

The detection and quantification of hybridization events were carried out with an ArrayWoRx^{e} scanner (Applied Precision, USA) using a 565 nm emission filter for the Cy3 signal. To maximize the dynamic range of microarrays and allow quantification of both high and low copy expressed genes, arrays were scanned using different exposure times. After image acquisition, the scanned 16-bit image was used to quantify the signal intensities with 'ImaGene® 4.1' software (BioDiscovery®, USA). The fluorescent intensity of each DNA spot (average intensity of each pixel present within the spot) was calculated using local mean background subtraction. A signal was accepted if the average intensity following background subtraction was at least 2.5-fold higher than its local background. Very bright element intensities (saturated signals, highly expressed genes) were deemed unsuitable for accurate quantitation because they underestimated the intensity ratios. They were excluded from further analysis.

Normalization and data analysis

A two-step normalization procedure was performed prior to data analysis. First, a correction of values was carried out by using a factor calculated from the intensity ratios of the internal standard mixture. The corresponding probes of the internal standards are located at six different areas of the microarray to determine the local background and evaluate the homogeneity of the microarray. Second, because the first step does not reflect the purity and quality of the RNA, a correction of values was carried out by using a factor calculated from the intensity ratios of housekeeping genes. Only those housekeeping genes for which the expression was not changed after treatment with toxin were taken into account for the normalization. The variance of the normalized set of housekeeping genes (except those affected by the tested condition) was used to generate an estimate of expected variance, leading to a predicted confidence interval for testing the significance of the gene expression ratios obtained. Ratios outside the 95% confidence interval were determined to be significantly different.

Results and Discussion

Exposure of *in vitro* cultured rat hepatocytes to 0.05 μM and 50 μM TMs resulted in 38 and 34 genes respectively that were expressed at significantly different levels compared to untreated hepatocyte controls. Statistical analysis of the 51 genes (housekeeping genes excluded) presented on the DualChip™ rat hepato revealed that 24 (47%) genes were up-regulated, 14 genes (27%) were down-regulated, and 13 (39%) remained stable following treatment with TMs at 0.05 μM. Dosed at a TMs concentration of 50 μM, 26 genes (50%) were up-regulated, eight genes (16%) were down-regulated, and 17 genes (34%) remained stable. Figure 1 shows the hybridization data of control and 0.05 μM- and 50 μM-treated samples on DualChip™ rat hepato at 24 hours. Figure 2 shows the corresponding scatter plot.

The most highly induced genes consistent within both treatments compared to untreated cells were: *FN, ecoa, MnSOD, mt_Co1* and *Osteon.* Their functions, including oxidative metabolism and detoxification of radicals which are toxic to biological systems (*MnSOD*- superoxide dismutase), drug metabolism (*mt_Co1* – cytochrome c oxidase*),* lyase activity/ β oxidation (*ecoa*-enoyl CoA) and maintaining the integrity of the extracellular matrix (*FN*-fibronectin and *Osteon*-Osteonectin), are consistent with expected up-regulation following a toxic insult. The most significantly down-regulated gene was *JNK3* that functions as a protein kinase activated as part of a stress response. This observation implies that *JNK3* kinase activity does not have a direct involvement in the hepatocellular response to TMs at the stage of TMs-induced hepatocellular stress at the time of cell harvesting.

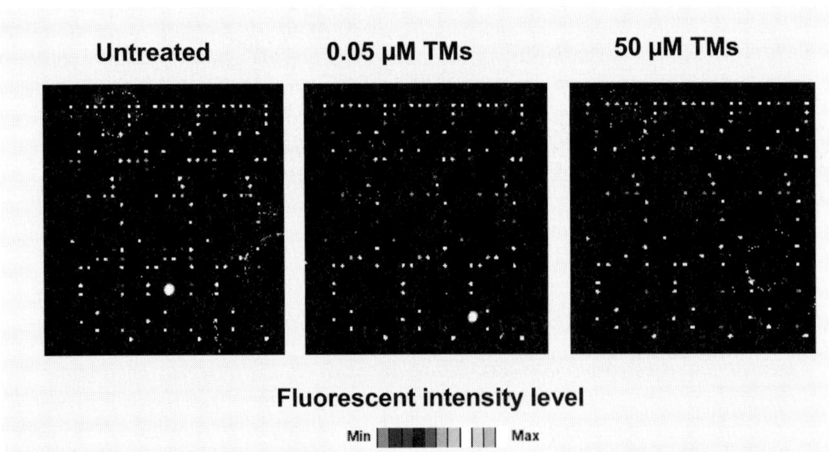

Figure 1. DualChip™ rat hepato hybridized with cDNA obtained from total RNA extracted from untreated rat hepatocytes (control) and hepatocytes treated with 0.05 and 50 μM TMs. Fluorescence is represented in false color scale and corresponds to the gene expression levels.

The difference in gene expression profiles between 0.05 μM and 50 μM was limited to only a few genes. An anti-apoptotic gene *Bclx* was shown to be significantly up-regulated in 50 μM-treated cells, but remained unchanged in the 0.05 μM sample. This result is consistent with previous studies demonstrating acute TMs treatment causes apoptosis (Finnie *et al.* 2004). Also *c-jun*, another stress response kinase, was shown to be down-regulated in 0.05 μM-treated cells, but was unchanged in the 50 μM. Furthermore, *FN* and *ecoa* (two of the five most highly expressed genes) demonstrated significantly greater induction in the 50 μM dose compared to 0.05 μM. This may indicate that the induction of *Bclx*, *FN* and *ecoa* is a dose-dependent cellular response to TMs.

Conclusions

This toxicogenomic approach to defining the effect of exposure of *in vitro* cultured rat hepatocytes to tunicamycins will complement previous work that has focused primarily on toxicology, pathology and pathogenesis associated with acute exposure to TMs.

Future work using significantly lower dose rates and other hepatotoxins will assess the specificity of the gene regulation changes for tunicaminyluracil toxicity and the potential utility of the gene profile as an early biomarker of exposure to the tunicaminyluracil glycolipid compounds.

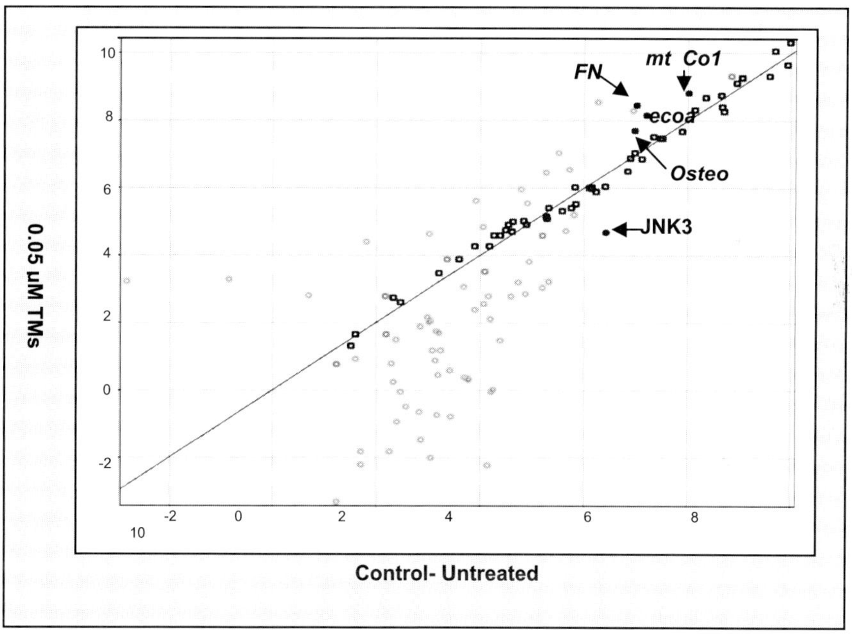

Figure 2. Scatter plot of normalized fluorescent intensity values from DualChip™ rat hepato hybridized with cDNA obtained from mRNA extracted from untreated and treated hepatocytes with 0.05 μM TMs at a medium PMT setting. Control values (x-axis) were plotted against TMs dosed values (y-axis). The arrows represent the most highly up-regulated and down-regulated genes consistent between all 3 PMT settings and both TMs doses.

References

De Longueville F, Surry D, Meneses-Lorente G, Bertholet V, Talbot V, Evrard S, Chandelier N, Pike A, Worboys P, Rasson JP, Le Bourdelles B, and Remacle J (2002). Gene expression profiling of drug metabolism and toxicology markers using a low-density DNA microarray. *Biochemical Pharmacology* 64:137–149.

Finnie JW (2001). Effect of tunicamycin on hepatocytes *in vitro*. *Journal of Comparative Pathology* 125:318–321.

Finnie JW and Jago MV (1985). Experimental production of annual ryegrass toxicity with tunicamycin. *Australian Veterinary Journal* 62:248–249.

Finnie JW, Read SH, and Swift JG (2004). Apoptosis in liver damage produced by tunicamycin. *Australian Veterinary Journal* 82:87–90.

Huang Q, Dunn II RT, Jayadev S, Di Sorbo O, Pack FD, Farr SB, Stoll RE, and Blanchard KT (2001). Assessment of cisplatin-induced nephrotoxicity by microarray technology. *Toxicology Science* 63:196–207.

Jago MV and Culvenor CCJ (1987). Tunicamycin and corynetoxin poisoning in sheep. *Australian Veterinary Journal* 64:232–235.

Jago MV, Payne AL, Peterson JE, and Bagust TJ (1983). Inhibition of glycosylation by corynetoxins, the causative agent of annual ryegrass toxicity: a comparison with tunicamycin. *Chemical Biological Interactions* 45:223–234.

Yeoh GC, Vogel P, Daly J, Brighton VJ, Light A, and Petterson DS (1984). Effect of corynetoxins isolated from parasitised annual ryegrass on albumin and transferrin synthesis and secretion by cultured fetal rat hepatocytes. *Experimental Cellular Research* 151:421–432.

Chapter 44

A Case of Acute Water Hemlock (*Cicuta maculata*) Poisoning and Death in Cattle after Ingestion of Green Seed Heads

K.E. Panter[1], D.R. Gardner[1], D. Holstege[2], and B.L. Stegelmeier[1]

[1] *USDA-ARS Poisonous Plant Research Laboratory, Logan, UT 84341, USA;*
[2] *California Veterinary Diagnostic Laboratory Systems, Toxicology, West Health Sciences Drive, Davis, CA 95616, USA*

Introduction

Water hemlocks (*Cicuta* spp.) are perennial forbs of the Umbelliferae family. Distinguishing features of water hemlock include a thickened tuber with many parsnip-like roots at the base of a segmented hollow green stem. When the tuber or lower stem is cut, many cross striations or a matrix of compressed nodes and internodes are revealed. Usually, drops of a yellowish, oily exudate soon appear on the cut surface and this is believed to contain high levels of toxins. Upon exposure to air these compounds are quickly oxidized to a brownish substance.

All species of *Cicuta* grow in similar wet habitats of swamps, marshes, active streams or where adequate moisture will support growth. Water hemlock often grows directly in the water or on the banks of shallow streams, creeks or lakes.

Apparently, water hemlock is toxic to all species of livestock and humans (Knutsen and Paszkowski 1984; Panter *et al.* 1988). Cattle readily eat new-growth water hemlock in the early spring, often without death losses unless tubers or excess quantities of vegetative material are ingested (K.E. Panter, personal observation 1990). As water hemlock grows, relative toxicity declines until flowers and seeds develop. It is believed that the flowers and seeds of water hemlock sequester large amounts of toxins and this is supported by theories and evidence that plants use secondary metabolites (toxins) for defense mechanisms (Wink and Schimmer 1999). According to the evidence, plant organs that are important for survival and reproduction such as flowers, fruits and seeds are nearly always a rich source of secondary compounds.

Four species are recognized in North America including *Cicuta douglassii*, *C. maculata*, *C. virosa* and *C. bulbifera* (Mulligan 1979). All species are believed to be toxic to livestock; however, only the tubers and early growth-plants have been implicated in poisoning of cattle (Smith and Lewis 1987). The typical scenario describing water hemlock poisoning is early spring grazing where the water hemlock is in the early vegetative stage while grasses and other forage are limited. Cattle grazing along stream beds or in marshy areas are likely to expose roots or tubers, resulting in ingestion of highly toxic tubers and the more toxic early-growth plant. Only a portion of a single tuber is enough to kill a 100 kg sheep. Tubers are quite palatable and most toxic early in the spring. In sheep, as little as 1 mg kg^{-1} (wet weight) of tuber was lethal (Panter *et al.* 1996).

There is a narrow 'safety' margin (with regard to water hemlock toxicoses) between dosages exhibiting little or no overt clinical effects and the induction of grand mal seizures. Little effect is noted until the dose is high enough to stimulate the central nervous system. Once that dose is reached, very little additional material will induce severe grand mal seizures resulting in death. If these seizures persist for more than a few minutes the animal usually dies from seizure-induced respiratory failure. Depending on the dose, seizures are often intermittent, becoming more intense and prolonged with time until death occurs. Generally, diagnostic lesions and changes in serum biochemistry do not occur unless there are multiple seizures and the animal survives, in which case the lesions are secondary to the grand mal seizures (Panter *et al.* 1996). For example, lesions and serum changes are limited to damage of heart muscle and long skeletal muscle groups. Significant changes in AST, LDH and CK occur within days after the seizure activity if the animal survives. However, in acutely poisoned animals that die in a relatively short period (<1 day) serum blood changes are minimal and histological changes in skeletal muscle fibers are only subtle and include variation in myofiber diameter with slight increases of satellite and myofiber nuclei (Panter *et al.* 1996). No lesions were observed in heart muscle fibers in sheep acutely poisoned that died in 1–2 hours. In contrast, severity of lesions in skeletal and heart muscle fibers was extensive and serum concentrations of AST, LDH and CK were elevated 3–7 days after poisoning in recovering sheep showing multiple grand mal seizures. Severity of lesions and biochemical changes were directly related to the number and severity of seizures before recovery. Treatment of ewes with barbiturates at the time of onset of seizure activity prevented death and subsequent histologic and serum biochemical changes (Panter *et al.* 1996).

This field case reports for the first time suspected acute poisoning and death in cattle ingesting immature seed heads of water hemlock.

Materials and Methods

History

Investigation of the pasture after the cows died revealed that the creek had been heavily infested with water hemlock and that all the water hemlock seed heads had been eaten. It appeared that the cows walked up the shallow gravelly creek bed and grazed some of the grasses but ate all the hemlock seed heads and most of the hemlock vegetative parts.

Because of the sudden death and the appearance of the cattle the owner suspected possible malicious involvement. The local sheriff's department and a local veterinarian investigated the incident and after it was determined that there was no apparent criminal activity, the Poisonous Plant Research Laboratory was contacted. Because of the acute nature of the deaths, poisonous plants were suspected and water hemlock and poison hemlock were high on the list. Field investigation of the site revealed heavy infestations of water hemlock in the late flower/early seed stage in ungrazed pasture along the creek. There was an obvious and significant contrast between the pasture where poisoning occurred (no hemlock seed heads observed) versus the ungrazed pastures. Closer examination determined that all of the hemlock seed heads and most of the vegetative portions of the plant had been eaten within a few days. The cows had been in this pasture less than a week and were in good physical condition at the time. There was no lack of feed as most of the forage was 0.5 to 1 m in height.

Sample collection

Three cows were necropsied; however, because of the summer heat and lapsed time between death and necropsy, the cows were bloated and tissues were severely autolyzed. Rumen samples, blood, liver, kidney and heart were collected and submitted on ice to the Utah State Veterinary Diagnostic Laboratory (Logan, Utah) for examination and histopathology. Plant specimens including stem, leaves, flowers and seed heads were submitted to the Intermountain Herbarium (Logan, Utah) for positive identification. Green seed, leaves/stem and tuber were collected for toxicity testing and chemical analysis. Seed and tuber were analyzed by the California Veterinary Diagnostic Laboratory Systems, Toxicology Department and the Poisonous Plant Research Laboratory for evidence of cicutoxin-like compounds.

Chemical analysis

Extraction

A 1.0 g sample of fresh tuber and seed was weighed into 250 mL square french homogenization vessels with Teflon lined caps (Fisher Scientific). Dichloromethane (25 mL) was added, and samples were homogenized (Polytron Model PT 10/35, Brinkman Instruments Inc.) for 1 min at 50% power. Extracts were centrifuged at 90 G (IEC CU-5000 centrifuge, International Equipment Co., USA) for 5 min.

Derivatization

A 1 mL aliquot of the extract was placed into a 3 mL reaction vial. The extract was evaporated to dryness with a stream of N_2 without using heat. Ethyl acetate (750 uL) was immediately mixed into the vial. The extract–ethyl acetate mixture was transferred to an auto-sampler vial and 250 uL BSTFA was added. The vial was capped and heated at 70°C for 15 min. The vial was cooled, and the derivatized extract was injected on the GC/MS.

Mouse bioassay

Thirty grams of green seed or 30 g of fresh tubers collected from plants in an adjacent pasture upstream from where poisoning occurred were mixed with 80 g of water and pureed in a household blender. The crushed mixture was strained through cheese cloth and the crude water extract orally dosed to mature Swiss Webster mice. The crude seed and tuber extracts were dosed at equivalent volumes for a rapid and simple comparison of the toxicity. The mice weighed between 30 and 40 g. Data and complete details of the mouse bioassay will be reported elsewhere.

Results

Tissue examination and histopathology did not implicate any disease condition. No gross or microscopic lesions were diagnostic in the tissues examined. Examination of the rumen content was also unremarkable.

Chemical analysis of tuber and green seed from water hemlock collected 3 days after the death of the cows revealed levels of cicutoxin and cicutol equal to or higher than archive samples from highly toxic tubers used in an earlier study (Panter *et al.* 1996). The green seed contained 0.01, 9.5 and 22.5 times more cicutoxin, cicutol-1 and cicutol-2, respectively (Figure 1), when compared to the highly toxic archive sample. Chemical analysis of the tubers from ingested plants contained 4.6, 9.8 and 18.8 times more cicutoxin, cicutol-1 and cicutol-2, respectively (Figure 2), compared to the archive sample. The tubers from the archive sample caused death in sheep when fed at 1 mg/kg wet weight in a single dose.

Oral dosing to mice of crude water extracts from both tubers and seeds from plants collected immediately upstream from the poisoning resulted in toxicity and death in mice. Toxicity was slightly higher in the groups dosed with tuber extracts compared to the green seed extracts; however, both were considered to be highly toxic. Clinical signs began with pilo-erection, tail flicking, muscle twitching and tremors, progressing quickly to grand mal seizures and death.

Discussion

Water hemlock-induced poisoning is economically important to livestock producers and in most cases a dead animal and associated evidence of ingested water hemlock is the only evidence. Understanding the plant=s potential for toxicity and controlling its invasion into pastures and stream beds is the best method of preventing livestock losses. Water hemlock is easily controlled with broadleaf herbicides; however, care should be used when spraying herbicides along streams or water courses and herbicides will often increase palatability of some poisonous plants. While most cases of water hemlock poisoning occur from ingestion of tubers or early-growth plant, we believe this is the first reported case of acute poisoning and death in cattle from grazing immature seed. Although green seed of water hemlock=s close relative, poison-hemlock (*Conium maculatum*) is known to have potent toxic and teratogenic activity (K.E. Panter, unpublished data), toxicity of water hemlock seed was unknown.

The toxins in water hemlock are long-chain highly unsaturated diols. A simplified analytical method for detecting cicutoxin in rumen samples of poisoned animals was reported (Smith and Lewis 1987). However, in this case the toxins were not detectable in the rumen samples collected from these cattle. We speculate that the toxins rapidly decreased in the rumen because of enzymatic activity enhanced by the hot summer environmental conditions. Further research is needed to determine how long after death these toxins can be detected in rumen, liver or kidney samples.

Cicutoxin ($C_{17}H_{22}0_2$) is a potent neurotoxin, whereas the alcohol cicutol ($C_{17}H_{22}0$) is reported to be non-toxic. This case report would suggest there are other derivatives of cicutol that are highly toxic. Further research is required to characterize these toxins and determine relative toxicity.

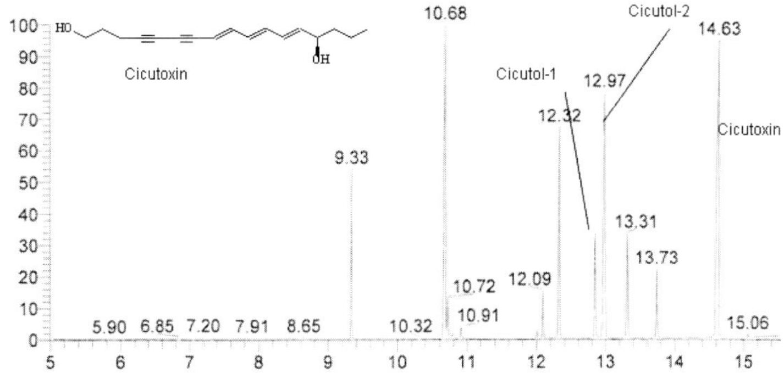

Figure 1. Chromatogram of water hemlock seed.

The pharmacologic action of cicutoxin is one of CNS stimulation and in mammals it probably stimulates the cortex, with the most pronounced effects on the midbrain and medulla as described with picrotoxin (Grundy and Howarth 1956; Goodman and Gilman 1960). The mouse bioassay demonstrated the toxicity of this water hemlock seed although a direct comparison to cattle is difficult to extrapolate. More research is required to fully evaluate the chemistry and toxicology of green water hemlock seed.

Field investigation, observations of the rapid and complete grazing of the seed heads from plants, chemical analysis of plant tubers and seeds, and mouse bioassay confirmation are consistent with water hemlock poisoning. We believe that had the seed been ripe (2 to 3 weeks later) it probably would not have caused poisoning as these types of seeds are hard and much less easily digested. Unfortunately for the cattle and the rancher we suspect the circumstances and conditions were just right for this poisoning to occur.

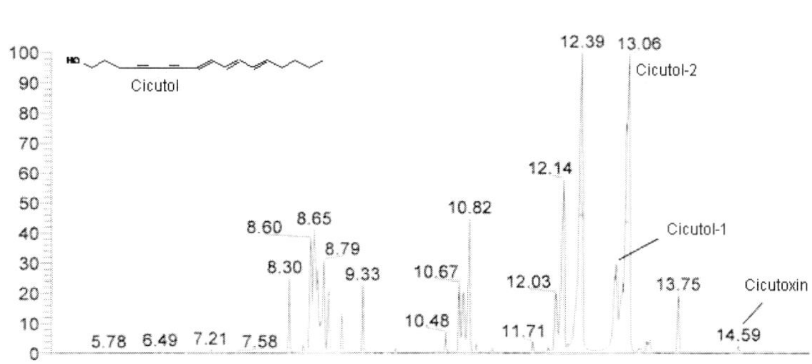

Figure 2. Chromatogram of water hemlock tuber from same plant as Figure 1.

References

Goodman LS and Gilman A (1960). *The Pharmacological Basis of Therapeutics*, pp. 325–326. Macmillan Publishing Company, New York.

Grundy HF and Howarth F (1956). Pharmacological studies on hemlock water dropwort. *British Journal of Pharmacology and Chemotherapy* 11:225–230.

Knutsen OH and Paszkowski P (1984). New aspects in the treatment of water hemlock poisoning. *Clinical Toxicology* 22:157–166.

Mulligan GA (1979). The genus *Cicuta* in North American. *Canadian Journal of Botany* 58:1755–1767.

Panter KE, Keeler RF, and Baker DC (1988). Toxicoses in livestock from the hemlocks (*Conium* and *Cicuta* spp.). *Journal of Animal Science* 66:2407–2413.

Panter KE, Baker DC, and Kechele PO (1996). Water hemlock (*Cicuta douglassii*) toxicoses in sheep: pathologic description and prevention of lesions and death. *Journal of Veterinary Diagnostic Investigations* 8:474–480.

Smith RA and Lewis D (1987). *Cicuta* toxicosis in cattle: Case history and simplified analytical method. *Veterinary and Human Toxicology* 29(3):240–241.

Wink M and Schimmer (1999). Modes of action of defensive secondary metabolites. In *Functions of Plant Secondary Metabolites and Their Exploitation in Biotechnology*, Vol 3 (M Wink, ed.), pp. 17–133. CRC Press, Boca Raton, Florida.

KIDNEY

GI

IMMUNOLOGY

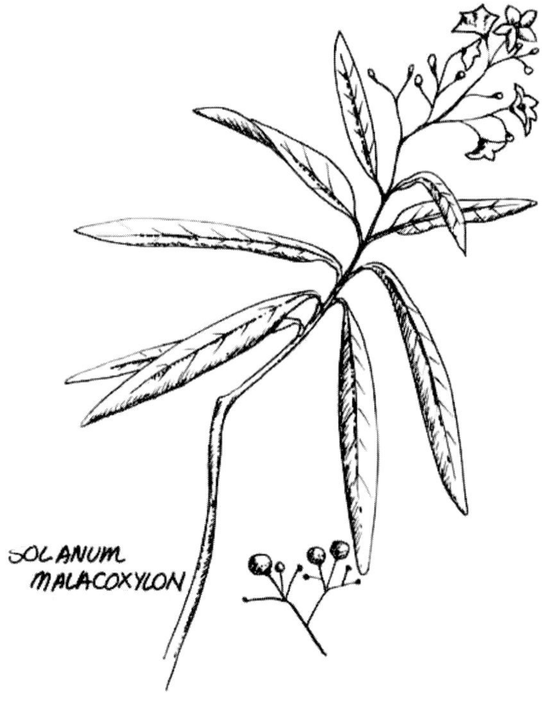

Chapter 45

Toxicity, Pathophysiology and Pathology in Sheep Following Dosing of the Nephrotoxic Plant *Nolletia gariepina* (DC) Mattf.

C.J. Botha, R.A. Meintjes, and L. Prozesky

Faculty of Veterinary Science, University of Pretoria, Private Bag X04, Onderstepoort, 0110 South Africa

Introduction

Farmers in the Kuruman district of the Northern Cape Province, Republic of South Africa have experienced mortalities in their free-ranging cattle herds. The mortalities, approximately 200 head of cattle, were ascribed to nephrotoxicity induced by a previously unknown poisonous plant (Du Plessis *et al.* 2004). Clinically, weakness, malaise, and perineal and brisket edema were observed. Necropsy findings included severe anasarca, ascites, perirenal edema and nephrosis, and mild hepatosis. Microscopic examination revealed extensive degeneration and necrosis of tubular epithelial cells in the renal cortex, accompanied by large numbers of protein and, sometimes, cellular casts in the tubular lumens. Dosing trials in cattle confirmed *Nolletia gariepina* (DC) Mattf. as the causative plant (Du Plessis 2004); sheep also were shown to be susceptible.

The primary aim of the current trial was to elucidate the pathophysiological changes in renal function in sheep induced by *N. gariepina*. A secondary objective was to evaluate the toxic effects and pathological findings following administration of this nephrotoxic plant to sheep.

Materials and Methods

Nolletia gariepina was collected on a farm in the Kuruman district, Northern Cape. The plant material was air dried, finely milled and frozen (at –10°C) before being used. Nine mutton Merino wethers, 9–18 months of age, were individually placed in steel metabolic crates in an experimental animal housing facility. Lucerne hay and drinking water were available *ad lib.* After a control period the sheep were deprived of feed for 24 hours and then dosed with dried, milled plant material (1.5–3 g kg^{-1} body weight) by stomach tube. Daily clinical examinations were conducted to evaluate the animals' health. Feed and water intake and urine volume were recorded daily.

An aliquot of urine was collected for conventional urinalysis (Combur urinary dip stick, Merck). Urinary creatinine, sodium (Na) and potassium (K) concentrations, gamma glutamyltransferase (GGT) activity and osmolarity were also determined.

Blood samples were collected for determination of hematocrit, osmolarity and plasma protein, Na, K, creatinine and urea concentrations. In addition, GGT and aspartate aminotransferase (AST) activities were determined to establish if any hepatic damage had occurred.

From the above variables the following were derived: (1) glomerular filtration rate (GFR) (Nawaz and Shah 1984, Bastl *et al.* 1985); and (2) fractional excretion of sodium (FE_{Na}) (Bastl *et al.* 1985)

The carcasses of sheep that died or were euthanased were submitted for necropsy. Tissue specimens (approximately 10 mm^3) from several organs were collected and fixed by immersion in 10% neutral buffered formalin for at least 5 to 7 days. Specimens were routinely processed for light microscopy and stained with hematoxylin and eosin (HE).

Renal tissue specimens were collected from two cases within minutes of death, cut into 1 mm^2 blocks and fixed in labeled glass vials containing 2.5% glutaraldehyde in 4% 0.1 M Millonig's buffer. Samples were routinely processed for electron microscopy using a transmission electron microscope and photographed.

Results

Toxicity

Four of seven animals that were dosed with 3 g kg^{-1} *N. gariepina* plant material died acutely within 24 hours. One sheep, which was depressed, recumbent and exhibited respiratory distress (hyperventilation, groaning, frothing at the mouth) was euthanased 8 hours after being dosed with 3 g kg^{-1} milled plant material. Only two animals, which had received 3 g kg^{-1} plant material, survived for longer periods, but were depressed, anorexic and had weak rumen movements and tachycardia.

Administration of half the dose (1.5 g kg^{-1}) to two wethers elicited no clinical effects. More than 3 months later the same two wethers were dosed again with 2.8 g kg^{-1} and at this dose, moderate to severe clinical signs developed.

Pathophysiology

The most noticeable aberrations in blood chemistry were the remarkable increases in serum urea (ranged from 21–65 mmol L^{-1}) and creatinine (ranged from 292–786 μmol L^{-1}) concentrations in sheep that survived for 24 hours or longer. A moderate increase in AST activity (up to 157 U L^{-1}) occurred in some sheep. There was no noticeable increase in GGT activity.

Following dosage of the sheep with *N. gariepina*, even at doses of 1.5 g kg^{-1}, urine production in general increased. In some cases, however, the increase was preceded by virtual anuria on the day immediately following dosage. Semi-quantitative urinalysis (urinary dip stick) revealed mild proteinuria and presence of hemoglobin in the urine after dosing. Dramatic decreases in GFR were calculated in sheep which survived at least 1 day post-dosing when the plant material was dosed in excess of 1.5 g kg^{-1}. During the control phase no sodium could be detected in the urine, but within 24 hours of dosing the *Nolletia* plant material at doses of 2.8 g kg^{-1} and above, urinary sodium concentrations rose. The fractional excretion of potassium increased, even after receiving relatively low doses (1.5 g kg^{-1}) of *Nolletia* plant material. Urine osmolarity decreased markedly in most of the animals that did not die within 24 hours of dosing the plant material. Although no change in serum GGT activity was noticed following dosing with *Nolletia* plant material, urine GGT activity tended to rise and then fall within a week.

Pathology

Macroscopic pathology

A severe nephrosis was present in all the animals. The kidneys exhibited various degrees of swelling, bulged slightly on cut surface and ranged in color from light brown to greyish-white. We also found widespread serosal hemorrhages, mild accumulation of fluid in the body cavities, moderate hepatosis characterized by mild swelling of the liver with rounded edges, slightly pale color and a taut capsule, and lung congestion and edema.

Light microscopy

A severe nephrosis was the most striking lesion present. The epithelium of the proximal convoluted tubules (PCT) in the cortex was the most severely affected. No specific lesions were visible in the glomeruli. Mild hepatocellular swelling with accumulation of multiple small vacuoles in the cytoplasm of hepatocytes (indicating hydropic degeneration) was the most prominent lesion in most animals. No specific lesions were visible in the portal tracts and bile ducts.

Transmission electron microscopy

The most severe changes occurred in the PCT of both animals. Individual cells lining the tubules could not be identified because of loss of the cell membranes. Affected tubules were demarcated by the intact basement membrane. The usually conspicuous brush border and apical canaliculi of the epithelial cells had disappeared. The cytoplasmic organelles and nuclei were disorganized and dispersed as a loose mass within the tubular lumen.

Discussion

Following ingestion of *N. gariepina* there appears to be a threshold before toxicity will be induced. When 3 g kg^{-1} of dried, milled plant material was dosed orally, five of the seven sheep died acutely or were euthanased for humane reasons. On the other hand, administration of 1.5 g kg^{-1} elicited no clinical signs. During the reported field outbreaks, the cattle were deprived of feed for 24–48 hours and, following their release, they presumably consumed large quantities of the plant material (Du Plessis *et al.* 2004; Du Plessis 2004) and probably exceeded the threshold level. The acute mortalities observed in this study could be ascribed to other toxins besides the nephrotoxin(s) contained by the plant. Analysis of the plant material for cyanogenic potential gave negative results.

In the animals which did not die acutely, kidney damage was severe enough to induce marked increases in the plasma concentrations of creatinine and urea as well as a low GFR.

As no specific damage to the glomerulus was described on histopathology and there is no reason to believe that the azotemia was pre-renal, the drastic fall in GFR was due to blockage of proximal tubules by cellular debris and protein casts. The reverse pressure across the filtration membrane due to such blockage would effectively stop the filtration process.

The appearance of sodium in the urine and the rise of potassium concentration in the urine as well as the increase in urine volume, despite the marked fall in GFR, could be ascribed to two factors. The flow of filtrate in non-damaged nephrons may have been so rapid that there was little chance for the reabsorption of sodium, potassium and water. More likely, however, in damaged nephrons (proximal tubular epithelial damage with loss of brush border), which still permitted filtrate flow, sodium, potassium and water reabsorption would be compromised. The reduction in urine osmolarity indicates that, in spite of the natriuresis and kaliuresis, more water relative to electrolytes was lost.

Gamma glutamyltransferase is present in the brush border of the proximal convoluted tubule where it facilitates transfer of glutamate into the tubular cells (Welbourne and Mathews 1999). A rise in urine GGT activities when there is proximal tubular damage is therefore in accordance with the pathological findings in this section of the nephron tubule as was seen in this trial.

The pathological lesions in the experimental sheep correspond to a large extent with the lesions described in the experimentally produced cases in cattle, but were much less pronounced compared to the lesions in the field outbreak in cattle (Du Plessis 2004). In both species under experimental conditions gross lesions included mild to moderate accumulation of fluid in body cavities, severe nephrosis and a moderate hepatosis. Contrary to the situation in the experimentally produced cases in cattle, perirenal edema was not a striking feature in sheep, whereas widespread serosal hemorrhages were present in the latter.

Nephrosis was characterized by degeneration and necrosis of the PCT, and affected tubules contained large numbers of protein casts in the lumens. Transmission electron microscopy confirmed that the basement membrane of affected tubules was still intact. No structural visible lesions were evident in the glomeruli.

The PCT epithelium is the most susceptible to both toxic and hypoxic injury because of its high metabolic rate (Confer and Panciera 2001). Nephrotoxic acute tubular necrosis (ATN) can be distinguished from ischemic ATN by two histological features *viz.* extensive necrosis of predominantly PCT epithelium and intact tubular basement membranes. Toxic damage to tubular epithelium has a better prognosis if the animal survives the initial toxic insult, as the basement membranes that are necessary for epithelial regeneration are intact (Maxie 1993).

Conclusion

Dosing *N. gariepina* plant material to sheep in doses exceeding 2.8 g kg^{-1} resulted in acute death of most animals and marked changes in the clinical chemistry of the survivors, characterized by a sudden, marked decrease in GFR, azotemia, elevated fractional excretion of Na and K, and increased urinary GGT activity.

Sheep which died or were euthanased exhibited severe nephrosis that was histologically characterized by necrosis of predominantly lining epithelial cells of the PCT and the presence of protein and cellular casts.

References

Bastl CP, Rudnick MR, and Nairns RG (1985). Assessment of renal function: characteristics of the functional and organic forms of acute renal failure. In *The Kidney: Physiology and Pathophysiology* (DW Seldin and G Giebisch, eds), pp. 1823–1824. New York Press, New York.

Confer AW and Panciera RJ (2001). The urinary system. In *Thomson's Special Veterinary Pathology*, 3rd ed. (MD McGavin, WW Carlton, and JF Zachary, eds), pp. 245–252. Mosby Inc., Missouri.

Du Plessis EC (2004). Pathological investigation of the nephrotoxic effects of the shrub *Nolletia gariepina* (DC) Mattf. in cattle. MMedVet (Path) dissertation, University of Pretoria.

Du Plessis EC, Joubert JPJ, Prozesky L, Naudé TW, Herman P, and Van Der Westhuizen GBM (2004). Nephrotic syndrome in cattle caused by the shrub *Nolletia gariepina* in the Kalahari sandveld of southern Africa. In *Poisonous Plants and Related Toxins* (T Acamovic, CS Stewart, and TW Pennycott, eds), pp. 76–78. CABI Publishing, Wallingford, Oxon, UK.

Maxie MG (1993). The urinary system. In *Pathology of Domestic Animals*, 4th ed. (KVF Jubb, PC Kennedy, and N Palmer, eds), pp. 465–468. Academic Press Inc., California.

Nawaz M and Shah BH (1984). Renal clearance of endogenous creatinine and urea in sheep during summer and winter. *Research in Veterinary Science* 36:220–224.

Welbourne TZ and Mathews JC (1999). Glutamate transport and renal function. *American Journal of Physiology* 277 (4 Pt2), F501–F505.

Chapter 46

Nephrotoxicity of Easter Lily (*Lilium longiflorum*)

J.O. Hall

Utah State University, Utah Veterinary Diagnostic Laboratory, 950 East 1400 North, Logan, UT 84322-5700, USA

Introduction

Easter lily (*Lilium longiflorum*) is a common ornamental houseplant of the Liliaceae family that is widely displayed during the Easter holiday period and is readily available for companion animal ingestion. The first report of the Easter lily toxicosis was published in 1990 (Hall 1990) and reported at ACVIM (Hall 1992), but the toxic principle is still unknown. Easter lily is still listed as a 'Non-Toxic Ingestion' by Poisindex® (1996), the poison identification system used by most human poison centers.

The National Animal Poison Control Center (NAPCC) specializes in the identification, diagnosis, and treatment of animal poisonings and it maintains computer records of each case. Examination of NAPCC case records (1987–1991) suggests that Easter lily is a toxic plant of major significance for cats. Since first described, there have been other reports of *Lilium* sp. poisoning in cats (Gathers 1991; Mullaney *et al.* 1993; Carson *et al.* 1994; Gulledge *et al.* 1997; Langston 2002; Rumbeiha *et al.* 2004).

The objective of this work is to outline the evidence that led to the identification of Easter lily as a plant toxic to cats. This work will include NAPCC case evaluation, clinicopathology, histopathology, treatment recommendations, and treatment outcomes.

Materials and Methods

NAPCC database evaluation

The NAPCC case database was evaluated by comparing the relative frequency of clinical signs for cases involving Easter lily ingestion by cats with that of the total case database. This database had in excess of 100,000 case records. In addition, individual NAPCC cases of Easter lily ingestion by cats were reviewed for details and callers were contacted for additional case information. Treatment recommendations and outcomes were evaluated via cases coming in to the NAPCC over time.

Animal dosing

Under the approval and supervision of the University of Illinois College of Veterinary Medicine Office of Laboratory Animal Care, a healthy 1-year-old intact male cat, six adult male Sprague Dawley rats, and two male New Zealand rabbits were used to investigate the toxic syndrome. The cat was allowed 10 days to acclimate to the facility and had feed and water provided *ad libitum*. Pretreatment CBC and serum chemistry values were obtained on day 7 of the acclimation period, then daily after dosing. Urine was collected via bladder expression. After an overnight fast the cat was dosed via stomach tube with a puree of fresh Easter lily leaves at a dose equivalent to 2% of the cat's body weight. After 7 days post-dosing, the cat was humanely euthanized with an overdose of pentobarbital, necropsied, and tissues were collected for histopathological evaluation. These tissues were compared with those received from practitioners in field cases. Tissues were fixed in 10% neutral buffered formalin, imbedded in paraffin, sectioned at 4 □m, and stained with hematoxylin and eosin for light microscopy.

The rats were allowed ≥10 days to acclimate to the facility and had feed and water provided *ad libitum*. After an overnight fast, two control rats were dosed via gavage needle with 2 mL of distilled water, two were dosed once with 2 mL of pureed Easter lily leaves, and two rats were similarly dosed with 2 mL twice daily for 2 days. The rats were monitored for 7 days post-dosing. Blood samples were obtained under deep pentobarbital sedation at the end of the study. The rats were humanely euthanized via an overdose of pentobarbital. Tissues collected were processed as previously described.

The rabbits were individually housed and allowed ≥10 days to acclimate to the facility. Feed and water were provided *ad libitum*. Dosing was achieved by providing each rabbit an Easter lily plant, which they readily consumed. Within 36 hours, each rabbit consumed plant material equivalent to ≥5% of their body weight. Blood was drawn via ear vein venipuncture for CBC and serum chemistry analyses prior to dosing and at the termination of the study. The rabbits were humanely euthanized by an overdose of pentobarbital at 7 days. Tissues collected were processed as previously described.

Results

NAPCC database evaluation

In April 1990 five cases of renal failure involving Easter lily ingestion by cats were reported to the NAPCC. As the plant was not recognized as toxic the first two cases were dismissed as coincidental. But, when the third, fourth, and fifth cases had similar exposure, clinical signs, and case progression it was thought to be much more than chance.

A review of the NAPCC data from 1987 to 1989 was performed. This case data supported the hypothesis that Easter lily ingestion by cats results in nephrosis and renal failure. Clinical signs identified in cats that ingested Easter lily included depression, anorexia, vomiting, anuria, and renal failure. All of these signs were reported with greater frequency than that of the overall database, although a majority of the cases were received several hours post-ingestion and no follow-up contacts had been made. Cats were the only species for which renal failure was a significant finding. Even when dogs ingested very large amounts the only clinical effect was a digestive upset.

Attempts were made to follow-up cases of Easter lily ingestion by cats that were reported to the NAPCC in 1989, but successful contacts were made in only seven cases. For these cases renal failure was a frequent finding. Four of the seven cats died or were euthanized due to renal failure. One cat developed renal failure, but underwent peritoneal dialysis for 10 days and regained kidney function. Two cats vomited leaf material acutely and did not develop subsequent clinical disease.

The reported amount of the Easter lily ingested was relatively small. Ingestions of one to 15 leaves were reported to the NAPCC, with most being one to five leaves. The cats that died in renal failure had ingested from two to 15 leaves. The cat that ate about 15 leaves had evidence of eight to ten leaves in its vomitus. In most of the cases, finding relatively intact leaves in the vomitus verified plant ingestion.

The seven cases from 1989 and the five cases in 1990 were very suggestive that Easter lily is nephrotoxic to cats. Ten of these 12 cats (83%) developed renal failure and nine of the 12 died or were euthanized. The age, sex, weight, and outcome of the 13 cases received by the NAPCC in 1990 and 1991 are provided in Table 1.

Table 1. Age, sex, weight, and outcome of the 13 cases of Easter lily ingestion received by the NAPCC in 1990 and 1991.

Animal	Age	Sex	Weight (lbs)	Outcome
1	6 years	female	15	Renal failure; euthanized
2	3 years	female	12	Renal failure; euthanized
3	4 years	male	14	Renal failure; euthanized
4	3 years	female	8.5	Renal failure; died
5	11 years	spayed female	8	Renal failure; died
6	7 years	male castrate	18	Renal failure; euthanized
7	11 years	male castrate	N/A[1]	Renal failure; euthanized
8	1 year	spayed female	6	Renal failure; euthanized
9	6 year	spayed female	15	Renal failure; euthanized
10	2.5 years	spayed female	8	No clinical signs. Emesis induced and successful within 45 min post-ingestion
11	12 years	spayed female	10	No clinical signs. Decontamination within 2–4 hours post-ingestion
12	N/A[1]	N/A[1]	N/A[1]	Vomited on its own 2–4 hours post-ingestion. No further clinical signs
13	1 year	female	12	Severe vomiting, depression, and anorexia acutely; fluid diuresed; no subsequent clinical signs

[1] N/A indicates the data were not provided

Clinical syndrome

Cats that ingested Easter lily developed clinical lethargy, anorexia, vomiting, anuria, and renal failure. Generally, the cats exhibited vomiting and salivating by 1–6 hours post-ingestion, but these signs resolved by approximately 12–18 hours. Also, the cats became lethargic and completely anorectic within 1–6 hours and remained so throughout the syndrome. Polyuria was present by 10–18 hours and continued through dehydration to the development of anuria. Other than the moderate to severe lethargy, anorexia, and polyuria no other clinical signs occurred from 12–18 hours to about 36–48 hours. By 36–48 hours the cats were very lethargic, inactive, dehydrated, and many began vomiting again.

Anuric renal failure was evident by 36–72 hours post-exposure in five cats in 1989, five in 1990, and four in 1991. These cats were anuric even after fluid replacement. There was a progressive weakening to a moribund condition as their uremia worsened. These cats were either euthanized or died by 4–7 days. Two additional cats that died were anuric and presumably in renal failure, but no diagnostic tests were performed.

Other clinical signs that were reported varied in occurrence. Three cats were reported to be very weak, but these cats were also in the terminal syndrome stages. One of the weak cats also had a lack of pupillary light reflexes and mydriasis. Two cats had diarrhea that accompanied the vomiting in the acute phase. Also, two cats began hiding from their owners as they progressed into the uremic phase of the syndrome.

Clinical pathology

Hematology and clinical chemistry parameters tested were not consistent among the reported cases. Complete blood counts (CBC) in the clinically affected cats revealed only a stress leukogram. Poisoned cats had biochemical changes suggestive of renal disease, including increases in serum urea nitrogen, creatinine, phosphorus, and potassium (Table 2). Serum creatinine values generally increased more markedly than did urea nitrogen. Mild sporadic increases were also seen in SGPT, SGOT, and CPK.

Due to the complete nature of the anuria, urine was obtained from only two cats. In both the urine specific gravity was 1.019. Increased glucose (++ and 1000 mg/dL) and protein (++ and 1700 mg/dL) were measured. One cat had 12–20 coarse granular casts per low power field and 15–20 red blood cells per high power field in the urine sediment. In both cats, the blood glucose was normal when the urine sample was obtained.

Histopathology

Tissues were collected from several clinical cases of Easter lily intoxication in 1990 and 1991. The only gross lesion reportedly observed in these cats was perirenal edema in two cats. In three cases only kidney was submitted. In two cases, kidney, liver, heart, lung, pancreas, and spleen were submitted. Other than vascular congestion of the lung and liver, only the kidney contained significant histological lesions (Figure 1). All cats examined had renal tubular lesions that included acute renal tubular necrosis with loss of tubular epithelium. The lesions only differed from case to case in severity. The tubular and collecting duct lumens contained eosinophilic granular and proteinaceous debris. In cats with less severe lesions the tubular epithelial necrosis was largely confined to the proximal tubules. However, in cats with more severely affected tubules, the epithelial necrosis involved both the proximal and distal tubular epithelium. The remaining renal tubular cells were flattened and partially covered the denuded basement membrane. With PAS stain, intact tubular basement membranes were observed. Also, mitotic figures were occasionally encountered in the affected tubules.

Table 2. Serum chemistry alterations reported for cases of Easter lily ingestion by cats that were received by the NAPCC in 1990 and 1991.

Animal[1]	Day[2]	Creatinine	K	P	BUN	other
1	3	17.9	N/A[3]	N/A	79.0	
2	4	23.0	N/A	16.0	340.0	
3	5	18.0	N/A	12.0	148.0	SGPT=429 SGOT=88
	6	23.0	N/A	N/A	N/A	N/A
4	5	37.0	9.4	Increased[4]	150.0	
5	6	16.3	N/A	N/A	170.0	SGPT=230 SGOT=13
6	5	24.8	6.7	11.5	215.0	CPK=1043
	6	32.5	8.0	11.6	261.9	N/A
7	3	15.0	4.6	N/A	138.0	
	4	16.0	5.0	N/A	153.0	
8	4	16.0	N/A	16.0	78	
9	-	N/A	N/A	N/A	N/A	
10–13	-	N/A	N/A	N/A	N/A	All Lived

[1] Listed in the same order as in Table 1
[2] Days post-ingestion
[3] N/A= Data not available
[4] Numerical value unavailable

Dosing studies

Despite eliminating >95% of the dosed material with the first vomiting episode, the acute clinical syndrome of Easter lily toxicosis in the cat mimicked that documented in the field cases. Profuse salivation and vomiting began at 45 min post-dose. The salivation and vomiting continued for 4 hours. Additionally, the cat was less responsive to entrance into the room through day 4, and was lethargic and anorectic through day 6.

Clinical evidence of renal effects was present by 12 hours, when the cat was polyuric and was dehydrated. A urine sample obtained at that time contained 1000 mg dL^{-1} glucose, 500 mg dL^{-1} protein, and urine sediment that had 10–15 tubular casts per low power field. The tubular casts had cellular characteristics and discernible nuclei. As per our laboratory animal protocol, subcutaneous fluids were administered to correct the dehydration. By 24 hours, the cat was not clinically dehydrated and the urine contained one to two granular casts per low power field, 50 mg dL^{-1} glucose, and 30 mg dL^{-1} protein.

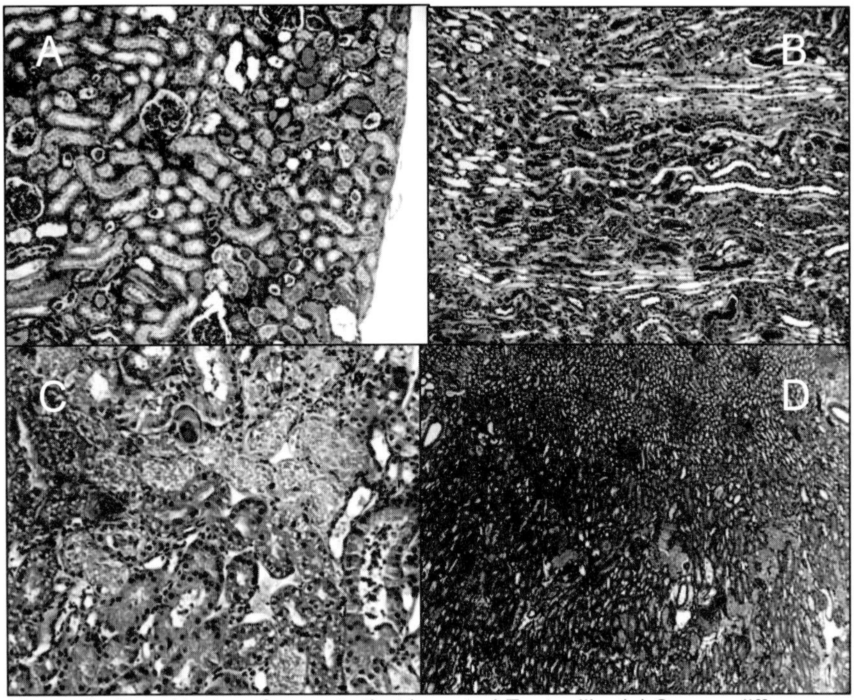

Figure 1. Renal lesions from cats that ingested Easter lily. (a) Severe diffuse renal tubular necrosis in a field case. (b) Renal tubular casts in a field case. (c) Moderate renal tubular necrosis in a dosed cat. (d) Renal tubular casts in a dosed cat.

Serum chemistry values were altered, but less severely than reported for NAPCC cases. Urea nitrogen, creatinine, potassium, magnesium, and phosphorus gradually increased from the pretreatment concentrations to maximum values on day 5 with the pre-dose and highest values of 17.3–35.6 mg dL^{-1}, 1.4–2.3 mg dL^{-1}, 4.2–5.4 mEq L^{-1}, 2.8–4.3 mg dL^{-1}, and 5.3–7.5 mg dL^{-1}, respectively. Serum glucose was normal, except at 12 hours when it was 208 mg dL^{-1}. By day 7, the serum chemistry values neared pre-dosing values. Hematology results revealed a stress leukogram from 12 hours through day 7.

Lesions were identified only in the kidney and were similar, but less severe, than those of field cases (Figure 1). Acute tubular necrosis was present in approximately 10–20% of the proximal tubules. In addition, proteinaceous casts were observed in the collecting ducts.

Attempts to reproduce the nephrotoxic syndrome in rats and rabbits were unsuccessful. Neither exhibited any detectable adverse effect from Easter lily ingestion. With rats and rabbits no clinicopathologic or histopathologic abnormalities were seen.

Treatment

Since complete renal shutdown was not observed in the NAPCC cases in which cats vomited the leaves soon after ingestion or the dosed cat that was rehydrated, it was thought that treatment should focus on early decontamination and enhanced renal perfusion. With acute renal failure being somewhat biphasic, the acute insult followed by renal ischemia secondary to dehydration, the use of fluid diuresis to keep the kidneys well perfused was thought to be very important. Early decontamination and providing fluid diuresis prior to renal shut-down was anticipated to result in a favorable outcome.

The decontamination procedures developed include emesis and activated charcoal with a saline cathartic for decontamination, as well as fluid diuresis with IV fluids at two to three times maintenance for 24–48 hours. With this treatment protocol, Easter lily-associated renal failures and deaths were reduced. All cats treated with fluid diuresis prior to renal shut-down survived, although some did not receive other treatments.

Once cats developed Easter lily-induced renal failure, the effectiveness of treatment was limited. In only one case was treatment successful after the onset of renal failure. In this case, the cat underwent peritoneal dialysis for 10 days and had regained normal renal function by 2 weeks, as determined by urinalysis, urine output, and serum chemistries. Other common treatments to initiate urine output in renal failure cats (dopamine, furosemide, thiazide diuretics, dextrose, and mannitol) were not effective.

Conclusion

The first identification and characterization of a syndrome requires extensive proof in order to prevent false information from being ingrained into the literature. With the information derived from these cases it was shown that Easter lily is uniquely nephrotoxic to cats. Essential treatment includes early decontamination and more importantly fluid diuresis prior to the onset of renal failure, as found with the dosed cat. There does not appear to be any age, weight, or sex predilection for Easter lily-associated toxicoses in cats. Histologic findings of intact basement membranes and mitotic figures in renal tubules, as well as the one case in which peritoneal dialysis was successful, suggest that Easter lily-exposed cats may be able to regenerate the damaged tubular epithelium and regain renal function.

References

Carson TL, Sanderson TP, and Halbur PG (1994). Acute nephrotoxicosis in cats following ingestion of lily (*Lilium* sp.) (abstract). *Proceedings Annual Meeting American Association Veterinary Laboratory Diagnosticians* 37:43.

Gathers TM (1991). Acute renal failure in a cat, was it tiger lily? *NAPINet Report* 3(2):February 1.

Gulledge L, Boos D, and Wachsstock R (1997). Acute renal failure in a cat secondary to tiger lily (*Lilium tigrinum*) toxicity. *Feline Practice* 25:38–39.

Hall JO (1990). Are Easter lilies toxic to cats? *NAPINet Report* 3(2):July 16.

Hall JO (1992). Nephrotoxicity of Easter lily (*Lilium longiflorum*) when ingested by the cat. *Proceedings Annual Meeting American College of Veterinary Internal Medicine* 6:121.

Langston CE (2002). Acute renal failure caused by lily ingestion in six cats. *Journal of the American Veterinary Medical Association* 220:49–52.

Mullaney TP, Slanker MR, and Poppenga RH (1993). Easter lily associated nephrotoxicity in cats (abstract). North Central Conference of Veterinary Laboratory Diagnosticians.

National Animal Poison Control Center (NAPCC) (1987–1991). Computerized Case Database. University of Illinois College of Veterinary Medicine.

Poisindex® CD-ROM (1996). Easter Lily. Denver: Microdex Inc.

Rumbeiha WK, Francis JA, Fitzgerald SD, Nair MG, Holan K, Bugyei KA, and Simmons H (2004). A comprehensive study of Easter lily poisoning in cats. *Journal of Veterinary Diagnostic Investigation* 16:527–541.

Chapter 47

Phytate: an Anti-Nutritive Factor in Animal Diets

W.L. Bryden[1], P.H. Selle[2], V. Ravindran[3], and T. Acamovic[4]

[1]School of Animal Studies, University of Queensland, Gatton Qld 4343, Australia; [2]Faculty of Veterinary Science, University of Sydney, Camden NSW 2570, Australia; [3]Monogastric Research Centre, Institute of Food, Nutrition and Human Health, Massey University, Palmerston North New Zealand; [4]Avian Science Research Centre, Scottish Agricultural College, Ayr, KA6 5HW, Scotland, UK

Introduction

Phytate, the mixed salt of phytic acid (*myo*-inositol hexaphosphate; Figure1), is a ubiquitous component of plant material, in which the majority (>70%) of plant phosphorus (P) can be bound to the inositol ring. This polyanionic molecule may complex with a variety of cationic elements, such as calcium, zinc and magnesium, thereby reducing utilization. Similarly monovalent cations such as sodium and potassium interact with phytate, influencing their absorption and excretion and potentially influencing water intake and balance. Monogastrics can only partially utilize phytate-P because they possess insufficient inherent phytase activity to hydrolyze phytate and liberate P. Increased phytate-P in animal diets increases environmental pollution by the loss of undigested phosphorus in effluent from animal units. Phytase feed enzymes were commercially developed some 15 years ago to increase phytate-P availability and decrease P excretion (Lenis and Jongbloed 1999). There were some limitations in the use of older phytases because of lack of thermal stability but newer enzymes have, however, become more specific and thermally stable and less expensive. While the focus of phytase research has been P orientated, the enzyme has provided new insights into the anti-nutritive properties of phytate. In particular, the interaction of phytate with dietary proteins, carbohydrates and fats may have significant economic consequences in animal nutrition. Some recent work has demonstrated that phytase also influences the excretion of endogenous material as measured by sialic acid and this may also have significant consequences on nutrient and energy utilization and gut health. Several studies have suggested that phytase has a positive effect on protein digestibility in pigs and poultry and, in broiler chickens, also enhances energy utilization.

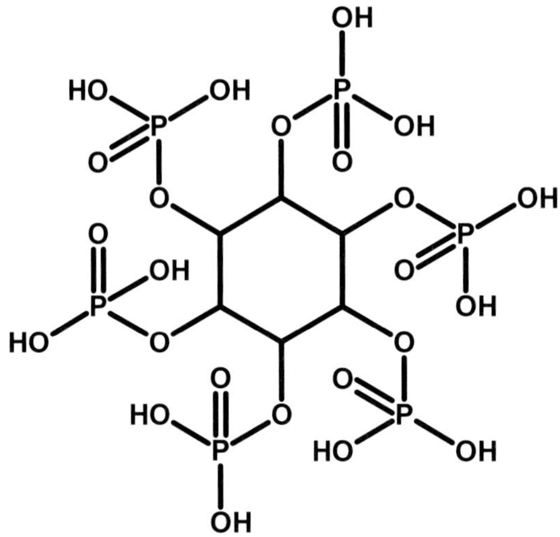

Figure 1. Inositol phosphate; phytate.

'Protein Effect' of Phytate and Phytase

Ravindran *et al.* (1999a) and Rutherfurd *et al.* (2002) reported positive responses in apparent and true ileal digestibility of amino acids in a range of individual feed ingredients following phytase supplementation. It is noteworthy that both studies show that wheat is more responsive to phytase than maize. Collectively, the data indicate that phytase increased the average ileal digestibility of indispensable amino acids by 11.5% (0.785 to 0.875) in wheat, but that in maize the response was only 3.6% (0.826 to 0.856). Moreover, in three studies (Ravindran *et al.* 2000, 2001; Selle *et al.* 2003) where phytase was included in complete broiler diets based on wheat–sorghum blends, responses in amino acid digestibility to phytase were positive. Collectively, phytase increased the average digestibility of indispensable amino acids by 4.0% (0.791 to 0.822). The propensity for protein to be bound by phytate as binary protein–phytate complexes is thought to be variable (Champagne 1988) and there is *in vitro* evidence indicating phytate has the capacity to bind with wheat protein but this may not be the case with maize.

Predictably, a larger number of broiler studies have been completed with maize-based diets, but here there is an interesting dichotomy in relation to the dietary marker used. In the two studies (Rutherfurd *et al.* 2004; Ravindran *et al.* 2005) where titanium oxide was used as the marker, phytase increased the average digestibility of seven indispensable amino acids by 4.3% (0.816 to 0.851). It was also observed that the inherent digestibility of the amino acids and percentage increases to phytase inclusion were significantly correlated in the two studies; where r = 0.939 and r = 0.788, respectively. In addition, Rutherfurd *et al.* (2004) found that phytase enhanced both the apparent and true ileal digestibility of amino acids. However, when chromic oxide was used as the marker in maize-based diets (examples include reports by Zhang *et al.* (1999), Camden *et al.* (2001) and Onyango *et al.* (2005)), it is noticeable that broiler responses in amino acid digestibility to phytase were considerably less than when titanium oxide is used. Indeed, in the eight studies the average response in digestibility to phytase of the seven amino acids was only 1.4%, from 0.850 to 0.862. The more pronounced and consistent responses to phytase observed with titanium oxide may indicate that chromic oxide is not a suitable marker for studies involving exogenous phytase. Jagger *et al.* (1992) concluded that titanium oxide should be preferred to chromic oxide as a dietary marker in digestibility studies.

In pigs, positive responses in ileal digestibility of protein and amino acids following dietary phytase supplementation have been reported by Officer and Batterham (1992) and Kornegay *et al.* (1998) where ileal digesta samples were taken from intact (anaesthetized or slaughtered) pigs. In contrast, in several studies where ileal digesta samples were taken from cannulated animals, responses to phytase have been negligible, including those of Mroz *et al.* (1994) and Kemme *et al.* (1999). As discussed by Selle *et al.* (2000), this apparent anomaly requires resolution and a number of possible contributing factors have been considered. It may be that, in addition to the sampling procedure followed, the fact that cannulated pigs are usually fed on a restricted, twice-daily basis whereas intact pigs are usually fed *ad libitum* has relevance.

While the indigestible marker used in digestibility assays may explain some of the variability noted by some authors, there are a number of other factors that may also contribute significantly to the variability. Selle *et al.* (2006) have reviewed those factors, which may include: dietary composition and ingredients; dietary levels of Ca and non-phytate P; dietary electrolyte balance; inherent digestibility of the amino acids of the dietary ingredients; sources and concentration of phytate in the diet; protein and amino acid concentrations of test diets; and the inclusion level and type of phytase added.

‘Energy Effect’ of Phytate and Phytase

In the majority of broiler studies completed at the University of Sydney, exogenous phytase has consistently increased the apparent metabolizable energy (AME) of broiler diets, which is in accord with previous studies by Farrell *et al.* (1992). Ravindran *et al.* (2000) found that the AME response to added phytase was influenced by dietary levels of phosphorus. Supplemental phytase increased AME values from 13.36 to 13.54 MJ kg^{-1} dry matter in phosphorus-deficient diets and from 12.66 to 13.38 MJ kg^{-1} dry matter in diets containing adequate phosphorus. Enhanced protein digestibility would contribute the energy response but it appears that increased utilisation of energy from lipids and starch is also involved.

It was suggested by Thompson (1988) that phytate has the capacity to complex with starch. Despite the acceptance of this suggestion there is, however, a paucity of supportive evidence for the formation of starch–phytate complexes. Therefore, it is interesting that Fuh and Chiang (2001) proposed that extruding rice bran induced complex formation between phytate and starch because extraction rates of phytate from extruded rice bran were increased from 61.4% to 73.4% by pre-treatment with α-amylase. This finding provides some indirect support for the suggestion by Thompson (1988). Alternatively, phytate is a potent α-amylase inhibitor (Sharma *et al.* 1978) and this may play a role in the control of energy reserves during seed germination (Desphande and Cheryan 1984) but it is not yet clear if phytate inhibition of α-amylase in the gut of pigs and poultry is of nutritional importance. The chelation of Ca by phytate may also influence the ability of α-amylase to hydrolyze starch within the GI tract since α-amylase is dependent on Ca as a co-factor. The possible involvement of insoluble Ca-phytates in the formation of metallic soaps in the gut lumen has been proposed as a mechanism whereby phytate and phytase influence fat digestion (Ravindran *et al.* 2000). The effect of phytate and phytase may be greater on net energy for poultry since endogenous losses tend to decrease in the presence of phytase in diets. Endogenous excretion is energy expensive and may not be clearly seen as changes in AME.

Mechanisms of Anti-nutritive Action

The capacity of phytate to bind proteins at low pH to form binary protein–phytate complexes was described by Cosgrove (1966). It is likely that binary protein–phytate complexes are formed *de novo* in the stomach and fore-stomach in pigs and poultry, respectively. However, as exogenous phytase is mainly active in the stomach and crop, prior hydrolysis of phytate would partially prevent complex formation (Selle *et al.* 2000). Protein–phytate complexes are resistant to pepsin digestion, and the subsequent amino acid digestion in the small intestine may be impeded by configurational changes to complexed protein induced by aggregation with phytate. This may be the main mechanism whereby phytate interferes with the digestion of dietary protein/amino acids. It is also probable that phytate increases the flow of endogenous amino acids (Cowieson *et al.* 2004), which would also decrease apparent ileal digestibility. Interestingly, these workers found that phytate was associated with substantially increased Na^+ secretion into the gut lumen, presumably to buffer the polyanionic phytate molecule, which was ameliorated by phytase.

The ability of phytate to complex with other molecules provides a number of possible mechanisms for the observed effects of phytase on amino acid and energy utilization in monogastric animals. The positive effect of phytase on improving a nutrient's utilization is the result of the hydrolysis of phytate complexes by the enzyme and the release of the phytate-bound nutrient for absorption and subsequent utilization. Possible enzyme–substrate complexes that may contribute to the observed effects include: native phytate–protein complex in plant feed stuffs; *de novo* formation of phytate–protein complex within the gut; formation of phytate–free amino acid complexes in mixed diets/gut; reduction in enzyme activity by the formation of phytate enzyme–protein complexes; reduction of the interaction of phytate with the gastro-intestinal tract within the animal.

It is likely that the interaction between phytase and phytate in the gut is complex and involves to varying degrees the different mechanisms proposed above. It should be remembered that phytase feed enzymes degrade only a proportion of dietary phytate and in broilers this may be as low as 30–35% in the ileum (Camden *et al.* 2001; Tamim *et al.* 2004; Cowieson *et al.* 2006). Therefore, complete phytate degradation should increase amino acid digestibility. This is supported by Newkirk and Classen (2001) who, by pre-treatment, 'dephytinized' the canola meal component (300 g kg^{-1}) of a corn/soy broiler diet and substantially increased the apparent ileal digestibility of amino acids. Moreover, it is likely that the efficacy of phytase may be further enhanced by the simultaneous use of other exogenous enzymes, which complement their activity, increase substrate access and/or absorption of liberated nutrients. It has been demonstrated, for example, that the simultaneous inclusion of phytase and glycanase in diets generates a synergistic increase in digestibility of amino acids (Ravindran *et al.* 1999b; Selle *et al.* 2003). This further illustrates the anti-nutritive effects of phytate and the possible avenues that will need to be further clarified if the anti-nutritive effects of phytate are to be ameliorated. It should be noted that phytases allow the release of nutrients, thus altering the balance of these with respect to each other, and it is necessary to consider the judicial re-formulation of diets in the presence of phytases to avoid anti-nutritional effects occurring.

References

Camden BJ, Morel PCH, Thomas DV, Ravindran V, and Bedford MR (2001). Effectiveness of exogenous microbial phytase in improving bioavailabilities of phosphorus and other nutrients in maize-soya-bean meal diets for broilers. *Animal Science* 73:289–297.

Champagne ET (1988). Effects of pH on mineral-phytate, protein-mineral-phytate and mineral-fiber interactions. Possible consequences of atrophic gastritis on mineral bioavailability from high fiber foods. *Journal of the American College of Nutrition* 7:499–508.

Cosgrove DJ (1966). The chemistry and biochemistry of inositol polyphosphates. *Reviews of Pure and Applied Chemistry* 16:209–224.

Cowieson AJ, Acamovic T, and Bedford MR (2004). The effects of phytase and phytic acid on the loss of endogenous amino acids and minerals from broiler chickens. *British Poultry Science* 45:101–108.

Cowieson AJ, Acamovic T, and Bedford MR (2006). Supplementation of corn-soy-based diets with an *Escherichia coli*-derived phytase: effects on broiler chick performance and the digestibility of amino acids and metabolisability of minerals and energy. *Poultry Science*85(8):1389–1397.

Desphande SS and Cheryan M (1984). Effects of phytic acid and divalent cations and their interactions on alpha-amylase activity. *Journal of Food Science* 49:516–519.

Farrell DJ, du Preez JJ, Bongarts M, Betts M, Sudaman A, Thomson E, and Ball W (1992). The improvement in phosphorus availability when phytase is added to broiler diets. *Proceedings of the Australian Poultry Science Symposium* 4:116–119.

Fuh W-S and Chiang B-H (2001). Dephytinisation of rice bran and manufacturing a new food ingredient. *Journal of the Science of Food and Agriculture* 81:1419–1425.

Jagger SJ, Wiseman J, Cole DJA, and Craigon J (1992). Evaluation of inert markers for the determination of ileal and faecal apparent digestibility values in the pig. *British Journal of Nutrition* 68:729–739.

Kemme PA, Jongbloed AW, Mroz Z, Kogut J, and Beynen AC (1999). Digestibility of nutrients in growing finishing pigs is affected by *Aspergillus niger* phytase, phytate and lactic acid levels. 1. Apparent ileal digestibility of amino acids. *Livestock Production Science* 58:107–117.

Kornegay ET, Zhang Z, and Denbow DM (1998). Influence of phytase and diet composition on phosphorus and amino acid digestibilities, and phosphorus and nitrogen excretion in swine. *BASF Technical Symposium preceding Carolina Swine Nutrition Conference*, pp. 125–155.

Lenis NP and Jongbloed AW (1999). New technologies in low pollution swine diets: diet manipulation and use of synthetic amino acids, phytase and phase feeding for reduction of nitrogen and phosphorus excretion and ammonia emission – Review. *Asian-Australasian Journal of Animal Sciences* 12:305–327.

Mroz Z, Jongbloed AW, and Kemme P (1994). Apparent digestibility and retention of nutrients bound to phytate complexes as influenced by microbial phytase and feeding regimen in pigs. *Journal of Animal Science* 72:126–132.

Newkirk RW and Classen HL (2001). The non-mineral nutritional impact of phytate in canola meal for chicks. *Animal Feed Science and Technology* 91:115–128.

Officer DI and Batterham ES (1992). Enzyme supplementation of Linola™ meal for grower pigs. *Proceedings of the Australian Society of Animal Production* 19:288.

Onyango EM, Bedford MR, and Adeola O (2005). Efficacy of an evolved *Escherichia coli* phytase in diets for broiler chicks. *Poultry Science* 84:248–255.

Ravindran V, Cabahug S, and Bryden WL (1999a). Influence of microbial phytase on apparent ileal amino acid digestibility in feedstuffs for broilers. *Poultry Science* 78:699–706.

Ravindran V, Selle PH, and Bryden WL (1999b). Effects of phytase supplementation, individually and in combinations, with glycanase on the nutritive value of wheat and barley. *Poultry Science* 78:699–706.

Ravindran V, Cabahug S, Ravindran G, Selle PH, and Bryden WL (2000). Response of broiler chickens to microbial phytase supplementation as influenced by dietary phytic acid and non-phytate phosphorus levels. II. Effects on apparent metabolisable energy, nutrient digestibility and nutrient retention. *British Poultry Science* 41:193–200.

Ravindran V, Selle PH, Ravindran G, Morel PCH, Kies AK, and Bryden WL (2001). Microbial phytase improves performance, apparent metabolizable energy and ileal amino acid digestibility of broilers fed a lysine-deficient diet. *Poultry Science* 80:338–344.

Ravindran V, Sands JS, Hruby M, and Partridge GG (2005). Influence of Phyzme™ XP phytase on nutrient utilisation in broilers fed diets containing graded levels of phytate. *Proceedings of the Australian Poultry Science Symposium* 17:238–241.

Rutherfurd SM, Chung TK, and Moughan PJ (2002). The effect of microbial phytase on ileal phosphorus and amino acid digestibility in the broiler chicken. *British Poultry Science* 43:598–606.

Rutherfurd SM, Chung TK, Morel PCH, and Moughan PJ (2004). Effect of microbial phytase on ileal digestibility of phytate phosphorus, total phosphorus, and amino acids in a low phosphorus diet for broilers. *Poultry Science* 83:61–68.

Selle PH, Ravindran V, Caldwell RA, and Bryden WL (2000). Phytate and phytase: Consequences for protein utilisation. *Nutrition Research Reviews* 13:255–278.

Selle PH, Ravindran V, Ravindran G, Pittolo PH, and Bryden WL (2003). Influence of phytase and xylanase supplementation on growth performance and nutrient utilisation of broilers offered wheat-based diets. *Asian-Australasian Journal of Animal Sciences* 16:394–402.

Selle PH, Ravindran V, Bryden WL, and Scott TA (2006). Influence of dietary phytate and exogenous phytase on amino acid digestibility in poultry: A review. *Journal of Poultry Science* 43:89–103.

Sharma CB, Goel M, and Irshad M (1978). Myoinositol hexaphosphate as a potential inhibitor of alpha-amylases. *Phytochemistry* 17:201–204.

Tamim NM, Angel R, and Christman M (2004). Influence of dietary calcium, and phytase on phytate phosphorus hydrolysis in broiler chickens. *Poultry Science* 83:1358–1367.

Thompson LU (1988). Antinutrients and blood glucose. *Food Technology* 42:123–131.

Zhang X, Roland DA, McDaniel GR, and Rao SK (1999). Effect of Natuphos® phytase supplementation to feed on performance and ileal digestibility of protein and amino acids in broilers. *Poultry Science* 78:1567–1572.

Chapter 48

Lectin Binding Pattern in the Small Intestine of Rabbits Intoxicated with *Solanum glaucophyllum*

C.N. Zanuzzi[1,2,3], C.G. Barbeito[1,2,3], P.A. Fontana[1], E.L. Portiansky[1,3], and E.J. Gimeno[1,3]

[1] Institute of Pathology, [2] Department of Histology and Embryology, School of Veterinary Sciences, National University of La Plata, 60 y 118, (1900) La Plata, Buenos Aires, Argentina; [3] Members of CONICET (National Research Council)

Introduction

Enzootic calcinosis (EC) in ruminants is a chronic plant intoxication characterized by the calcification of soft tissues and loss of body condition. *Solanum glaucophyllum* (*Sg*) is one of the plants that has been recognized as responsible for inducing chronic systemic calcinosis (Gimeno 2000). It has high levels of 1,25-dihydroxyvitamin D_3 as glycoside derivatives and 25-hydroxyvitamin D_3 as well. The plant predominates in the low-lying and wet land of the poorly drained eastern and central areas of Buenos Aires and littoral provinces of Argentina, and it can also be found in Paraguay, Uruguay, and in the south of Brazil (Gimeno 2000).

Affected bovines are generally more than 2 years old. The symptoms of the disease include loss of appetite and weight, with stiffness and a painful gait. Calcium and inorganic phosphorus levels are higher than in normal animals and this leads to the deposition of calcified material. At necropsy, the animals show advanced cachexia and calcification of the vascular system, lungs and other tissues (Gimeno 2000).

In the last years vitamin D receptors have been identified in multiple tissues and organs (Colston *et al.* 2000). Vitamin D is not only involved in calcium homeostasis but also has been implicated in such essential and varying processes as cell differentiation and proliferation. Owing to its steroidal nature, 1,25-dihydroxyvitamin D_3 induces its biological activities through specific receptors. These receptors interact with the vitamin D response elements of the DNA in which translation results in multiple cellular functions in different organs. The activation of the vitamin D receptor can increase or diminish this gene transcription and, consequently, modify the secretion of the products coded by them (Dusso and Brown 1998).

The small intestine is one of the target organs of vitamin D, in which it stimulates calcium absorption (Dusso and Brown 1998). In addition to this contribution to mineral homeostasis, vitamin D_3 also affects intestinal proliferation and differentiation by inducing the synthesis of compounds known as polyamines, which regulate both processes (Suda *et al.* 1990). High doses of vitamin D_3 inhibit cell proliferation in the colonic neoplasias, whereas lower amounts stimulate epithelial differentiation (Newmark and Lipkin 1992). However, cell differentiation in plant-induced hypervitaminosis D has been only partially evaluated in domestic animals (Barros *et al.* 1981; Barros and Gimeno 2000; Gimeno *et al.* 2004).

The main cause of anorexia and cachexia in clinically affected animals remains unknown. Proliferation, differentiation and cellular death are naturally occurring processes of the intestine (Leblond 1981). Taking into account these biologic characteristics and a possible association of the intestine with the loss of body condition, we decided to study this organ as a useful model for understanding the effects of the hypervitaminosis D.

Lectins are carbohydrate-binding proteins which can be used to localize and identify specific terminal carbohydrate residues. Lectin binding pattern has a great value as a cell differentiation marker (Damjanov 1987; Gimeno and Barbeito 2004).

Since there are various disadvantages of working with bovines for experimental purposes, such as economical and management difficulties, rabbits were selected in this study as experimental models due to their high susceptibility to the *Sg* intoxication as has been shown in previous studies (Dallorso *et al.* 2001; Gimeno *et al.* 2004).

The purpose of this study is to analyze the intestinal carbohydrate expression under the effect of the hypervitaminosis D induced by *Sg* intoxication, using lectinhistochemistry (LHC).

Material and Methods

Five male 3-month-old New Zealand rabbits were orally intoxicated with a dose of 125 mg kg^{-1} day^{-1} of *Sg*, twice a week for a month. Four clinically normal animals were used as controls. Samples of duodenum, jejunum and ileum were taken and fixed in 10% buffered formaldehyde. Tissues were dehydrated using an increasing graded alcohol scale and embedded in paraffin. Five μm sections were stained with hematoxylin and eosin. After dewaxing with xylene the slides were incubated with 0.03% hydrogen peroxide in methanol for 30 min at room temperature to inhibit endogenous peroxidase activity. Then, slides were passed through an increasing graded alcohol scale, rinsed three times in phosphate-buffered saline (PBS) and treated with bovine serum albumin (BSA; 1% in PBS) for 30 min. Sections were then incubated with biotinylated lectins for 2 hours. The following seven lectins (Lectin Kit BK 1000 Vector, Burlingame, CA, USA) with different carbohydrate specificity were used: Con-A (*Concanavalia ensiformes*, specifically binding α-D-Man and α-D-Glc); DBA (*Dolichus biflorus*, with binding specificity to α-D-GalNAc); SBA (*Glicine maximus*, binding specificity to α-D-GalNAc, β-D-galNAc and α and β-Gal); PNA (*Arachis hypogea*, that specifically binds β-D-Gal and (1-3) GalNAc); RCA-1 (*Ricinus communis*-1, binding specificity β-D-Gal and α-D-Gal); UEA-1 (*Ulex europaeus*-1, binding specificity α-L-Fuc); and WGA (*Triticum vulgaris,* binding specificity α-D GlcNAc and NeuNAc). Sections were washed three times with PBS and then incubated with the detection system LSAB® (DAKO, Carpinteria, CA, USA) for 20 min. The employed chromogen was 3,3´diaminobenzidine tetrahydrocloride (DAB). All the slides incubated with the same lectins were processed for the same time. The dark, golden brown DAB hydrogen peroxide reaction product showed the positively stained structures. Mayer's hematoxylin was used for counterstaining.

We studied the lectin binding pattern of the goblet cells, the glycocalix and apical cytoplasm of the epithelial cells at the bottom and in the middle area of the intestinal crypts or Lieberkühn glands, and those cells overlying the intestinal villi. Lectin labeling of goblet cells, both in the intestinal crypts and villi, was analyzed as well. The intensity of staining was graded using a semiquantitative scale with 0 representing no detectable labeling; 1: weak; 2: moderate and 3: strong. The morphology of the cells was assessed on hematoxylin- and eosin-stained sections while the presence of goblet mucin was confirmed with the periodic acid Schiff (PAS) histochemical technique.

Results

Differences were found in the lectin pattern of the intoxicated group in comparison with controls. In the duodenal sections the most interesting finding was a reduction in the intensity of lectin DBA labeling in the apical granular cytoplasm of some pyramidal cells at the bottom of the crypts (Table 1).

Incubation with the lectin PNA in the duodenum and jejunum revealed a heterogeneous pattern among the control animals since some of them exhibited a strong intensity in the apical glycocalix of the cells of the whole crypts, whereas others had a fainter labeling, especially in the middle segment. Nevertheless, in almost all the intoxicated rabbits we observed a strong staining of the glycocalix of the whole crypts and of some cells at the bottom of the jejunal crypts (Figure 1).

In the ileum of the intoxicated animals the apical glycocalix of the whole crypt was more intensely labeled with the lectin PNA and RCA-1 than the control group (Table 1). Individual cells in the base of some crypts were strongly labeled with the lectin UEA-I, both in the control and intoxicated animals, mostly in the jejunum and ileum.

Discussion

Cell differentiation and maturation in tissues undergoing constant renewal like the intestinal epithelium can be efficiently studied by LHC (Barros *et al.* 1981; Barros and Gimeno 2000). We used this technique to assess modifications of glycoconjugate expression in intoxicated and control animals. Some changes in the lectin-binding pattern were observed in the intoxicated rabbits.

Lectin binding pattern in several non-inflammatory and inflammatory lesions is altered as it has been shown at different intestinal sites from various diseases and species. As an example, WGA and Con-A are increased in the ileal glycocalix of pigs infected by *Escherichia coli* 987 P, whereas SBA, PNA and UEA-I are decreased (Sohn and Chae 2000). A particular non-inflammatory alteration of young rabbits, named mucoid enteropathy, is characterized by an increase of goblet cell reactivities to DBA, SBA, UEA, WGA and PNA (Itagaki *et al.* 1994). In the rabbit enterotoxemia changes are observed in the glycoconjugate composition of the brush border and in the Golgi apparatus of the cecum expressed by the reduction of SBA and RCA-I labeling (Jelinek 1996).

Table 1. Lectin binding pattern in the small intestine of control and *Sg*-intoxicated rabbits.

Lectin binding intensity*															
Sites		WGA		Con-A		DBA		SBA		PNA		RCA-I		UEA-I	
		C	I	C	I	C	I	C	I	C	I	C	I	C	1
Duodenum															
Goblet cells		0–2	0–1	0–2	0	0–2	2	1–2	2	0–2	0	1	1–2	0	0
Villi	Gc†	2–3	2	2–3	2	2–3	3	2–3	2–3	1–3	1–2	2	3	0	0
	Ac‡.	0–2	0–1	0–1	1–2	1	0–1	0	0–1	0	0	0	0–1	0	0
Middle crypts	Gc	2–3	2–3	0–2	0–3	2–3	3	2–3	3	1–3	2–3	2	2–3	0	0
	Ac	0–2	0–1	0–1	0–2	1	0–1	0–1	0	0–1	0	0	0–1	0	0
Bottom crypts	Gc	2–3	2–3	1–3	2–3	2–3	3	2–3	3	2–3	2–3	2	2–3	0	0
	Ac	0–2	1–2	0–1	1	2–3	0–1	0	0	0–1	0–1	0	1	0	0
Jejunum															
Goblet cells		1	1	0–1	0–1	0–1	2	1–2	1–2	0	0–1	0–2	1–2	0	1–2
Villi	Gc	3	1–3	2–3	1–3	2–3	3	2–3	3	0–3	2	3	3	0–3	2
	Ac	1	0	1	1	0–1	1	0–1	0	0	0	0–1	1	0	0
Middle crypts	Gc	3	2–3	2–3	3	2–3	3	2–3	3	0–3	3	3	3	0–3	0
	Ac	1	1–2	0–1	1	0–1	1–2	1	0–1	0	0–1	0	0	0	0
Bottom crypts	Gc	3	2–3	0–1	2–3	3	3	3	3	2–3	3	3	3	0–3	2
	Ac	2–3	1–2	0–1	1	2–3°	1–2	2–3°	1–2°	0	1–2	0	0	0–3°	0
Ileum															
Goblet cells		0–2	1–2	0	0	0–2	1–2	0–2	2–3	0–1	0	0–1	1–2	0–2	0–2
Villi	Gc	2–3	3	3	2–3	2–3	3	3	3	0	1	2	3	0	0
	Ac	1	0	1	1	1	1	0–2	0–1	0	0	0	0–1	0	0
Middle crypts	Gc	1–2	3	3	0–3	3	3	2–3	2–3	0	3	2	3	0	0
	Ac	1–2	0–1	1	0–2	0–2	1–2	0–2	0–2	0	0	0	0	0	0
Bottom crypts	Gc	1–3	3	3	3	0–3	3	2–3	3	2	3	2	2–3	0–2	2–3
	Ac	1–2	0–1	1–3	0–1	0–3	1–2	0–3	0–1	0	0	0	0	2–3°	2°

Control (C) and *Sg*-intoxicated (I) rabbits. * Scored as 0: nil; 1: weak; 2: moderate; 3: strong. † Gc: glycocalix. ‡ Ac: apical cytoplasm.
° Only some cells

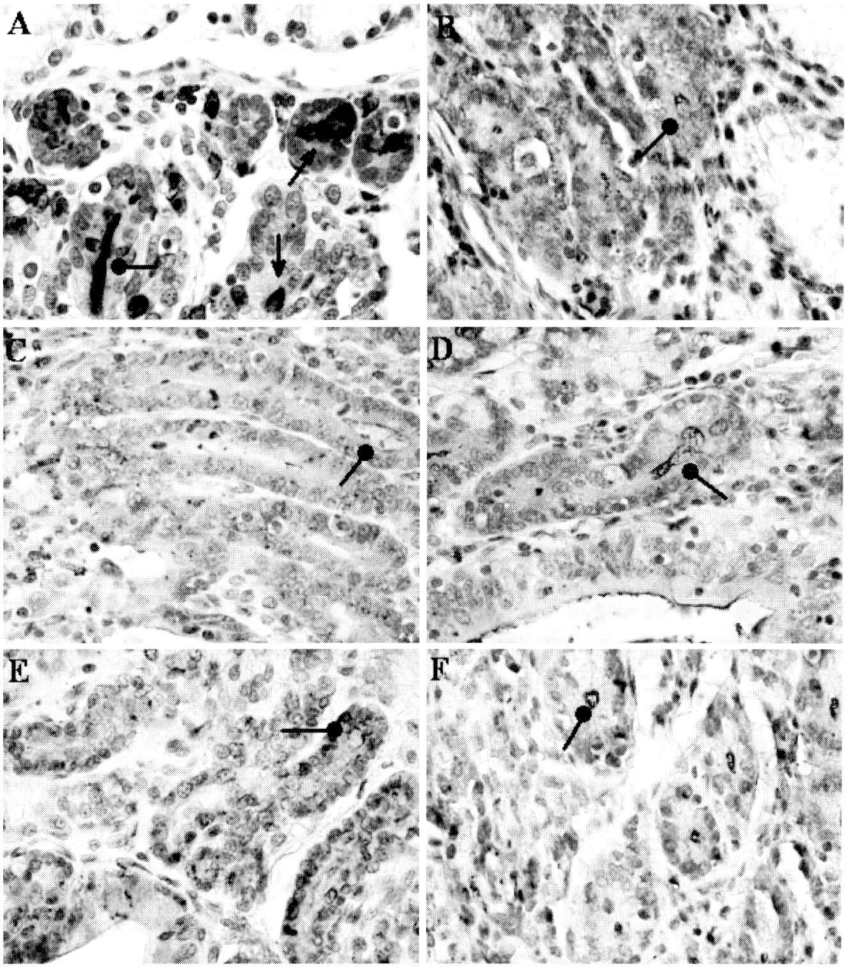

Figure 1. Lectinhistochemical pattern in small intestine of *Sg*-intoxicated rabbits. **A**: DBA binding in duodenum of control animals. Positive staining was observed in individual cells at the bottom of the crypts (arrows) and enterocytes glycocalix (circle ending arrows). **B**: DBA binding in duodenum of intoxicated rabbits. Only binding in the glycocalix was detected. **C**: PNA binding in jejunum of control animals. **D**: PNA binding in jejunum of *Sg*-intoxicated rabbits. **E**: PNA binding in ileum of controls. **F**: PNA binding in ileum of *Sg*-intoxicated animals.
PNA binding at jejunum and ileum was detected at the bottom of the crypts in control animals while in the entire crypt glycocalix of the *Sg*-intoxicated rabbits. 400 x.

In our study we found that the lectin DBA binding was fainter in the cytoplasm of some duodenal cells at the bottom of the crypts in the intoxicated animals. Although the sections and structures analyzed were different, the decrease in DBA labeling was noted in chronic inflammatory diseases such as ulcerative colitis and Crohn's disease, and it was attributed to a reduction in the expression of α-D-GalNAc (Cooper *et al.* 1987). Paneth cells, which are located at the bottom of the crypts and function as defense cells in the small intestine, can be strongly labeled with DBA, SBA and UEA-I in mice (Falk *et al.* 1994). The DBA positive cells of the control group may be Paneth cells. The increase in PNA binding has also been previously observed in different species and pathological conditions, as in the ileum of the paratuberculosis-affected cattle when compared with normal animals (Massone *et al.* 1991). Similar changes have been described in Crohn's disease, active ulcerative colitis and dysplasia, exhibiting the same labeling as is present in colon cancer, adenomas and fetal colons (Cooper *et al.* 1987). These differences have been explained as a loss in the addition of terminal neuraminic acid that leads to the exposure of carbohydrates which are normally masked. Possibly, the same is occurring in the *Sg*-intoxicated animals.

The reasons for the changes observed in the lectin-binding pattern of this study are not well understood. Vitamin D modifies the gene expression and may alter the synthesis of some enzymes involved in modifying cell saccharide composition, as glycosiltransferases and glycosidases. These modifications can be revealed using lectins as cell differentiation markers. We should remember that carbohydrates are considered a secondary product of the expression of genes which play a pivotal role in various cell processes (Gimeno and Barbeito 2004). The intestinal epithelial cells are involved in enzymatic digestion of food by the peptidases of the glycocalix as well as absorptive functions and lubrication of the alimentary tract (Leblond 1981). Abnormal carbohydrate expression could alter these processes and be reflected as a loss in weight and body condition, as observed in *Sg*-intoxicated animals.

Acknowledgements

The authors would like to thank Mrs Rosa Villegas and Mr Ruben Mario for their technical assistance. Financial support was provided in part by grants from the 'Agencia Nacional de Promoción Científica', the 'Academia Nacional de Agronomía y Veterinaria' and the 'Consejo Nacional de Investigaciones Científicas (CONICET)', Argentina.

References

Barros SS and Gimeno EJ (2000). Cell differentiation and bone protein synthesis in the lung of sheep affected by spontaneous calcinosis. *Journal of Comparative Pathology* 123:270–277.

Barros SS, Tabone E, Dos Santos M, Andujar M, and Grimaud JA (1981). Histopathological and ultrastructural alterations in the aorta in experimental *Solanum malacoxylon* poisoning. *Virchow Archives (Cell Pathologie)* 35:167–175.

Colston K, Pirianov G, and Mansi J (2000). Non-classical actions of vitamin D: mechanisms and clinical aspects. In *Vitamin D Endocrine System: Structural, Biological, Genetic And Clinical Aspects* (AW Norman, R Bouillon and M Thomaset, eds), pp. 477–484. University of California Press.

Cooper HS, Farano P, and Coapman RA (1987). Peanut lectin binding sites in colons of patients with ulcerative colitis. *Archives of Pathology and Laboratory Medicine* 111:270–275.

Dallorso ME, Gil S, Pawlak E, Lema F, and Márquez A (2001). 1,25-$(OH)_2$ vitamin D concentration in the plasma of *Solanum glaucophyllum* intoxicated rabbits. *Australian Veterinary Journal* 79:419–423.

Damjanov I (1987). Lectin cytochemistry and histochemistry. *Laboratory Investigations* 57:5–20.

Dusso AS and Brown AJ (1998). Mechanisms of vitamin D action and its regulation. *American Journal of Kidney Diseases* 32:S13–S24.

Falk, P, Roth KA, and Gordon JI (1994). Lectins are sensitive tools for defining the differentiation programs of mouse gut epithelial cell lineages. *Gastrointestinal and Liver Physiology* 29:G987–G1003.

Gimeno EJ (2000). Calcinosis enzoótica En Rumiantes: Un problema vigente de la ganadería nacional. *Anales de la Academia Nacional de Agronomía y Veterinaria* 54:202–234.

Gimeno EJ and Barbeito CG (2004). Glicobiología, una nueva dimensión para el estudio de la biología y de la patología. *Anales de la Academia Nacional de Agronomía y Veterinaria* 57:1–34.

Gimeno EJ, Portiansky EL, Gomar MS, Costa EF, Massone AR, Alonso CR, Dallorso ME, and Barros SS (2004). Calcinosis in ruminants due to plant poisoning: Contributions on the pathogenesis. In *Poisonous Plants and Related Toxins* (T Acamovic, CS Stewart, and TW Pennycott, eds), pp. 84–89. CABI Publishing, Wallingford, Oxon, UK.

Leblond C (1981). The life history of cells in renewing systems. *American Journal of Anatomy* 160:113–158.

Itagaki S, Perfumo CJ, Petrucelli MA, and Doi K (1994). Lectinhistochemical changes of goblet cell mucin in rabbit mucoid enteropathy. *Laboratory Animal Science* 44:82–84.

Jelinek F (1996). Changes of glycid composition and binding of some lectins in the caecum of rabbits with experimentally induced enterotoxaemia. *Veterinary Medicine (Praha)* 41:219–223.

Massone AR, Itagaki S, Doi K, and Gimeno EJ (1991). Lectinhistochemical study on normal and paratuberculosis-affected bovine ileum. *Journal of Veterinary Medical Sciences* 53:761–763.

Newmark H and Lipkin M (1992). Calcium, vitamin D, and colon cancer. *Cancer Research*. 52:2067–2070.

Sohn YS and Chae C (2000). Lectin binding capacity of glycoconjugates in *E. coli* 09: K103: NM, 987 P^+ST^+-infected porcine lower small intestines. *Journal of Veterinary Medical Science* 62:543–547.

Suda T, Shinki T, and Takahashi N (1990). The role of vitamin D in bone and intestinal cell differentiation. *Annual Reviews of Nutrition* 10:195–211.

Chapter 49

A Preliminary Investigation on Changes in Esophageal Diameter and Function of Sheep Recovering from Vermeersiekte

L.D. Snyman[1], A. Olivier-Carstens[2], R. A. Schultz[1], and L. Labuschagne[1]

[1]Division of Toxicology, Onderstepoort Veterinary Institute, Private bag X5, Onderstepoort 0110, South Africa; [2]Department of Companion Animal Clinical Studies, Faculty of Veterinary Science, UP, Private bag X4, Onderstepoort 0110, South Africa

Introduction

Vermeersiekte, a plant poisoning of ruminants, is caused by ingestion of various *Geigeria* spp. The most important of these, *Geigeria ornativa*, causes widespread vermeersiekte in the dry western parts of South Africa. In sheep vermeersiekte is characterized by stiffness, paresis, paralysis, and by regurgitation of rumen contents through the mouth and nose. At necropsy, marked dilatation of the esophagus is sometimes evident (Kellerman *et al.* 1996). The prognosis is good if stock is promptly removed from the toxic pasture. However, if affected animals remain on the pasture for an extended period, recovery may be slow and some sheep may never reach their full production potential again. This might be due to inefficient functioning of the esophagus as mild dilatation of the esophagus in two vermeersiekte sheep was still evident 55 days after withdrawal from *G. ornativa* (Van Heerden *et al.* 1993). The objective of this study was to investigate the changes in esophageal diameter and function during recovery from vermeersiekte in an attempt to explain the slow improvement to full production.

Materials and Methods

Vermeersiekte was induced in three sheep with *G. ornativa* (Table 1). *Geigeria ornativa* was ingested by Sheep 4188, 4468 and 4242 for 26 days, 42 days and 6 months, respectively (Table 1), after which the animals had no further access to the plants. Animals were periodically examined to determine the:

- esophageal diameter index (esophagus diameter/length of the 7th thoracic vertebral body) by means of survey radiography at the maximal width; and

- esophageal function by following a swallowed bolus of food, mixed with contrast medium, by means of fluoroscopy (Olivier-Carstens *et al.* 2002).

Table 1. Intoxication of sheep with *Geigeria ornativa*

Treatment	Sheep number		
	4188	4468	4242
Dose (g kg^{-1} bw day^{-1})	2	3.1	The sheep grazed on a *G. ornativa*-infested field for 6 months during which time the sheep was temporarily removed from the toxic field to recover on a number of occasions
Dosing period (days)	26	42	
Days before clinical signs became evident and disappeared	21–34	21–58	

Results and Discussion

The maximal esophageal diameter indexes for the sheep during recovery from vermeersiekte are indicated in Figure 1. Withdrawal from *G. ornativa* resulted in a decrease in the esophageal diameter index over a period of approximately 200 days to values approaching normal, indicating a slow anatomical recovery of the esophagus.

The esophageal function of the three sheep during recovery from vermeersiekte is given in Table 2. The esophageal function of sheep 4188 was only moderately affected and returned to normal between 345 and 562 days after withdrawal from vermeerbos. Sheep 4468 improved from severely to moderately affected between days 56 and 159 after withdrawal from vermeerbos, but did not show any further improvement until day 610 when the last measurement was performed. Sheep 4242 was extremely affected and improved to severely affected after day 98 following withdrawal from vermeerbos and then remained unchanged until day 1070, the last day on which the esophageal function was determined.

Conclusion

The slow and incomplete recovery of the esophageal diameter and function over these extended periods in all probability will result in reduced feed intake production and might offer some explanation for the poor improvement to full production following vermeersiekte.

Acknowledgements

Funding was provided by the North West Province (Department of Agriculture, Conservation, Environment and Tourism) and the Northern Cape Province (Department of Agriculture and Land Reform).

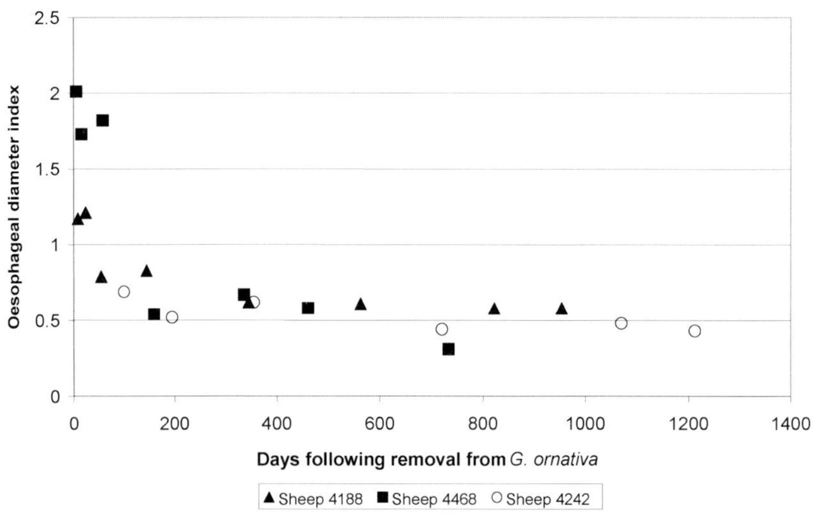

Figure 1. Maximum esophageal diameter index of sheep during recovery from vermeersiekte.

Table 2. Degree of impairment of sheep esophageal function after different periods of removal from *Geigeria ornativa*.

Sheep 4188		Sheep 4468		Sheep 4242	
Day after removal	Esophageal function	Day after removal	Esophageal function	Day after removal	Esophageal function
8	Moderate[a]	0	Severe[b]	98	Extreme[c]
23	Moderate	15	Severe	195	Severe
53	Moderate	56	Severe	355	Severe
144	Moderate	159	Moderate	720	Severe
345	Moderate	335	Moderate	1070	Severe
562	Normal	460	Moderate		
822	Normal	610	Moderate		
953	Normal				

[a]Moderately affected – Feed bolus did not pass directly into the rumen
[b]Severely affected – Feed was partially retained in the esophagus
[c]Extremely affected – Feed accumulated in the thoracic esophagus

References

Kellerman TS, Naude TW, and Fourie N (1996). The distribution, diagnosis and estimated economic impact of plant poisonings and mycotoxicosis in South Africa. *Onderstepoort Journal of Veterinary Research* 63:65–90.

Olivier-Carstens A, Snyman LD, Schultz RA, and Labuschagne L (2002). Radiological and fluoroscopic oesophageal evaluation of the disease progress of sheep dosed with *Geigeria ornativa*: a pilot study. In *Proceedings of the 9th Annual Conference of the European Association of Veterinary Diagnostic Imaging*, pp. 71–72. 24–27th July, Archena (Murcia) Spain.

Van Heerden J, Van Der Lugt JJ, and Durante E (1993). Experimental vermeersiekte (*Geigeria ornativa* O. Hoffm. poisoning) in sheep. 1: An evaluation of diagnostic aids and an assessment of the preventive effect of ethoxyquin. *Journal of the South African Veterinary Association* 64:76–81.

Chapter 50

Cutleaf Nightshade (*Solanum triflorum* Nutt.) Toxicity in Horses and Hamsters

B.L. Stegelmeier, S.T. Lee, L.F. James, D.R. Gardner, K.E. Panter, M.H. Ralphs, and J.A. Pfister

USDA-ARS Poisonous Plant Research Laboratory, Logan, UT 84341, USA

Introduction

Solanum triflorum Nutt. (cutleaf nightshade) is an annual herb that grows in disturbed sites throughout North America (Figure 1; Welsh *et al.* 2003). Historical reports of poisoning in horses and cattle are brief with little description of the lesions (Pammel 1921). Although neurologic signs have been observed in cases of field poisoning, the primary described change of *Solanum* toxicity in experimental studies has been gastrointestinal irritation. In previous studies, it has been shown that the *Solanum* glycoalkaloids including α-chaconine and α-solanine and the solasodine-type glycoalkaloids cause gastrointestinal necrosis (Baker *et al.* 1987, 1989). There are conflicting reports if *Solanum* toxins inhibit acetylcholinesterase and cause neurologic disease (Baker *et al.* 1991). Recently we investigated several incidents of possible cutleaf nightshade poisoning in horses. The affected horses developed signs of cholinergic stimulation after they were observed eating the plant. Affected horses had severe salivation, frequent urination, diarrhea and colic. In all cases the horses recovered with no detectable permanent sequelae of poisoning. The purpose of this study was to develop a small animal model of poisoning and if possible, identify the neurotoxin.

Methods

Solanum triflorum Nuttall (Intermountain Herbarium Voucher #238848, N42°06.64′ W112°00.91′) was collected, frozen, freeze dried and finely ground. Thirty Syrian hamsters were randomly divided into five groups and dosed with 0, 50, 100, 150 and 200 mg of ground plant suspended in 2 mL of water four times a day for 7 days. The hamsters were monitored four times each day for neurologic changes and pupil diameters were measured using a transparent ruler. After dosing, the animals were euthanized, blood was collected and the hamsters were necropsied. Tissues were collected, fixed and prepared for histologic studies. Serum was collected and analyzed for electrolyte and biochemical changes. The serum biochemistries and pupil diameters were analyzed using analysis of variance with $P \leq 0.05$ interpreted as being significant. Where differences occurred the means were separated using Duncan's multiple range test (Proc GLM, SAS Institute, Cary NC).

Figure 1. *Solanum triflorum* Nutt. is a reclining annual weed found in disturbed sites in western and central North America. The leaves are narrow, lobed and pinnate. The flowers are white and the 1 cm berries are green.

Results and Discussion

None of the treated hamsters developed cholinergic-related clinical signs and there were no significant differences in pupil diameter between the groups. Various *Solanum* species have been associated with both cholinergic and anticholinergic toxicity (Mukherjee *et al.* 1961; Baker *et al.* 1991; Ceha *et al.* 1997). However, these effects appear to be sporadic as reproducing toxicity and isolating the toxin has been difficult. Keeler and Baker studied the toxicity of several *Solanum* species in which they isolated and characterized various steroidal alkaloids. Several of these alkaloids have been shown to be teratogenic in rodent models, but in clinical conditions in livestock, most caused severe gastroenteritis (Baker *et al.* 1989, 1991; Keeler *et al.* 1990). Controlled experimental studies have not been able to reproduce neurologic toxicity as no signs or chemical changes suggestive of neurologic toxicity were observed. Brain cholinesterase activities in hamsters treated with *Solanum sarrachoides, S. melongena, S. eleagnifolium or S. dulcamara* were not significantly inhibited (Baker *et al.* 1989). Certainly the horses we observed in the clinical poisonings had signs of cholinergic stimulation (Figure 2). The salivation was copious and all of the horses lacked oral lesions or pain that can be associated with mechanical- or trauma-induced salivation. Additionally the poisoned horses quickly recovered (within several hours) when exposure was discontinued. The lack of salivation or other neurologic signs in this study suggests that the toxin only affects horses or that it may be present in only some plant phenotypes or locations. Alternatively it may be easily degraded when the plants are collected, dried or ground. Additional work is needed to resolve this problem.

Figure 2. Horse with copious salivation after eating cutleaf nightshade.

After several days of dosing, several of the high dose hamsters developed diarrhea and small labial ulcerations. These animals had sporadic biochemical changes consistent with electrolyte loss and lipid mobilization (Table 1). At necropsy, the hamsters dosed with 150 and 200 mg had segmental dilation of the intestine and stomach. Affected bowels were dilated with gas and mucoid exudates. The mucosa of some animals was red and edematous. Histologically these animals had focally extensive hemorrhagic and necrotizing gastroenteritis (Figure 3). No other significant histologic lesions were identified.

The hemorrhagic enteritis was most severe in the ileum but it involved the entire small intestine and glandular portion of the stomach. The distribution and severity was similar to those described by sprout material of *Solanum tuberosum* (Baker *et al.* 1989). This toxicity was primarily attributed to the solanidane alkaloids similar to those identified in *Solanum dulcamara* and *S. eleagnifolium* toxicity (Baker *et al.* 1989). It is likely that *S. triflorum* contains similar glycoalkaloids.

Summary

These findings suggest that in rodents, cutleaf nightshade toxicity is due to the direct toxic effects of *Solanum* glycoalkaloids. Additional studies are needed to determine if cutleaf nightshade contains cholinergic toxins and when and to what species those alkaloids are toxic.

Table 1. Serum biochemistries of hamsters treated with 0, 100 and 200 mg ground *Solanum triflorum* Nuttall four times daily for 7 days.

	0 mg	100 mg	200 mg
Serum glucose (mg/dL)	174±72	166±45	222±97
Serum urea nitrogen (mg/dL)	23±2	19±4	23±7
Serum creatinine (mg/dL)	0.35±0.05	0.31±0.04	0.39±0.14
Total protein (g/dL)	6.7±0.3	6.8±0.4	6.2±0.7
Serum albumin (g/dL)	2.7±0.1	2.8±0.2	2.6±0.3
Alkaline phosphatase (U/L)	90±17	79±6	94±24
Aspartate aminotransferase (U/L)	78±46	64±29	96±54
Sodium (mequ/L)	139±3	139±4	134±5
Potassium (mequ/L)	8.3±0.5	8.2±0.5	7.7±1.4
Chloride (mequ/L)	97±2	97±4	95±7
Carbon dioxide (mequ/L)	36.3±2.8	36.1±4.7	29.6±9.0

Data are presented as means±standard deviation.

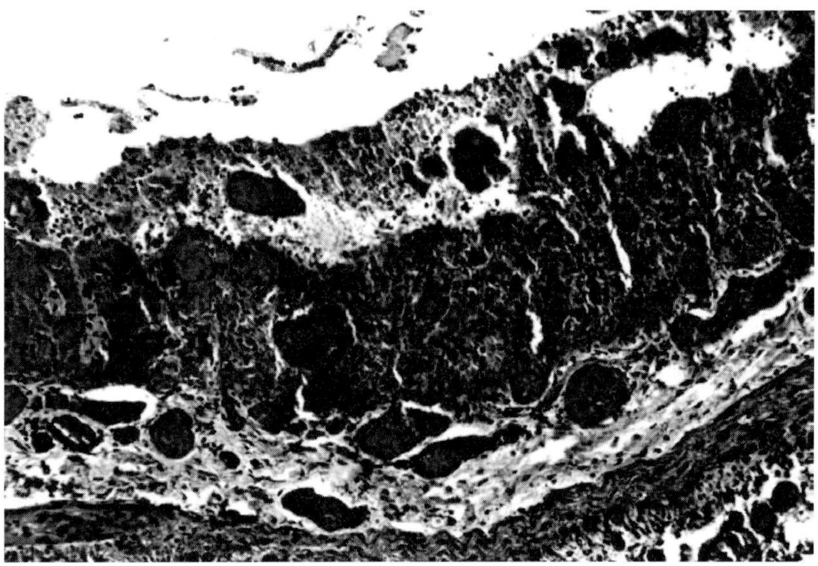

Figure 3. Photomicrograph of the ileum of a hamster treated with 200 mg ground *Solanum triflorum* Nutt. for 7 days. Note the extensive necrotizing and hemorrhagic enteritis. The submucosa is edematous and congested.

References

Baker D, Keeler R, and Gaffield W (1987). Lesions of potato sprout and extracted potato sprout alkaloid toxicity in Syrian hamsters. *Journal of Toxicology and Clinical Toxicology* 25:199–208.

Baker DC, Keeler RF, and Gaffield W (1989). Pathology in hamsters administered *Solanum* plant species that contain steroidal alkaloids. *Toxicon* 27:1331–1337.

Baker DC, Keeler RF, and Gaffield W (1991). Toxicosis from steroidal alkaloids of *Solanum* species. In *Handbook of Natural Toxins*, Vol. 6 (RF Keeler and AT Tu, eds), pp. 71–81. Marcel Dekker Inc., New York.

Ceha LJ, Presperin C, Young E, Allswede M, and Erickson T (1997). Anticholinergic toxicity from nightshade berry poisoning responsive to physostigmine. *Journal of Emergency Medicine* 15:65–69.

Keeler RF, Baker DC, and Gaffield W (1990). Spirosolane-containing *Solanum* species and induction of congenital craniofacial malformations. *Toxicon* 28:873–884.

Mukherjee SR, Dey PK, and Paul S (1961). Cholinergic alkaloid in *Solanum nigrum* Linn. *Journal Francais de Medecine et Chirurgie Thoraciques* 4:136–144.

Pammel LH (1921). Three flowered nightshade poisoning. *Veterinary Medicine* 16:46.

Welsh SL, Atwood ND, Goodrich S, and Higgins LC, eds (2003). *A Utah Flora* Third Edition, p. 643. Brigham Young University Press, Provo, Utah.

Chapter 51

Effect of *Senna occidentalis* Seeds in Broiler Chickens: a Preliminary Immunotoxicologic Study

I.M. Hueza, A.O. Latorre, D.P. Mariano-Souza, L.E.R. Raspantini, P.C. Maiorka, and S.L. Górniak

Research Center for Veterinary Toxicology (CEPTOX), Department of Pathology, Faculty of Veterinary Medicine and Zootechny, University of São Paulo, CEP 05508-900, Brazil

Introduction

Senna occidentalis (*So*) is a weed of the Caesalpinoideae family found ubiquitously in tropical and subtropical regions of the world, and is a contaminant of soybean, maize, wheat and other cereal crops (Lal and Gupta 1973). Although most harvested cereals are mechanically cleaned and selected before being processed, if present *So* seeds always contaminate the final product, including animal rations, because of their similarity in size and density to crop grains. The main toxic principles of *So* seed have been identified as dianthrons (Haraguchi *et al.* 1996).

Regardless of the animal species in both natural and experimental intoxication, the ingestion of large amounts of this plant produces ataxia, muscle weakness and tremors, body weight loss and ultimately death (O'Hara *et al.* 1969; Barth *et al.* 1994).

Skeletal muscle degeneration is the predominant lesion found in the majority of animals intoxicated with *So* seed; however, other lesions have been described, such as degenerative myopathy of myocardial muscle (Tasaka *et al.* 2000), congestion and pulmonary edema (Irigoyen *et al.* 1991), and hepatic cell hypertrophy and necrosis (Tasaka *et al.* 2000). In addition, broiler chicks treated with the external tegument of the seed showed significant alterations in the relative weight of lymphoid organs, mainly the bursa of Fabricius, and their histological structure (Silva *et al.* 2003).

Moreover, studies conducted by Haraguchi *et al.* (2003) with low concentrations of *So* seeds mixed in broiler chick rations showed that these animals were very sensitive to the toxic effects of dianthrons. Taken together, these studies indicate that *So* toxicity might impact the immune system of chickens. Thus, the purpose of the present study was to evaluate the possible immunotoxic effects of low concentrations of *So* seeds in broiler chickens.

Materials and Methods

Ripe *So* seeds were collected from plants grown at the Biological Institute of São Paulo. After harvesting, the seeds were first frozen in liquid nitrogen and then immediately ground and incorporated into the chicken ration at different concentrations.

Sixty one-day-old Hubbard commercial broiler chicks were randomly assigned to four groups, including one control and three experimental groups treated with commercial ration mixed with 0.25, 0.50 and 0.75% of *So* seeds, for 28 days. At the end of the experimental period, lymphoid organs were collected to evaluate their relative weights and histology. Peritoneal macrophages (Mø) were harvested to determine Mø activity (spreading, phagocytosis, and release of hydrogen peroxide (H_2O_2) and nitric oxide (NO)). An additional 32 one-day-old Hubbard broiler chicks were randomly assigned to four groups and treated with 0.0, 0.25, 0.50 and 0.75% of *So* seeds, for 35 days. These chicks were vaccinated 14 days after birth with a commercial anti-Newcastle vaccine, and at the end of the experimental period lymphoid organs were collected to evaluate their relative weights and histology.

Macrophage activity

Forty-eight hours before sacrifice, the 28 1-day-old birds in the first experiment received an intraperitoneal injection of 5.0 mL kg^{-1} of Sephadex solution (3%) in order to promote peritoneal Mø activation. The methods used to study Mø spreading and phagocytosis were based on those described by Rabinovitch and De Stefano (1973) with some modifications. Briefly, a total of 200 cells per slide were counted for each bird and the Mø spreading index (SI) and phagocytosis index (PI) were calculated as follows: SI or PI = number of spreading Mø or phagocytic activity x 100 ÷ 200 adherent cells counted (i.e. SI = percentage of spreading Mø and PI = percentage of Mø with phagocytized zimosan particles). The mean of four counts obtained from the two slides of each bird was used to express SI and PI.

Spontaneous and phorbol myristate-acetate solution (PMA)-induced H_2O_2 release by Mø was measured by the method of Russo *et al.* (1989). H_2O_2 concentration was calculated from absorbance measurements. Spontaneous and PMA-induced H_2O_2 were repeated four times for each bird in each group and the mean value of the four counts was used to determine H_2O_2 concentration.

NO production was indirectly measured by assessment of the nitrite concentration, the stable end product of NO synthase-generated reactive nitrogen intermediates. Nitrite concentration was determined using a colorimetric assay (Green *et al.* 1982). In brief, 150 μL of the cultured Mø supernatants were incubated with 300 μL of a 1:1 mixture of 1% sulfanilamide in 30% acetic acid and 0.1% *N*-1-naphtyl-ethylenediamine dihydrochloride in 60% acetic acid at room temperature. After 5 min of incubation in the dark, absorbance was read at 540 nm. Concentrations were calculated according to a standard curve prepared by serial dilutions of sodium nitrite.

Experimental and control data were analyzed statistically by analysis of variance (ANOVA); significant F tests were followed by Dunnett's *post hoc* test to compare control and treated groups at $P<0.05$. Data are expressed as mean ± SEM.

Results

The spreading and the phagocytic activity of peritoneal Mø from treated birds were not affected by *So* seeds at any level in the ration when compared with control birds. The administration of *So* seeds for 28 days, however, increased the spontaneous and the PMA-induced H_2O_2 release by peritoneal Mø in birds in the 0.75% treatment when compared with cells from control animals ($P<0.05$) (Figure1). No difference was observed in the release of NO by peritoneal Mø in any of the birds.

Only chicks in the 0.75% treatment given for 28 and 35 days showed a significantly reduced weight in the bursa of Fabricius (Figure2) ($P<0.05$). No significant differences were observed in the weight of other lymphoid organs (i.e. thymus and spleen), in treated birds compared with the control group.

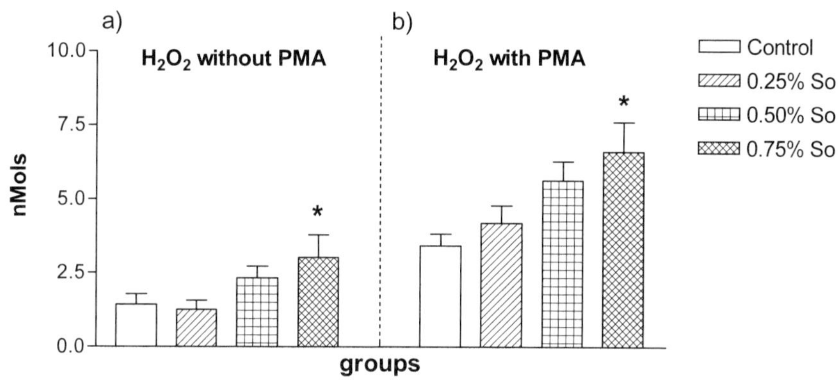

Figure 1. Effect of administration of *Senna occidentalis* seeds on spontaneous (A) and PMA-induced H_2O_2 release (B) by peritoneal macrophages in broiler chicks given rations containing 0.0, 0.25, 0.50 and 0.75% *Senna occidentalis* seeds for 28 days. Results represent means ± SEM.
* Significantly different from control ($P<0.05$).

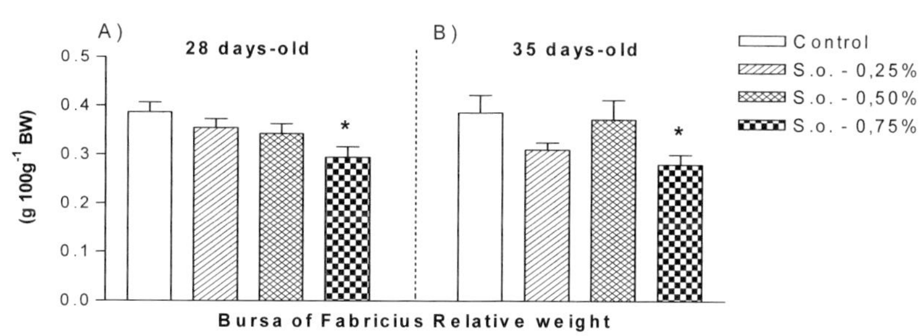

Figure 2. Effect of administration of *Senna occidentalis* seeds on relative weight of the bursa of Fabricius in broiler chicks given rations containing 0.0, 0.25, 0.50 and 0.75% *Senna occidentalis* seeds for 28 days (A) and 35 days (B). Results represent means ± SEM.
* Significantly different from control ($P<0.05$).

Histological examination of the bursa of Fabricius from birds in the 0.75% treatment for 28 days revealed reduced follicular areas, and within follicles there were depleted cellular areas with the marked presence of Mø (Figure 3).

Discussion

The most important lesions caused by *So* intoxication are skeletal and cardiac muscle degeneration. Electron microscopy of affected muscle fibers revealed the presence of mitochondrial damage, reflective of the mitochondrial myopathy promoted by *So.* The mechanism of action of *So* toxins on mitochondrial metabolism is still unknown; however, a study conducted by Lewis and Shigamoto (1989) with *Senna obtusifolia* showed that *So* can cause impairment of mitochondrial metabolism, inhibiting NADH oxidation and electron transport in mitochondria, resulting in uncoupling of oxidative phosphorylation.

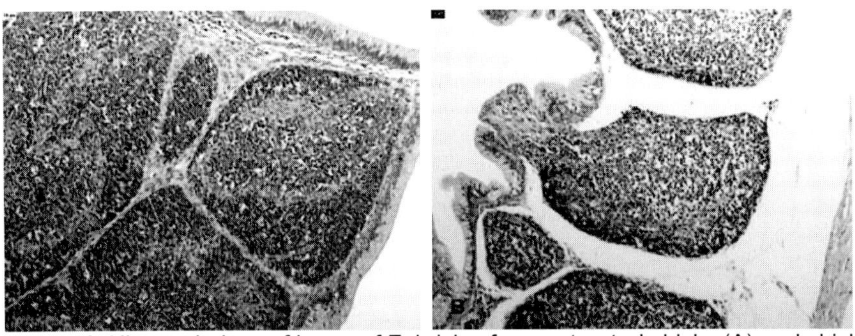

Figure 3. Histopathology of bursa of Fabricius from untreated chicks (A) and chicks fed a ration with 0.75% *Senna occidentalis* seeds (B) for 28 days. The bursa of Fabricius of treated chicks showed reduced follicle area, presence of depleted cell areas in the cortical and in the medullar regions, with the marked presence of internal macrophages. H&E.

The mitochondrial respiratory chain uses electron flow to drive adenosine triphosphate (ATP) synthesis. Briefly, the electrons are passed to the terminal electron acceptor oxygen which is then reduced to water. Associated with the movement of electrons along the electron chain transport is the movement of protons (H^+) from the mitochondrial matrix into the intramembrane spaces. The movement of H^+ through the ATPase provides energy to support ATP synthesis. If a site-specific defect exists in electron transport, electrons will leak from the respiratory chain and cause a univalent reduction of oxygen that results in the formation of superoxide (O_2^-) which is converted to H_2O_2 in the presence of superoxide dismutase.

The present study clearly shows an increased H_2O_2 production by peritoneal Mø of birds treated with 0.75% *So* seeds, but no further enhanced effects were observed in other Mø activities. This suggests that H_2O_2 results from the mitochondria and not from any possible pro-inflammatory effect of *So* intoxication. Indeed, *So* seeds are employed in several popular medicinal preparations used for different therapeutic purposes including anti-inflammatory drugs (Sadique *et al.* 1987).

The mitochondrion, besides being the organelle in which oxidative phosphorylation and ATP production occur, is the locale where several important aspects of apoptosis occur. We hypothesize that some caspase activators, such as cytochrome *c*, could be released from the mitochondria to the cytosol during disturbance of the mitochondrial transmembrane potential, leading to activation of pro-apoptotic genes, and eventual disruptions in electron transport and uncoupling oxidative phosphorylation. We further hypothesize that the reduced weight of the bursa of Fabricius and associated histological changes (e.g. reduced follicles with depleted cellular area, including the marked presence of Mø) result from an apoptotic process triggered by the plant toxin(s). Conversely, spleen and thymus were not affected by *So*, even in unvaccinated birds. Therefore, more work will be required to reconcile these results. Future experiments focusing on mitochondrial participation in both H_2O_2 production and the possible apoptosis process in B lymphocytes will be carried out in our laboratory to obtain more information about the mechanism of *So* intoxication in mitochondria.

Acknowledgements

This study was supported by grants from Fundação de Amparo à Pesquisa do Estado de SãoPaulo – FAPESP, Brazil (Proc. No. 02/00080-0).

References

Barth AT, Kommers GD, Salles MS, Wouters F, and Barros L (1994). Coffee senna (*Senna occidentalis*) poisoning in cattle in Brazil. *Veterinary and Human Toxicology* 36: 541–545.

Green LC, Wagner DA, Glogowski J, Skipper PL, Wishnok JS, and Tannenbaum SR (1982). Analysis of nitrite and nitrate in biological fluids, *Analytical Biochemistry* 126:131–138.

Haraguchi M, Górniak SL, Dagli, MLZ, and Raspantini PCF (1996). *Determinacão dos constituintes químicos das frações tóxicas de fedegoso (Senna occidentalis (L.)* Annals of Reunião Annual da Sociedade Brasileira de Química 19, Poços de Caldas.

Haraguchi M, Dagli MLZ, Raspantini PCF, and Górniak SL (2003). The effects of low doses of *Senna occidentalis* seeds on broiler chickens. *Veterinary Research Communications* 27:321–328.

Irigoyen LF, Graça DL, and Barros CSL (1991). Intoxicação experimental por *Cassia occidentalis* (Leg. Caes) em equinos. *Pesquisa Veterinária Brasileira* 11:35–44.

Lal J and Gupta PC (1973). Anthraquinone glycoside from the seeds of *Cassia occidentalis* Linn. *Experientia* 29:42–143.

Lewis DC and Shigamoto T (1989). Effects of *Cassia occidentalis* (sicklepod) extracts and anthraquinones on muscle mitochondrial function. *Toxicon* 27:519–529.

O'Hara PJ, Pierce, KR, and Read WK (1969). Degenerative myopathy associated with ingestion of *Cassia occidentalis* and pathologic features of the experimentally induced disease. *American Journal of Veterinary Research* 30:2173–2180.

Rabinovitch M and De Stefano MJ (1973). Macrophage spreading *in vitro*: I. Inducers of spreading. *Experimental Cell Research* 77:323–334.

Russo M, Teixeira HC, Marcondes MCG, and Barbuto JAN (1989). Superoxide independent hydrogen peroxide release by activated macrophages. *Brazilian Journal of Medical Biological Research* 22:1271–1273.

Sadique J, Chandra T, Thenmozhi V, and Elango V (1987). Biochemical modes of action of *Cassia occidentalis* and *Cardiospermum halicacabum* in inflammation. *Journal of Ethnopharmacology* 19:201–212.

Silva TC, Górniak SL, Oloris SCS, Raspantini PC, Haraguchi M, and Dagli MLZ (2003). Effects of *Senna occidentalis* on chick bursa of Fabricius. *Avian Pathology* 32:633–637.

Tasaka AC, Weg R, Calore EE, Sinhorini IL, Dagli MLZ, Haraguchi M, and Górniak SL (2000). Toxicity testing of *Senna occidentalis* seeds in rabbits. *Veterinary Research Communications* 24:573–582.

Chapter 52

Chronic Canine Aflatoxin and Management of an Epidemic

T. Garland[1] and J.C. Reagor[2]

[1]*Indiana Board of Animal Health, 805 Beachway Dr., Suite 50, Indianapolis, IN 46224, USA;* [2]*Department of Toxicology, Texas Veterinary Medical Diagnostic Laboratory, 1 Sippel Road, College Station, TX 77843, USA*

Introduction

An epizootic of canine hepatitis, later termed 'Hepatitis X', in the southeastern United States was described by Seibold and Bailey (1952). This disease was classified as being mostly subacute with some acute and some chronic cases based on histopathological observations and clinical histories. The condition was characterized by severe liver damage, which resulted in depression, anorexia and icterus without fever. Many animals had hemorrhages in various tissues and organs. The liver was described as being yellowish, greenish yellow or reddish yellow. The hepatic parenchyma cells showed a marked fatty change and various amounts of bile duct proliferation and hepatic cell regeneration. This hepatic pathology was reproduced by Newberne *et al.* (1955) by feeding suspect lots of dog food.

Acute aflatoxin B_1 (AFB_1) toxicity was shown to be the etiology of Hepatitis X in 1966 (Newberne *et al.* 1966). Dosing with 0.2 mg AFB_1 per day for 15 days followed by 13 days of observation reproduced moderate to severe hepatic lesions. Dogs consuming food containing 500 parts per billion (ppb) aflatoxin for 35 days before euthanasia had minimal to moderate hepatic lesions but no clinical signs. A single intraperitoneal injection of AFB_1 at 1 mg kg^{-1} resulted in the death of two out of three animals within 14 days. These animals had moderate to severe hepatic lesions, but no significant changes in the kidneys other than cholemic nephrosis.

The pathological findings of a natural outbreak of aflatoxicosis in dogs were reported where the dog food contained 100 to 300 ppb AFB_1 (Bastianello *et al.* 1987). A small-scale study (Kellerman *et al.* 1988) where two animals per group were fed a diet containing either 250 or 500 ppb AFB_1 reproduced the disease. The low level required approximately 6 months to cause death, while the higher level required only 2 months. These levels are much less than previous reports and are similar to the levels found in the epidemic reported here.

Outbreak

In late October of 1998, The Texas Veterinary Medical Laboratory (TVMDL) received a group of unrelated cases in which otherwise healthy dogs went into acute hepatic failure and died. Clinical signs included anorexia, depression, icterus, sudden death, melena, coagulopathy, vomiting and diarrhea. This was chronic disease that presented acutely. Several owners reported their animal had not seemed well at the feeding time and did not eat that meal and was dead by the next. In animals exhibiting vomiting and diarrhea, blood was often reported in the vomitus and/or the feces.

The most profound abnormality in the clinical chemistry profile of severely affected animals was the ALT (alanine aminotransferase, formerly SGPT) concentration, which was often over 2000 IU L^{-1} (normal 13 to 75 IU L^{-1}). The serum alkaline phosphatase (SAP) concentration was often up to 400 IU L^{-1} (normal 12 to 122 IU L^{-1}). Total bilirubin level was consistently elevated up to 3 mg percent (normal 0.1 to 0.4 mg percent) and GGT (gamma-glutamyltransferase) level was moderately elevated up to 45 IU L^{-1} (normal less than 10 IU/L). Animals with very high GGT (50 to 100 IU L^{-1}) levels were consistently diagnosed as something other than aflatoxicosis. Changes in concentrations of AST (aspartate aminotransferase, formerly SGOT) and CK (creatine kinase) were variable.

Histologic lesions observed in liver sections of all affected animals were similar (Hoffman *et al.* 1999). Hepatic lipidosis, cholestasis and biliary hyperplasia were observed in every section. The lipidosis was characterized by vesicles ranging from 1 to 2 μm in diameter with some clear spaces up to 15 μm in diameter. The cholestasis was usually intrahepatocellular and some bile canaliculi contained plugs of bile. The biliary hyperplasia was characterized by numerous haphazardly arranged bile ducts proliferating within portal areas. Lesions observed in many of the liver sections included hepatic necrosis, nodular regeneration, minimal fibrosis and inflammation. Cholemic nephrosis was commonly observed.

Liver failure can result from a number of agents, including anticoagulants, phosphine, mushrooms, cycads, blue-green algae and iron. These were eliminated as the cause of liver failure in these cases. Dog food from two of the early cases contained about 150 ppb aflatoxin. Intense case-by-case investigation by TVMDL revealed all of the initially unexplained deaths occurred in dogs consuming foods made by the same manufacturer. All of the several dog food formulations were prepared on either 1 day in mid-July and/or 1 day in mid-August at one particular plant. Contamination occurred on possibly 2 days out of 90 days that food was recalled. The company operated on extreme caution and recalled more food than just the likely contaminated lots.

Recall

On November 2, immediately following the diagnosis, the dog food manufacturer made a complete, voluntary recall for all their dog food made from mid-July through August, in the particular plant where the AFB_1 tainted food originated. This plant manufactured approximately 25% of all the dog food sold in Texas during that time period. The manufacturer published a toll free phone number and hired several veterinarians to return phone calls to pet owners and their veterinarians. The manufacturer covered the cost for veterinary examinations, clinical chemistry profiles and CBCs (complete blood counts) of sick dogs consuming the recalled product, as well as treatment of animals with confirmed liver disease. The last confirmed death of 55 total occurred on November 22, 3 weeks after the recall was initiated.

The company recalled about 90 days of product, although contamination occurred on only 1 day in approximately mid-July and/or 1 day in mid-August. The company recalled more product than the likely affected lots of food as an act of extreme caution. This company had a sterling record for testing its raw products, but aflatoxin is not uniform within corn and to test all the product would mean there would be no corn for making of pet food. Additionally, the company has dramatically increased its testing and quality assurance protocols.

The canine seems to be sensitive to a chronic, low-level intake of AFB_1. These animals exhibited an acute clinical syndrome after chronic consumption of the contaminated food for 90 to 120 days. Although most people do not purchase enough food to last that length of time, and particularly from one lot of food, some people did purchase in bulk, and some people actually got the same lot of contaminated food, by accident from different stores, purely by chance based upon their shopping habits. It is unlikely that all the dogs affected consumed tainted food that entire duration of time. There is no way to determine how much aflatoxin an individual animal consumed. There may have been some bags containing a large amount of toxin, while other bags contained lower levels or none at all. However, of 188 food samples assayed by TVMDL, only one contained over 300 ppb AFB_1 (370 ppb). Most samples associated with sick animals contained between 100 and 300 ppb. It may be only a few animals were unlucky enough to be continually fed contaminated food or only the most sensitive of the canine population were clinically affected. It required 6 months for Naudé (Kellerman *et al.* 1988) to produce death in the experimental trial with approximately this level of aflatoxin.

Recall Management

The manufacturer issued the recall without veterinary assistance and the toll free number became overwhelmed with customer calls. They hired a veterinary toxicologist to come and educate them, assess the magnitude of the calls, and assist the company in formulating a plan to handle the situation. Facts relating to canine aflatoxicosis were sent to the private veterinarians as a fax through the TVMDL and the Texas Veterinary Medical Association, and in electronic format through the TexVetMed list server. The company hired local veterinarians to assist with customer calls and under company directives empowered the recall team to assist the client in 'becoming whole again.'

Overwhelmingly, the company's attitude and motto was 'Do the right thing.' In attempting to accomplish this, they provided guidelines and limited training to the veterinarian with regard to grief counseling and active listening procedures. Furthermore, each veterinarian was given training in the clinical signs related to canine aflatoxicosis. Using established company guidelines, they empowered the veterinarian to make decisions regarding reimbursement for medical expenses or for the replacement of an animal. The guidelines involved the veterinarians using clinical judgment regarding actual signs of illness in the animal, proper time frame of the disease, evidence of consumption of potentially contaminated food, and logical medical treatment and/or reasonable replacement costs.

Within days it became apparent the recall office needed to be located in the state where the outbreak was occurring, instead of the state where the manufacturer's home office was located. At the suggestion of the consulting veterinary toxicologist and the manager of the consumer relations division, a recall team of veterinarians and veterinary technicians was assembled in Texas within 48 hours of leaving the home office. By establishing the recall office in Texas, the customers and private veterinarians developed more of a sense of talking to someone 'local' who had an understanding of the problem, instead of some corporate entity. Furthermore, having the recall office in the same town as the TVMDL facilitated timely return of laboratory reports that were picked up each day by the veterinary toxicologist. The TVMDL reports were matched with customer records to facilitate assessing each case.

The manufacturer established an account at the TVMDL for private veterinarians to use when sending clinical samples for testing, and paid for the testing on the pet instead of the veterinarian charging the customer. The manufacturer agreed to pay for clinical chemistries, a CBC and a fecal examination. This was an attempt to either incriminate aflatoxin or to rule out other diseases. Also, an outbreak of a very virulent strain of parvo virus occurred prior to and throughout the outbreak.

The temporary recall office utilized local veterinarians and veterinary technicians. They were provided the same technical information and customer care guidelines as those in the manufacturer's main office. The toll free recall information telephone number was transferred to Texas, and an office staff was established to assist in maintaining the volumes of records generated. Each customer had a record generated when they called the hotline. The record consisted of basic customer information, clinical history of the pet or pets involved, and permission to consult with the veterinarian who treated the pet. Extensive notes were generated as the medical staff visited with customers. Confusion occurred when a pet owner would call and then the pet owner's family member would call with a different name. Similar confusion occurred when the veterinarian would call with a duplicate pet owner's name. Sometimes these two callers were assigned to different veterinarians or veterinary technicians. The recall team developed a half-hour meeting at the end of each 6-hour shift to debrief other team members and clear up double complaints on the same animals.

The recall office operated two shifts of 6 hours each with both the customer relations manager and the veterinary toxicologist present throughout the day. Veterinarians and technicians worked flexible hours so there was often overlap in the personnel manning the shifts. This promoted continuity between the shifts.

Case Studies and Treatments

During the recall, many owners carried pets to their veterinarian for evaluation, which included a chemistry profile. Several non-clinical animals had indications of liver disease. A typical case where follow-up profiles are available is one where there was a two adult dog household. Two neutered beagles, one female and one male, had been consuming contaminated food and were slightly depressed according to the owner. The owner took them to the veterinarian because of advertisements and news articles regarding the recall. Examination was unremarkable; the CBCs were essentially normal, but the GGT concentration was elevated in both (17 and 42 IU L^{-1}). They were not treated other than removal of the contaminated food. Sixty-three days later they returned to the veterinary clinic for a follow-up exam and at that time the GGT was 3 IU L^{-1} in both animals.

One case involved a dog in crisis taken to the veterinarian. The patient had a profoundly high ALT and moderately high total bilirubin, GGT and AST. The dog had profuse diarrhea with dark blood (melena) and vomit with some blood. As the animal continued to worsen and appeared to be sloughing the gastrointestinal track, the dog was put on balanced fluids, electrolytes, choline and cysteine. By the next day, the animal had begun to seizure. The attending veterinarian controlled the seizures and upon consultation with the veterinary toxicologist put the dog on Vitamin K_1 and continued the previous therapy. The rationale for the Vitamin K_1 was to re-establish clotting factors potentially diminished by an aflatoxin affected liver. It was reasoned the animal was bleeding into the cranial vault, causing the seizures. The seizures were medically controlled, and within 72 hours of initiation of this treatment the seizures and the blood loss stopped. The patient was maintained for 5 days on Vitamin $K_{1,}$ and was gradually weaned off of the remainder of the supportive therapy. Clinical chemistries at the time of discharge, 15 days after admission, revealed that the total bilirubin, GGT and AST were within the normal range. The ALT was slightly high, but clinically the dog was normal, except for weight loss.

Many owners believed the recall was the opportunity to have someone else pay for routine preventative medical care for their pets. Those who had not vaccinated against parvo virus wanted treatment for their animal when it was unrelated to the aflatoxin outbreak. Such was the case of one customer who was certain her pet had been poisoned by the dog food. A history revealed the dog had never been vaccinated for anything except rabies. Despite intensive care the dog died. The owner was so insistent her dog was a victim of aflatoxin that a necropsy was requested. Necropsy revealed lesions associated with parvo virus, and the virus was isolated from the gastrointestinal tract of the dog. Understandably, owners in this type of situation were distraught, but this was not a dog food recall issue. Necropsy was a valuable tool in helping to rule out aflatoxicosis or other diseases, thereby obtaining an accurate diagnosis.

Likewise there were owners who were excellent care-givers. The stress of pregnancy and whelping sometimes made the clinical disease apparent. One such case involved a very well cared for dog which had recently whelped. The pet was vomiting, developed diarrhea and was visibly jaundiced. Despite intensive therapy the dog died. Histopathological lesions at necropsy confirmed aflatoxicosis. The puppies in this case did fine. However, in some cases the female dog was not clinically ill but the nursing puppies up to about 8 or 10 weeks of age were affected.

Mitigating Circumstances

Drought and unusually high temperatures during the summer of 1998 in Texas and the southern United States resulted in a large amount of aflatoxin in the corn harvested in this area (sometimes above 1000 ppb). This episode was not the result of a company not checking incoming ingredients for the toxin, but rather the toxin was missed in the corn as the result of sampling. Corn was purchased with the guarantee it contained less than 20 ppb aflatoxin and each truck was checked for aflatoxin. Assays of the finished product from many batches by the company indicated there were probably only two truck-loads of corn which contained a hazardous level of the toxin. Analyses of the recalled products submitted by the public to TVMDL were in agreement as about 89.4% of the samples contained less than 20 ppb AFB_1.

Conclusion

This food-related outbreak was contained in a timely fashion and is a good example of how veterinarians, diagnosticians, regulators and industry, working together, brought a relatively quick end to this incident. A similar situation went unchecked in South Africa (Bastianello *et al.* 1987) in the mid-80s and the animals continued to die for about 3 months, versus the Texas outbreak that was contained in 3 weeks. The South Africa papers certainly helped make the diagnosis and convince industry that unintentional contamination within the feed ingredients was the source of the problem.

References

Bastianello SS, Nesbit JW, Williams MC, and Lang AL (1987). Pathological findings in a natural outbreak of aflatoxicosis in dogs. *Onderstepoort Journal of Veterinary Research* 54:635–640.

Hoffman RJ, Barr AC, and Reagor JC (1999). Histologic lesions of chronic canine aflatoxicosis. *Proceedings of the IX International Symposium of the World Association of Laboratory Diagnosticians*, p. 69. College Station, Texas.

Kellerman TS, Coetzer JAW, and Naudé TW (1988). *Plant Poisonings and Mycotoxicoses of Livestock in Southern Africa*, p. 4. Oxford University Press, Cape Town.

Newberne JW, Bailey WS, and Seibold HR (1955). Notes on a recent outbreak and experimental reproduction of hepatitis X in dogs. *Journal of the American Veterinary Medical Association* 127:59–62.

Newberne PM, Russo R, and Wogan GN (1966). Acute toxicity of aflatoxin B_1 in the dog. *Pathologia Veterinaria* 3:331–340.

Seibold HR and Bailey WS (1952). An epizootic of hepatitis in the dog. *Journal of the American Veterinary Medical Association* 121:201–206.

Chapter 53

Toxigenic *Fusarium* Species on Annual Medic Pastures and Reduced Feed Intake by Sheep in Western Australia

M.J. Barbetti[1] and J.G. Allen[2]

[1]School of Plant Biology, Faculty of Natural and Agricultural Sciences, The University of Western Australia, 35 Stirling Highway, Crawley WA 6009, Australia; [2]Animal Health Laboratories, Department of Agriculture, Locked Bag No. 4, Bentley Delivery Centre WA 6983, Australia

Introduction

In southern Australia annual *Medicago* species are widely utilized (Chatterton and Chatterton 1996) and are an important component of agricultural systems involving self-regenerating pastures. In these systems they provide both green feed for grazing animals in the winter/spring growing season, and significant amounts of pods and seed as feed during the drier months. Feed refusal or reduced feed intake was reported in sheep on seven farms grazing senesced stands of annual *Medicago* spp. during the summer of 1989 in the wheatbelt area of Western Australia and a subsequent feeding experiment confirmed that up to 50% of sheep offered an *ad libitum* diet of *Medicago polymorpha* var. *brevispina* cv. Serena pods from one of these farms completely rejected them as a feed source (Barbetti and Allen 2005). Toxigenic *Fusarium* spp. have been shown to be associated with annual *Medicago* spp. in South Africa, and some of the isolated *Fusarium* spp. produced diacetoxyscirpenol (Lamprecht *et al.* 1986), a mycotoxin known to cause decreased feed consumption (Villar and Carson 2004). A study was undertaken by Barbetti and Allen (2005) to determine the incidence of *Fusarium* spp. associated with annual *Medicago* seed pods in Western Australia, in particular those associated with *M. polymorpha* var. *brevispina* pods where feed refusal or reduced feed intake by sheep had been evident or suspected. *Fusarium acuminatum*, *F. avenaceum*, *F. equiseti*, *F. chlamydosporum* and *F. graminearum* were all isolated from pods collected from sites where feed intake had been a problem, with the incidence of infection by *F. acuminatum* being as high as 83% (Barbetti and Allen 2005). Since there are approximately 1 million ha of annual *Medicago* pastures present in Western Australia (You *et al.* 1999), feed refusal in stock grazing them was considered a serious issue and warranted investigation of a possible toxigenic cause.

This paper reports a preliminary study of the toxigenicity of the *Fusarium* spp. isolated by Barbetti and Allen (2005) from annual *Medicago* pods associated with reduced feed intakes by sheep in Western Australia. For interest, four additional *Fusarium* isolates collected from *Trifolium subterraneum* and *Hordeum leporinum* in Western Australia were also included in this toxigenicity study.

Materials and Methods

Isolates and culture methods

Methods used to obtain the *Fusarium* isolates from annual *Medicago* pods were as published earlier by Barbetti and Allen (2005). In brief, 200 pods from each collection site were either given a vigorous wash in sterile distilled water, or placed in 70% ethyl alcohol for 30 sec, blotted dry, plated to potato dextrose agar (PDA) containing 25 ppm of aureomycin hydrochloride, and incubated at room temperature (approximately 25°C). All *Fusarium* cultures were subcultured to PDA and carnation leaf agar (Fisher *et al.* 1982), and incubated at 20°C under a light bank containing four 40-W cool white fluorescent tubes and one 40-W black light tube (Philips TL 40 W.8p RS F40 BLB) for a 12-hour photoperiod each day for up to 42 days. *Fusarium* isolates were identified using the guidelines of Burgess *et al.* (1988) and representative cultures sent to the International Mycological Institute, Kew, England for confirmation of identification. All isolates were preserved using lyophilization and have been lodged in the Mycology Collection of the Department of Agriculture Western Australia, South Perth, Western Australia.

A total of 16 *Fusarium* spp. isolates, consisting of four of *F. acuminatum,* five of *F. avenaceum,* two of *F. chlamydosporum,* two of *F. equiseti* and three of *F. graminearum,* were chosen for toxigenicity testing (Table 1). Twelve of these 16 isolates were obtained from annual *Medicago* pods using the methods described above, while three of the *Fusarium avenaceum* isolates were isolated from diseased foliage of *T. subterraneum*, and one of the *Fusarium graminearum* isolates had been isolated from a diseased crown of *H. leporinum*.

To prepare the *Fusarium* isolates for toxicity tests, five colonized PDA plugs of each isolate (4 mm in diameter) were used to inoculate 500 ml conical flasks containing sterilized millet seeds (*Panicum mileaceum* L.). Sterilized millet seed was prepared by soaking 100 g of seeds overnight in distilled water in bottles, excess water was then drained and the millet seeds autoclaved at 121°C for 20 min on 3 consecutive days. Inoculated millet seeds were incubated at 25°C on a laboratory bench for 14 days and were shaken daily to facilitate uniform colonization of the millet seed. No uninoculated millet seeds were prepared as a control.

Extraction procedure

100 g of colonized millet of each isolate was blended with 250 mL of ethyl acetate for 3 min. The resultant slurry was placed in a shaking water bath for 1 hour and then the ethyl acetate decanted off. This procedure was then repeated using 90% aqueous methanol and again using 50% aqueous methanol. The three solvent fractions were combined, filtered through Whatman No. 1 filter paper, and then evaporated down to 10 mL by rotary evaporation in a water bath at 50°C. The resultant extracts had their pH adjusted to 7.3 by the addition of a few drops of NaOH, and contained the extract from 10 g of culture in each mL.

Table 1. Details of 16 *Fusarium* species isolates used in the toxigenicity study. In addition to the 12 isolates from pods of annual *Medicago* spp., three isolates from foliage of *Trifolium subterraneum* and one from foliage of *Hordeum leporinum* were also included.

Fusarium spp.	WAC No.[a]	IMI No.[b]	Isolated from genus/species	Isolated from
F. acuminatum	5715	342250	*M. polymorpha*	pods
F. acuminatum	5716	342244	*M. polymorpha*	pods
F. acuminatum	5718	342251	*M. polymorpha*	pods
F. acuminatum	5738	342246	*M. polymorpha*	pods
F. avenaceum	7912	232897	*T. subterraneum*	leaves
F. avenaceum	7914	234383	*T. subterraneum*	leaves
F. avenaceum	7962	262019	*T. subterraneum*	stems
F. avenaceum	5585	339114	*M. polymorpha*	pods
F. avenaceum	5588	339115	*M. polymorpha*	pods
F. chlamydosporum	5710	350954	*M. polymorpha*	pods
F. chlamydosporum	5720	342257	*M. polymorpha*	pods
F. equiseti	5681	342260	*M. polymorpha*	pods
F. equiseti	5682	342253	*M. polymorpha*	pods
F. graminearum	5680	342262	*M. polymorpha*	pods
F. graminearum	5707	350956	*M. polymorpha*	pods
F. graminearum	6698	354674	*H. leporinum*	crown

[a] Western Australian Department of Agriculture Collection Number.
[b] International Mycological Institute Number.

Toxigenicity tests

Young male rats weighing between 35 and 45 g ($\bar{X} = 39$ g) were used. For each isolate, 1 mL of concentrated extract, 1 mL of a 1 in 2 dilution of the concentrated extract, or 1 mL of a 1 in 10 dilution of the concentrated extract was injected intraperitoneally into each of three rats. The resultant doses were equivalent to 256, 128, and 26 g culture kg^{-1} liveweight, respectively. The rats were observed regularly, and convulsing and/or moribund animals were euthanased. Uninjected control rats were also maintained for comparison with those injected. This experiment was conducted in 1993 in accordance with the relevant ethics approval processes of the Department of Agriculture Western Australia.

Results

All but one of the *Fusarium* isolates chosen for this toxigenicity study displayed at least some toxicity (Table 2). Overall, 77% of rats injected with the highest dose died, and half of those that died did so within 4 hours; the rest within 22 hours. Rats receiving the concentrated extract from *F. acuminatum* WAC5715 from annual *Medicago* pods died within 5 min. These rats and some others (e.g. rats receiving the concentrated extract from *Fusarium equiseti* WAC5682 from annual *Medicago* pods or *F. avenaceum* WAC7914 from *T. subterraneum*) were seen convulsing immediately prior to death. Many of the rats that survived appeared depressed for several hours after injection. No unusual signs or behavior were observed in the uninjected control rats.

Conclusion

This preliminary study demonstrated that isolates of *F. acuminatum*, *F. aveanaceum*, *F. equiseti*, *F. chlamydosporum* and *F. graminearum* from annual *Medicago* spp., *F. aveanaceum* from *T. subterraneum* and *F. graminearum* from *H. leporinum* in Western Australia were toxigenic to rats. The range of results obtained, and particularly the lack of toxigenicity of one of the *F. graminearum* isolates, suggests the extraction procedure was not extracting toxic materials intrinsically present in the millet, nor that there were harmful residues of the extraction solvents in the extracts. Hence, we believe that the absence of any extract from uninoculated millet is of little if any significance to the outcomes of this study.

Table 2. The number of deaths of young adult male rats from three dose rates (three rats per dose rate) of extracts of 16 *Fusarium* spp. isolates injected intraperitoneally

Fusarium spp.	WAC No.[a]	No. deaths at dose rate (max. = 3)		
		256 g/kg	128 g/kg	26 g/kg
F. acuminatum	5715	3	0	0
F. acuminatum	5716	2	0	0
F. acuminatum	5718	3	0	0
F. acuminatum	5738	2	0	0
F. avenaceum	7912	2	0	0
F. avenaceum	7914	3	3	3
F. avenaceum	7962	3	3	0
F. avenaceum	5585	3	0	0
F. avenaceum	5588	1	0	0
F. chlamydosporum	5710	3	0	0
F. chlamydosporum	5720	1	1	0
F. equiseti	5681	3	3	0
F. equiseti	5682	3	0	0
F. graminearum	5680	3	0	0
F. graminearum	5707	0	0	0
F. graminearum	6698	2	0	0

[a] Western Australian Department of Agriculture Collection Number

Provided that these *Fusarium* spp. produce the same mycotoxins in annual *Medicago* pods in the field as they did on millet cultures under controlled environmental conditions, and that the mycotoxins that were toxic to rats are also toxic to sheep, then the conclusion may be drawn that these *Fusarium* spp. were very likely the cause of the feed refusal and/or reduced feed intake seen in sheep in Western Australia.

Four of the *Fusarium* spp. we tested and found to be toxigenic, namely *F. acuminatum*, *F. aveanaceum*, *F. equiseti* and *F. graminearum*, have also been isolated from annual *Medicago* seeds in South Africa and shown to produce mycotoxins lethally toxic to day-old ducklings (Lamprecht *et al.* 1986). In addition the South African isolates of *F. acuminatum* and *F. aveanaceum* were lethal to sheep, and both these *Fusarium* spp. produced mainly the mycotoxin moniliformin (Lamprecht *et al.* 1986). In trials where sheep were dosed with either cultures of *F. acuminatum* or *F. aveanaceum*, or crystalline moniliformin, the main pathological change in sheep that died was degeneration of the proximal tubules of the kidneys (Lamprecht *et al.* 1986). This is extremely interesting because each year the

Animal Health Laboratories in Western Australia observes occasional unexplained cases of nephrosis in sheep during summer (JG Allen, unpublished data).

It is noteworthy that the two most toxigenic *F. avenaceum* isolates (WAC7914, WAC7962) in our study were in fact isolated from *T. subterraneum* foliage, and that a mildly toxic isolate of *F. graminearum* was isolated from foliage of *H. leporinum*. This suggests that wider investigations need to be undertaken into the association of toxigenic *Fusarium* spp. with a greater range of pasture species than just annual *Medicago* spp. in Western Australia.

References

Barbetti MJ and Allen JG (2005). Association of *Fusarium* species, with potential for mycotoxicosis, on pods of annual *Medicago* in Western Australia. *Australian Journal of Agricultural Research* 56:279–284.

Burgess LW, Liddell CM, and Summerell BA (1988). *Laboratory Manual for* Fusarium *Research*, p. 156. University of Sydney, Sydney, New South Wales.

Chatterton L and Chatterton B (1996). *Sustainable Dryland Farming*. Cambridge University Press, Cambridge.

Fisher NL, Burgess LW, Toussoun TA, and Nelson PE (1982). Carnation leaves as a substrate and for preserving cultures of *Fusarium* species. *Phytopathology* 72:151–153.

Lamprecht SC, Marasas WFO, Thiel PG, Schneider DJ, and Knox-Davies PS (1986). Incidence and toxigenicity of seedborne *Fusarium* species from annual *Medicago* species in South Africa. *Phytopathology* 76:1040–1042.

Villar D and Carson TL (2004). Tricothecene mycotoxins. In *Clinical Veterinary Toxicology* (KH Plumlee, ed), pp. 270–273. Mosby, St. Louis, Missouri.

You MP, Sivasithamparam K, Riley IT, and Barbetti MJ (1999). Relationship of rainfall, cultural practices, soil and plant nutrients and seedling survival with root disease and parasitic nematode numbers in annual *Medicago* spp. pastures. *Australian Journal of Agricultural Research* 50:977–983.

Chapter 54

Evaluation of the Possible Immunomodulatory Effects of *Ipomoea carnea*

A.O. Latorre[1], I.M. Hueza [1], M. Haraguchi [2], and S.L. Górniak [1]

[1]*Research Center for Veterinary Toxicology (CEPTOX), Department of Pathology, Faculty of Veterinary Medicine and Animal Science, University of São Paulo, Av. Prof. Orlando Marques de Paiva, 87, São Paulo, CEP 05508-900, Brazil;* [2] *Biological Institute of São Paulo, SP, CEP 04014-002, Brazil*

Introduction

The indolizidine alkaloid swainsonine (SW) is the main toxin found in *Ipomoea carnea*, a shrub which grows in Brazil and other tropical countries. This compound is a potent inhibitor of lysosomal α-mannosidase, which promotes lysosomal accumulation of incompletely processed oligosaccharides, leading to loss of cellular function and ultimately cell death (Stegelmeier *et al.* 1998). SW also inhibits Golgi mannosidase II, which is involved in N-linked glycoprotein processing (Elbein 1989). The resulting alteration in glycoprotein synthesis causes impairment and dysfunction of cell adhesion molecules and various membrane receptors (Stegelmeier *et al.* 1998). This effect may be responsible for immune system alterations; in fact, SW has potential immunomodulatory properties on different cell lines such as NK lymphocytes (Humphries *et al.* 1988) and macrophages (Das *et al.* 1995; Hueza *et al.* 2003). In addition, SW also has been suggested to protect bone marrow cells against the toxic effects of chemotherapeutic agents and radiation (Oredipe *et al.* 2003). Thus, the purposes of the present study were to investigate, in Swiss outbreed mice, the possible effect of SW on lymphoid tissues and to observe the immunomodulatory effects of SW and/or the aqueous fraction (AF) of *I. carnea* on peritoneal macrophage activity (e.g. macrophage spreading, phagocytosis, and hydrogen peroxide release).

Materials and Methods

Ipomoea carnea leaves were collected from the Research Center for Veterinary Toxicology (CEPTOX), University of São Paulo (USP) – Pirassununga, Brazil, in May 2002 and an aqueous fraction (AF) was obtained as previously described by Hueza *et al.* (2003). SW was isolated from *I. carnea* as previously described by Haraguchi *et al.* (2003) and analyzed to verify that the purity was 98%.

Evaluation of spreading and phagocytosis of peritoneal macrophages

A modification of the method described by Rabinovitch and De Stefano (1973) was used to study macrophage spreading and phagocytosis. Briefly, a total of 200 cells per slide were counted for each mouse and the macrophage spreading index (SI) and phagocytosis index (PI) were calculated as follows: SI or PI = number of spreading macrophages or phagocytic activity x 100 ÷ 200 adherent cells counted, *i.e.*, SI = percentage of spreading macrophages and PI = percentage of macrophages with phagocytized zymosan particles. The mean of four counts obtained from the two slides of each mouse was used to express the SI or PI.

Hydrogen peroxide release

Spontaneous and phorbol myristate-acetate solution (PMA)-induced H_2O_2 release by macrophages were measured by the method of Russo *et al.* (1989). H_2O_2 concentration was calculated from absorbance measurements. Spontaneous and PMA-induced H_2O_2 production experiments were repeated four times in each mouse in each group and the mean value of the four counts was used to determine H_2O_2 concentration.

Bone marrow, thymus and spleen evaluation

Thymus and spleen were harvested and weighed. A single cell suspension was prepared by crushing spleen between two pieces of ground glass. Bone marrow cell suspensions were obtained by flushing the femur cavity with ice-cold RPMI-1640 medium using a sterile syringe with a 26-gauge needle. Cell viability and cell number were determined using an hemacytometer.

Ipomoea carnea treatment

Two experiments were done. In the first, macrophage activity was determined using 32 mice randomly placed into four groups: one control (Co1) and three experimental groups (A1, A3 and A7). The experimental groups were treated by gavage for 14 consecutive days with AF: 1.0 (A1), 3.0 (A3) and 7.0 (A7) $g{\cdot}kg^{-1}$ of extracted *I. carnea* leaves (dry weight). The Co1 group received only tap water during the same period. Seven days before the end of the experimental period all animals received an intraperitoneal (ip) injection of *Bacillus* Calmette Guerin (BCG) in order to promote peritoneal macrophage activation.

In the second experiment, bone marrow cellularity and thymus and spleen cell viability were determined. Thirty mice were randomly assigned to three groups: one control (Co2) and two experimental groups (B1 and B3) that were treated by gavage for 14 consecutive days with the AF of *I. carnea*: 1.0 (A1) and 3.0 (A3) $g\ kg^{-1}$. The Co2 group received only tap water.

Swainsonine treatment

In the first experiment, macrophage activity was determined with 20 mice randomly placed in two groups: one control (C1) and one experimental group (Sw1) that received twice-daily ip injections (1.0 mg kg^{-1} per day) of SW solution for 8 days. The daily injections of SW were spaced 8 hours apart. Five days before the end of the experimental period all animals received an ip injection of thioglycolate (3%) to elicit macrophages. In the second experiment, bone marrow cellularity and thymus and spleen cell viability were assessed as previously described. Twenty mice were randomly divided into two groups: one control (C2) and one experimental group (Sw2). The SW treatment described above was employed. Animals from C2 group received only the vehicle, i.e. PBS, for the same period.

Histopathological evaluation

On the last day of the experiments, all animals were humanely euthanized and necropsied for sample collection of central nervous system, thyroid, sternum bone marrow, lung, liver, pancreas, spleen and kidneys. The samples were fixed in 4% neutral buffered formalin, embedded in paraffin, sectioned into 5 μm thickness, and stained with hematoxylin and eosin (H&E).

Statistical analysis

Experimental and control data were analyzed statistically by analysis of variance (ANOVA). Significant F tests were followed by either Students's *t*-test (one treated group and one control) or Dunnett's *post hoc* test to compare $\geq$ two treated groups against a control. Data are expressed as mean ± SEM.

Results

The peritoneal macrophages of mice treated with AF did not show any differences in their activity. Similarly, no statistical differences ($P>0.05$) were observed in both cellularity and relative weight of lymphoid tissues when treated animals were compared to controls. The histopathology revealed the presence of vacuolar lesions on pancreas, liver and kidneys of mice treated with 7.0 g kg^{-1} of AF of *I. carnea* (Figure 1).

SW did not significantly affect phagocytosis. Conversely, we observed an enhancement of both spreading activity and PMA-induced H_2O_2 production of peritoneal macrophages in mice treated with SW ($P<0.05$) (Table 1). No statistical differences were observed in lymphoid weight and cellularity of mice from all groups (Table 2).

Discussion and Conclusion

There are several *in vitro* investigations showing that SW stimulates the immune system in different effector cells, such as murine NK cells (Humphries *et al.* 1988), lymphokine activated killer (LAK)-cell (Bowlin *et al.* 1989) and peritoneal macrophages (Das *et al.* 1995). In addition, studies performed in rats given AF of *I. carnea* revealed that macrophages of these animals showed enhanced phagocytosis and H_2O_2 production (Hueza *et al.* 2003). In the present study SW treatment had stimulatory effects on macrophage spreading and PMA-induced H_2O_2 production by peritoneal cells, although no stimulatory effects were observed in those mice treated with AF of *I. carnea*. *Ipomoea carnea* contains not only SW but also the nortropane alkaloids calystegines B_2 and C_1 (De Balogh *et al.* 1999). Calystegines are potent inhibitors of β-glucosidase, and calystegine B_2 is also a potent inhibitor of α-galactosidase enzymes (Asano *et al.* 2000), suggesting that this lack of stimulatory effects may be linked to the effects of the nortropane alkaloids.

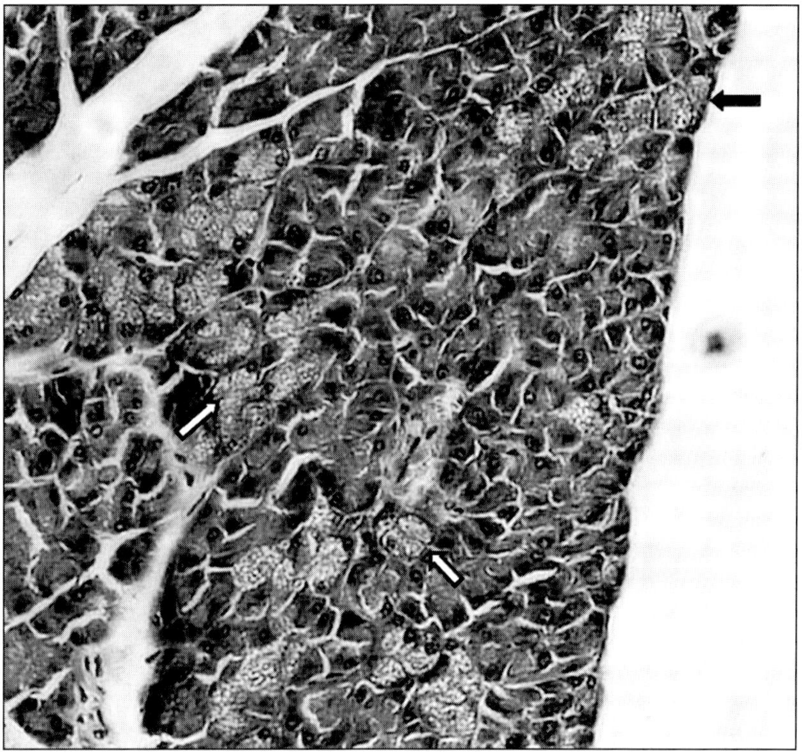

Figure 1. Pancreas tissue from mouse treated by gavage with aqueous fraction equivalent to 7.0 $g \cdot kg^{-1}$ bw from dry leaves of *Ipomoea carnea* during 14 days. Note the presence of vacuolar lesions on acinar cells (arrows). H&E, 10x.

Table 1. Effects of SW administrated to mice on macrophage activities: spreading, phagocytosis and spontaneous and PMA-induced H_2O_2 release.

Groups	PI %	SI %	H_2O_2 (nMols)	PMA-induced H_2O_2 (nMols)
C1 (n=10)	76.7	47.9	0	0.2
SW1 (n=10)	79.2	61.5*	0	0.5*

SI = percentage of spreading macrophages; PI = percentage of macrophages with phagocytized zymosan particles.
*$P<0.05$. Student's *t* test.

Oredipe *et al.* (2003) showed unequivocally that SW protects bone marrow cells and the hematopoietic system against toxic effects of chemotherapeutic agents and radiation. In the Oredipe *et al.* (2003) study, SW promoted an increase in bone marrow cellularity on inbred mice (C57BL/6; C3H-HEN; Balb/C and DBA-2). Conversely, we found no differences in cell variables from lymphoid organs (*i.e.*, bone marrow and spleen) in control and experimental mice. Because we used Swiss mice as experimental animals (*i.e.*, an outbreed strain), these results also suggest that wild strains may be less sensitive than inbreed mice to the effects of SW in bone marrow.

Table 2. Effects of SW administrated to mice on bone marrow and spleen cellularity and also on lymphoid organ weights (thymus and spleen).

Groups	Bone marrow Cells x 10^7	Spleen Cells x 10^7	Thymus g 100^{-1} g bw	Spleen g 100^{-1} g bw
C2 (n=10)	5.4	15.3	0.08	0.31
SW2 (n=10)	5.7	15.8	0.08	0.35

In conclusion, contrary to earlier studies with SW, the experiments conducted with *I. carnea* AF in Swiss mice did not show alterations in the immunomodulatory response. Hence, future experiments will be conducted in order to verify if this discrepancy between results is related to the other toxic compounds in *I. carnea* (i.e. nortropane alkaloids), or to animal species (inbred and outbred) or perhaps an interaction between these two factors.

Acknowledgements

This study was supported by grants from Fundação de Amparo à Pesquisa do Estado de São Paulo – FAPESP, Brazil (Proc. 02/02758-3).

References

Asano N, Nash RJ, Molyneux RJ, and Fleet GWJ (2000). Sugar-mimic glycosidase inhibitors: natural occurrence, biological activity and prospects for therapeutic application. *Tetrahedron Asymmetry* 11(8):1645–1680.

Bowlin TL, McKown BJ, Kang MS, and Sunkara PS (1989). Potentiation of human lymphokine-activated killer cell activity by swainsonine, an inhibitor of glycoprotein processing. *Cancer Research* 49(15):4109–4113.

Das PC, Roberts JD, White SL, and Olden K (1995). Activation of resident tissue-specific macrophages by swainsonine. *Oncology Research* 7(9):425–433.

De Balogh KIM, Dimande AP, van der Lugt JJ, Molyneux RJ, Naudé TW, and Welman WG (1999). A lysosomal storage disease induced by *Ipomoea carnea* in goats in Mozambique. *Journal of Veterinary Diagnostic Investigation* 11(3):266–273.

Elbein AD (1989). The effects of plant indolizidine alkaloids and related compounds in glycoprotein processing. In *Swainsonine and Related Glycosidase Inhibitors* (LF

James, AD Elbein, RJ Molyneux and CD Warren, eds), pp. 87–155. Iowa State University Press, Ames.

Haraguchi M, Górniak SL, Ikeda K, Minami Y, Kato A, Watson AA, Nash RJ, Molyneux RJ, and Asano N (2003). Alkaloidal components in the poisonous plant, *Ipomoea carnea* (Convolvulaceae). *Journal of Agriculture and Food Chemistry* 51(17):4995–5000.

Hueza IM, Fonseca ESM, Paulino CA, Haraguchi M, and Górniak SL (2003). Evaluation of immunomodulatory activity of *Ipomoea carnea* on peritoneal cells of rats. *Journal of Ethnopharmacology* 87(2-3):181–186.

Humphries MJ, Matsumoto K, White R, Molyneux RJ, and Olden K (1988). Augmentation of murine natural killer cell activity by swainsonine, a new antimetastatic immunomodulator. *Cancer Research* 48(6):1410–1415.

Oredipe OA, Furbert-Harris PM, Green WR, White SL, Olden K, Laniyan I, Parish-Gause D, Vaughn T, Griffin WM, and Sridhar R (2003). Swainsonine stimulates bone marrow cell proliferation and differentiation in different strains of inbred mice. *Pharmacological Research* 47(1):69–74.

Rabinovitch M and De Stefano MJ (1973). Macrophage spreading *in vitro*: I. inducers of spreading. *Experimental Cell Research* 77(1):323–334.

Russo M, Teixeira HC, Marcondes MCG, and Barbuto JA (1989). Superoxide-independent hydrogen peroxide release by activated macrophages. *Brazilian Journal of Medical and Biological Research* 22(10):1271–1273.

Stegelmeier BL, Snyder PD, James LF, Panter KE, Molyneux RJ, Ralphs MH, and Pfister JA (1998). The immunologic and toxic effects of locoweed (*Astragalus lentiginosus*) intoxication in cattle. In *Plant Toxins: Proceedings of the Fifth International Conference on Poisonous Plants* (TR Garland, MR Bailey, KE Panter and PR Dorling, eds), pp. 285–290. CABI Publishing, Wallingford, Oxon, UK.

Chapter 55

Structural and Functional Changes in Organs and Cells of the Immune System in *Solanum glaucophyllum*-intoxicated Heifers

P.A. Fontana[1], E.J. Gimeno[1,3], G. Laguens[2], C.G. Barbeito[1,3], W. De Girolamo[2], E.F. Costa[1], S. Coronato[2], and E.L. Portiansky[1,3]

[1]Institute of Pathology, School of Veterinary Sciences, National University of La Plata (UNLP), 60 and 118, (1900) La Plata, Buenos Aires, Argentina; [2]Department of Pathology B, School of Medical Sciences UNLP; [3]Members of CONICET (National Scientific Research Council)

Introduction

Enzootic calcinosis (EC) is an intoxication of bovine livestock caused by the chronic ingestion of *Solanum glaucophyllum* (*Sg*) also known as 'Duraznillo blanco' (Collier 1927; Worker and Carrillo 1967; Puche and Bingley 1995; Mello 2003). *Solanum glaucophyllum* is a plant with calcinogenic activity that contains high levels of glycoside conjugates with 1,25-dihydroxyvitamin D_3 or calcitriol (Corradino and Wasserman 1974).

The 1,25-dihydroxyvitamin D_3, an activated form of Vitamin D, is mainly related with calcium and bone metabolism. Excess calcitriol stimulates calcium absorption thus inducing hypercalcemia. Since the excessively absorbed mineral cannot be physiologically accommodated, there is an increase in its deposition in soft tissues, resulting in calcinosis (Puche and Bingley 1995). It was suggested that 1,25-dihydroxyvitamin D_3 induces mesenchymal cell differentiation and the synthesis of a calcifiable matrix (Barros and Gimeno 2000). Calcium salts are deposited preferentially at the large vessels, heart, lungs, kidneys, tendons and ligaments.

Clinical signs of intoxicated animals include stiffness and painful gait, reduction of weight with loss of body condition, raised pulse and increase respiratory rate as well as decreased vitality (Worker and Carrillo 1967). These symptoms can be partially explained by the lesions found in intoxicated bovines.

However, some of the EC signs still remain unclear. Calcitriol has, in addition to its well-known functions on mineral metabolism, several effects on growth and differentiation of many cell types and marked immunomodulatory properties (Manolagas *et al.* 1985; De Luca and Cantorna 2001; Mathieu *et al.* 2004). These biological effects are mediated by the vitamin D receptor. The presence of this receptor has been found in most cell types of the immune system, particularly in antigen presenting cells (APC), such as macrophages (Provvedini *et al.* 1983) and dendritic cells (DC) (Brennan *et al.* 1987), as well as in both

$CD4^+$ and $CD8^+$ T lymphocytes (Veldman *et al.* 2000). Here we describe the alterations in the morphological and functional characteristics of the organs related to the immune system in *Sg*-intoxicated cattle.

Materials and Methods

Plan

The *Sg* leaves were collected from EC incidence zones. The leaves were air dried, powdered and orally administrated with tap water in individual doses of 25 g kg^{-1} twice a week.

Animals – Experimental design

Two-year-old Aberdeen Angus heifers were dosed with during 15, 30 and 60 days (experimental groups). A group was dosed for 15 days and sacrificed at 60 days post intoxication (recovery group). Control animals only received tap water during the same period and were sacrificed at 0 and 60 days. Weekly control tests were performed to determine EC classical signs. The animals were fed with hay and water *ad libitum.*

Cells and tissue samples – Techniques

Sterile peritoneal fluid containing macrophages was extracted immediately after sacrifice of the animals to perform a phagocytosis test that corroborates the functional capacity of macrophages. Briefly, peritoneal fluid macrophages of each animal were incubated with a suspension of *Candida* sp. Two hundred cells per slide were counted and the phagocytosis index (PI) calculated (PI= percentage of macrophages with phagocytic activity).

Fresh samples as well as frozen sections of subiliac lymph nodes were taken for immunodetection of DC in spots, using rabbit anti-cow monoclonal S100 protein antibody (DakoCytomation, Carpinteria, CA, USA). The immunohistochemical detection system was a dextran polymer-based method (Universal EnVision®System, DakoCytomation); 3-amino-9-ethyl-carbazol was employed as a chromogen while hematoxylin was used for counterstaining. The slides were mounted in aqueous medium Aquamount® (Lerner Laboratories, Pittsburgh, PA, USA).

Samples of spleen, thymus and subiliac and prescapular lymph nodes were fixed in buffered formalin paraffin embedded and cut in 5 µm-thick section. The slides were routinely stained for general observations. Morphometric analysis was performed in all processed sections. The lectinhistochemical study, to determinate the carbohydrate expression pattern of tissues, was also performed with seven biotinylated lectins: PNA, UEA-I, CON-A, SBA, RCA-I, WGA (VectorLab, Burlingame, CA, USA). The reactions were amplified with ABC complex. The 3,3′diaminobenzidine tetrahydrocloride (DAB)(DakoCytomation) was used as a chromogen and hematoxylin used for counterstaining. The intensity of the structures observed was graded using a semiquantitative scale from 0 to 3.

Results

The phagocytic activity of the peritoneal macrophages was progressively reduced in the intoxicated animals with time.

The main histological findings in lymph nodes were a decrease in the cortex width and a reduction in the lymph nodules and germinal centers (GC) size, the latest presenting lymphocyte and tingible-body macrophages depletion (Figure 1 A–B). Moreover, the medullar region showed dilatation of sinus and a decrease in the number of its cells. In the recovery group, a marked follicular hypertrophy with GC containing numerous debris-laden macrophages and sinus histiocytosis was observed. The thymus showed atrophy with reduction of the parenchyma and a relative increase of adipose and connective tissue per unit area. In addition, we also observed a progressive reduction of the cortical region. In the spleen no particular changes were observed in the number of lymphoid follicles nor in their size.

Lectinhistochemical studies revealed that normal lymphatic tissues had a great reactivity with PNA (Figure 1 C–D). The remaining lectins reacted with supporting tissues of the lymphatic organs but not on cells as PNA did. PNA bound to T- and B-dependent areas in spleen and lymph nodes and to the thymic cortex. In *Sg*-intoxicated animals even though the intensity of staining never decreased in all studied organs, the lectinhistochemical stained area decreased with intoxication time. Recovery group animals showed similar results to those observed in control heifers. A significant decrease in the number and size of DC was observed in isolated cell cultures as well in the lymph node sections (Figure 1 E–F).

Discussion

We describe some functional and structural modifications of the cell belonging to and associated with the immune system in *Sg*-intoxicated heifers that contribute to the negative influence of calcitriol on cellular immune responses.

The progressive reduction of macrophages phagocytosis activity observed in intoxicated animals was consistent with *in vitro* studies where calcitriol substantially decreased endocytosis of cytokine-treated monocytes (Berer *et al.* 2000). However, other authors (Hubel *et al.* 1991; Mathieu *et al.* 2000) showed different results perhaps due to different experimental conditions. Moreover, we found a decrease in the number and size of DC in isolated cultures and tissues with the S-100 protein. APC and notably DC are key targets of calcitriol, which may induce tolerogenic DC by inhibiting differentiation and maturation of these cells (Penna and Adorini 2000; Adorini *et al.* 2004). DC conducts the response of lymphocytes by secreting chemokines with different specificity and function. Several studies have shown that *in vitro* treatment of DC with 1,25-dihydroxyvitamin D_3 leads to a downregulated expression of different costimulatory molecules, with decreased IL-12 and enhanced IL-10 production. This results in a marked decrease in T cell stimulatory capacity, with negative consequences for T cell activation. Although direct effects on T lymphocytes have also been observed, it is mainly by this indirect pathway that calcitriol modulates the fate and function of $CD4^+$ T lymphocytes (Mathieu *et al.* 2004). In this way, B lymphocyte number can also be controlled since these cells are dependent on the proliferation of $CD4^+$ (Liu and Arpin 1997).

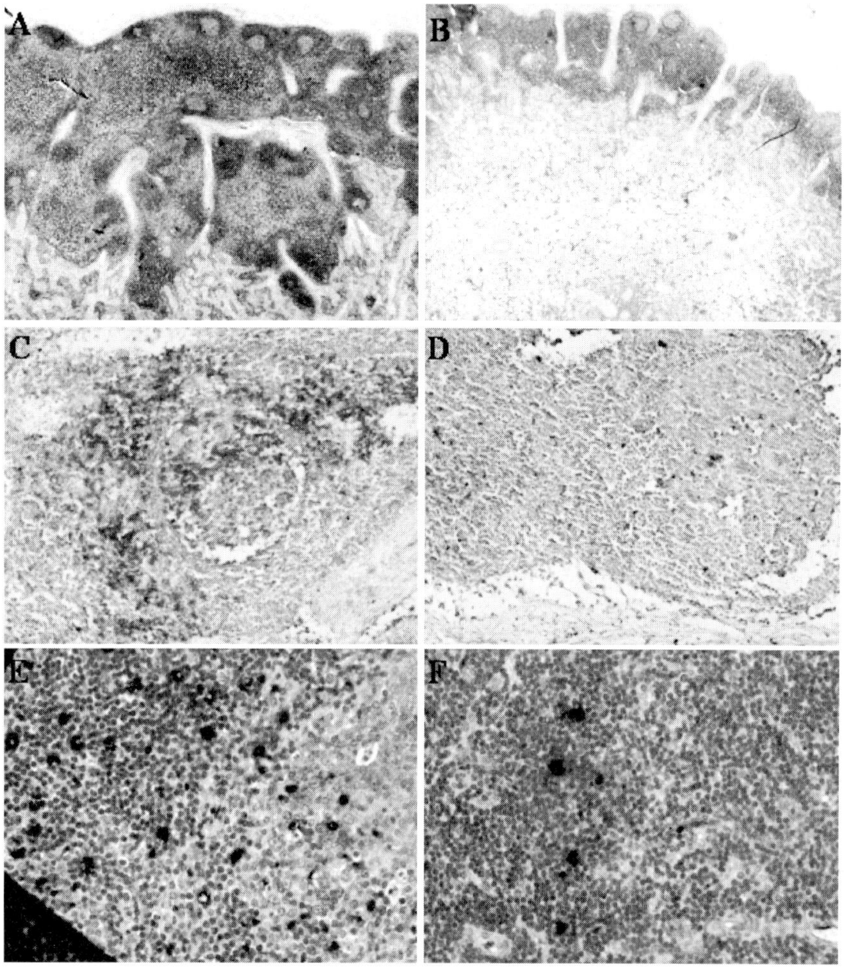

Figure 1. Effect of *Sg* on cattle lymph nodes. **A**: Normal lymph node. **B**: Dosed heifers (60 days) showing a marked reduction in the cortex region. Magnification 12.5x. **C**: Lectinhistochemical staining with PNA of control animals showing a widespread labeling of B- and T-dependent zones. **D**: Decrease in the area and intensity of staining of 60-day *Sg*-intoxicated animals, lymph node incubated with PNA. Magnification 100x. **E**: Immunohistochemistry anti-S100 in normal animals. **F**: The same antibody against *Sg*-dosed heifers. Note the difference in staining and number of DC detected. Magnification 400x.

This mechanism may explain our findings in lymphoid tissues showing a reduction in cortex width and in the number of lymphoid and non-lymphoid cells. It also explains the decrease in the lectinhistochemical stained area of PNA, a lectin with probed affinity to several cell populations of the immune system, in T- and B-dependent zones (Rose and Malchiodi 1981). The progressive decrease of cortical thymic areas with increased dose may also be related to these findings, although hypercalcemia and not 1,25-dihydroxyvitamin D_3 itself could be the main cause (Mohamed *et al.* 1996).

This study demonstrates the negative influence on the immune system and related cells that occur in EC. This may explain the enhanced sensitivity to infectious disease of the *Sg*-intoxicated animals (Costa and Gomar, personal communication) and consequently, some of the non-specific signs of this illness that are not clear.

Acknowledgements

The authors would like to thank Mrs Rosa Villegas and Mr Ruben Mario for their technical assistance. Financial support was provided in part by grants from 'Agencia Nacional de Promoción Científica', 'Consejo Nacional de Investigaciones Científicas (CONICET)' and 'Academia Nacional de Agronomía y Veterinaria', Argentina.

References

Adorini L, Penna G, Giarratana N, Roncari A, Amuchastegui S, Daniel KC, and Uskokovic M (2004). Dendritic cells as key targets for immunomodulation by Vitamin D receptor ligands. *Journal of Steroid Biochemistry and Molecular Biology* 89-90:437–441.

Barros SS and Gimeno EJ (2000). Cell differentiation and bone protein synthesis in the lungs of sheep with spontaneous calcinosis. *Journal of Comparative Pathology* 123:270–277.

Berer A, Stockl J, Majdic O, Wagner T, Kollars M, Lechner K, Geissler K, and Oehler L (2000). 1,25-dihydroxyvitamin D_3 inhibits dendritic cells differentiation and maturation in vitro. *Experimental Hematology* 28:575–583.

Brennan A, Katz DR, Nunn JD, Barker S, Hewison M, Fraher LJ, and O'Riordan JL (1987). Dendritic cells from human tissues express receptors for the immunoregulatory Vitamin D3 metabolite, dihydroxychole-calciferol. *Immunology* 61:457–461.

Collier WA (1927). Zür Kenntnis einer als Enteque bezeichneten Krankheit der Rind in der Provinz Buenos Aires. *Zeitschrift für Infektionskrankheiten, Parasitäre Krankeiten und Hyginie der Haustiere* 31:81–92.

Corradino RA and Wasserman RH (1974). 1,25-dihydroxycholecalciferol-like activity of *Solanum malacoxylon* extract on calcium transport. *Nature* 252:716–718.

De Luca HF and Cantorna MT (2001). Vitamin D: its role and uses in immunology. *FASEB Journal* 15:2579–2585.

Hubel E, Kiefer T, Weber J, Mettang T, and Kuhlmann U (1991). In vivo effect of 1,25-dihydroxyvitamin D_3 on phagocyte function in hemodialysis patients. *Kidney International* 40:927–933.

Liu Y-J and Arpin C (1997). Germinal center development. *Immunological Reviews* 156:111–126.

Manolagas SC, Provvedini DM, and Tsoukas CD (1985). Interactions of 1,25-dihydroxyvitamin D_3 and the immune system. *Molecular and Cellular Endocrinology* 43:113–122.

Mathieu C, Van Etten E, Kato S, Verstuyf A, Laureys J, Depovere J, Valckx D, and Bouillon R (2000). In vitro and in vivo analysis of the immune system of VDR-KO mice. In *Vitamin D Endocrine System: Structural, Biological, Genetic and Clinical Aspects* (AW Norman, R Bouillon, and M Thomaset, eds), pp. 555–559. University of California Press.

Mathieu C, Van Etten E, Decallonne B, Guilietti A, Gysemans C, Bouillon R, and Overbergh R (2004). Vitamin D and 1,25-dihydroxyvitamin D_3 as modulators in the immune system. *Journal of Steroid Biochemistry and Molecular Biology* 89–90:449–452.

Mello JR (2003). Calcinosis. Calcinogenic plants. *Toxicon* 41:1–12.

Mohamed MI, Beckman MJ, Meehan J, and De Luca HF (1996). Effect of 1,25-dihydroxyvitamin D_3 on mouse thymus: role of extracellular calcium. *Biochemistry and Biophysical Acta* 1289:275–283.

Penna G and Adorini L (2000). 1,25-dihydroxyvitamin D_3 inhibits differentiation, maturation, activation and survival of dendritic cells leading to impaired alloreactive T cell activation. *Journal of Immunology* 164:2405–2411.

Provvedini DM, Tsoukas CD, Deftos LJ, and Manolagas SC (1983). 1,25-dihydroxyvitamin D_3 receptors in human leukocytes. *Science* 221:1181–1183.

Puche R and Bingley J (1995). *Solanum glaucophyllum* Deft. In *Calcinosis of Cattle in Argentina*, First English Ed. (Rosario, ed), pp. 23–32. Universidad Nacional de Rosario.

Rose ML and Malchiodi F (1981). Binding of peanut lectin to thymic cortex and germinal centres of lymphoid tissue. *Immunology* 42:583–591.

Veldman CM, Cantorna MT, and De Luca HF (2000). Expression of 1,25 dihydroxyvitamin D_3 receptor in the immune system. *Archives of Biochemistry and Biophysics* 374:334–338.

Worker NA and Carrillo BJ (1967) 'Enteque seco', calcification and wasting in grazing animals in Argentina. *Nature* 215:72–74.

MANAGEMENT

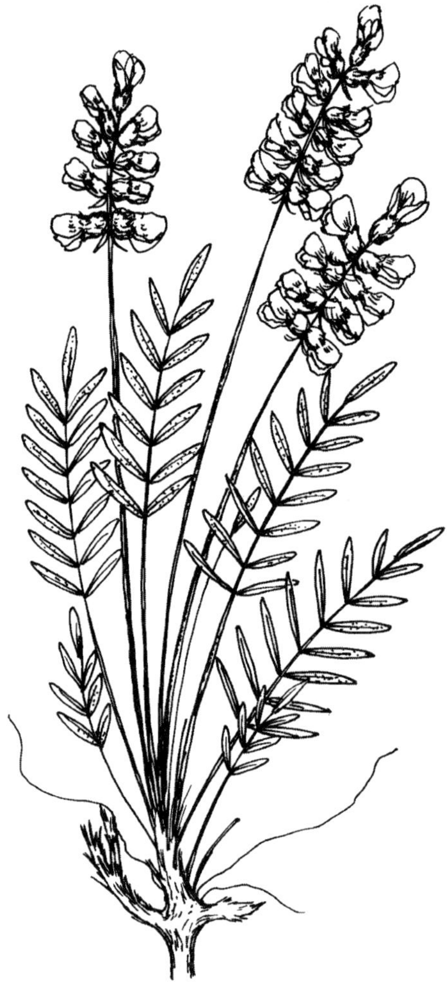

OXYTROPIS SERICEA

Chapter 56

Implications of Linking Herbivore Experience, Varied Diets, and Plant Biochemical Diversity for Affecting Ingestion of Plants with Toxins by Herbivores

F.D. Provenza

Department of Forest, Range, and Wildlife Sciences, Utah State University, Logan, Utah, 84322-5230, USA

Introduction

Historically, we believed only a subset of plants – poisonous plants – contained compounds potentially toxic to herbivores, but during the past 25 years we have come to appreciate that all plants contain toxins, even those we grow in our gardens (Provenza 2003). Thus, animals foraging on pastures and rangelands must cope with toxins every day of their lives. Foraging behavior is a function of its consequences and ingesting foods results in postingestive consequences that affect the kinds and likelihoods of future foraging behaviors (Provenza and Villalba 2006). Animals likely experience the postingestive consequences of their behaviors, for instance well-being following a nutrient boost or illness following nutrient excesses or toxicosis, through body mapping by the central nervous system that links life-sustaining processes to feelings (Damasio 2003). Nerve endings in every region of the internal milieu provide detailed mapping simultaneously of the current state of cells that feed back to the central nervous system. At the most fundamental levels, this involves metabolic regulation, basic reflexes, and immune responses. At somewhat higher levels, organisms experience the consequences of their behaviors as a continuum from pain (aversive) to pleasure (positive). At yet higher levels, drives and motivations (incentives) emanate from the consequences of past behaviors that influence the likelihood of future behaviors. At the highest levels, the experience of emotions and ultimately feelings arise from different degrees of well-being emanating from ongoing interactions among cells, organs, individuals, social and biophysical environments. Given this degree of body mapping and the importance of postingestive feedback in well-being, why do animals die from over-ingesting plants that contain toxins?

The Toxin Challenge and Nutrient–Toxin Interactions

Any substance can be harmful or beneficial depending on the dose and the physiological condition of an animal. At too high doses, nutrients are toxic, whereas at the appropriate dose, toxins can have medicinal benefits (Provenza and Villalba 2006). Moreover, as opposed to killing animals outright, most toxins set limits on the amount of a particular food an animal can eat through feedback mechanisms that link cells and organs of the body with the palate (Provenza 1995). The net result is that animals must eat a variety of foods containing different kinds and amounts of toxins to meet their needs for nutrients (Freeland and Janzen 1974).

Biochemical diversity is critical because nutrients and toxins interact with one another – nutrients with nutrients, nutrients with toxins, and toxins with toxins (Provenza *et al.* 2003a). The amount of toxin an animal can ingest depends on the kinds and amounts of toxins and nutrients in the forages on offer. The general mechanism of detoxification involves converting more toxic lipophilic compounds to less toxic water-soluble compounds that can be excreted in the urine (Cheeke 1998). These transformations require nutrients such as protein and energy that otherwise would be available for maintenance and production (Illius and Jessop 1995, 1996).

Animals fed adequately can eat more foods with toxins. Sheep eat more food containing LiCl as the energy content of their diet increases (Wang and Provenza 1997). Supplemental energy and protein increase the ability of sheep and goats to eat foods with terpenes (Banner *et al.* 2000; Villalba *et al.* 2002a), tannins (Villalba *et al.* 2002b), and saponins (Williams *et al.* 1992; Martinez *et al.* 1993). In contrast, herbivores eat less food with toxins when nutrients such as sodium are inadequate (Freeland *et al.* 1985; Freeland and Choquenot 1990). Nutritional needs evidently depend on the particular toxin. Lambs infused with terpenes, nitrates, tannins, or lithium chloride selected diets with higher ratios of protein:energy, whereas lambs infused with cyanide preferred foods with lower ratios of protein:energy (Villalba *et al.* 2002c).

Finally, while the aforementioned findings support the thesis that feeding decisions depend on an animal's capacity to detoxify plant toxins, they do not provide direct tests, which have been difficult because we typically do not know the specific mechanisms the body uses to detoxify toxins. Recently, more direct tests have been conducted with brushtail possums (Marsh *et al.* 2005). Possums supplemented with glycine metabolize benzoic acid faster, and in response eat more; they detoxify benzoic acid primarily by conjugating it with glycine to form benzoyl glycine (hippuric acid). Moreover, given a choice possums select a diet with both benzoate and glycine over diets with a high concentration of just one of these supplements. Their ability to regulate intake of benzoate and glycine offered separately or mixed together suggests they recognize amino acid or benzoate excesses and modify their feeding accordingly.

Food intake and preferences depend on consequences manifest as the interrelationship between a food's flavor and its postingestive feedback (Provenza 1995). The senses of smell, taste, and sight enable animals to discriminate among foods and provide pleasant sensations – liking for a food's flavor – associated with eating. Postingestive feedback from nutrients and toxins calibrates sensory experiences – like or dislike – in accord with a food's utility to the body. Feedback from the 'body' to the 'senses' is critical for health and well-being. Bodies are integrated societies of cells that interact one with another and with the external environment through feedback mediated by nerves, neurotransmitters, and hormones. Feedback from the body to the palate is how societies of cells and organs influence which foods, and how much of those foods, are eaten by an animal.

So how can this help to explain why animals eat poisonous plants? Breakdowns in flavor–feedback interactions are one way to explain why animals die from over-ingesting poisonous plants. They are likely to occur under a number of conditions. When aversive feedback from toxins in foods does not occur, or is delayed beyond 10 to 12 hours, the likelihood of an animal associating a food with its consequences is greatly diminished (Burritt and Provenza 1991). D-con kills mice by providing immediate positive postingestive feedback from nutrients and delayed (several days) negative effects from toxins. This is likely to be true with any poisonous plant that causes delayed poisoning and with plants that cause teratogenic effects (Provenza 1997). There also is evidence that nausea is critical in the formation of food aversions, and any food that contains compounds that circumvent the emetic system is likely to be ingested (Garcia *et al.* 1985; Garcia 1989; Provenza *et al.* 1994; Dziba *et al.* 2005). Thus, ailments such as photosensitization that affect the skin and bloat that causes distention are not likely to condition food aversions. There are other factors and situations, described in what follows, that can either increase or decrease the likelihood of poisoning.

Uniqueness of the Individual and the Spice of Life

Food preferences depend on how individuals are built morphologically and how they function physiologically, and foraging decisions are affected by differences in organ mass and how animals metabolize nutrients and toxins (Konarzeski and Diamond 1994). Importantly, marked variation is common even among closely related animals in needs for nutrients and abilities to cope with toxins. From the standpoint of nutrients, for instance, the same dose of sodium propionate that conditions preferences in some lambs, conditions aversions in others (Villalba and Provenza 1996). Likewise, lambs given a choice of barley (high energy) and alfalfa (high protein) vary greatly in their preferences (Scott and Provenza 1999). The same is true with toxins. Doses of tannins that condition aversions in some goats do not deter others (Provenza *et al.* 1990), and some sheep fed high levels of *Galega officinalis* failed to show any symptoms of toxicosis, whereas others were killed by a low dose (Keeler *et al.* 1988). Differences in ability to tolerate the same dose of a toxin and in metabolism of nutrients and toxins guarantees no one diet will be ideal for every individual. This has two implications for poisonous plants.

First, it allows people to select for animals that are locally adapted. An animal's foraging behavior is a function of its evolutionary history, genetically expressed, in concert with its uniquely individualistic history of the social and biophysical environments where it was conceived and reared. Thus, selecting for local adaptation is a way to match animals to local landscapes. Differences in abilities to metabolize nutrients and toxins are the basis for selection, be that by nature or man, in ways that match animals to landscapes. Though no studies have been conducted, that process likely occurs in situations where animals are challenged to eat diets high in toxins for prolonged periods. For instance, some moose populations in Norway winter at high elevations, rather than along the coast where conditions are more favorable, due to predation by humans that began thousands of years ago (Andersen 1991). Moose that winter at high elevations eat large amounts of conifer trees high in various secondary metabolites. Undoubtedly, there has been strong selective pressure for individuals with the ability to tolerate ingesting these metabolites. Similar selective pressure has been imposed by people whose cattle forage on diets high in endophyte-infected tall fescue, high in alkaloids, and several anecdotal accounts suggest some producers have selected for animals with the ability to ingest the alkaloids in fescue.

Second, diets and habitats that allow animals to select among alternatives enable individuals to better meet needs for nutrients and to better cope with toxins. For instance, herders in France use understanding of the uniqueness of the individual and biochemical diversity to stimulate ingestion of different plants and more fully use the range of plants available by herding in grazing circuits (Provenza *et al.* 2003a). The circuit includes a moderation phase, which provides sheep access to plants that are abundant but not highly preferred to calm a hungry flock; the next phase is a main course for the bulk of the meal with plants of moderate abundance and preference; then comes a booster phase of highly preferred plants for added diversity; and finally a dessert phase of palatable plants that complement previously eaten forages. Daily grazing circuits are designed to stimulate and satisfy an animal's appetite for different nutrients, and they enable animals to maximize intake of nutrients and regulate intake of different toxins.

Experience, Variety, and the Toxin Challenge

Experience affects an animal's propensity to eat different foods. When herbivores can select only the most preferred plants, they are not likely to learn to mix foods high in nutrients with foods that contain toxins, but when they are encouraged to eat all plants in an area they are more likely to learn to eat mixtures of plants that mitigate toxicity (Provenza *et al.* 2003a). Experienced animals who have learned to eat a variety of foods that differ in nutrients and toxins do so even when nutritious alternatives are available, whereas naive animals familiar only with the nutritious alternatives eat only that subset of familiar foods (Villalba *et al.* 2004).

Importantly, different systems of grazing management influence how animals learn to forage (Provenza *et al.* 2003a). Continuous grazing at low stock densities encourages selectivity and reduces diet and habitat breadth, whereas short-duration grazing at high stock densities increases diet and habitat breadth. Thus, what was traditionally considered proper grazing management – rotational grazing at low stock densities – may have trained generations of livestock to 'eat the best and leave the rest' rather than to 'mix the best with the rest' thus inadvertently accelerating an increase in the abundance of plant species that are less desirable.

For example, Ray Banister, who manages 7200 acres of land in eastern Montana, has modified his grazing management from reliance on rotational grazing to boom–bust grazing involving intense periods of grazing during the growing season followed by two growing seasons of rest (Ray Bannister, personal communication). Such intense grazing ensures that less palatable and weedy plants do not acquire a competitive advantage over more palatable species. The ranch has some of the highest vegetation cover and diversity in eastern Montana (Joe Fidel, Natural Resource Conservation Service, personal communication). The change from rotational to boom–bust grazing meant cattle could no longer select only the most preferred plants. During the 3-year transition his cows learned to eat shrubs such as snowberry and sagebrush, which are usually considered unpalatable due to their levels of toxins, and they evidently learned to eat palatable and unpalatable species simultaneously to mitigate the aversive effects of toxins.

The foreman at the Rex Ranches in Nebraska reports the same thing for leafy spurge (John Young, personal communication). While most people know sheep and goats eat leafy spurge readily, few believe that cattle will eat it too. According to John Young, four factors are critical: (1) the Rex Ranches use high stock densities for short periods to encourage cattle to eat a variety of plants; (2) leafy spurge occurs in patches, it does not dominate the landscape, so cattle do not have to eat only spurge; (3) cattle have a variety of other plants from which to choose – leafy spurge is only one part of their diet; and (4) the Rex Ranches keep their replacement heifers so young animals learn to eat leafy spurge from their mothers – it has become part of the culture. Undoubtedly, there are critical thresholds of herbivore knowledge and plant diversity below which plant diversity and animal performance decline at ever increasing rates, and above which diversity and performance can beget greater diversity and enhanced performance (Provenza *et al.* 2003b).

Conclusions

Animals must cope with toxins every day of their lives, and for the most part they are successful at selecting diets that minimize toxicity when they: (1) are on an adequate plane of nutrition; (2) have a variety of foods from which to choose; and (3) have learned how to mix and match their diets from the foods available in an environment (Provenza *et al.* 1992, 2003a). Diets and habitats that allow animals to select among alternatives enable individuals to better meet needs for nutrients and to better cope with toxins. All plants contain toxins, and the amount of toxin an animal can ingest depends on the kinds and amounts of nutrients and toxins in the forages on offer. Thus, individuals can better meet their needs for nutrients and regulate their intake of toxins when offered a variety of foods that differ in nutrients and toxins than when constrained to a single food, even if the food is 'nutritionally balanced.' Food intake and preference also depend on differences in how individual animals are built morphologically and how they function physiologically, and marked variation is common even among closely related animals in needs for nutrients and abilities to cope with toxins. Thus, feeding and grazing practices that allow producers to capitalize on the individuality of animals are likely to improve performance of the herd by enabling the uniqueness of individuals to become manifest. Finally, experiences play a key role in an animal's propensity to learn to eat different foods. When herbivores are allowed to eat only the most preferred plants, they are not likely to learn to mix foods high in nutrients with foods that contain toxins. Conversely, herbivores encouraged to eat all plants in an area are more likely to learn to eat mixtures of plants that mitigate toxicity. Importantly, different systems of grazing management influence how animals learn to forage. Continuous grazing at low stock densities encourages selectivity and reduces diet and habitat breadth, whereas short-duration grazing at high stock densities increases diet and habitat breadth.

References

Andersen R (1991). Habitat deterioration and the migratory behaviour of moose (*Alces alces* L.) in Norway. *Journal of Applied Ecology* 28:102–108.

Banner RE, Rogosic J, Burritt EA, and Provenza FD (2000). Supplemental barley and activated charcoal increase intake of sagebrush by lambs. *Journal of Range Management* 53:415–420.

Burritt EA and Provenza FD (1991). Ability of lambs to learn with a delay between food ingestion and consequences given meals containing novel and familiar foods. *Applied Animal Behaviour Science* 32:179–189.

Cheeke PR (1998). *Natural Toxicants in Feeds, Forages, and Poisonous Plants.* Interstate Publishing Inc., Danville, Illinois.

Damasio A (2003). *Looking for Spinoza: Joy, Sorrow, and the Feeling Brain*. Harcourt, Inc., New York.

Dziba LE, Hall JO, and Provenza FD (2005). Feeding behavior of lambs in relation to kinetics of 1,8-cineole dosed intravenously and into the rumen. *Journal of Chemical Ecology* 32:391–408.

Freeland WJ and Choquenot D (1990). Determinants of herbivore carrying capacity: Plants, nutrients, and *Equus asinus* in northern Australia. *Ecology* 71:589–597.

Freeland WJ and Janzen DH (1974). Strategies in herbivory by mammals: The role of plant secondary compounds. *American Naturalist* 108:269–286.

Freeland WJ, Calcott PH, and Anderson LR (1985). Tannins and saponin: interaction in herbivore diets. *Biochemical Systematics and Ecology* 13:189–193.

Garcia J (1989). Food for Tolman: cognition and cathexis in concert. In *Aversion, Avoidance and Anxiety* (T Archer and L Nilsson, eds), pp. 45–85. Hillsdale, New Jersey.

Garcia J, Lasiter PA, Bermudez-Rattoni F, and Deems, DA (1985). A general theory of aversion learning. In *Experimental Assessments and Clinical Applications of Conditioned Food Aversions*. (NS Braveman and P Bronstein, eds), pp. 8–21. New York Academy of Sciences, New York.

Illius AW and Jessop NS (1995). Modeling metabolic costs of allelochemical ingestion by foraging herbivores. *Journal of Chemical Ecology* 21:693–719.

Illius AW and Jessop NS (1996). Metabolic constraints on voluntary intake in ruminants. *Journal of Animal Science* 74:3052–3062.

Keeler RF, Baker DC, and Evans JO (1988). Individual animal susceptibility and its relationship to induced adaptation or tolerance in sheep to *Galega officinalis*. *Letters in Veterinary and Human Toxicology* 30:420–423.

Konarzewski M and Diamond J (1994). Peak sustained metabolic rate and its individual variation in cold-stressed mice. *Physiological Zoology* 67:1186–1212.

Marsh KJ, Wallis IR, and Foley WJ (2005). Detoxification rates constrain feeding in common brushtail possums (*Trichosurus vulpecula*). *Ecology* 86:2847–2859.

Martinez JH, Ross TT, Becker KA, and Smith GS (1993). Ingested dry snakeweed foliage did not impair reproduction in ewes and heifers during late gestation. *Proceedings of the Western Section American Society of Animal Science* 44:32–35.

Provenza FD (1995). Postingestive feedback as an elementary determinant of food preference and intake in ruminants. *Journal of Range Management* 48:2–17.

Provenza FD (1997). Feeding behavior of animals in response to plant toxicants. In *CRC Handbook of Plant and Fungal Toxicants*. (JPF D'Mello, ed), pp. 231–242. CRC Press Inc., Boca Raton, Florida.

Provenza FD (2003). Twenty-five years of paradox in plant-herbivore interactions and 'sustainable' grazing management. *Rangelands* 25:4–15.

Provenza FD and Villalba JJ (2006). Foraging in domestic vertebrates: Linking the internal and external milieu. In *Feeding in Domestic Vertebrates: From Structure to Function*. (VL Bels, ed), pp. 210–240. CABI Publishing, Wallingford, UK.

Provenza FD, Burritt EA, Clausen TP, Bryant JP, Reichardt PB, and Distel RA (1990). Conditioned flavor aversion: a mechanism for goats to avoid condensed tannins in blackbrush. *American Naturalist* 136:810–828.

Provenza FD, Pfister JA, and Cheney CD (1992). Mechanisms of learning in diet selection with reference to phytotoxicosis in herbivores. *Journal of Range Management* 45:36–45.

Provenza FD, Ortega-Reyes L, Scott CB, Lynch JJ, and Burritt EA (1994). Antiemetic drugs attenuate food aversions in sheep. *Journal of Animal Science* 72:1989–1994.

Provenza FD, Villalba JJ, Dziba LE, Atwood SB, and Banner RE (2003a). Linking herbivore experience, varied diets, and plant biochemical diversity. *Small Ruminant Research* 49:257–274.

Provenza FD, Villalba JJ, and Bryant JP (2003b). Foraging by herbivores: Linking the biochemical diversity of plants with herbivore culture and landscape diversity. In *Landscape Ecology and Resource Management: Linking Theory with Practice.* (JA Bissonette, and I Storch, eds), pp. 387–421. Island Press, New York.

Scott LL and Provenza FD (1999). Variation in food selection among lambs: effects of basal diet and foods offered in a meal. *Journal of Animal Science* 77:2391–2397.

Villalba JJ and Provenza FD (1996). Preference for flavored wheat straw by lambs conditioned with intraruminal administrations of sodium propionate. *Journal of Animal Science* 74: 2362–2368.

Villalba JJ, Provenza FD, and Banner RE (2002a). Influence of macronutrients and activated charcoal on utilization of sagebrush by sheep and goats. *Journal of Animal Science* 80:2099–2109.

Villalba JJ, Provenza FD, and Banner RE (2002b). Influence of macronutrients and polyethylene glycol on intake of a quebracho tannin diet by sheep and goats. *Journal of Animal Science* 80:3154–3164.

Villalba JJ, Provenza FD, and Bryant JP (2002c). Consequences of nutrient-toxin interactions for herbivore selectivity: benefits or detriments for plants? *Oikos* 97:282–292.

Villalba JJ, Provenza FD and Han G (2004). Experience influences diet mixing by herbivores: Implications for plant biochemical diversity. *Oikos* 107:100–109.

Wang J and Provenza FD (1997). Food deprivation affects preference of sheep for foods varying in nutrients and a toxin. *Journal of Chemical Ecology* 22:2011–2021.

Williams JL, Campos D, Ross TT, Becker KA, Martinez JM, Oetting BC, and Smith GS (1992). Snakeweed (*Gutierrezia* spp.) Toxicosis in beef heifers. *Proceedings of the Western Section American Society of Animal Science* 43:67–69.

Chapter 57

Fungal Endophytes of Locoweed: an Overview

R. Creamer, J. McLain-Romero, and E. Oldrup

Department of Entomology, Plant Pathology, and Weed Science, New Mexico State University, Las Cruces, NM 88003, USA

Poisoning due to consumption of locoweeds causes significant losses to grazing animals in the western USA (James and Panter 1989). Locoweeds are perennial flowering legumes found in arid and semiarid regions. The plants are prolific seed producers, and the seed can remain viable after many years in the soil (James *et al.* 1999). Common species of locoweed endemic to the western USA include *Oxytropis lambertii*, *O. sericea*, *Astragalus mollissimus*, and *A. lentiginosus* (Allred 1991). The aerial plant parts of locoweeds are toxic, even when dried. However, not all *Astragalus* species are toxic and the level of toxicity varies by specific plant population. For example, *Astragalus crassiparus* is not toxic and some populations of *Oxytropis lambertii* are not toxic (Smith *et al.* 1992; Ralphs *et al.* 2002).

Toxic *Astragalus* and *Oxytropis* locoweed plants contain swainsonine, a polyhydroxy alkaloid that causes locoism symptoms by specifically inhibiting alpha-mannosidases (Dorling *et al.* 1980; Molyneux and James 1982; Elbein 1989). Locoism is characterized by a staggered gait, lack of muscular coordination, and difficulty eating and drinking (James 1970; James *et al.* 1981). Loss of mannosidase activity can ultimately lead to cellular vacuolization and death (Stegelmeier *et al.* 1999). The swainsonine biosynthesis pathway has been partially determined from two fungi, *Metarhizium anisopliae* (Sim and Perry 1997) and *Rhizoctonia leguminicola* (Wickwire *et al.* 1990). The latter fungus was incorrectly taxonomically placed, and is actually an Ascomycete (McGinn *et al.* 1999). The swainsonine biosynthesis pathway has also been partially determined from *Astragalus* and *Ipomoea* plants (Harris *et al.* 1987, 1988; Pastuszak *et al.* 1990; Haraguchi *et al.* 2003).

Locoweed Endophytes

Our group has isolated and characterized endophytic fungi from toxic *Oxytropis* and *Astragalus* plants (Braun *et al.* 2003). The isolated fungi appear white to brown with a black under surface when grown on agar media. The fungi grow slowly (0.03–0.34 mm day^{-1}) with thin septate mycelia (Braun *et al.* 2003). Conidial production is sporadic in some isolates and absent in others. Conidia are dark with very dark transverse septa, but vary somewhat in size and shape. The fungi are considered endophytes because they do not produce external mycelia on the plants or spores within the plants, nor cause any symptoms of disease, and are seed transmitted (Braun 1999). The fungi can be isolated from the stems, seeds, leaves, and flowers of field collected plants.

These endophytic fungi produce swainsonine in culture media, even after repeated transfers for several years after isolation from plants. Gardner *et al.* (2004) found that low swainsonine levels (<0.001% swainsonine g dry wt^{-1}) were correlated with endophyte absence and high swainsonine levels (>0.001% swainsonine g dry wt^{-1}) were correlated with endophyte presence in a comparison of 16 populations of *O. lambertii*. We found that endophyte cultures that produce high levels of swainsonine were isolated from locoweed field populations with higher swainsonine content (Braun *et al.* 2003), which provided initial evidence that fungal genetics plays a role in swainsonine production.

We have shown that endophytes alone can cause symptoms of locoism. Rats that were fed locoweed endophytes or locoweed showed reduced feeding and increased alkaline phosphatase and aspartate aminotransferase enzyme levels compared to controls (McLain-Romero *et al.* 2004a). Similar vacuolization was found in pancreatic, renal, and hepatic tissues of both locoweed- and endophyte-fed rats.

Alkaloid-producing fungal endophytes are widespread among grasses (Powell and Petroski 1992), but have rarely been reported for legumes. The grass endophytes usually inhabit the leaves, stems, flowers, and seeds of the plants (Faeth and Fagan 2002) and are maternally transferred through the seed. These fungal endophytes also cause livestock poisoning due to forage consumption (Bacon *et al.* 1977). Benefits to the grasses derived from endophyte infections include increased tolerance to heat and drought stress, improved vigor, and resistance to some pathogens (Petroski *et al.* 1990; Siegel 1993). Benefits to the grass endophytes include improved dissemination (through seed) and access to nutrients and water (Siegel *et al.* 1987; Clay 1988; Knoch *et al.* 1993).

Endophyte Genetics

The locoweed endophytes have been taxonomically classified based on the conidial morphology and nucleic acid sequence analysis of the ITS (internal transcribed sequences of the ribosomal DNA) and beta-tubulin encoding regions as *Embellisia* species (Braun *et al.* 2003). *Embellisia* falls within the family Pleosporaceae, and is closely related to *Alternaria*, with the type of conidial septa as the primary differentiating factor. The genus *Embellisia* is not very clearly defined and some species within the genera presently are more closely related to fungi with the genera of *Alternaria* and *Ulocladium*. Morphological studies on additional fungal endophyte isolates and sequence analysis from the glyceraldehyde phosphate dehydrogenase (*gpd*) encoding region of the isolates have shown that these fungi form a clade distinct from other *Embellisia* (Pryor *et al.* 2004). Additional analyses of all described members of *Embellisia* are ongoing to better determine the taxonomic placement of the endophytes.

RAPD (random amplified polymorphic DNA) analyses and nucleic acid sequencing ITS and *gpd* regions of larger numbers of endophyte isolates from a wider geographic area has shown that *Embellisia* isolated from *Oxytropis* locoweeds are genetically similar with the same RAPD banding patterns and nearly identical sequences (Belfon and Creamer 2003). The similarity was found among isolates from several different western states. In contrast *Embellisia* isolated from *Astragalus* locoweeds vary in RAPD banding pattern and nucleic acid sequence from the fungus isolates from *Oxytropis*. The fungal isolates from *Astragalus* are highly genetically diverse, exhibiting different RAPD banding patterns from isolates collected only a few miles apart, as well as by species of *Astragalus* (Belfon and Creamer 2003).

The recent work comparing RFLP (restriction fragment length polymorphism) patterns of chloroplast genes among *Oxytropis* and varieties of *A. mollissimus* showed that *O. lambertii* and *Oxytropis sericea* are nearly genetically identical, while *Astragalus* were quite genetically variable, even among varieties of *A. mollissimus* (Kulshreshtha *et al.* 2004). The limited genetic variability of *Oxytropis* plants and endophytes and the high level of genetic variability among *Astragalus* plants and endophytes suggest the possibility of a long evolutionary relationship between fungus and plant.

Endophyte-free Locoweeds

Seed from locoweed have been shown to contain swainsonine and give rise to toxic plants. We found that the *Embellisia* endophytes can be found in the seed coat of toxic *Astragalus* and *Oxytropis* plants (Romero *et al.* 2002). Plants that are germinated from the whole seed contain *Embellisia* and swainsonine. When the seed coats are separated from the embryo, *Embellisia* can be isolated from the coat, but not the embryo. When plants are germinated from the embryo alone and grown on tissue culture media, they are fungus-free as determined by PCR and culturing and do not contain or produce swainsonine as determined by LC-MS (Gardner *et al.* 2001; McLain-Romero *et al.* 2004b; Oldrup 2005). This shows that the fungus alone is responsible for the toxicity of locoweed plants.

Tissue culture media is essential for the culturing, because mortality of plants germinated without seed coats is very high within a month if grown in soil mix or sand. We have localized the fungus to the aluerone and parenchyma layers of the seed coat using microscopy and PCR (McLain-Romero *et al.* 2004b). We have essentially produced non-toxic locoweed plants from toxic seed.

Endophyte colonization of the seed coat may aid in seed germination. Grass endophytes are thought to increase germination and seed survival (Clay 1988; Wolock-Madej and Clay 1991; Knoch *et al.* 1993). The hard outer seed coat of many locoweed species can cause a low germination rate without scarification (Lewis 1973). However, we have found that seed several decades old germinates well and produces viable endophytic fungi.

Environmental Effects on Endophyte/Locoweed System

The growth of *O. sericea* plants and the *Embellisia* isolated from the plants is influenced by the growth medium and environmental conditions. *Oxytropis sericea* plants (infected with *Embellisia*) and *Embellisia* cultures grown on plant tissue culture media (Reyes-Vera *et al.* 2004) showed a threefold increase in swainsonine levels (Gardner *et al.* 2001) when the pH of the media was decreased from 5.5 to pH 4.5 (Oldrup 2005). An increase to pH 8.5 prevented any fungal growth. Increased swainsonine production was associated with a decrease in media pH also for *M. anisopliae* (Sim and Perry 1997).

Addition of sufficient polyethylene glycol (PEG) to the media to induce moderate water deficiency in plants caused a fivefold increase in swainsonine levels of both *O. sericea* plants (infected with *Embellisia*) and *Embellisia* cultures (Oldrup 2005). This agrees with the preliminary field studies of Purvines and Ralphs (1999), which found that below normal precipitation was associated with increased swainsonine levels in locoweed plants. Altering other environmental parameters including increasing temperature to 32°C, and decreasing the N, P, and K to deficiency levels, did not affect the swainsonine levels of plants or fungus compared to controls (Oldrup 2005).

The addition of PEG to the media also caused a significant increase in the dry weight of fungal-infected *O. sericea*, but not in the dry weight of non-infected plants (Oldrup 2005). The increase was not due to an increase in the fungal mass, because the same PEG treatment caused a significant decrease in the dry weight of *Embellisia* cultures. Thus, like the grass endophytes, the locoweed endophytes aid the plant in withstanding drought conditions.

The precise relationship of locoweeds and their fungal endophytes is not well understood. *Embellisia* appears to function as a mutualist, with the plant also gaining from the interaction. The fungus produces swainsonine, which can deter herbivory of the locoweed, and helps with water stress. The endophyte gains a place to live and easy access to nutrients and water, and dispersal through seed. However, additional studies are needed to better define the plant–microbe interaction.

Summary

The toxicity of locoweeds is due to their fungal endophytes. Most of the trends associated with locoweed toxicity are in reality due to the endophyte or the effect of the environment on the fungus. Although the taxonomy of the endophyte has not been clearly defined, the relationship between locoweeds and their endophytes is obviously a long one where both components have adapted to the association. Much of the earlier literature dealing with locoweeds needs to be re-evaluated, keeping in mind that the fungus is the toxic agent. The pathway of swainsonine biosynthesis in locoweed plants should be looked at again. Further, the role of the endophyte in disease needs to be considered in range management decisions.

References

Allred KW (1991). Locoweeds and livestock poisoning. *New Mexico Cooperative Extension Service Publication* 400 B-15.

Bacon CW, Porter JK, Robbins JD, and Luttrell ES (1977). *Epichloe typina* from toxic tall fescue grasses. *Applied Environmental Microbiology* 34:576–581.

Belfon R and Creamer R (2003). RAPD-PCR analysis of fungal endophytes of locoweed. *Phytopathology* 93:S7.

Braun K (1999). Fungal endophyte infection and swainsonine toxicity in locoweed. MS thesis, New Mexico State University. Las Cruces.

Braun K, Romero J, Liddell C, and Creamer R (2003). Production of swainsonine by fungal endophytes of locoweed. *Mycological Research* 107:980–988.

Clay K (1988). Clavicipitaceous fungal endophytes of grasses: coevolution and the change from parasitism to mutualism. In *Coevolution of Fungi with Plants and Animals* (KA Pirozynski and DL Kawksworth, eds), pp. 79–105. Academic Press, New York.

Dorling PR, Huxtable CR, and Colegate SM (1980). Inhibition of lysosomal α-mannosidase by swainsonine, an indolizidine alkaloid isolated from *Swainsona canescens*. *Biochemical Journal* 191:649–651.

Elbein AD (1989). The effects of plant indolizidine alkaloids and related compounds on glycoprotein processing. In *Swainsonine and Related Glycosidease Inhibitors* (LF James, AD Elbein, RJ Molyneux, and CD Warren, eds), pp. 155–187. Iowa State University Press, Ames.

Faeth SH and Fagan WF (2002). Fungal endophytes: common host plant symbionts but uncommon mutualists. *Integrative and Comparative Biology* 42:360–368.

Gardner DR, Molyneux RJ, and Ralphs MH (2001). Analysis of swainsonine extraction methods, detection, and measurement in populations of locoweeds (*Oxytropis* spp.). *Journal of Agricultural and Food Chemistry* 49:4573–4580.

Gardner DR, Romero J, Ralphs MH, and Creamer R (2004). Correlation of an endophytic fungus (*Alternaria* spp.) with the presence of swainsonine in Lambert locoweed (*Oxytropis lambertii*). In *Poisonous Plants and Related Toxins* (T Acamovic, CS Stewart, and TW Pennycott, eds), pp. 32–38. CABI Publishing, Wallingford, Oxon, UK.

Haraguchi M, Gorniak SL, Ikeda K, Minami Y, Kato A, Watson A, Nash R, Molyneux RJ, and Asano N (2003). Aklaloidal components in the poisonous plant, *Ipomoea carnea* (Convolvulaceae). *Journal Agricultural and Food Chemistry* 51:4995–5000.

Harris DM, Campbell BC, Molyneux RJ, and Harris TM (1988). Biosynthesis of swainsonine in the diablo locoweed (*Astragalus oxyphysus*). *Tetrahedron Letters* 29:4815–4818.

Harris TM, Harris CM, Hill JE, Ungemach FS, Broquist HP, and Wickwire BM (1987). (1S,2R,8aS)-1,2-Dihydroxy-indolizidine formation by *Rhizoctonia leguminicola*, the fungus that produces slaframine and swainsonine. *Journal Organic Chemistry* 52:3094–3098.

James LF (1970). Physiopathologic changes in locoweed poisoning of livestock. *American Journal of Veterinary Research* 33:663–672.

James LF and Panter KE (1989). Locoweed poisoning in livestock. In *Swainsonine and Related Glycosidase Inhibitors* (LF James, AD Elbein, RJ Molyneux, and CDWarren, eds), pp. 23–28. Iowa State University Press, Ames.

James LF, Bennett KL, Parker KG, Keeler RF, Binns W, and Lindsay B (1981). Syndromes of *Astragalus* poisoning in livestock. *Journal of the American Veterinary Medical Association* 178:146–150.

James LF, Panter KE, Stegelmeier BL, Ralphs MH, Pfister JA, and Gardner DR (1999). *Astragalus* and *Oxytropis* poison livestock with different toxins. In *Locoweed Research* (TM Sterling and DC Thompson, eds), pp. 10–11. New Mexico Cooperative Extension Service Research Report 730.

Knoch TR, Faeth SH, and Arnott DL (1993). Endophytic fungi after foraging and dispersal by desert seed-harvesting ants. *Oecologia* 95:470–473.

Kulshreshtha S, Creamer R, and Sterling S (2004). Phylogenetic relationships among New Mexico *Astragalus mollissimus* varieties and *Oxytropis* species by restriction fragment analysis. *Weed Science* 52:984–988.

Lewis J (1973). Longevity of crop and weed seeds: survival after 20 years in soil. *Weed Research* 13:179–191.

McGinn ME, Liddell CM, and Fernandez-Pavia SP (1999). *Rhizoctonia leguminicola* and its reclassification as an ascomycete. *Phytopathology* 89:S51.

McLain-Romero J, Creamer R, Zepeda, Strickland J, and Bell J (2004a). The toxicosis of *Embellisia* fungi from locoweed (*Oxytropis lambertii*) is similar to locoweed toxicosis in rat. *Journal of Animal Science* 82:2169–2174.

McLain-Romero J, Padilla A, and Creamer R (2004b). Microscopic localization of *Embellisia* in seed and seedlings of southern specklepod locoweed (*Astragalus lentiginosus*). *Phytopathology* 94:S69

Molyneux RJ and James LF (1982). Loco intoxication: indolizidine alkaloids of spotted locoweed (*Astragalus lentiginosus*). *Science* 216:190–191.

Oldrup WO (2005). Determining the influence of environmental parameters on swainsonine production in *Oxytropis sericea* and its fungal endophyte. MS thesis. New Mexico State University, Las Cruces, NM.

Pastuszak I, Molyneux RJ, James LF, and Elbein AD (1990). Lentiginosine, a dihydroxyindolizidine alkaloid that inhibits amyloglucosidase. *Biochemistry* 29:1886–1891.

Petroski R, Dornbos D, and Powell R (1990). Ecological and physiological aspects of host-specificity in endophytic fungi. In *Endophytic Fungi in Grasses and Woody Plants: Systematics, Ecology, and Evolution* (S Redlin and C Carris, eds), pp. 87–100. American Phytopathological Society, St. Paul, Minnesota.

Powell RG and Petroski RJ (1992). Alkaloid toxins in endophyte-infected grasses. *Natural Toxins* 1:163–170.

Pryor BM, Romero J, and Creamer R (2004). Morphological and molecular characterization of a new species of *Embellisia* isolated from locoweed. *Proceedings of the Annual Meeting of Mycological Society of America*, Asheville, NC.

Purvines J and Ralphs MH (1999). Drought-stressed locoweed contains more swainsonine. In *Locoweed Research* (TM Sterling and DC Thompson, eds), pp. 34–35. New Mexico Cooperative Extension Service Research Report 730.

Ralphs MH, Welsh SL, and Gardner DR (2002). Distribution of the locoweed toxin swainsonine in populations of *Oxytropis lambertii*. *Journal of Chemical Ecology* 28:701–707.

Reyes-Vera I, Hubstenberger J, and Barrow J (2004). Hyperhydricity reversal and clonal mass propagation of four-wing saltbush [*Atriplex canescens* (Pursh) Nutt.] cultivated *in vitro*. *In Vitro Cellular and Developmental Biology-Plant* 40:55-A.

Romero J, Creamer R, Ralphs MH, and Gardner DR (2002). Association of a fungal endophyte with seed tissue and locoweed toxicity. *Phytopathology* 92:S70.

Siegel MR (1993). *Acremonium* endophytes: our current state of knowledge and future directions for research. *Agriculture Ecosystems and Environment* 44:301–321.

Siegel MR, Latch GCM, and Johnson MC (1987). Fungal endophytes of grasses. *Annual Review of Phytopathology* 25:293–315.

Sim KL and Perry D (1997). Analysis of swainsonine and its early metabolic precursors in cultures of *Metarhizium anisopliae*. *Glycoconjugate Journal* 14:661–668.

Smith GS, Allred KW, and Kiehl DE (1992). Swainsonine content of New Mexican locoweeds. *Proceedings, Western Section, American Society of Animal Science* 43:405–407.

Stegelmeier BL, James LF, Panter KE, Gardner DR, Pfister JA, Ralphs MH, and Molyneux RJ (1999). Dose response of sheep poisoned with locoweed (*Oxytropis sericea*). *Journal of Veterinary Diagnostic Investigation* 11:446–454.

Wickwire BM, Wagner C, and Broquist HP (1990). Pipecolic acid biosynthesis in *Rhizoctonia leguminicola*. II. Saccharopine oxidase: a unique flavin enzyme. *Journal of Biological Chemistry* 265:14748–14753.

Wolock-Madej C and Clay K (1991). Avian seed preference and weight loss experiment: the role of fungal-infected fescue seeds. *Oecologia* 88:296–302.

Chapter 58

Effect of Fungicide on Swainsonine Production in White Locoweed (*Oxytropis sericea*)

J.R. Valdez Barillas[1], R.D. Child[1], and M.H. Ralphs[2]

[1]*Forest, Rangeland, and Watershed Stewardship Department, Colorado State University, Fort Collins, CO 80523, USA;* [2]*USDA-ARS Poisonous Plant Research Laboratory, Logan, UT 84341, USA*

Introduction

White locoweed (*Oxytropis sericea*) is a perennial legume toxic to livestock throughout western North America (Ralphs and James 1999). Livestock intoxication is attributed to the alkaloid swainsonine, produced by the plant in association with the fungal endophyte *Embellisia* sp. (Braun *et al.* 2003; McLain-Romero *et al.* 2004). Swainsonine inhibits lysosomal and Golgi mannosidases, resulting in retention of glycoproteins and oligosaccharides, which compromises cellular function and eventually causes cell death (Harris *et al.* 1988; Panter *et al.* 1999; Stegelmeier *et al.* 1999).

The effects of the presence of an endophyte on alkaloid production, plant growth, and plant survival have been studied in endophyte-infected and uninfected plants using fungicides to regulate infection. Faeth and Fagan (2002) used propiconazole to produce *Neotyphodium* sp.-free Arizona fescue plants. Bayman *et al.* (2002) used Benomyl and propiconazole to determine the effects of endophytic fungi in the growth and survival of the orchid *Lepanthes rupestris*. Fungal endophyte control in both seed and established plants of tall fescue was achieved by using Benomyl and Bayleton (Backman *et al.* 1983). However, swainsonine production and endophyte infection in white locoweed have only been studied in wild populations.

The objective of this research was to determine the effects of fungicide on swainsonine production in white locoweed plants. We hypothesized that fungicide treatments on the seeds and the leaves would affect plant growth and decrease swainsonine production in white locoweed.

Materials and Methods

The study was conducted at the Plant Growth Facilities at Colorado State University from October 2003 to September 2004. The plants used in this study were started from white locoweed scarified seed provided by the USDA-ARS Poisonous Plant Research

Laboratory in Logan, Utah. One batch of seeds was treated with 1.5 g L^{-1} of Cleary's 3336 (thiophanate ethyl) and a second batch remained untreated.

Twenty-four of the fungicide-treated and 24 untreated seeds were planted in PVC pots filled with sand. Seedlings were watered regularly and treated once a week with Miracle-Gro (15-30-15). Sixteen weeks after germination, 12 plants grown from fungicide-treated seeds and 12 plants grown from fungicide-untreated seeds received a foliar fungicide application. The other 12 plants grown from fungicide-treated seeds and the 12 plants grown from fungicide-untreated seeds remained untreated

Twenty-four weeks after germination each of the 48 plants was measured to estimate average leaf length, average number of leaves, and the number of clusters (equivalent of a tiller in grasses). Leaves from each plant were clipped 1cm above the crown, and oven dried at 55°C for 72 hours in paper bags to determine leaf biomass. Dried samples were ground in a Wiley mill, using a 1mm mesh, for swainsonine analysis at the USDA-ARS Poisonous Plant Research Laboratory.

The procedure used for the HPLC-MS assay for swainsonine in white locoweed samples was based on the methods published by Gardner *et al.* (2004). Standards for this analysis were prepared using ground *Astragalus lentiginosus* leaves, previously assayed for swainsonine content following the same method applied to the white locoweed plant material.

The statistical calculations using a matrix of the data were run on SAS 8.02 GLM procedure. Significant differences in the number of leaves, number of clusters, leaf biomass, and swainsonine concentration among fungicide treatments were estimated using a one-way ANOVA.

Results

Fungicide treatments had an effect on plant growth and reduced, but did not eliminate, the swainsonine content in white locoweed. Leaf biomass was different between treatments ($F=2.91$ $P=0.0493$). Plants that received seed and seed/foliar fungicide had lower ($t=2.17$ $P=0.03778$) biomass (1.46g SE ± 0.119 and 1.47g SE ±0.283) than the controls (2.23g SE ±0.302) and the foliar-treated plants (2.15g SE ±0.244).

Control plants and those treated with foliar fungicide application had greater ($t=2.92$ $P=0.0065$) swainsonine content (0.153% SE ±0.026) than the plants that received seed and seed/foliar fungicide application (0.044% SE ±0.027 and 0.084% SE ±0.035). Plants that received seed fungicide had the lowest concentrations of swainsonine compared to the fungicide-untreated controls (Table 1).

Table 1. Mean leaf biomass and swainsonine concentrations in white locoweed plants treated with fungicide.

Treatment	Number of plants	Mean leaf biomass (g)	SE	Mean swainsonine (%)	SE
Control	7	2.232 [a]	0.302	0.153 [a]	0.026
Foliar	11	2.150 [a]	0.244	0.150 [a]	0.017
Seed/Foliar	8	1.467 [b]	0.283	0.084 [ab]	0.035
Seed	9	1.461 [b]	0.119	0.044 [b]	0.027

Letters indicate significance ($P < 0.05$).

Conclusions

Fungicide application on white locoweed seeds indirectly reduced swainsonine levels in the plant by reducing *Embellisia* sp. infection. In addition, the fungicide application on the seed was associated with lower foliage biomass, demonstrating a possible negative effect of fungicide on white locoweed seedling growth, or alternatively, the positive effects of the endophyte on plant growth.

These results elucidate the benefits of a fungicide application on the seed of white locoweed by lowering plant toxicity and growth. Due to the large seed bank of white locoweed and its geographic range, fungicide application would not be economically feasible. In addition, fungicide application of this magnitude may have severe ecological consequences by eliminating beneficial fungi in ecosystems where white locoweed is widely distributed. However, experimental application of fungicides is a promising tool to manipulate fungal infection when studying swainsonine production in white locoweed.

References

Backman PA, Williams MJ, and Pedersen JF (1983). Control of the fungal endophyte *Acremonium coenophialum* in seed and established plants of tall fescue. In *Proceedings in Forage and Turfgrass Endophyte Workshop*, pp. 77–85. Oregon State University Extension, Corvallis.

Bayman P, Gonzales EJ, Fumero JJ, and Tremblay RL (2002). Are fungi necessary? How fungicides affect growth and survival of the orchid *Lepanthes rupestris* in the field. *Journal of Ecology* 90:1002–1008.

Braun K, Romero J, Liddell C, and Creamer R (2003). Production of swainsonine by fungal endophytes of locoweed. *Mycological Research* 107(8):980–988.

Faeth SH and Fagan WF (2002). Fungal endophytes: common host plant symbionts but uncommon mutualist. *Integral and Comparative Biology* 42:360–368.

Gardner DR, Romero J, Ralphs MH, and Creamer R (2004). Correlation of an endophytic fungus (*Alternaria* spp.) with the presence of swainsonine in lambert locoweed (*Oxytropis lambertii*). In *Poisonous Plants and Related Toxins* (T Acamovic, CS Stewart, and TW Pennycott, eds), pp. 32–38. CABI Publishing, Wallingford, Oxon, UK.

Harris CM, Schneider MJ, Ungemach FS, Hill JE, and Harris TM (1988). Biosynthesis of the toxic indolizidine alkaloids slaframine and swainsonine in *Rhizoctonia leguminicola*: metabolism of 1-hydroxyindolizidines. *Journal of the American Chemical Society* 110:940–949.

McLain-Romero J, Creamer R, Zepeda H, Strickland J, and Bell G (2004). The toxicosis of *Embellisia* fungi from locoweed (*Oxytropis Lambertii*) is similar to locoweed toxicosis in rats. *Journal of Animal Science* 82:2169–2174.

Panter KE, James LF, Stegelmeier BL, Ralphs MH, and Pfister JA (1999). Locoweeds: effects on reproduction in livestock. *Journal of Natural Toxins* 8(1):53–62.

Ralphs MH and James LF (1999). Locoweed grazing. *Journal of Natural Toxins* 8(1):47–51.

Stegelmeier BL, James LF, Panter KE, Ralphs MH, Gardner DR, Molyneux RJ, and Pfister JA (1999). The pathogenesis and toxicokinetics of white locoweed (*Astragalus* and *Oxytropis* spp.) poisoning in livestock. *Journal of Natural Toxins* 8(1):35–45.

Chapter 59

Locoweed Research in Northeastern New Mexico

D. Graham

New Mexico State University Cooperative Extension Service, Union County, Clayton, NM 88415, USA

Introduction

Locoweed research in northeastern New Mexico encompasses some 11 million hectares (27 million acres) of native rangeland that supports an annual $250 million animal agriculture industry, the area's primary economic base (NMDA 2005). Elevations range from 1219 m (4000 ft) to 3048 m (10,000ft) (U.S. Geological Survey), with annual precipitation ranging from 147 mL (9 in) to 492 mL (30 in) (National Oceanic and Atmospheric Administration 2005). The dominant forage is blue grama *(Bouteloua gracillis)* and western wheatgrass *(Agropyron smithii* Rydb.) (Gay and Dwyer 2005) growing primarily on volcanic soils (USDA Soil Survey 1981).

Northeastern New Mexico hosts a wide variety of *Astragalus* and *Oxytropis* plants; but it is fortunate that there are only two locoweeds, *Oxytropis sericea* and *Astragalus mollissimus* (Fox *et al.* 1998), that have proven to be economically significant with a calculated loss of approximately $10 million in the late 1980s and early 1990s (Torell *et al.* 1998).

Livestock operations in northeastern New Mexico for the most part are seasonal grazing of stocker cattle, imported from every state in the United States, Canada, and Mexico. These cattle have a different history and background in grazing habits, diseases, and previous management. Operations include native year-round cow/calf operations that market all calves at the end of the grazing season. Some hold over their yearling cattle and graze them in conjunction with the year-round cow/calf operation, and some import stocker cattle to graze along with the year round cow/calf operation on a seasonal basis.

Concerns, observations, and basic research on locoweeds have been ongoing for some time with the first printed account of the weed and its effects appearing in 1873 according to a paper presented to the New Mexico Association for the Advancement of Science, November 4, 1937 by Dr John D. Clark from the University of New Mexico. Prior to this date, the only toxin was believed to be 'barium and/or selenium', which was proven wrong by G.S. Fraps, Chief Chemist of the Texas Agricultural Experiment Station, and E.C. Carlyle, Assistant Chemist of the Division of Chemistry at the Texas Agricultural

Experiment Station, and announced in 1936. Dr Fraps isolated a compound he called 'locoine'. Since this time, research literature was sparse and difficult to find and what was available was very contradictory.

Research

In 1994, an outcry from livestock producers was heard at a regional meeting held in Clayton, New Mexico, by the New Mexico congressional delegation, and money was appropriated for locoweed research. The authority and the money were awarded to the USDA Poisonous Plant Research Laboratory (Logan, Utah) and New Mexico State University (Las Cruces) to increase the level and intensity of research on locoweeds. There has been research on all levels, in the field and in the laboratories, by an interdisciplinary team of scientists from these institutions working together to develop economical and environmentally safe management methods to reduce locoweed to acceptable levels on New Mexico rangeland, as well as to minimize its toxic effects on livestock. Current research is addressing various fungi and their roles in locoweed toxicity. Researchers are looking for ways to identify loco toxicity in animals at an earlier stage of intoxication. They are studying the quantity and quality of swainsonine required for toxic effects, and data related to tolerance levels. Researchers are searching for field tests to determine swainsonine ingestion. Grazing studies continue in an effort to determine the effects of maternal feeding habits that influence locoweed ingestion, the effects and extent of social facilitation, and when and why animals graze locoweed.

Animal management

- Cattle can graze locoweed-infested pastures for 9 to 10 days and recover from locoweed exposure within 3 weeks. Furthermore, there is no cumulative effect of repeating the short-term exposure to locoweed up to three times.
- Supplementation of copper, Rumensen®, protein, mineral packages, or hormone implants has no effect on the course of locoweed toxicosis.
- Grazing studies demonstrate that cattle do readily graze purple locoweed (*A. mollissimus*) and white locoweed (*O. sericea*) in March, April and May, but usually cease grazing these plants in June when warm season grasses start rapid growth and locoweed matures.
- Animals eating locoweed influence others to start eating it (social facilitation).
- Locoweed ingestion causes numerous and severe effects on livestock reproduction. Abortion in pregnant animals will occur after 20 to 30 days of grazing. Besides reduced reproductive performance, other problems include emaciation, reduced lactation, inability to breed, delayed puberty, reduced maternal/fetal behavior, lengthened postpartum interval and reduced growth rates in offspring.
- Denying livestock access to locoweed during critical spring and fall periods when it is the principal green feed available is the simplest management solution to prevent intoxication.
- Results from a 2000 grazing trial indicated Brangus cattle started eating locoweed before either Hereford or Charolais. Gross observations of Brangus cattle suggest

that they are very social and may encourage each other to select locoweed because of social facilitation.

Plant management

- Purple locoweeds *(A. mollissimus)* are typically short lived (< 4 years), flourishing in wet years but easily die out with drought and/or in combination with insect damage.
- Purple locoweed *(A. mollissimus)* rarely persists more than a few years in a given area because of recurring droughts and from damage inflicted by the larvae of the four-lined locoweed weevil. All attempts to rear these beneficial insects in the laboratory for later release into the field have been unsuccessful, so presently there is no known way to augment their activity.
- White locoweed (*O. sericea*) is more stable than purple locoweed (*A. mollissimus*) and causes the most widespread poisoning problems in northeastern New Mexico. Its population also cycles as evidenced by die offs during droughts in 1996, 2000, 2002, 2003, and new propagation of new seedlings in autumn 1997, 2000, and 2004. Locoweeds are prolific seed producers with 600 to 4000 seeds/m^2 retained within the soil.
- White locoweed (*O. sericea*) germinates when late summer rains are heavy followed by a wet fall and winter.
- The biology of many native insects that damage locoweeds is known but propagation and release is not yet possible.
- Locoweeds sprayed in early vegetative and bloom stages are easily controlled by herbicides at a cost of about $14 to $18 per acre.
- Herbicides can eliminate a current locoweed crop, but seed remains in the soil for a long period of time. Spraying the same area for three locoweed generations has resulted in lower populations over time but complete elimination of plants has not yet occurred.
- Herbicide activity is enhanced when absorption time is lengthened. Greatest herbicide uptake occurs when the relative humidity is high, wind speed is low, and air temperature is moderate.

Toxicology and animal health

- Locoweed's toxic component is a natural constituent of the plant, and is named swainsonine.
- The chemical structure of swainsonine is not complex and is quite similar to simple sugars, such as mannose and glucose, which it appears to mimic.
- Swainsonine is very water soluble and therefore distributes rapidly to many parts of the animals' body. It is also rapidly excreted, primarily in the urine. But in lactating animals, a portion of it is transferred to the milk, so that nursing young can become 'locoed.' This fast excretion rate suggests that occasional locoweed consumption for short periods is unlikely to have serious adverse effects. Continuous consumption, even at relatively low levels, can result in locoweed poisoning. Therefore, it is important to remove animals from locoweed-infested land as soon as they are observed grazing the plant.
- Presently, there is no known way to determine how intoxicated an animal might be.

- Many histologic lesions and loss of function seen in poisoned animals are regained shortly after locoweed consumption is discontinued. However, some neurologic changes are irreversible and permanent after long-term ingestion.
- Research indicates that animals that have shown signs of intoxication and are then removed from locoweed will eventually reach body and weight condition in the feedlot, but will be 30 to 60 days behind peers that have not been intoxicated.

Grazing Management

The diversity in topography, elevation, and precipitation, as well as individual producer situations, dictates diversity in grazing management options for use in rangelands with locoweed infestations. There is not one best method or management system. Based on the available research, demonstration results, and observations, several management plans have been developed to help producers cope with grazing locoweed-infested rangeland.

- Abstention from grazing any livestock in locoweed-infested pastures is ideal but usually impractical and economically impossible.
- Timed grazing is a good option since livestock should be denied access to locoweed when it is green and lush, especially in spring and autumn when grass is dormant. Animals should not be allowed to graze pastures until warm season grasses are growing and locoweed is maturing; grazing animals should also be removed in autumn when locoweeds begin new growth and grasses are maturing.
- Flash grazing or intensive grazing for 9 to 10 days, then removing animals from the locoweed area for a 3 to 6 week period is an effective management tool. A regular rotational system that forces animals to uniformly graze all plants in a pasture, including locoweeds, can induce locoism.
- Social facilitation or peer pressure strategy is used extensively. In smaller herds, offering fresh cut locoweed in a confined area prior to turning out to a locoweed-infested range is practical for naïve animals. Observers should watch animals for a 5 min period and classify any that consume the loco as eaters and never allow them to graze where locoweed is prevalent. This 'eat and pull' management strategy, whereby livestock are watched closely and removed if they are observed eating locoweeds, is a highly recommended practice and should be continued throughout the grazing period.
- Conditioned food aversion is also a practice that can be used on naïve animals using the availability of fresh locoweed, pulling the eaters and using a drench of lithium chloride (LiCl) as an effective emetic.
- Herbicide control of locoweeds can be achieved by several rangeland herbicides, thus maintaining locoweed-free pastures for a time. However, several studies have shown that new plants will germinate, probably within 5 years of spraying. Repeated spraying has not completely eliminated locoweed. This practice is very expensive (approximately $15 per acre) and cannot be economically practiced on an entire ranch. This practice can be used to create locoweed-free pastures for a successful rotational grazing program.

Conclusion

The research conducted in northeastern New Mexico is a success. The results have been used extensively to allow effective management of livestock around locoweed infestations. The locoweed intoxication problem cannot and will not be handled in a simple way, and a planned management system based on the individual operation is the best option to date. There is no truth to the myth that native cattle will be less likely to consume locoweed in any given year. Winter grazing strategies using alfalfa hay play no role in the consumption of locoweed, and there is no supplement or feed additive that can either prevent or enhance the consumption of locoweed. It is this author's opinion that 'social facilitation' is the key to major locoism problems. It has been observed that, in any group of animals, only a small percentage will be natural eaters (< 15%) and that if these cattle are removed or averted, the total locoism problem will be greatly reduced.

Acknowledgements

Special thanks are given to the USDA-ARS Poisonous Plant Research Laboratory and the New Mexico State University researchers and crew for their support and work in the area.

References

Fox WE, Allred KW, and Roalson EH (1998). *A Guide to the Common Locoweeds and Milkvetches of New Mexico*. New Mexico Agriculture Experiment Station and Cooperative Extension Service Circular 557, New Mexico State University, Las Cruces.

Gay CC Jr and Dwyer DD (revised 2005). *New Mexico Range Plants*. New Mexico Agriculture Experiment Station and Cooperative Extension Service Circular 374, New Mexico State University, Las Cruces.

New Mexico Department of Agriculture (2005). *New Mexico Agricultural Statistics, 2004*. Las Cruces, New Mexico.

Torell LA, Hawkes JM, and Stromei TD (1998). *Range Livestock Cost and Return Estimates for New Mexico, 1996*. New Mexico State University Agriculture Experiment Station Research Report 726.

Chapter 60

Herbicidal Control of Locoweed

K. McDaniel[1], T. Sterling[2], and S. Ivey[1]

[1]Department of Animal and Range Sciences, New Mexico State University, Las Cruces, NM 88003, USA; [2]Department of Entomology, Plant Pathology and Weed Sciences, New Mexico State University, Las Cruces, NM 88003, USA

Introduction

Plants commonly referred to as locoweeds or milkvetches are among 373 species of *Astragalus* and 22 species of *Oxytropis* that are indigenous to North America (Fox *et al.* 1993; Sterling and Jochem 1995). Of these, some are poisonous and targeted by livestock producers for herbicide control, especially within the mountains, plains, and prairie regions of the intermountain west. This paper considers a few common species of *Astragalus* and *Oxytropis* included in major herbicide control programs conducted on western US rangelands.

The Problem

The perennial locoweeds – crazyweeds (*Oxytropis* spp.) and loco or milkvetches (*Astragalus* spp.) – are members of the legume (Fabaceae) family. This large group of legumes is taxonomically very difficult to identify, even by the experienced botanist (Kulshreshthna *et al.* 2004). Locoweed toxicity is complicated as some members of these taxa produce locoism in horses, cattle, sheep, and elk, but others are useful forage plants. Locoism, blind staggers, alkali disease, and a syndrome characterized by respiratory and coordination difficulty have all been attributed to species of *Astragalus* and *Oxytropis*. Three distinct syndromes are now recognized by virtue of the different toxins that various species of these plants may accumulate. In general, toxic locoweed species contain indolizidine alkaloids (swainsonine), whereas milkvetches contain nitroglycosides (miserotoxin). In addition, some *Astragalus* species accumulate toxic levels of selenium. Some species may contain more than one of the toxins and consequently may cause a combination of clinical effects in animals.

Locoweeds are particularly abundant on rangeland pastures once cultivated or disturbed. Livestock owners with locoweed try their best to keep animals from consuming the plant, and sometimes take measures to control the weed. Ralphs *et al.* (1993, 1994a, b, 1997a, b) suggested various ways to prevent cattle from eating locoweed, such as reducing stocking rates, limiting grazing to times when grasses are actively growing, and adversely conditioning animals so they will not seek the plants. Locoism is most commonly reported by ranchers in early spring or late autumn when locoweed is vegetative rather than in summer during flowering. This is partially because although locoweed remains lush and green throughout most of the year, livestock are less apt to graze the plant when prominent warm season grasses are green and healthy in summer (Ralphs *et al.* 1993).

The Plant

In general, locoweeds grow up to 3 ft (1 m) in height, have branching stems from a stout crown, and possess an extensive taproot (Fox *et al.* 1993). Leaves are alternate, pinnately compound, and each leaflet is usually elliptical or oval with minute hairs. The racemes are usually produced at the ends of the branches and are densely covered with white to purple pea-like flowers, depending on the species. The leguminous seed pods vary considerably in shape and contain many kidney bean-shaped seeds. The pod is often conspicuously enlarged, which allows seeds inside to rattle when ripe; hence the common names, rattlebox and rattleweed, are often assigned. The keel petal of *Oxytropis* tends to be long and pointed whereas that of *Astragalus* tends to be blunter.

It is important when considering locoweed control to identify the species correctly and to appreciate that each has certain unique environmental and genetic influences that have a bearing on the plant's life history. Plant longevity is an extremely important consideration when deciding if spraying should or should not be considered. For example, woolly loco (*Astragalus mollissimus*) is particularly prone to feeding damage by the four-lined locoweed weevil (*Cleonidius trivittatus*) and other insects, thus it is considered to be a relatively short-lived species (Thompson 1999). By comparison, white locoweed (*Oxytropis sericea*) is relatively long-lived and according to Torell *et al.* (2000), there are greater economic benefits from control of this species than spraying woolly loco.

Locoweed Control Literature

Before the introduction of herbicides, hand grubbing was the primary method for locoweed removal, and is still practiced where it grows sparsely in pastures. Broadcast and individual plant sprays with phenoxy herbicides, such as 2,4-D, have been common herbicide control methods since the 1950s and, while effective, locoweed death is often erratic (Bohmont 1952; Payne and Warden 1952). In a review of herbicide control of locoweed provided by Ralphs and Ueckert (1988) and Ralphs *et al.* (1991), they concluded that newer herbicides, such as picloram, dicamba, clopyralid, metsulfuron, and triclopyr, provide more consistent control than 2,4-D, especially when they are applied under less than optimum environmental conditions.

Recent research reporting herbicide control results for different locoweed species is scant. Most work reports control results from spraying during the bud to bloom growth stages (Ralphs and Ueckert 1988), and information concerning control during earlier or later vegetative stages is limited. Depending on elevation, soil moisture, temperature, and other local environmental influences, locoweed phenology can vary widely both locally or regionally and the plant's growth stage may affect herbicide control results.

Sterling and Jochem (1995) compared uptake, translocation and metabolism of picloram and metsulfuron by silky crazyweed (or white locoweed) and woolly loco. Compared to woolly loco, silky crazyweed was more than ten times as sensitive to increasing rates of either herbicide. The two locoweed species absorbed similar amounts of both herbicides, and there was no difference in translocation or metabolism of the chemicals within the plants. The differential response of these two species to picloram and metsulfuron was attributed to the sensitivity of the plants within their sites of action.

Herbicide Choices

When the proper herbicide is applied during optimum growing and environmental conditions, control of a majority of *Astragalus* and *Oxytropis* species can be achieved (Ralphs and Ueckert 1988). Numerous herbicides were compared in research trials conducted in northeastern New Mexico on white locoweed and woolly loco from 1990 to 2000 (McDaniel 1999) (Table 1). Applications were timed to coincide with three locoweed growth stages: early vegetative (April to mid-May), bloom or flowering (late May and June), and late vegetative (September to mid-October). Averaged across years and spray locations, herbicide treatments given in Table 1 were rated as excellent (>95% control), good (85 to 94% control), fair (70 to 84% control) and poor (<70% control). These categories are arbitrary but reflect livestock producers' expectations for eliminating these highly toxic plants. Product performance cannot be guaranteed and failures do occur about 3 to 10% of the time.

Table 1. Herbicides that provided white locoweed control in test conducted in northeastern New Mexico, 1990 to 2000.

Herbicide	Dosage	Locoweed growth stage		
		Early vegetative[a]	Bloom[b]	After flower-Late vegetative[c]
	(lb ai/ac)			
Picloram	0.375	E[d]	E	E
Picloram + 2,4-D (1:4)	1.25	E	E	E
Picloram + clopyralid (1:1)	0.250	E	E	E
Picloram + dicamba (1:1)	0.25	F	G	G
Clopyralid	0.25	G	E	E
Clopyralid + dicamba	0.125 + 0.25	G	E	E
Dicamba	0.50	P	G	G
Dicamba + 2,4-D	2.0	G	G	G
2,4-D ester	2.0	F	G	P
Metsulfuron	0.375 oz/ac	F	E	E

[a] Spraying for early vegetation was between April 1 and May 15.
[b] Spraying for bloom was between June 1 and June 15.
[c] Spraying for late vegetation was between September 1 and October 15.
[d] E=Excellent control; G=Good control; F=Fair control; P=Poor control

Picloram (0.375 lb acre^{-1}) applied alone or mixed with 2,4-D (1:4 ratio at 1.25 lb acre^{-1}) provides excellent control of white and woolly loco when applied across all growth stages (Table 1). By mixing dicamba or clopyralid with picloram, herbicide cost can be reduced and still provide satisfactory locoweed control. Metsulfuron and clopyralid applied alone or mixed with other herbicides also give excellent control when sprayed during flowering and after flowering. A major limitation to locoweed control is propagation of new plants after herbicide treatment. Research suggests that follow-up control should be anticipated.

Spraying Guidelines

Environmental conditions at the time of spraying are a critical consideration when applying foliar active herbicide. Locoweed control by various herbicides applied during the plant's early vegetative, bloom, and late vegetative growth stages in northeastern New Mexico (Table 1) indicated the following conditions are ideal when spraying: high relative humidity (above 50%); low wind speed (<8 mph); and air temperatures that are moderate (near 60 to 70°F). These conditions keep droplets wet longer, thereby increasing herbicide absorption into the leaves (Sterling and Jochem 1995). Soil that is moderately moist with a temperature above 55°F at 6 inch depth ensures that plants are growing vigorously.

Longevity of Locoweed Control with Herbicide Treatments

To estimate (and to a certain extent, guess) how long herbicide control will last is influenced by two separate but equally important considerations: first, how effective the initial herbicide treatment was; and second, how often the threshold environmental conditions occur that promote or trigger locoweed to propagate. Herbicide treatments, especially those that include picloram in the mixture, kill most if not all locoweed roots within 2 months of spraying and areas remain locoweed-free until new plants germinate from seed (McDaniel 1999). Herbicides used to control locoweed generally do not affect seed already in the soil, nor do the chemicals prevent these seeds from germinating. The environmental requirements for germination differ by locoweed species and while a few new plants may be added to a local population in a given year, typically conditions suited for widespread propagation occur infrequently, about once or twice a decade.

To investigate if white locoweed and its soil seed source could be eliminated from an area by repeated spraying, a series of experiments were conducted in northeastern New Mexico. Picloram treatments sprayed in 1992 killed 100% of white locoweed and no new plants were observed in research plots until summer 1997 (Figure 1). Above-normal rainfall in this region throughout most of the summer apparently triggered white locoweed to germinate, presumably from seed that was stored in the soil for at least 5 years. Plots previously sprayed had fewer seedlings than areas not sprayed, but the number of returning plants was still fairly high, averaging about 1 per m^2 (compared to 1.6 per m^2 in non-sprayed areas). All previously sprayed research plots were treated with picloram in autumn 1997, and again shortly after new plants were noted in 1999, and 2004. After four herbicide treatments applied over a 13-year period it is still not clear if white locoweed seed has been completely eliminated from the soil. However, there is a definite trend of reduced locoweed germination in herbicide-treated areas, which suggests that perhaps the seed bank is being depleted.

Alternatives to Herbicide Spraying

Considering ways to manage livestock to prevent locoweed poisoning should always be considered before investing in herbicide spraying (McDaniel and Ralphs 2001). The simplest management solution is to deny cattle access to locoweed, especially during critical spring and fall periods when it is the primary green feed available (Ralphs *et al.* 1994a). This of course is not always possible.

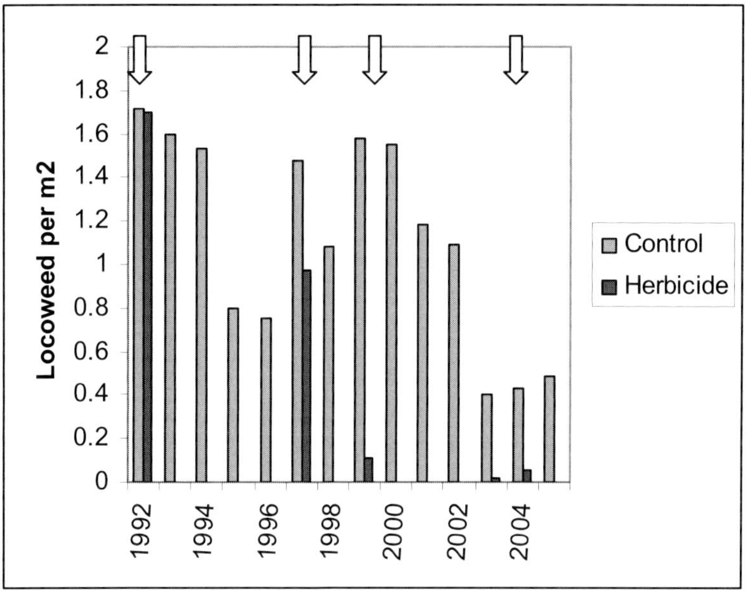

Figure 1. White locoweed density (plants per m^2) in non-sprayed (control) and herbicide-sprayed research plots in northeastern New Mexico. Arrows indicate times when herbicide plots were (re)sprayed.

Using herbicide to control locoweed in strategic locations is an approach used by some to create locoweed-free pastures for critical times (Torell *et al.* 2000). The idea is simple. First, pastures or portions of existing pastures that have little or no locoweed are identified and their use is earmarked for grazing during critical times. Spot spraying can be used to eliminate isolated plants and pockets of locoweed in these pastures. These locoweed-free pastures also serve a purpose in the event animals become poisoned, providing a safe zone for recovering livestock.

References

Bohmont DW (1952). Chemical control of poisonous range plants. *Wyoming Agricultural Experiment Station Bulletin* 313, University of Wyoming, Laramie.

Fox WE, Allred KW, and Roalson EH (1993). A guide to common locoweeds and milkvetches of New Mexico. *College of Agriculture and Home Economics Circular* 557, New Mexico State University, Las Cruces.

Kulshreshthna S, Creamer R, and Sterling TM (2004). Phylogenetic relationships among New Mexico *Astragalus mollissimus* varieties and *Oxytropis* species by restriction fragment analysis. *Weed Science* 52:984–988.

McDaniel KC (1999). Controlling locoweed with herbicides. *Locoweed Research Updates and Highlights, Agricultural Experiment Station Research Report* 730, New Mexico State University, Las Cruces.

McDaniel KC and Ralphs MH (2001). What we know about locoweed management. *Proceedings of the Livestock Research Briefs and Cattle Growers Short Course*, March 29–30, 2001. New Mexico State University, Las Cruces.

Payne GH and Warden RL (1952). Chemical control of white loco. *Montana Agricultural Experiment Station Mimeo Circular* 65, Montana State University, Bozeman.

Ralphs MH and Ueckert DN (1988). Herbicide control of locoweeds: A review. *Weed Technology* 2:460–465.

Ralphs MH, Whitson TD, and Ueckert DN (1991). Herbicide control of poisonous plants. *Rangeland* 13:73–77.

Ralphs MH, Graham JD, Molyneux RJ, and James LF (1993). Seasonal grazing of locoweeds in northeastern New Mexico. *Journal of Range Management* 46:416–420.

Ralphs MH, Graham JD, and James LF (1994a). Cattle grazing white locoweed in New Mexico: influence of grazing pressure and phenological growth state. *Journal of Range Management* 47:270–274.

Ralphs MH, Graham JD, and James LF (1994b). Social facilitation influences cattle to graze locoweed. *Journal of Range Management* 47:123–126.

Ralphs MH, Graham JD, Galyean ML, and James LF (1997a). Creating aversions to locoweed in naive and familiar cattle. *Journal of Range Management* 50:361–366.

Ralphs MH, Graham JD, Galyean ML, and James LF (1997b). Influence of over-wintering feed regimen on consumption of locoweed by steers. *Journal of Range Management* 50:250–252.

Sterling TM and Jochem HS (1995). Uptake, translocation, and metabolism of picloram and metsulfuron methyl by two locoweed species. *Weed Science* 43:13–17.

Thompson DC (1999). Common locoweed-feeding insects. *Locoweed Research Updates and Highlights, Agricultural Experiment Station Research Report* 730, New Mexico State University, Las Cruces.

Torell LA, Owen LP, McDaniel KC, and Graham D (2000). Perceptions and economic losses from locoweed in northeastern New Mexico. *Journal of Range Management* 53:376–383.

Chapter 61

The Comparative Pathology of Locoweed Poisoning in Livestock, Wildlife and Rodents

B.L. Stegelmeier, S.T. Lee, L.F. James, D.R. Gardner, K.E. Panter, M.H. Ralphs, and J.A. Pfister

USDA-ARS Poisonous Plants Research Laboratory, Logan, UT 84341, USA

Introduction

Over 500 *Astragalus* and *Oxytropis* species and varieties have been identified in North America (Welsh *et al.* 2003). Many are nutritious forages; however, a few are toxic. Of the toxic species about 20 are known as locoweeds. Locoweeds contain the endophyte-produced toxin, swainsonine, which with prolonged ingestion produces wasting and neurologic disease. Extended exposure to locoweed inhibits cellular α-mannosidases and Golgi mannosidase II, resulting in cytoplasmic vacuolation, accumulations of incompletely metabolized hybrid oligosaccharides and glycoproteins (Stegelmeier *et al.* 1999a). Though all animals appear to be susceptible to poisoning, there are species-specific differences in locoweed-induced neurologic disease and the distribution and severity of histologic lesions. The purpose of this study is to compare the distribution, severity and nature of locoweed poisoning in rodents, sheep, cattle, horses and mule deer.

Methods

Clinical, biochemical and histologic studies from various locoweed studies were reviewed, compiled and evaluated. This includes field studies in horses, sheep and cows where the locoweed dose was estimated by analyzing the plant and estimating daily consumption using bite counts (Stegelmeier *et al.* 1995a; Pfister *et al.* 2003). Pen studies were also included where rats, hamsters, sheep, cattle, and mule deer were fed locoweed pellets. The swainsonine dose in these studies was estimated from the amounts of pellet that each animal ingested (Stegelmeier *et al.* 1995b, 2005). Additionally, several more controlled dose-response studies are included. In these studies rats, mice, hamsters, sheep and cattle were dosed with known quantities of swainsonine or locoweed (Stegelmeier *et al.* 1985, 1995b, 1998). As these studies varied in dose, duration and number, all comparisons are subjective with no statistical comparisons.

Results and Discussion

Clinically, sheep, cattle and horses develop similar changes to locoweed poisoning. Though less is known about the minimal dose that affects horses, at swainsonine doses above 0.2 mg kg^{-1} day^{-1} sheep and cattle develop clinical signs after 10–12 days. The initial changes are mild reluctance to move and loss of appetite. With continued exposure these animals become more reluctant to move and eat. The initial neurologic changes are subtle intention tremors that become only noticeable when the animals are forced to move. These intention tremors become more severe until obvious proprioceptive deficits are evident at about 30 days. After 30 days the animal's coat begins to appear dull and most animals have glassy appearing eyes. With continued exposure all animals lose condition and weight. Cattle often develop a pendulous abdomen and some may develop congestive heart failure. Pregnant animals often abort and some develop hydrops amnii. Both the fetuses and neonates of intoxicated dams are poisoned. Doses estimated from clinical cases or calculated from field studies suggest horses are uniquely sensitive to locoweed poisoning as many develop clinical signs at doses much less than 0.1 mg swainsonine kg^{-1} bw day^{-1}. Additionally horses readily eat most locoweeds – even when other abundant green forages are available. The clinical signs of poisoning in horses are similar to those in sheep and cattle; however, when stimulated, startled, or stressed horses often develop behavioral changes including aggression, maniacal fits, uncontrollable trembling and seizures. When poisoned animals are removed from locoweed, nearly all of the clinical signs of poisoning resolve within a couple of weeks. However, with long-duration exposures some animals continue to loose weight and become emaciated. Some seemingly recovered animals may have residual proprioceptive deficits that are best demonstrated as a reluctance to back up or walk down hill. Additionally, with stress or sudden stimuli some animals may demonstrate tremors or mild seizures similar to those seen when they were poisoned. These permanent neurologic changes indicate that, although poisoned animals might seem to recover and they might be used as breeding animals, they should never be used as work animals.

Deer and rodents are relatively resistant to developing clinical signs of locoweed poisoning. Relatively high doses of swainsonine (>8 mg kg^{-1} bw day^{-1}) with durations of nearly 30 days are required for mice, rats and hamsters to develop mild intention tremors and subtle proprioceptive deficits. Doses near, at or above 2 mg cause rats and hamsters to gain less weight than controls. Similarly deer do not develop many of the neurologic signs that are common in livestock. The clinical signs in poisoned deer are mostly rough and straggly coats, failure to shed, and remarkable weight loss. This makes locoweed poisoning in deer visually indistinguishable from other wasting diseases.

All animal species are susceptible to the hematologic effects of locoweed poisoning. All animals studied developed a non-regenerative anemia with about 30 days of poisoning. All also had lymphocyte and monocytes vacuolation that was first detected after about 7 days. This vacuolation persists for several weeks after swainsonine exposure is discontinued. Immunologic studies in cattle showed that poisoning causes an initial stimulatory effect in lymphocyte blastogenesis and immunoglobulin production; however, after several weeks of poisoning, locoweed reduces both cellular and humeral immunologic responses.

Several of the serum biochemical changes and swainsonine concentrations of poisoning are shown in Table 1. Horses, cattle and sheep tended to have serum swainsonine concentrations above 300 ng mL^{-1}. More resistant rodents and deer have much lower serum concentrations, which were most often below 200 ng mL^{-1} even though they were often dosed at much higher rates than livestock. These differences are most likely due to toxicokinetic differences between species, though the variation between studies of the same species with similar dosing suggest that direct comparisons are questionable. Aspartate aminotransferase activity (AST) has been proposed as a good indicator of locoweed poisoning. Although AST was elevated above activities seen in control groups in most species, the increased activities were minimal and rarely two to three times higher than normal. As such modest elevations are often associated with minor stressful events such as trailering or physical exertion, this change has little use as a clinical indicator of poisoning.

In all species serum swainsonine clearance is relatively rapid ($T_{½}$<24 hrs). As swainsonine is rapidly cleared, it is of little use as an indicator of locoweed poisoning if exposure has stopped. Glycosylation of serum proteins appears to be a better long-term indicator of duration and extent of poisoning. Work continues to develop a consistent serum biomarker that can be used to monitor poisoning and better predict the fate of poisoned animals.

There are large species-specific differences in both the distribution and severity of locoweed-induced histologic lesions. Locoweed-poisoned deer developed extensive pancreatic lesions (Figure 1). Though locoweed-induced pancreatic lesions are common in cows, sheep and rodents, they are relatively mild in horses. This extensive damage of the exocrine pancreas certainly accounts for the remarkable wasting and loss of condition that is seen in poisoned deer. The chronic wasting and emaciation locoweed produces in deer are easily clinically confused with other wasting diseases of deer, including spongiform disease. Sadly, the government-sponsored monitoring and post mortem examinations for chronic wasting disease rarely include complete examinations. Thus they are unlikely to detect locoweed-poisoned animals or to provide a better indicator of the incidence of poisoning in wild North America cervids.

Table 1. Serum swainsonine concentration and biochemical changes from various locoweed studies.

Study	Species	Dose[a]	Time	SW[b]	AST[c]	ALP[d]
A. lentiginosus; field study	Horse	5 mg/kg	45 d	433±41	160±43	
(Pfister *et al.* 2003)	Cow	1 mg/kg	30 d	170±95	210±5	
Swainsonine; osmotic pump	Hamster	15 mg/kg	28 d	685±121		
(Stegelmeier *et al.* 1995b)	Rat			878±240		
A. mollissimus; rodent pellet	Hamster	7 mg/kg	28 d	185±32		
(Stegelmeier *et al.* 1995b)	Rat			170±50		
A. lentiginosus; alfalfa pellet (Stegelmeier *et al.* 1985)	Sheep	1 mg/kg	21 d	576±70		
O. sericea; gavage study	Cows	0.8 mg/kg	14 d	204±42		
(Stegelmeier *et al* 1995a)	Sheep	1 mg/kg		432±64		
O. sericea; gavage study (Stegelmeier *et al* 1999b)	Sheep	1 mg/kg	30 d	369±72	317±66	945±740
A. lentiginosus; gavage study (Stegelmeier *et al.* 1998)	Cow	2.25 mg/kg	45 d			
O. sericea; alfalfa pellet (Stegelmeier *et al.* 2005)	Deer	1.5 mg/kg	180 d	125±35	327±185	156±160

Data were compiled from various studies that are referenced within the table (mean±sd).
[a] Dose is reported as mg swainsonine kg^{-1} body weight day^{-1}
[b] Serum swainsonine (ng/mL)
[c] Aspartate aminotransferase activity (U/L)
[d] Alkaline phosphatase activity (U/L)

Nearly all animals with the exception of mule deer develop extensive thyroid follicular vacuolation. Thyroid dysfunction has also been documented as decreased serum T3 and T4 concentrations. Thyroid lesions of deer and horses are minimal (Figure 2). Figure 2 also demonstrates the extensive vacuolation of the thyroid follicular epithelium that is seen in sheep, cattle and horses. Thyroid lesions in rodents are moderate, but much less severe than the livestock species.

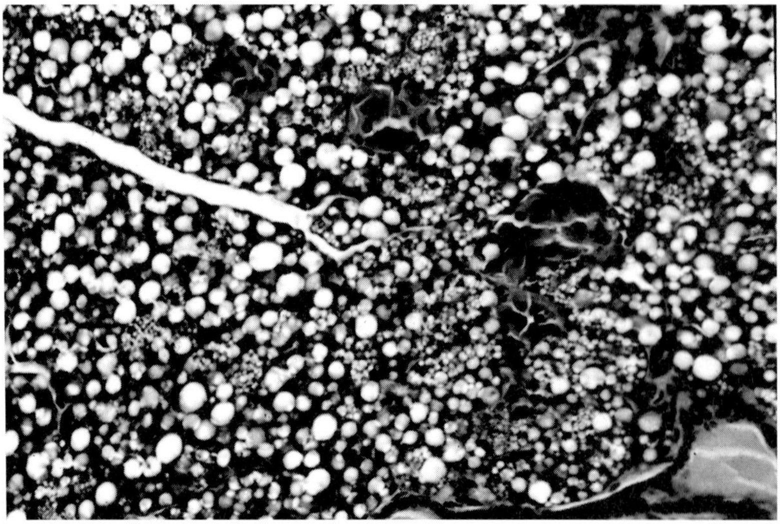

Figure 1. Photomicrograph of the pancreas of a deer poisoned with locoweed for 180 days. Note the extensive vacuolation and degeneration of the exocrine pancreas with relative sparing of the endocrine islet.

Mule deer and rodents are relatively resistant to neurologic lesions. Sheep, cattle and horses quickly develop vacuolation of cerebellar Perkinje cells, the cerebellar granular neurons, and many of the large neurons through the cerebellar and basal nuclei. With prolonged intoxication (>45 days) nearly all neurologic cells including many of the glial cells become vacuolated. Deer and rodents require much higher doses to develop neurologic lesions. In rats doses of nearly 10 mg swainsonine kg^{-1} bw are required to develop Perkinje cell lesions similar to those seen in sheep. With prolonged intoxication, deer Perkinje cells also become vacuolated; however, they rarely develop lesions in the cerebellar granular neurons. In sheep, horses and cattle these granular neurons appear to be highly susceptible to poisoning as they quickly become vacuolated, degenerative and pyknotic with only weeks of exposure. The lesions in these neurons seem to be closely related to the clinically irreversible neurologic disease. With prolonged poisoning, deer develop neuronal vacuolation, but the distribution of neurologic lesions is distinct from other species. The most severe locoweed-induced neurologic lesion in mule deer is extensive neuronal necrosis and axonal dystrophy of the gracilis and cuneate nuclei (Figure 3). Such extensive necrosis has not been seen in other species.

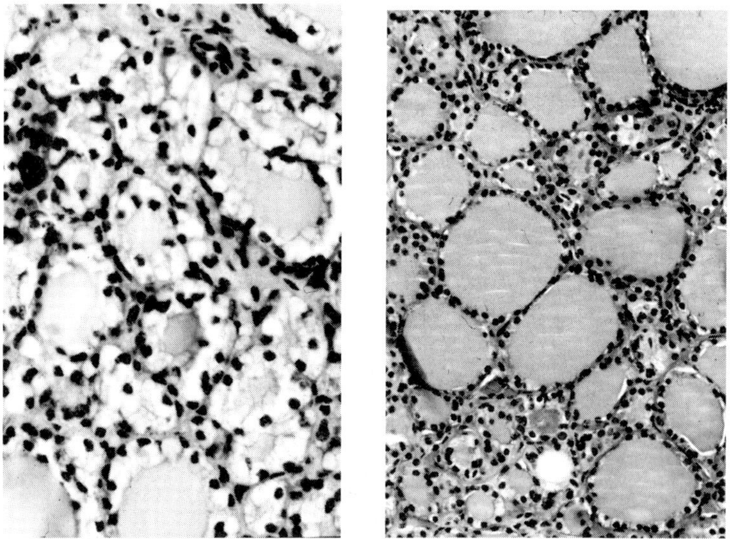

Figure 2. Photomicrograph of thyroid from a locoweed-poisoned sheep (left) and horse (right). Compare the extensive vacuolation and degeneration of the follicular epithelium in the sheep when compared to the minimal vacuolation seen in the horse.

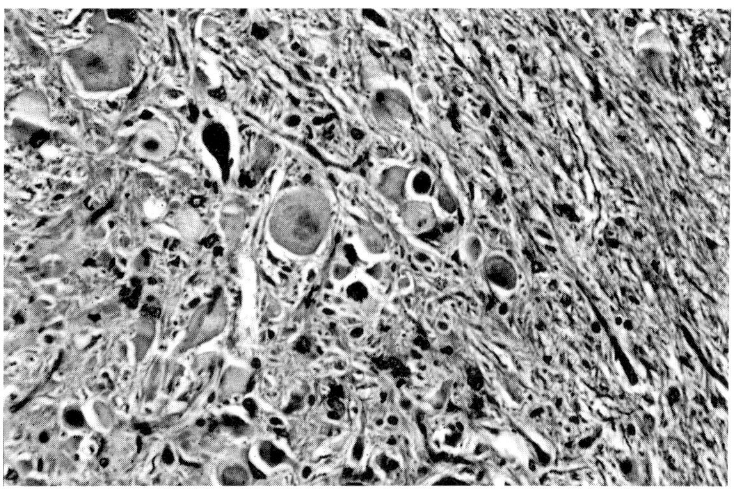

Figure 3. Photomicrograph of severe neuronal degeneration and axonal degeneration in the nucleus cuneatus of a deer poisoned with locoweed for 180 days.

Though species differences in toxicokinetics are likely, swainsonine is highly soluble and in most cases, absorption and most excretion appears to be passive. This makes it less likely as an explanation for all of the species-specific lesions. Another likely explanation is that susceptibility is directly linked to swainsonine affinity to the various different mannosidases and its ability to inhibit them. Though closely conserved, several different mannosidases have been identified in humans, rodents, dogs and cows (Park *et al.* 2005).

Tissue-specific mannosidase expression may also explain the distribution of locoweed-induced lesions. Expression of Golgi mannosidase II in rat tissues was found to vary greatly with most expressed in certain epithelia (Dong *et al.* 2000). This expression appears to be highest in tissues that have prominent locoweed-induced vacuolation. As other plant toxins have varying affinity for different animal receptors, it is likely that these different mannosidases have unique swainsonine binding affinities as well as unique expression patterns.

Summary

In summary, all species examined in these studies are susceptible to locoweed poisoning. However, there are species-specific differences in sensitivity to poisoning and the distribution and severity of locoweed-induced lesions. Horses appear to be most sensitive as they are highly susceptible to the effects of locoweed. Cattle and sheep have similar susceptibility and rodents and mule deer are relatively resistant to poisoning.

References

Dong A, Zuber C, Spiro MJ, Spiro RG, and Roth J (2000). Immunohistochemical evaluation of endomannosidase distribution in rat tissues: evidence for cell type-specific expression. *Histochemistry and Cell Biology* 114:461–467.

Park C, Meng L, Stanton LH, Collins RE, Mast SW, Yi K, Strachan H, and Moremen KW (2005). Characterization of a human core-specific lysosomal alpha 1-6 mannosidase involved in N-glycan catabolism. *Journal of Biological Chemistry* 280:37204–37216.

Pfister JA, Stegelmeier BL, Gardner DR, and James LF (2003). Grazing of spotted locoweed (*Astragalus lentiginosus*) by cattle and horses in Arizona. *Journal of Animal Science* 81:2285–2293.

Stegelmeier BL, James LF, Panter KE, and Molyneux RJ (1985). Tissue and serum swainsonine concentrations in sheep ingesting *Astragalus lentiginosus* (locoweed). *Veterinary and Human Toxicology* 37:336–339.

Stegelmeier BL, James LF, Panter KE, and Molyneux RJ (1995a). Serum swainsonine concentration and alpha-mannosidase activity in cattle and sheep ingesting *Oxytropis sericea* and *Astragalus lentiginosus* (locoweeds). *American Journal of Veterinary Research* 56:149–154.

Stegelmeier BL, Molyneux RJ, Elbein AD, and James LF (1995b). The lesions of locoweed (*Astragalus mollissimus*), swainsonine, and castanospermine in rats. *Veterinary Pathology* 32:289–298.

Stegelmeier BL, Snyder PD, James LF, Panter KE, Molyneux RJ, Gardner DR, Ralphs MH, and Pfister JA (1998). The immunologic and toxic effects of locoweed (*Astragalus lentiginosus*) intoxication in cattle. In *Toxic Plants and Other Natural Toxicants* (TR Garland and AC Barr, eds), pp. 285–290. CABI Publishing, Wallingford, Oxon, UK.

Stegelmeier BL, James LF, Panter KE, Ralphs MH, Gardner DR, Molyneux RJ, and Pfister JA (1999a). The pathogenesis and toxicokinetics of locoweed (*Astragalus* and *Oxytropis* spp.) poisoning in livestock. *Journal of Natural Toxins* 8:35–46.

Stegelmeier BL, James LF, Panter KE, Gardner DR, Pfister JA, Ralphs MH, and Molyneux RJ (1999b). The response of sheep poisoned with different doses of locoweed (*Oxytropis sericea*). *Journal of Veterinary Diagnostic Investigations* 11:446–454.

Stegelmeier BL, James LF, Gardner DR, Panter KE, Lee ST, Ralphs MH, Pfister JA, and Spraker TR (2005). Locoweed (*Oxytropis sericea*)-induced lesions in mule deer (*Odocoileius hemiionus*). *Veterinary Pathology* 42:90–102.

Welsh SL, Atwood ND, Goodrich S, and Higgins LC, eds (2003). *A Utah Flora* Third Edition. Brigham Young University Press, Provo, Utah.

Chapter 62

White Locoweed (*Oxytropis sericea* Nutt.) Ingestion by Naïve Stockers: Diet Preference Patterns of Natural "Eaters" and "Avoiders"

K.T. Jackson[1], A.F.Cibils[1], J.D. Graham[4], W.R. Gould[2], and C.D. Allison[3]

[1]*Department of Animal and Range Sciences,* [2]*Experimental Statistics Program,* [3]*Extension Animal Sciences and Natural Resources, New Mexico State University, Las Cruces, NM 88003,USA;* [4]*County Extension Director, Cooperative Extension Service, New Mexico State University, Union County Extension Office, Clayton, NM 88415,USA*

Introduction

Grazers have evolved complex feeding strategies and detoxification mechanisms to cope with toxic plants (Freeland and Janzen 1974; Provenza *et al.* 1992; Launchbaugh *et al.* 2001). Mammals possess a suite of physiological and behavioral adaptations, collectively known as the gut defense system, that enable them to learn to avoid toxic foods (García and Holder 1985; Provenza 1995). A few plant toxins such as alkaloids, however, are able to circumvent the gut defense system and induce severe physiological disorders that can lead to an animal's death. It is not surprising, therefore, that alkaloids are the phytotoxins responsible for the majority of rangeland-raised livestock poisoning losses worldwide (Pfister *et al.* 2001). In North America alone, 20 to 30% of perennial herbaceous species contain alkaloids (Pfister *et al.* 2001); locoweeds (*Astragalus* spp. and *Oxytropis* spp.) are important members of this group of plant species and can synthesize swainsonine, an indolizidine alkaloid that inhibits critical enzymatic processes in the herbivores that ingest them (Pfister *et al.* 2001; Panter *et al.* 2002).

Livestock that consume white locoweed commonly exhibit nervousness, loss of proprioception, ataxia, anorexia, and disruption of the reproductive functions; animals often show signs of aggression, solitary behavior and wasting, ending in death if toxicosis is severe (Panter *et al.* 1999). If exposure to locoweed is brief, grazing animals usually recover completely; however, if a certain threshold of intoxication is crossed, permanent neurological damage occurs (Panter *et al.* 1999).

Circumstantial observations suggest that most cattle that are unfamiliar with white locoweed will avoid consuming it upon first encounter (locoweed avoiders) and will continue to avoid it unless they are induced by peers to consume it (Ralphs *et al.* 1994). However, the majority of herds include a few individuals (5–10% of the herd) which will readily consume this plant (locoweed eaters) despite its novelty. Large animal-to-animal variation in levels of ingestion of plants that contain toxins has been well documented in other grazer and plant species. For example, sheep fed mesquite (*Prosopis glandulosa* Torr.) leaves rich in phenolics (Baptista and Launchbaugh 2001) or goats fed ashe and redberry juniper (*Juniperus ashei* Buch. and *J. pinchotii* Sudw) containing high levels of terpenes (Pritz *et al.* 1997) exhibit high animal-to-animal variation in voluntary intake.

Graham (unpublished data) developed a simple on-farm test to classify naïve stockers into categories on the basis of their propensity to consume locoweed. He found that fasted naïve animals (raised in locoweed-free environments) that avoided consuming freshly cut plants offered in a pen continued to avoid consuming locoweed in a pasture throughout the spring. Conversely, fasted naïve yearlings that readily consumed the freshly cut locoweed plants in the pens continued to consume this plant when released into a pasture. We used Graham's protocol to classify a group of naïve heifers so as to study the dietary preferences of locoweed eaters and avoiders in a controlled feeding trial. The objective of this study was to determine whether differences in locoweed preference exhibited by heifers classified as eaters or avoiders were contingent upon the quality of alternative food choices.

Materials and Methods

We conducted our study at New Mexico State University's Clayton Livestock Research Center (CLRC) during the month of June 2004. Sixteen cross-bred heifers averaging 156.4 ± 4.54 kg that were naïve to white locoweed were used in this study. A series of three pre-trials were conducted to select eight locoweed eaters and eight locoweed avoiders from a pool of 32 animals. Animals that sampled locoweed during pre-trials were classified as locoweed eaters whereas those that avoided it were classified as locoweed avoiders.

All animal handling protocols were approved by New Mexico State University's Institutional Animal Care and Use Committee. Animals were penned in pairs, had access to drinking water, and were fed a daily *ad libitum* bunk ration that consisted of 50% corn and 50% alfalfa and wheat hay mix. Adaptation to experimental foods was done by placing two rubber bowls with 300 g of ground alfalfa hay (5 days) or ground wheat straw (5 days) in every pen before daily feeding. This allowed each heifer to become familiarized with foods and containers later used during trials. Locoweed used in this experiment was harvested in the mature pod stage at a site near Gladstone, NM, air-dried, and ground through a 25 mm screen. Samples were sent to USDA-ARS Poisonous Plant Research Laboratory to determine swainsonine concentration.

Locoweed eaters and avoiders were randomly assigned to one of two treatment groups: (a) WS: locoweed offered with either high or low availability of low quality food (ground wheat straw with approximately CP: 3.5%, NDF: 78.9%, TDN: 41%; NRC 2000); or (b) AL: locoweed offered with either high or low availability of a high quality food (ground alfalfa hay with approximately CP:18.6%, NDF:43.9%, TDN:60%; NRC 2000). All animals were exposed to a 35 m^2 (7 m x 5 m) arena with sixteen 6-liter rubber bowls placed 1 m apart, simulating a single patch of forage covering most of the central area of the arena. The arena was sufficiently removed from the holding pens to prevent heifers from observing the feeding trials. The location of bowls with locoweed and other foods was randomized and remained the same throughout trials.

Each heifer was exposed to the arena individually for 10 min daily. The order in which heifers were led to the arena was randomized to control for potential time of day effects. Food bowls contained 50g of alfalfa hay, wheat straw (depending on the treatment) or ground white locoweed. During the first 6 days, heifers were exposed to a food patch with low availability of locoweed (six bowls) and high availability of the alternative food (ten bowls of ground wheat straw or alfalfa hay), whereas in the following 6 days, food quantities were switched to simulate a food patch with high availability of locoweed (ten bowls) and low availability of other foods (six bowls). Locoweed and alternative food refusals were collected and weighed after each trial to calculate individual intake of each. Preference indices (ratio between percent locoweed consumed and percent locoweed available) were derived from the intake measurements.

Data were analyzed using a repeated measures mixed-effects model. The same heifers were measured over time, hence the use of repeated measures. Heifers were considered to be random effects. Mixed model analysis was conducted so as to evaluate a variety of models accounting for different covariance structures. Model selection was facilitated using information criterion methods (AIC; Akaike 1973) to select the most parsimonious model. Treatment effects were examined in the mixed modeling context using single degree of freedom contrasts in ANOVA type tests using version 9.1 of SAS (SAS 2006).

Results

Swainsonine content of air-dried and ground white locoweed used in this experiment was 0.11% on a DM basis. One heifer in the locoweed eater AL group (locoweed offered with high quality food choice) died during the last 6 days of trials due to unknown causes and was therefore excluded from the analysis.

Table 1. Mean individual intake (g) ± SEM for all foods offered during 10 min trials.

Treatment group	Locoweed	Alfalfa	Wheat straw	Total
Days 1–6				
AL Eater	37.4±12.3	89.5±18.1		126.9±29.9
AL Avoider	32.5± 9.9	115.1±20.8		147.7±30.1
WS Eater	67.9±9.7		84.1±16.3	152.0±31.0
WS Avoider	1.1±0.9		78.5±14.7	79.6±16.2
Days 7–12				
AL Eater	95.7±20.2	120.5±18.4		216.2±50.9
AL Avoider	82.7±20.4	103.9±17.2		186.6±42.5
WS Eater	137.1±20.2		65.4±18.1	202.5±43.6
WS Avoider	30.6±9.9		108.9±12.1	139.5±28.5

Days 1–6: High availability of alternative food choices (500 g); low availability of locoweed (300 g). Days 7–12: Low availability of alternative food choices (300 g); high availability of locoweed (500 g).

Locoweed eaters showed significantly higher preference for white locoweed than did avoiders when availability of the alternative food choice was high (62% of total food offered) during the first 6 days of pre-conditioning trials ($F_{(1,10)} = 6.56$, $\underline{P} = 0.03$; Figure 1). Locoweed preference was not affected by the quality of the alternative food offered (alfalfa or wheat straw) ($F_{(1,10)} = 0.87$, $\underline{P} = 0.37$). An interaction between alternative food and heifer category (locoweed eater or avoider) approached statistical significance ($F_{(1,10)} = 3.28$; $\underline{P} = 0.10$). Locoweed eaters tended to show a higher preference for locoweed when the quality of the alternative food was low (wheat straw). Conversely, locoweed avoiders tended to exhibit higher preference for locoweed when the quality of the alternative food was high (alfalfa) (Figure 1). Intake of alternative foods by locoweed eaters and avoiders was similar (Table 1).

During days 7–12, when availability of the alternative food choice was low (38% of total food offered), differences in locoweed preference persisted, but were only marginally detectable ($F_{(1,10)} = 4.69$, $\underline{P} = 0.06$; Figure 2) and were, again, not affected by the quality of the alternative food choice. Locoweed eaters and avoiders continued to exhibit a tendency to mix their diets differently (Figure 2). Again, intake of alternative foods was similar across treatments, except for locoweed eaters on WS treatment, which tended to consume less WS than their avoider counterparts (Table 1).

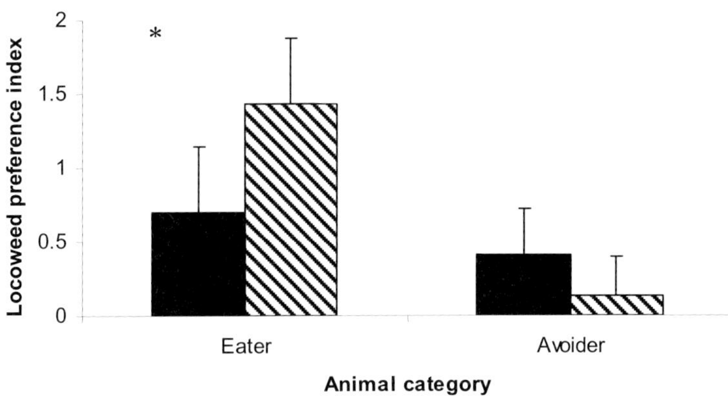

Figure 1. Locoweed preference of eaters and avoiders during days 1–6, when availability of alternative food choices was high (62% of total food offered). Heifers were offered either alfalfa hay (full bars) or wheat straw (hatched bars) as the alternative food choice. Error bars show S.E.M; asterisk indicates detectable preference difference between eaters and avoiders at $\underline{P} \leq 0.05$.

Conclusions

Locoweed preference of heifers classified as natural eaters was detectably higher than that of avoiders regardless of the quality and availability of alternative food choices. These results supported the validity of Graham's (unpublished data) classification protocol, which had not been previously tested under controlled conditions for consistency over time.

Locoweed eaters and avoiders exhibited a tendency to mix their diets differently. Eaters, on the one hand, appeared to behave as nutrient maximizers in that they chose higher quality forage regardless of its toxicity; white locoweed contains up to 17% crude protein according to Ralphs and Olsen (1987). This diet selection pattern appeared to fit Westoby's (1978) nutrient maximization model of diet selection. Avoiders, on the other hand, appeared to behave as toxin minimizers, ingesting more locoweed only when the alternate food choice (alfalfa) had a higher concentration of macronutrients (crude protein) that could potentially increase detoxification rates and reduce the risk of intoxication

(Provenza *et al.* 2003). This behavior appeared to fit predictions from Freeland and Janzen's (1974) diet selection model of toxin minimization. More research is needed to determine whether natural locoweed eaters and avoiders possess contrasting capabilities to detect or detoxify alkaloids and whether contrasting preference for locoweed of eaters and avoiders is heritable.

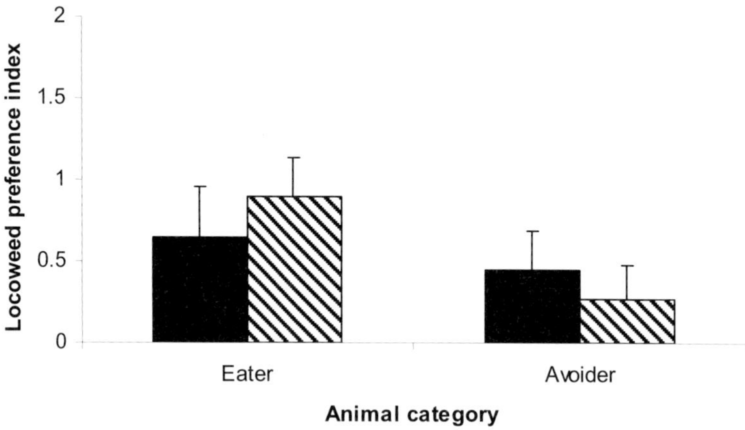

Figure 2. Locoweed preference of eaters and avoiders during days 7–12, when availability of alternative food choices was low (38% of total food offered). Heifers were offered either alfalfa hay (full bars) or wheat straw (hatched bars) as the alternative food choice. Error bars show S.E.M.

References

Akaike H (1973). Information theory as an extension of the maximum likelihood principle. In 2nd *International Symposium on Information Theory* (BN Pterov and F Csaksi, eds), pp. 267–281. Akadeimiai Kiado, Budapest, Hungary.

Baptista R and Launchbaugh KL (2001). Nutritive value and aversion of honey mesquite leaves to sheep. *Journal of Range Management* 54:82–88.

Freeland WJ, and Janzen DH (1974). Strategies in herbivory by mammals: The role of plant secondary compounds. *American Naturalist* 108:269–289.

Garcia J and Holder MD (1985). Time, space and value. *Human Neurobiology* 4:81–89.

Launchbaugh KL, Provenza FD, and Pfister JA (2001). Herbivore response to anti-quality factors in forages. *Journal of Range Management* 54:431–440.

National Research Council (2000). Composition of selected foods. In *Nutrient Requirements of Beef Cattle, Seventh Revised Edition* (National Research Council), pp. 133–148. National Academy Press, Washington, DC.

Panter KE, Ralphs MH, James LF, Stegelmeier BL, and Molyneux RJ (1999). Effects of locoweed (*Oxytropis sericea*) on reproduction in cows with a history of locoweed consumption. *Veterinary and Human Toxicology* 41:282–286.

Panter KE, James LF, Gardner DR, Ralphs MH, Pfister JA, Stegelmeier BL, and Lee ST (2002). Reproductive losses to poisonous plants: Influence of management strategies. *Journal of Range Management* 55:301–308.

Pfister JA, Panter KE, Gardner DR, Stegelmeier BL, Ralphs MH, Molyneux RJ, and Lee ST (2001). Alkaloids as anti-quality factors in plants on western U.S. rangelands. *Journal of Range Management* 54:447–461.

Pritz RK, Launchbaugh KL, and Taylor CA (1997). Effects of breed and dietary experience on juniper consumption by goats. *Journal of Range Management* 50:600–606.

Provenza FD (1995). Postingestive feedback as an elementary determinant of food preference and intake in ruminants. *Journal of Range Management* 48:2–17.

Provenza FD, Pfister JA, and Cheney CD (1992). Mechanisms of learning in diet selection with reference to phytotoxicosis in herbivores. *Journal of Range Management* 45:36–45.

Provenza FD, Villalba JJ, Dziba LE, Atwood SB, and Banner RE (2003). Linking herbivore experience, varied diets, and plant biochemical diversity. *Small Ruminant Research* 49:257–274.

Ralphs MH and Olsen JD (1987). Alkaloids and palatability of poisonous plants. In *Proceedings—Symposium on Plant-Herbivore Interactions* (FD Provenza, JT Flinders, and ED McArthur, comp), pp. 78–83. Gen. Tech. Rep. INT-222, USDA-FS Intermountain Research Station, Ogden, Utah.

Ralphs MH, Graham D, and James LF (1994). Social facilitation influences cattle to graze locoweed. *Journal of Range Management* 47:123–126.

SAS Institute (2006). *Statistical Analysis System*. SAS Institute Inc., Cary, NC.

Westoby M (1978). What are the biological bases of varied diets? *American Naturalist* 112:627–631.

Chapter 63

Water Deficit Effects on *Astragalus mollissimus* and *Oxytropis sericea* Swainsonine Content, Physiology, and Growth

A.D. Vallotton and T.M. Sterling

Department of Entomology, Plant Pathology, and Weed Science, New Mexico State University, Las Cruces, NM 88003, USA

Introduction

Locoweeds are part of a plant group comprised of *Astragalus* and *Oxytropis* species that contain the polyhyroxyindolizidine alkaloid swainsonine. When ingested, swainsonine acts as a α-mannosidase inhibitor, thereby preventing the complete metabolism of oligosaccharides. These partially metabolized sugars accumulate in intracellular lysosomes, causing cellular disruption and impairment, leading to erratic neurological function and ultimately livestock death (James and Panter 1989). Given significant economic losses due to locoweed poisoning in the western United States (Torell *et al.* 2000), a better understanding of environmental and genetic effects on swainsonine content is crucial for locoweed management.

In our initial work, swainsonine levels differed among field-collected *Astragalus mollissimus*, *Oxytropis sericea* and *O. lambertii* plants that had been grown in a common greenhouse environment, thereby suggesting that swainsonine content is genetically controlled (Vallotton and Sterling 2002). In a subsequent study, field samples of *Astragalus mollissimus* (*A.m.*) var. *bigelovii*, *A.m.* var. *mollissimus*, and *A.m.* var. *thompsonae* also contained different swainsonine levels, further indicating that genetics play a role in constitutive swainsonine content (Vallotton *et al.* 2004). However, given different environmental conditions at each of the sampling sites, the role of any environment by genetic interaction is unknown.

A major environmental factor affecting plant growth is water deficit stress, which impacts primary metabolic and physiological processes, as well as secondary plant metabolite production. Physiological impacts include reduced cell turgor, photosynthesis, transpiration, and stomatal conductance, leading to lower carbohydrate accumulation. Photosynthate tends to be channeled to root growth rather than to shoot growth (Hsiao 1973). While the role of water deficit as a causal agent of increased alkaloid content in leaves has been suggested (Waller and Nowacki 1978), the relationship between alkaloid content and water stress is directly modified by the severity of drought, nutrient status, plant phenological stage, tissue part, and the particular alkaloid compound produced (Saenz *et al.* 1993; Christiansen *et al.* 1997; Liu 2000; Baričevič *et al.* 1999).

The objective of this research was to determine the effect of water deficit in *A. mollissimus* and *O. sericea* response in terms of swainsonine levels, as well as physiological processes and overall plant growth.

Materials and Methods

In late spring of 2004, individual plants from flowering field populations of *A.m. bigelovii*, *A.m. mollissimus*, *A.m. thompsonae*, and *O. sericea* in New Mexico were dug up and transplanted into one gallon size pots filled with a 2:1:1 mixture of sand, perlite, and Metro Mix 360, respectively. *A.m. bigelovii* plants were collected from City of Rocks, NM (experiment 1) and the Burro Mountains, NM (experiment 2); *A.m. mollissimus* plants were collected from Farley, NM (experiment 1 and 2); *A.m. thompsonae* plants were collected from Farmington, NM (experiment 1 and 2); and *O. sericea* plants were collected from Maxwell, NM (experiment 1 and 2). All plants were then acclimated in a common greenhouse environment and replanted into larger 2½ gallon size pots. At about 4 months after collection, these plants were then used for two similar experiments each conducted over a 56-day time course, in which plants were exposed to two 21-day drought–recovery cycles, followed by a final 14-day drought period.

Initially, pot capacity for each species was established and target moisture levels determined. At treatment initiation, all plants were weighed daily using a portable high capacity balance to obtain gravimetric moisture loss, then re-watered to obtain target moisture levels for both droughted and control plants within each species. During drought cycles, water-deficit levels for droughted plants were established at 40% soil moisture content as compared to well-watered plants. During recovery cycles, all plants received the same moisture levels (100%) within a given species.

For each of the two experiments, gas exchange parameters and relative water content, as well as swainsonine concentration, were measured before and after each treatment period. Fifty-six days after treatment, total dry weight, root to shoot ratio, and water use per g dry weight per day were also determined. Data shown here are for the final measurement period. Gas exchange parameters were measured on fully mature, healthy leaves using a Licor LI-6400 photosynthesis meter set at a flow rate of 400 μmol CO_2 and quantum flux of 1000 μmol m^{-2} s^{-1} PAR. The same leaves were used to determine relative water content (RWC). Fresh leaves were weighed, submerged in deionized water for 16 hours and re-weighed for turgid weights, then dried at 60°C for 72 hours and weighed to obtain dry weights. Young, fully expanded leaves were also obtained for swainsonine analysis. Samples were dried , ground, extracted using resin-filled cation exchange tubes, and then analyzed using a Varian CP-3800 gas chromatograph equipped with a Saturn 2200 MS detector. Methodology followed Gardner *et al.* (2001). In addition, whole plants were harvested, in which leaves, stems, and roots were cut then separated. Roots were washed using a Gillison hydropneumatic root elutriation system. Shoots and roots were dried at 60°C for 72 hours then weighed to obtain leaf, stem, and root weights. Total biomass and root to shoot ratios were determined. Experiments were completely randomized in a 4 x 2 factorial arrangement with five reps for each treatment group. SAS ANOVA was used to separate means for each response variable in each data set. When no experiment interaction was detected, data were pooled across experiments.

Results

Plant water status

There was no species by water interaction for relative water content (RWC). Leaf RWC was not different between water-deficit versus well-watered plants except in *A.m. thompsonae*, where RWC was reduced 10% under water stress. *Oxytropis sericea* and *A.m. thompsonae* had higher RWC as compared to *A.m. mollissimus* and *A.m. bigelovii.*

Gas exchange

Under well-watered conditions, photosynthesis (μmol m^{-2} s^{-1}) was reduced in *A.m. bigelovii*. In contrast, photosynthesis for drought versus well-watered plants increased in *A.m. bigelovii*, whereas it decreased in *A.m. mollissimus* and *A.m. thompsonae*, and was not different in *O. sericea*. Under well-watered conditions, transpiration (mmol m^{-2} s^{-1}) was greater in *A.m. thompsonae* than other species. Under water-deficit conditions, transpiration increased in *A.m. bigelovii*, but remained the same in *A.m. mollissimus*, *A.m. thompsonae*, and *O. sericea*. Regardless of water treatment, transpiration and stomatal conductance were twofold greater in *A.m. thompsonae* than the other species.

Plant water use efficiency

Under well-watered conditions, water use (mL) per g dry weight per day was greatest in *A.m. thompsonae* and least in *O. sericea*. A similar trend was observed in water-deficit plants. Across both water treatments, *O. sericea* was the most efficient water user, as compared to *A.m. thompsonae*, which was the least efficient water user. Across all species, for every g dry weight gained, less water was used in droughted plants.

Plant growth

Shoot and root dry weight 56 days after treatment initiation was not different between water treatments; however, root to shoot (R:S) ratio increased with water deficit stress across all species (P = 0.0134, n = 40). Under well-watered conditions, *A.m. bigelovii* had the lowest root to shoot (R:S) ratio. Under water-deficit conditions, R:S ratio increased in *A.m. bigelovii* and *O. sericea*. Regardless of water treatment, *A.m. mollissimus* and *A.m. thompsonae* R:S ratio was the same. Total dry weight (TDW) 56 days after treatment initiation was not different between water treatments for any of the species. There was a species main effect, in which *A.m. bigelovii* and *A.m. thompsonae* TDW were twofold greater than *A.m. mollissimus* TDW and fivefold greater than *O. sericea* TDW.

Swainsonine

In droughted versus well-watered plants, *A.m. bigelovii* and *A.m. mollissimus* had greater swainsonine, whereas *O. sericea* and *A.m. thompsonae* were not affected by water treatment. Regardless of water level, *A.m. mollissimus* produced between four- to 155-fold greater swainsonine than the other species.

Conclusions

While water deficit stress generally did not translate into reduced relative water content, total dry weight, and shoot and root dry weight responses, there were some differences detected between species in terms of water use efficiency and gas exchange parameters measured. These differences suggest that while the drought imposed affected physiological function, locoweed possesses strategies to acclimate to low water conditions.

Water deficit was severe enough to induce swainsonine production in at least two species, suggesting that swainsonine production in newly forming leaves can be sensitive to environment. Because there was no genetics by environment interaction for two species of locoweed (*A.m. thompsonae* and *O. sericea*), genetic differences also can play a role in swainsonine production. Considering all species, these data suggest that *A.m. bigelovii*, *A.m. mollissimus*, and *O. sericea* may be more adapted to water deficit as compared to *A.m. thompsonae.* Overall, *A. mollissimus* response to water deficit is more phenotypically plastic as compared to *O. sericea*, which used water efficiently, partitioned more carbon to root growth, and sustained consistent swainsonine levels.

References

Baričevič D, Umek A, Kreft S, Maticic B, and Zupančič A (1999). Effect of water stress and nitrogen fertilization on the content of hyoscyamine and scopolamine in the roots of deadly nightshade (*Atropa belladonna*). *Environmental and Experimental Botany* 42:17–24.

Christiansen JL, Jørnsgård B, Buskov S, and Olsen CE (1997). Effect of drought stress on content and composition of seed alkaloids in narrow-leafed lupin, *Lupinus angustifolius* L. *European Journal of Agronomy* 7:307–314.

Gardner DR, Molyneux RJ, and Ralphs MH (2001). Analysis of swainsonine: extraction methods, detection, and measurement in populations of locoweeds (*Oxytropis* spp.). *Journal of Agriculture and Food Chemistry* 49:4573–4580.

Hsiao TC (1973). Plant responses to water stress. *Annual Review Plant Physiology* 24:519–570.

James LF and Panter KE (1989). Locoweed poisoning in livestock. In *Swainsonine and Related Glycosidase Inhibitors* (LF James, AD Elbein, RJ Molyneux, and CD Warren, eds.), pp. 23–28. Iowa State University Press, Ames.

Liu Z (2000). Drought-induced in vivo synthesis of camptothecin in *Camptotheca acuminata* seedlings. *Physiologia Plantarum* 110: 483–488.

Saenz L, Santamaria JM, Villanueva MA, Loyola-Vargas VM, and Oropeza C (1993). Changes in the alkaloid content of plants of *Catharanthus roseus* L. (Don.) as a result of water stress and treatment with abscissic acid. *Journal of Plant Physiology* 142:244–247.

Torell AL, Owen LP, McDaniel KC, and Graham D (2000). Perceptions and economic losses from locoweed in northeastern New Mexico. *Journal of Range Management* 53:376–383.

Vallotton AD and Sterling TM (2002). Variation in swainsonine content among extraction methods and between locoweed genera. *Proceedings Western Society of Weed Science* 55:18.

Vallotton AD, Gardner DR, Ralphs MH, and Sterling TM (2004). Comparison of three methods for swainsonine detection and quantification in woolly loco. *Society Range Management, 57th Annual Meeting.*

Waller GR and Nowacki EK (1978). *Alkaloid Biology and Metabolism in Plants*, pp. 83–119. Plenum Press, New York.

Chapter 64

Rumen Function and the Ability of Goats to Consume Redberry Juniper (*Juniperus pinchottii)*

W.T. Dunson[1], C.B. Scott[1], E.S. Campbell[2], C.A. Taylor, Jr.[2], M.A. Carr[1], and T.R. Callaway[3]

[1]Department of Agriculture, Angelo State University, ASU Station Box 10888, San Angelo, TX 76909, USA; [2]Texas Agricultural Experiment Station, 7887 Hwy 87N, San Angelo, TX 76903, USA; [3] USDA ARS, Southern Plains Agricultural Research Center, 2881 F&B Rd., College Station, TX 77845, USA

Introduction

Redberry juniper (*Juniperus pinchotii*) is a basal-sprouting, evergreen conifer common in the southwestern US (Adams and Zanoni 1979). The plant is particularly problematic in central and western Texas where densities have increased dramatically in the last 30 years (Ansley *et al.* 1995). Infestation of rangelands by redberry juniper reduces forage production for livestock, decreases habitat quality for wildlife, and may adversely impact water yield and water quality (Dye *et al.* 1995; Taylor *et al.* 1997). Conventional brush control methods are typically impractical because of cost (Johnson *et al.* 1999).

Most species of livestock do not utilize redberry juniper because the terpenoids found in the plant cause aversive postingestive feedback (Riddle *et al.* 1996; Pritz *et al.* 1997). Even when forced to consume juniper in pen-feeding situations, intake rarely exceeds 30% of the diet (Straka *et al.* 2004). Redberry juniper intake is usually reported to be lower in free-ranging studies, at 0–13% of diet composition (Bryant 1977).

Some goats are apparently able to adapt to the terpenoids in juniper through exposure early in life (Bisson *et al.* 2001; Ellis *et al.* 2005). Apparently, gastrointestinal and hepatic physiology changes to the point that goats are able to detoxify and excrete terpenoids from the plant. The liver serves as the primary clearing organ for many toxins (Bidlack 1982; Weimer 1998; Nebbia 2001). Rumen microflora also aid in the degradation of some toxins (Jones and Megarrity 1986; Prins 1987; Allison *et al.* 1990; Majak 1992; Duncan *et al.* 2000).

The objective of this study was to determine if changes in the rumen microbial population are responsible for the detoxification and reduction of redberry juniper terpenoids in rumen fluid. If the mechanism of detoxification can be identified, it may be possible to select for or manipulate this process in an effort to increase redberry juniper utilization. For example, if rumen microbial populations are responsible for degradation of the terpenes found in redberry juniper, improving juniper consumption may simply involve feeding juniper in pens to improve acceptance.

Methods

Two experiments were conducted during the fall of 2003 at the Angelo State University Management, Instruction, and Research Center, San Angelo, TX, USA. The objective of both studies was to quantify differences in rumen microbial activity between goats fed juniper and goats fed alfalfa alone. Research objectives were addressed by placing goats in individual pens for 10–14 days and feeding juniper. Redberry juniper was harvested at the Texas Agricultural Experiment Station, Sonora, TX, USA, stripped from the stems, and stored at 0°C until feeding. Intake, serum metabolite levels, terpene disappearance rates, volatile fatty acid profiles, and ammonia levels were monitored and compared between treatments (fed juniper or fed alfalfa). Fresh water and calcium/phosphorous mineral with trace elements were provided *ad libitum* to all goats during testing in individual pens. All goats also received alfalfa pellets (2% bw) daily to meet maintenance requirements (NRC 1981).

Experiment 1

For the first experiment, 20 freshly weaned, mixed-sex Boer-Spanish cross goats (0=20 kg) (tengoats/treatment) and six freshly weaned, female, ruminally cannulated Boer-Spanish cross goats were placed in individual pens (4 X 5 m) after weaning. After a 14-day pen-adjustment pretrial where goats were fed only alfalfa (2% bw), 13 randomly selected goats were fed redberry juniper for 2 hours daily for 14 days. Three goats were randomly selected from the cannulated group and ten from the 20 goats without cannulas. Intake was monitored daily. Serum metabolite levels were monitored at day 0, 5, 10, and 15 to assess physiological status. In previous studies, dosing goats with juniper also affected serum constituent levels (serum aspartate transaminase, blood urea nitrogen, gamma glutamyltransferase, creatinine, and bilirubin) (Pritz *et al.* 1997; Bisson *et al.* 2001). Pritz *et al.* (1997) noted increases in AST above the physiologically normal range. Bisson *et al.* (2001) noted a statistically significant increase in AST levels that remained within biologically normal range (ISIS 1995). Changes in serum levels are indicative of toxicosis (Cornelius 1989; Cheeke 1998). It appears that a minor hepatic insult occurs when goats are exposed to juniper monoterpenes but the insult does not approach pathological levels. Blood was collected via jugular veinpuncture and serum was extracted by centrifugation, frozen, and sent to the Texas Veterinary Medical Diagnostic Lab in College Station, TX, USA for chemical analysis.

Rumen fluid was collected from the cannulated goats on day 10 to assess changes in rumen microbial populations and toxin degradation. Redberry juniper contains volatile oils (Gershenzon and Croteau 1991) which contain monoterpenes of several different chemical classes (Straka 2000). Four of these terpenes (myrcene, limonene, terpineol, ∀-pinene) were used in this experiment because of commercial availability and because three of them (myrcene, terpineol, ∀-pinene) have been negatively correlated with juniper intake (Riddle *et al.* 1996). Approximately 200 mL of rumen fluid was collected from each goat, mixed with rumen fluid from goats fed the same diet, and filtered through cheesecloth to remove particulates. Oxygen was displaced by CO_2 in triplicate glass tubes and terpenes were added individually (%/mL) and in combination (1 mL total) to 9 mL of a rumen fluid/media mixture (100 mL rumen fluid with 200 mL media) (Callaway *et al.* 1997). Oxygen was displaced again and the tubes were sealed and incubated for 24 hours at 39EC. Concentrations used (13% ∀-pinene, 58% limonene, 24% myrcene, 5% terpineol) in this study were based on terpene levels from juniper found growing on the same collection site (Owens *et al.* 1998). Tubes were uncapped after 24 hours and fluid was transferred to sterile 10 mL test tubes. Thereafter, tubes were centrifuged for 10 min at 7000 rpm. After spinning, the supernatant was transferred to three cryogenic tubes (2 mL each) and frozen for testing. Differences in volatile fatty acids (VFAs) and ammonia levels were determined using the procedure described by Callaway *et al.* (1997) and Chaney and Marbach (1962). *In vitro* terpene degradation was determined by gas chromatography. An HP 5790 instrument equipped with an HP-1 (non-polar) 0.32mm x 25m x 0.52 um column was used to measure monoterpenes prior to and following *in vitro* fermentation using a modification of the procedure as described by Owens *et al* (1998). We hypothesized that if rumen degradation of terpenes accounted for differences between animals in ability to consume juniper, then terpene disappearance rates, VFA levels, and ammonia levels (indicators of rumen microbial activity) would differ between goats fed juniper and goats fed only alfalfa.

Experiment 2

Immediately following the first experiment, approximately 325 mL of rumen fluid was collected from each cannulated goat. Fluid was thoroughly mixed with rumen fluid from goats fed the same diet and strained through cheesecloth to remove particulates. The ten goats not fed juniper in Experiment 1 were inoculated with rumen fluid from either cannulated goats fed juniper or cannulated goats fed only alfalfa. All ten naive goats were fed redberry juniper for 2 hours daily for 10 days thereafter and intake was monitored daily to determine if inoculation from goats fed juniper would increase rate of adaptation to the plants toxins faster than inoculation from naive goats.

Statistical analysis

Differences between treatments (fixed effect) for intake (both Experiments 1 and 2) and serum metabolite levels (Experiment 1) were assessed using repeated measures analysis of variance with individual goats (random effect) as replications nested within treatments and day as the repeated measure (Hicks 1993). Toxin degradation, VFA profiles, and ammonia levels were assessed using analysis of variance with goats nested within treatments (fed juniper or fed alfalfa alone) as replications. Differences among means were determined using the least significant difference (LSD) test when $\underline{P}<0.05$ (Gomez and Gomez 1984). Data was analyzed with the statistical package JMP (SAS 1994).

Results

Experiment 1

Alfalfa intake was similar between treatments and remained constant throughout Experiment 1 (Figure 1). Juniper intake increased daily from day 1 (0.38 g kg^{-1} bw) to day 11 (8.75 g kg^{-1} bw). Intake levels declined on day 12; intake on day 13 was similar to day 10 before increasing on day 14 (9.47 g kg^{-1} bw; Figure 2).

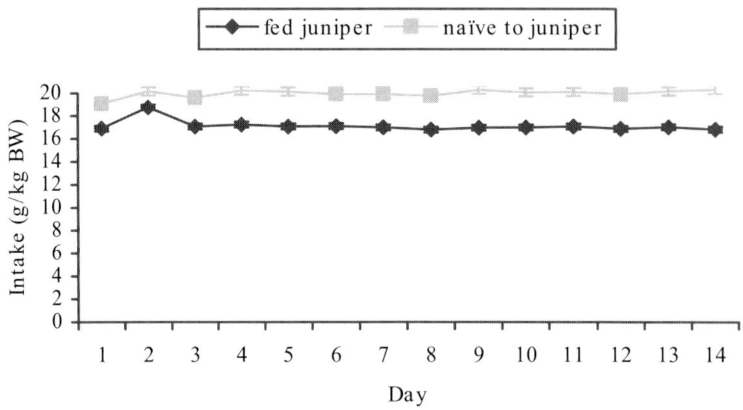

Figure 1. Intake (g kg^{-1} bw) of alfalfa across both treatments during Experiment 1. For both treatments, alfalfa was fed daily to meet maintenance requirements.

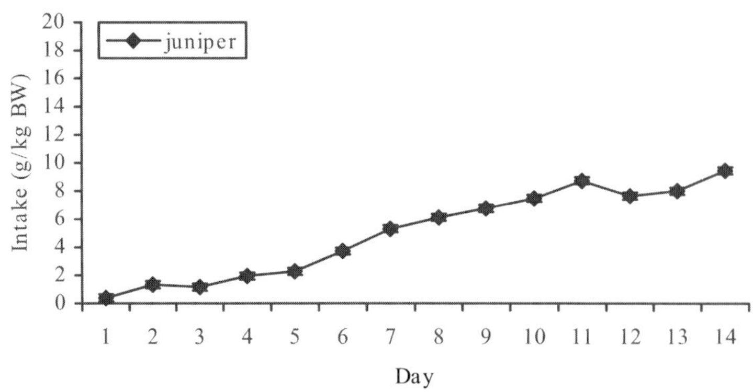

Figure 2. Intake (g kg^{-1} bw) of redberry juniper during Experiment 1. Goats were offered juniper for 2 hours daily for 14 days and intake was monitored daily.

Alfalfa intake was similar for cannulated goats and for non-cannulated (19.6∀0.2 g kg^{-1} bw versus 20.1∀0.1 g kg^{-1} bw). Conversely, juniper intake was lower for cannulated versus non-cannulated goats fed juniper (1.3∀0.1 g kg^{-1} bw versus 6.1∀0.5 g kg^{-1} bw). Juniper intake ranged fro 0.9 to 1.5 g kg^{-1} bw for cannulated goats while juniper intake ranged from 0.9 to 9.6 g kg^{-1} bw for non-cannulated goats. For four of the non-cannulated goats juniper made up over 40% of the diet.

Serum metabolite levels of gamma glutamyltransferase (GGT), aspartate transaminase (AST), and creatinine were elevated in goats consuming juniper (Table 1). A significant treatment x day interaction with creatinine, GGT, and AST existed. For goats fed juniper, metabolite levels increased across days of collection while levels remained constant for goats fed only alfalfa. Although serum metabolite levels were elevated in goats consuming juniper, none were outside the normal range for healthy goats (ISIS 1995).

Table 1. Serum metabolite levels for goats fed redberry juniper for 14 days and goats naive to juniper. Samples were collected on days 0, 5, 10, and 15 of the study. Serum levels on day 0 were used as a covariate to account for variation among individual goats.

	Treatment		
Serum metabolite	Fed Juniper	Fed Alfalfa	SEM
Creatinine (mg/dL)	0.60^{a}	0.54^{b}	0.01
AST (U/L)	62.57^{a}	61.26^{b}	1.93
GGT (U/L)	47.78^{a}	44.52^{b}	0.97
BUN (mg/dL)	22.54	23.62	0.51

$^{a-b}$Means within rows with different superscripts differ ($\underline{P}$<0.05)

In vitro toxin degradation of terpenes was similar between treatments (Table 2). Disappearance of individual terpenes, with the exception of myrcene, (limonene, terpineol, α-pinene), and a combination of all four, was similar in fluid from goats consuming juniper and fluid from naive goats.

Table 2. Remaining levels of terpenes after incubation in rumen fluid from goats either fed juniper or alfalfa alone. Rumen fluid was collected from cannulated goats on day 10, mixed with a rumen fluid media and incubated for 24 hours in an anaerobic environment.

Terpene	Amount added to fluid	Fed Juniper	Fed Alfalfa	P-value
Combined Terpenes	1 mL	0.82 mL	0.68 mL	0.60
α-pinene	130 μL	83.33 μL	50.02 μL	0.28
Limonene	580 μL	566.70 μL	316.70 μL	0.09
Myrcene	240 μL	107.86 μL	30.71 μL	0.01
Terpeniol	50 μL	13.52 μL	11.13 μL	0.80

Most volatile fatty acid levels were unaffected by feeding juniper. However, when all four terpenes were added in combination, acetic acid and isobutyric acid levels decreased for goats that had been consuming juniper. Acetic acid levels were also lower for goats consuming juniper when myrcene was added alone. Adding terpineol to rumen fluid also resulted in lower levels of acetic acid, butyric acid, and valeric acid. Ammonia levels were similar between goats fed juniper and goats fed alfalfa alone.

Experiment 2

Dosing with rumen fluid had no effect on subsequent juniper intake (treatment and treatment x day effects were not significantly different). Both goats dosed with rumen fluid from goats consuming juniper and from those dosed with rumen fluid from goats fed alfalfa alone consumed juniper at similar levels (Figure 3).

Discussion

Results of this study confirm Boer-Spanish cross goats will consume redberry juniper when fed juniper in pens after weaning (Bisson *et al.* 2001; Ellis *et al.* 2005). Juniper intake increased daily until day 11. At this point, goats were probably experiencing aversive postingestive feedback from the monoterpenes in juniper. Goats in Experiment 2 experienced similar results for juniper intake with levels increasing daily to day 8 (9.0 g kg^{-1} bw) before dropping on days 9 and 10 to 8.0 g kg^{-1} bw.

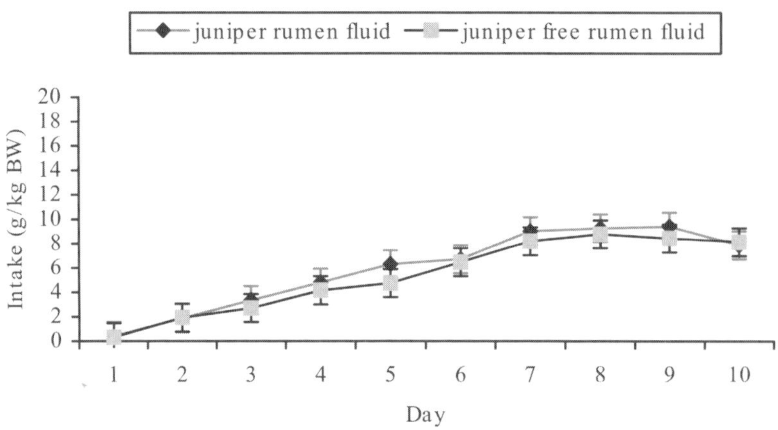

Figure 3. Intake (g kg^{-1} bw) of redberry juniper after inoculation with rumen fluid from goats fed juniper versus rumen fluid from goats fed only alfalfa during Experiment 2.

Analysis of serum metabolite levels suggests that goats may have experienced some effects from toxicosis from over-ingestion of juniper. Levels of certain blood serum metabolites can be an indicator of toxicosis when elevated beyond normal ranges (Cheeke 1998). In Experiment 1, levels of GGT, AST, and creatinine were elevated in goats consuming juniper. However, none were elevated beyond the normal range for healthy goats (ISIS 1995). Creatinine can be used as an indicator of kidney damage, but is such a sensitive indicator of muscle damage that, generally, only large increases are of clinical significance (Cornelius 1989). Aspartate transaminase (AST) and gamma glutamyltransferase (GGT) are both used for diagnosis of liver damage (Cornelius 1989), but elevated levels are irrelevant when still within the accepted range.

In vitro toxin degradation of terpenes, VFA profiles, and ammonia levels were measured in an effort to determine if a rumen microbial adaption to terpenes in juniper existed. Terpenes were added to rumen fluid from goats consuming juniper and from goats consuming alfalfa alone, incubated for 24 hours, and analyzed for toxin disappearance and changes in VFA and ammonia levels. Terpene disappearance rates were similar for goats fed juniper versus goats fed alfalfa alone. Similarly, ammonia levels were unaffected by treatments. These results suggest rumen microbial adaption does not appear to be the physiological processes allowing goats to consume juniper.

Some volatile fatty acid levels differed between treatments. When juniper intake exceeds 30% of the diet, microbial death occurs, which alters VFA levels (Straka 2000). Differences in VFA levels reported in this study were probably the result of removal of some of the ruminal microbial populations, and decreased degradation of terpenes in the rumen as illustrated in Table 2.

In Experiment 2, ten goats were inoculated with rumen fluid from either goats familiar with juniper or goats naive to juniper and juniper intake was monitored daily for 10 days. Juniper intake should have been higher when inoculated with fluid from familiar goats, especially if the rumen was adapting to the toxin. However, both goats dosed with rumen fluid from goats consuming juniper and from those consuming alfalfa alone consumed juniper at similar levels.

If toxins are not degraded into nontoxic compounds in the rumen, detoxification typically relies on liver function (Cheeke 1998). Many toxins are converted to nontoxic compounds by the cytochrome P-450-dependent mixed function oxidase (MFO) system and several conjugation enzymes that convert lipophilic compounds to more water soluble products which enhance excretion (Bidlack *et al.* 1986; Nebbia 2001). Given the lack of evidence supporting rumen degradation of monoterpenoids in this study, it appears the liver may be the primary site of detoxification in goats.

References

Adams RP and Zanoni TA (1979). The distribution, synonymy and taxonomy of three junipers of southwestern United States and northern Mexico. *Southwestern Naturalist* 24:323–329.

Allison MJ, Hammond AC, and Jones RJ (1990). Detection of ruminal bacteria that degrade toxic dihydroxypyridine compounds produced from mimosine. *Applied Environmental Microbiology* 56:590–594

Ansley RJ, Pinchak WE, and Ueckert DN (1995). Changes in redberry juniper distribution in northwest Texas (1948 to 1982). *Rangelands* 17:49–53.

Bidlack WR (1982). Toxicant metabolism and the role of nutrients. *Food Technology* 36:106–113.

Bidlack WR, Brown RC, and Mohan C (1986). Nutritional parameters that alter hepatic drug metabolism, conjugation, and toxicity. *Federation Proceedings* 45:142–148.

Bisson MG, Scott CB, and Taylor CA (2001). Activated charcoal and experience affect intake of juniper by goats. *Journal of Range Management* 54:274–278.

Bryant FC (1977). Botanical and nutritive content in diets of sheep, Angora goats, Spanish goats and white-tailed deer grazing a common pasture. PhD Dissertation, Texas A&M University, College Station.

Callaway TR, De Melo AMSC, and Russell, JB (1997) The effect of Nisin and Monensin on ruminal fermentations *in vitro*. *Current Issues in Microbiology* 35:90–96.

Chaney AL and Marbach EP (1962). Modified reagents for determination of urea and ammonia. *Clinical Chemistry* 8:130–132.

Cheeke P (1998). *Natural Toxicants in Feeds, Forages and Poisonous Plants*. Interstate Publishing, Danville, Illinois.

Cornelius CE (1989) Liver function. In *Clinical Biochemistry of Domestic Animals* (JJ Kaneko, ed), pp. 364–397. Academic Press, New York.

Duncan AJ, Frutos P, and Young SA (2000). The effect of rumen adaptation to oxalic acid on selection of oxalic-acid-rich plants by goats. *British Journal of Nutrition* 83:59–65.

Dye KL, II, Ueckert DN, and Whisenant SG (1995). Redberry juniper–herbaceous understory interactions. *Journal of Range Management* 48:100–107.

Ellis CR, Jones RE, Scott CB, Taylor CA, Walker JW, and Waldron DF (2005). Sire influence juniper consumption by goats. *Journal of Rangeland Ecology and Management* 58:324–328.

Gershenzon J and Croteau R (1991) Terpenoids. In *Herbivores: Their Interaction with Secondary Plant Metabolites* (GA Rosenthal and MR Berenbaum, eds), pp. 165–219. Academic Press Inc., San Diego, California.

Gomez KA and Gomez AA (1984). *Statistical Procedures for Agricultural Research*, pp. 187–207. John Wiley and Sons, New York.

Hicks CR (1993) *Fundamental Concepts in the Design of Trials*, pp. 173–199. Saunders College Publishing, New York.

ISIS Physiological Data Reference Values (1995).

Johnson P, Gerbolini A, Ethridge D, Britton C, and Ueckert D (1999). Economics of redberry juniper control in the Texas rolling plains. *Journal of Range Management* 52:569–574.

Jones RJ and Megarrity RG (1986). Successful transfer of DHP-degrading bacteria from Hawaiian goats to Australian ruminants to overcome the toxicity of *Leucaena*. *Australian Veterinary Journal* 63:259–262.

Majak W (1992). Further enhancement of 3-nitropropanol detoxification by ruminal bacteria in cattle. *Canadian Journal of Animal Science* 72:863–870.

Nebbia C (2001). Biotransformation enzymes as determinants of xenobiotic toxicity in domestic animals. *The Veterinary Journal* 161:238–252.

NRC (1981). *Nutrient Requirements of Goats*. National Academy Press, Washington D.C.

Owens MK, Straka EJ, Carroll CJ, and Taylor CA, Jr (1998). A comparison of techniques for extracting monoterpenoids from *Juniperus* (Cupressaceae) species. *Journal of Range Management* 51:540–544.

Prins RA (1987). Rumen microbial metabolism of plant secondary compounds. In *Physiological and Pharmacological Aspects of the Reticulo-rumen* (LAA Ooms, AD Degryse, and ASJPAM Van Miert, eds), pp. 199–225. Martinus Nijhoff Publishers, Dordrecht.

Pritz RK, Launchbaugh KL, and Taylor CA, Jr (1997). Effects of breed and dietary experience on juniper consumption by goats. *Journal of Range Management* 50:600–606.

Riddle RR, Taylor CA, Jr, Kothman MM, and Huston JE (1996). Volatile oil contents of ashe and redberry juniper and its relationship to preference by Angora and Spanish goats. *Journal of Range Management* 49:35–41.

SAS (1994). *JMP User's Guide*, Version 3.1. SAS Institute Inc., North Carolina.

Straka E, Scott CB, Taylor CA Jr, and Bailey EM (2004). Biological control of the toxic shrub juniper. In *Poisonous Plants and Related Toxins* (T Acamovic, CS Stewart, and TW Pennycott, eds.), pp. 436–442. CABI Publishing, Wallingford, Oxon, UK.

Straka EJ (2000). The physiological effects of monoterpenes on Spanish and Angora goats. PhD Dissertation, Texas A&M University College Station.

Taylor CA, Jr, Launchbaugh KL, Huston JE, and Straka E (1997). Improving the efficacy of goating for biological juniper management. In *Juniper Symposium*, San Angelo, Texas (CA Taylor, ed). Texas Agricultural Experiment Station Technical Report 97-1.

Weimer PJ (1998). Manipulating ruminal fermentation: a microbial ecological perspective. *Journal of Animal Science* 76:3114–3122.

Chapter 65

The Use of Fecal NIRS to Identify Levels of Consumption of the Toxic Shrub Juniper in a Selective Breeding Program with Goats

E.S. Campbell[1], B.S. Engdahl[2], C.J. Lupton[2], C.B. Scott[3], C.A. Taylor, Jr.[1], D.F. Waldron[2], J.W. Walker[2], and W.R. Whitworth[3]

[1] *Texas Agricultural Experiment Station at Sonora, P.O. Box 918, Sonora, TX 76950, USA;* [2] *Texas A&M University Agricultural Research and Extension Center, 7887 US HWY 87N, San Angelo, TX 76901, USA;* [3] *Angelo State University, Box 10888 ASU Station, San Angelo, TX 76909, USA*

Introduction

Juniper encroachment is a major ecological problem in the Edwards Plateau region of Texas. The primary species are *Juniperus pinchotii* (Sudworth) and *Juniperus ashei* (Buchholz).

Juniper infestation of Texas rangelands is an important dilemma because of its impact on livestock production, water yield and quality, wildlife habitats, and the prohibitive expense of conventional control methods. Juniper species (once minor proportions of the rangeland flora) have expanded their range. They now occupy most soil types and vegetation communities in central and west Texas. Livestock carrying capacity can be reduced by as much as 85% in the transition from grassland to a closed canopy cover of juniper.

The objective of this project is to investigate the use of selective breeding to improve the use of goats as biological control agents. The research has two components. The first is to investigate the use of fecal near-infrared reflectance spectroscopy (NIRS) technology for predicting juniper consumption in free-ranging goats. The second is to predict the genetic merit for juniper consumption for individual animals in two herds of goats (Boer x Spanish and Angora) in order to divergently select for high and low juniper consumption.

By selecting goats that exhibit an increased preference for juniper, we believe that we are selecting animals with a higher physiological capacity to denature the terpenoids in juniper through hepatic clearance. Metabolic fate of the terpenoids present in juniper involves hepatic biotransformation and subsequent excretion in the urine. Expression of hepatic microsomal enzymes is highly variable in nature due largely to the influence of genetics. Studies with sporidesmin toxicity in sheep have shown that resistant and susceptible ewes differ in their hepatic biotransformation capacity (Smith *et al.* 1980).

Also, breeding studies with mice have shown that lines that are either resistant or susceptible to sporidesmin can be created, and that resistance was a function of levels of expression of hepatic Phase II enzymes (Hohenboken and Blodgett 1997).

It was necessary to develop a quantitative tool that provided an accurate means of identifying superior juniper-consuming goats. Near Infrared Spectroscopy (NIRS) technology was adapted to predict diet quality (CP %, DOM %) of free-ranging cattle (Lyons and Stuth 1992), and goats (Leite and Stuth 1995) from fecal samples. Fecal NIRS was further refined to predict a specific botanical component of the diet by Walker *et al.* (1998), who estimated leafy spurge concentration in a sheep's diet. This technology posed a quick and easy way to potentially measure juniper consumption. There are limitations to this technique, however, and they are the difficulty of independent validation of predictions, and the need to consider predictions as interval scale data. This paper will provide an overview of the research roadmap we followed to develop NIRS as an effective tool to measure levels of a specific botanical component of a goat's diet under free-range conditions.

Calibration Equation Development

Two NIRS calibration studies were conducted to produce the current prediction equation. The first study compared two calibration data sets, one with two levels of juniper superimposed on eight diverse background forages, the other with ten levels of juniper and two background forages. Fecal spectra were used to develop prediction equations using modified partial least squares regression to predict percent juniper in the diets fed using the log (1/reflectance) for wavelengths between 1108 and 2492 nm. Equations developed with the calibration data set containing diverse background diets were more robust than the calibration data set with numerous juniper levels (Whitworth 2002). The second study estimated the effect of including samples with no juniper (zero fecals) on precision and accuracy of predictions. Both precision (R^2 increased from 0.76 to 0.94) and accuracy (slope increased from 0.66 to 0.88) of the equations were improved.

Repeatability of predictions, however, posed a challenge. Correlations of NIRS-predicted juniper between multiple samples taken from the same animals at different times varied from moderate to low. In the Spanish x Boer population correlations were highest between samples taken when forage quality was low due to season or drought (Table 1). In the Angora population correlation between successive samples on the same animal were low (Table 2), and followed the same trend seen in meat goats, i.e. correlations were highest between samples when intake was also high. Predictions from the four sampling dates showed the expected seasonal trends in juniper intake to be high in the winter and low in the summer. The lack of high correlation between successive measurements indicated a need to identify potential sources of variability which were contributing to poor repeatability.

Effect of Breed, Gender and Age

We first evaluated whether calibration equations developed from mature Boer x Spanish muttons (castrates) could accurately predict juniper in diets of other species, genders and ages of animals. In other words, to identify the contributions of breed, gender and age of goat on variation in fecal NIRS spectra and predicted juniper in the diet. One trial compared adult goats from two breeds (Angora and Boer x Spanish) and three genders (nanny (female), billy (male) and mutton goats) in a factorial design. The other trial compared adult and kid (<12 months of age) Angora goats from two genders (nanny and billy) in a factorial design. Eleven goats were fed in each treatment combination. Goats were pen fed a pelleted goat feed containing no juniper for 10 days followed by 10 days of a pelleted goat feed containing 14% juniper. Fecal samples were collected on days 7–10 for each diet. Fecal samples were prepared and scanned according to the method described by Walker *et al.* (1998).

Table 1. Pearson correlation coefficients (r) for correlation of seasonal fecal NIRS-predicted juniper intakes for Boer x goats at Sonora, Texas.

	Aug 03	Feb 04
Feb 03	0.11	0.53
Aug 03		0.03

Table 2. Pearson correlation coefficients (r) for fecal NIRS-predicted juniper intakes in different seasons for Angora goats at Read Ranch, Texas.

	Dec 03	Mar 04	May 04
Aug 03	0.12	0.18	0.07
Dec 03		0.10	−0.06
Mar 04			0.08

Spectra from feces of goats in the feeding trial were analyzed to determine if there were differences due to the effect of breed, gender or age on the spectral characteristics of the feces and on the predicted percent juniper in the diets using internal and external calibration equations. Comparison of the first five PCS (principle component score) scores between adult Angora and meat goats for diet, breed and sex (Table 3) showed that diet differed ($P < 0.001$) on PC1, which accounted for 71% of the variation in the original data set. Gender was the second most important factor affecting fecal spectral characteristics as shown by differences in PC2, PC3 and PC4 ($P < 0.03$), which accounted for a total of 26% of the variation in the original data. The breed effect was only significant for PC5, which accounted for 1% of variation. Analysis of the principle components for the effects of diet, age and gender in Angora goats showed an effect of diet (P=0.000) for PCS1, which accounted for 81.6% of the variation in the original data set. Age was the second most important factor affecting fecal spectral characteristics as shown by differences in PC2 and PC4 ($P < 0.034$), and accounted for 11% of the variation of the spectral data. The gender effect was only significant for PC3 ($P = 0.037$).

This analysis indicates that goats of different gender and age when fed the same diet can have different fecal spectra. The effect of breed on fecal spectra was only significant on PCS 5, indicating that it was relatively unimportant.

Analysis of fecal NIRS-predicted percentage of juniper in the diet from the internal and independent equations (Table 4) showed that predictions from the independent equation were about twice the actual percentage of juniper fed. Despite the upward bias for the independent equation, differences were readily detected ($\underline{P} < 0.001$), and the difference between the two diets was close to the actual differences, i.e. 13 and 18 percentage units for adult and Angora, respectively. As expected, predictions for the internal equation were near the actual values. For the internal equation the effect of diet accounted for 95% of the variance (i.e. ω^2) compared to only about 50% for the independent equation. Most of the remaining variation in the analysis of variance for the independent predictions was in the error. This is consistent with the low validation r^2 for the independent equation.

Table 3. Effect of treatments on principle component scores (PCS) calculated from reflectance data of feces from goats on feeding trial.

	PCS1	PCS2	PCS3	PCS4	PCS5
Breed & Gender Comparison					
	Percent of Variation				
Individual PCS Score	71.0	19.3	4.3	2.1	1.0
Cumulative Variation	71.0	90.3	94.6	96.7	97.7
	Probability of Greater F				
% Juniper in Diet	0.000	0.097	0.560	0.004	0.380
Breed	0.313	0.843	0.950	0.656	0.001
Gender	0.819	0.000	0.000	0.029	0.604
Angora Age & Gender Comparison					
	Percent of Variation				
Individual PCS Score	81.6	8.9	4.2	2.6	1.0
Cumulative Variation	81.6	90.5	94.7	97.3	98.3
	Probability of Greater F				
% Juniper in Diet	0.000	0.227	0.360	0.485	0.885
Age	0.262	0.033	0.229	0.001	0.252
Gender	0.267	0.505	0.038	0.592	0.498

The effect of gender for the adult data was significant for both the internal ($P = 0.048$) and independent ($P = 0.004$) equations (Table 4). The effect of gender was relatively less important compared to the diet effect ($\omega^2 = 0$ and 9% for internal and independent predictions, respectively). In the Angora data, gender was significant ($P = 0.003$; $\omega^2 = 7\%$) only for the independent predictions. The actual differences among genders were about 1 percentage unit for the internal equation but 7 percentage units for the independent equations. The effects of breed and age as predicted by both equations were not different. Thus, even though spectral characteristics of fecal samples from goats fed the same diet differed because of gender, analysis of variance of these predictions showed that the diet differences could be detected and the difference between the two diets in predicted percentage of juniper was close to the actual difference (Table 4). This finding is consistent with the conclusion of Walker *et al.* (2005) that fecal NIRS predictions of botanical composition of the diet from independent calibration equations represent an interval scale of measurement. An interval scale of measurement means that treatments can be ranked and the differences between treatments have meaning and are equal across the range of measurements, but there is not a true zero point.

Long-term Consumption Patterns

Previous studies have identified temporal variations in goats' consumption of juniper (Pritz *et al.* 1996). Daily, monthly and seasonal variation presents a problem with respect to fecal sampling. Samples may inadvertently be collected on a peak or trough day and not consistently represent juniper eating ability. This study sought to quantify those variations so that sampling techniques can be refined to collect fecal pellets representing that goat's consumptive capacity for juniper.

Table 4. Effect of breed, gender and age of goat on predicted percent juniper in the diet using either an internal or an external (independent) calibration equation[1]. Predicted fecal samples were from six animals in each breed, sex and age combination and were not used in either of the calibration data sets.

	Data Source for Calibration Equation	
	Internal[2]	External[3]
Breed & Sex Comparison		
% Juniper in Diet	P=0.000	P=0.000
0	0.5	9.3
14	14.1	22.4
Breed	P=0.365	P=0.492
Angora	7.5	15.3
Meat	7.2	16.4
Sex	P=0.048	P=0.004
Billy	6.8	18.4
Mutton	7.8	17.2
Nanny	7.3	11.9
Angora Age & Sex Comparison		
% Juniper in Diet	P=0.000	P=0.000
0	0.1	4.2
14	13.6	22.3
Age	P=0.583	P=0.383
Adult	7.0	14.1
Kid	6.7	12.3
Sex	P=0.494	P=0.003
Billy	7.0	16.5
Nanny	6.7	10.0

[1] $\omega^2 = (SS_{effect} - (df_{effect})(MS_{error})) / (MS_{error} + SS_{total})$ Omega squared is an estimate of the dependent variance accounted for by the independent variable in the population for a fixed effects model

[2] Internal modified partial least squares calibration equations were developed by randomly selecting four or five animals in each breed gender and age combination from the feeding trial to develop calibration equations.

[3] External modified partial least squares calibration equations were developed using data from feeding trials conducted in 1999, 2002 and zero fecals from goats grazing pastures with no juniper.

For 4 months 12 Angora nanny goats were maintained in a 40-acre pasture supplied with free-choice mineral and water. Six goats were selected from the higher consumption group, and six from the lower consumption group. Two times per week goats were gathered and penned for fecal collection. Animals were individually penned in a screen-bottom pen (50 x 90 cm) and fecal samples were collected from the screen. Rainfall and temperature were recorded.

A repeated measures analysis of variance was used to compare predicted juniper levels between the two treatment groups, high consumers and low consumers. Results identified a significant difference ($\underline{P} < 0.05$) between the high and low goats, with the difference being more pronounced in December and January compared to October and November. There was not a treatment by time effect. When the difference, mean square error, and probability were plotted (Figure 1), the difference between the groups became clearer. As difference increased going into the period of poor alternative forage availability, error terms decreased, and significance increased. This reinforced the selection of December and January as the optimal months for sample collection.

Heritability

The potential for genetic improvement of juniper consumption by selection is a function of heritability. In order to estimate heritability, fecal samples (N=789) were collected in January or February of 2002–2004 from a herd of Boer-cross goats with known pedigrees. A total of 417 different does were sampled. The does were the progeny of 64 different sires. The percent of juniper in the diet was predicted from fecal NIRS calibration equations. Juniper predictions were combined with the pedigree information to estimate heritability of juniper preference. The sire-model heritability estimate from using all records was 41%. Fecal samples from 159 Angora kids (from six sires) were collected over a 1-year period. The Angora sire-model heritability estimate was 19%.

The moderate level of heritability seen in the Boer x Spanish goats suggests that there is significant scope for genetic improvement by selection. The lower value in the Angoras is possibly due to less variation for juniper consumption in Angora goats or the relatively small sample of sires used in this first year. Angoras have historically been selected for fiber production and it is possible that selection for this characteristic has selected away from diversity of metabolic foraging strategies. Spanish goats evolved in this region and descended from goats brought to Texas by early settlers and Spanish explorers (Shelton 1990). As feral goats depend upon utilization of native browse species, including chemically defended plants, evolution would favor development of systems of metabolic compensation. Diet consumption studies reflect this as Spanish goats consume higher levels of juniper compared to Angora goats (Riddle *et al.* 1996).

Applications

The use of fecal NIRS to predict botanical composition of grazing ruminant diets offers new opportunities in the field of monitoring consumption of poisonous and/or noxious plants. Caution must be used, however, to understand the limitations of this technique. The two greatest limitations are that predictions cannot be easily independently validated and that predictions cannot be considered absolute and must be considered to represent interval scale data.

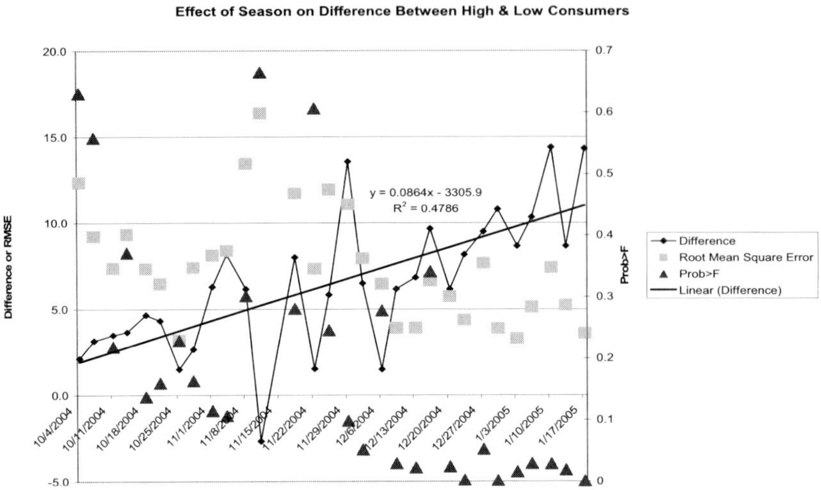

Figure 1. Effect of season on difference between high and low juniper-consuming angora goats.

Selective breeding to maximize the genetic capacity of goats to consume noxious plants may allow managers to custom build a biological tool specific to the ecological problem. Selective breeding has long been an acceptable means of improving a product, whether it be food, fiber, or reproduction. The results of this work show that selective breeding of goats also has the potential to be directed towards improving a trait such as brush control. If successful, benefits will be multi-fold: selectively bred goats will be physiologically more adept at utilizing their existing resource, at the same time providing an environmentally sustainable, low-cost means of keeping brush encroachment at bay. An anticipated result of this project will be development of source flocks of goats that are genetically superior for juniper consumption. Subsequently, we would release high-consuming genetics to producers for use in range improvement.

Fecal NIRS technology allows us to identify superior juniper-eating goats, to monitor their consumption of juniper, and to predict genetic merit for juniper consumption for use in breeding programs. This model is not limited to juniper and goats in the Edwards Plateau but has the potential to be applied to other situations involving the control of noxious plants with sheep and goats.

References

Hohenboken WD and Blodgett DJ (1997). Growth and physiological responses to toxicosis in lines of mice selected for resistance or susceptibility to endophyte-infected tall fescue in the diet. *Journal of Animal Science* 75:2165–2173.

Lyons RK and Stuth JW (1992). Fecal NIRS equations for predicting diet quality of free-ranging cattle. *Journal of Range Management* 45:238–244.
Leite ER and Stuth JW (1995). Fecal NIRS equations to assess diet quality of free-ranging goats. *Small Ruminant Research* 15:223–230.
Pritz RK, Launchbaugh KL, and Taylor CA Jr (1996). Effects of breed and dietary experience on juniper consumption by goats. *Journal of Range Management* 48:2–17.
Riddle RR, Taylor CA Jr, Kothmann MM, and Huston JE (1996). Volatile oil contents of ashe and redberry juniper and its relationship to preference by Angora and Spanish goats. *Journal of Range Management* 49:35–41.
Shelton M (1990). Goat production in the United States. In *Proceedings of the International Goat Production Symposium* (RC Gray, ed). Florida A&M University, Talahassee.
Smith BL, Stanbridge TA, and Embling PP (1980). Sheep breed differences in pentobarbitone sleeping time and response to experimental sporidesmin intoxication. *New Zealand Veterinary Journal* 28:35–36.
Walker JW, Clark DH, and McCoy SM (1998). Fecal NIRS for predicting percent leafy spurge in diet. *Journal of Range Management* 51:450–455.
Walker JW, Campbell ES, Engdahl BS, Lupton CJ, Scott CB, Taylor CA Jr., Waldron DF, and Whitworth WR (2005). Considerations for using fecal NIRS for predicting botanical composition of diets. *Annual Report of the Western Extension, Research, and Academic Coordinating Committee 039*, May 16–17, Houston, Texas, pp. 18–19.
Whitworth WR (2002). Estimating redberry and ashe juniper intake using fecal NIRS. MS Thesis, Angelo State University, San Angelo, Texas.

Chapter 66

Body Condition Affects Blood Alkaloid and Monoterpene Kinetics and Voluntary Intake of Chemically-Defended Plants by Livestock

K. Launchbaugh[1], J. Pfister[2], S. Lopez-Ortiz[3], and R. Frost[4]

[1]*Rangeland Ecology and Management, University, Moscow, ID 83844, USA;* [2]*USDA-ARS Poisonous Plant Research Laboratory, Logan, UT 84341, USA;* [3]*Colegio de Postgraduados, Veracruz, Mexico;* [4]*Animal and Range Sciences, Montana State University, Bozeman, MT 59717, USA*

Introduction

Poisonous plants are a substantial component of grazing lands worldwide. Higher losses to poisonous plants are often observed in times of drought or at high stocking rates. Increased incidences of fatality may occur because plants can be more toxic under these conditions or because limited access to alternative forages may induce livestock to eat excessive and lethal amounts of toxic plants. Limited access to nutritious forage alternatives could also limit chemical resources necessary for biotransformation and detoxification (Boyd and Campbell 1983; Illius and Jessop 1995). Furthermore, long-term nutrient deprivation results in loss of body weight and condition, which could then increase an animal's susceptibility to toxic plants by compromising their ability to detoxify and process toxic plant compounds (Foley *et al.* 1995). An herbivore's body condition may therefore strongly affect consumption of poisonous plants and detoxification of plant secondary metabolites (PSMs).

Body Condition Affects Consumption

To test the hypothesis that animals in low body condition may eat greater amounts of toxic plants we examined the consumption of chemically defended plants by animals in low and high body condition. Specifically, we examined the consumption of velvet lupine (*Lupinus leucophyllus*) and ponderosa pine (*Pinus ponderosa*) by cattle and consumption of red berry juniper (*Juniperus coahuilensis*) by goats.

Cattle consumption of lupine

Lupines are an important group of poisonous plants widely distributed throughout North America and cause major economic losses for the livestock industry. Velvet lupine, found in western North America, is one of many lupines that contain the teratogenic quinolizidine alkaloid, anagyrine. When pregnant cows eat toxic lupine during specific stages of gestation, their offspring often exhibit skeletal malformations, termed 'crooked calf disease' and/or cleft palate. Further, all livestock species are at risk for acute toxicity and death if lupine consumption is excessive (Panter *et al.* 1999). We hypothesized that cows in reduced body condition would ingest more velvet lupine than cows in elevated body condition.

This research was conducted with nine mature Hereford cows that were randomly assigned to treatments of low or high body condition (LBC or HBC, respectively). During a 60-day conditioning period, cattle were fed prescribed diets to reach desired body conditions. At the end of the conditioning period body condition scores were 4.2 ± 0.2 SE for the LBC and 6.6 ± 0.1 SE for HBC group (assessed on a 9-point scale with 9 high and 1 low).

Cows were moved to a 4 ha experimental pasture in eastern Washington dominated by annual grasses and perennial and annual forbs including velvet lupine to examine voluntary consumption of lupine. The field study was conducted at a time of rapid emergence and growth of lupine. Cattle were habituated to the close presence (1–2 m) of an observer. During an 18-day trial, the cows grazed as one group and one observer recorded bite counts of lupine and other forages (annual grass, perennial grass, grass-like plants, and forbs other than lupine) in 5-min observation periods beginning at 0500 each day and continuing during active grazing bouts until 2000. Typically each individual animal was observed for 20 to 30 min each day.

Herbaceous biomass was reduced by 80% during the grazing period. Cows in LBC averaged 8.8 ± 0.6% of bites as lupine over the trial compared to 6.3 ± 0.9% for the HBC cows ($\underline{P} = 0.08$). Days were different ($\underline{P} = 0.001$), and there was a day by treatment interaction ($\underline{P} = 0.02$) as low body condition animals began eating lupine before high condition cows. The number of bites of lupine increased exponentially at a decreasing rate with no differences between body condition treatments before reaching a maximum number of bites. Model estimates revealed that LBC cows reached a maximum of 3.2 bites min^{-1} on day 8, whereas cows in HBC reached 2.5 bites min^{-1} on day 13 (Figure 1).

Cattle consumption of ponderosa pine needles

Ponderosa pine needles cause abortion when grazed by pregnant cattle. Abortions generally occur in late fall and winter during the last trimester of pregnancy. Cattle generally graze pine needles during storms with increased snow, wind, cold, changes in feed, or hunger. The toxin in pine needles is the diterpene acid, isocupressic acid, and its ruminal metabolites.

We examined voluntary intake of ponderosa pine needles by cows in low and high body condition. Eight of the cows used in the lupine intake study described above were used to compare intake of ponderosa pine needles by cows in LBC (n=4) and HBC (n=4). Cows were fed appropriate diets to maintain prescribed body condition for 14 weeks. At the start of the pine needle trial, mean body weight and condition score were 689 ± 60 kg and 7.5 ± 0.2 BCS (on a scale of 1–9) for the high treatment, and 458 ± 18 kg and 3.5 ± 0.4

BCS for the low treatment. Body condition was assessed visually by an experienced observer from Utah State University who was blind to the treatments.

Cattle were placed into individual pens (4 x 4 m) in a heated barn with *ad libitum* access to water and salt. To avoid refusals, the diet was restricted to commercial alfalfa pellets fed at 1.25% of body weight (bw) during a 5-day acclimation period and the 14-day feeding trial. Pellets were fed at 0800 each morning, with any orts weighed the following morning. Pine needles were stored frozen, but thawed for feeding (hand-picked needles only, no woody material). Needles were fed whole in a separate feeding box each morning at 0800, and any refusals weighed. Cattle were fed pine needles *ad libitum*, and the feeding boxes were checked and replenished each evening as necessary to ensure a constant supply.

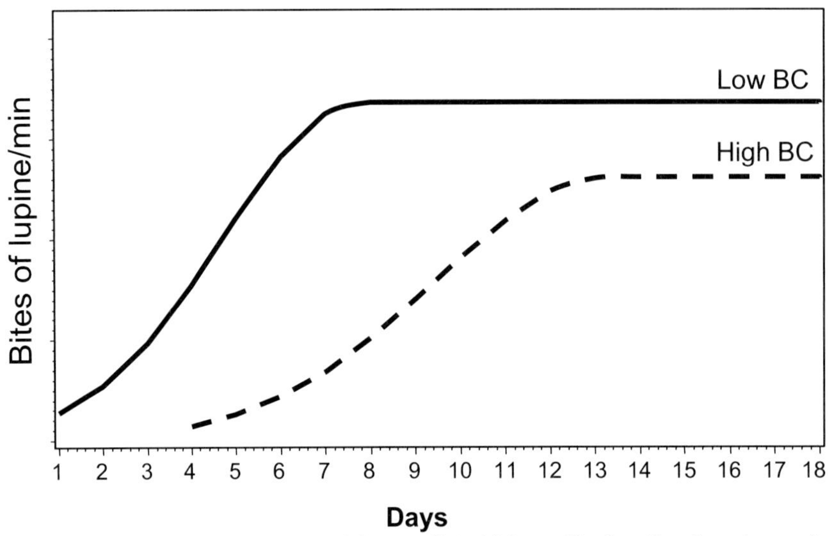

Figure 1. Exponential increase depicted for predicted bites of lupine (*Lupinus leucophyllus*) by cattle grazing in eastern Washington state, USA. Cows were in high and low body condition. The natural logarithm of observed bites of lupine fit a quadratic-plateau model.

Actual pine needle consumption was 691 ± 99 g day^{-1} for the high group compared to 2490 ± 116 g day^{-1} for the low condition group. There was a day effect and a day x treatment interaction ($\underline{P} < 0.001$), as daily consumption by cattle in low body condition averaged about 5.0 to 6.6 g kg^{-1} bw. Cattle in HBC ate a small amount of needles until day 10 of the trial, when their consumption peaked at 3.2 g kg^{-1} bw. High body condition cows lost 9 kg during the trial, whereas LBC animals gained 6 kg.

Consumption of juniper by goats

Redberry juniper is an invasive evergreen tree that is rapidly expanding throughout western and central Texas. This small tree is an aggressive competitor that has a dense branch pattern and lateral root system that can reduce the amount of water reaching the soil thereby decreasing the productivity of adjacent grasses and forbs and increasing the amount of bare ground. This results in a dramatic decline in herbaceous forage for livestock and wildlife. Goats will consume some juniper and are considered a potential option for juniper control through prescribed grazing. However, juniper intake is hindered by foliar monoterpene content. The objective of our research was to determine how body condition of goats would influence their consumption of juniper.

Research was conducted with 39 Boer × Spanish crossbred female goats at the Texas A&M University Agricultural Experiment Station near Sonora, TX. Animals were assigned to either a HBC or LBC treatment. For a 10 week conditioning period those animals in the HBC treatment were offered *ad libitum* access to an above-maintenance complete ration while those assigned to the low body condition treatment received 1.5% bw of a below-maintenance ration. Body condition was evaluated by weight and by visual and tactile examination weekly. At the end of the conditioning period those animals in the HBC group had an average score of 3.7 ± 0.1 while those in the LBC group had an average score of 1.4 + 0.1 (on a 1 to 5 scale).

Five days before the experiment, all animals were offered a daily ration of alfalfa pellets (2% BW) in the morning and afternoon. Fresh redberry juniper branches were harvested each morning of the 5-day trial. The leaves were stripped and offered in excess to each goat for 4 hours each morning. Juniper consumption was recorded.

Goats in LBC ate more ($\underline{P}$< 0.001) juniper than those in HBC (average 239 ± 23 g day^{-1} vs. 90 ± 21 g day^{-1}, respectively; Figure 2). Intake of both groups generally increased throughout the 5-day trial.

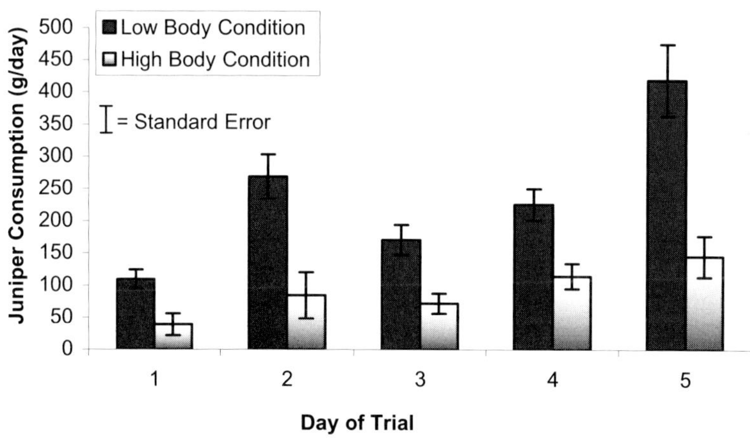

Figure 2. Mean daily intake of live redberry juniper (*Juniperus coahuilensis*) foliage by domestic goats in low (n=20) or high (n=20) body condition for 5 days.

Examining Metabolic Response Depending on Body Condition

Herbivores have developed a variety of mechanisms to counteract the negative digestive and metabolic effects of PSMs. The efficiency of these detoxification systems can be influenced by the internal environment of the herbivore. Well-fed animals, in good body condition and on a high plane of nutrition, may effectively process phytotoxins while nutrient-deprived animals may have limited ability to detoxify or excrete the toxins (Foley *et al.* 1995). The biotransformation of toxins requires nutrients in excess of those for maintenance (Illius and Jessop 1995) and the activity and mass of visceral organs involved in detoxification, including the liver, is often reduced in animals after a prolonged period of nutritional stress (Noziere *et al.* 1999).

We examined whether body condition would affect the disposition (absorption, metabolism, or elimination) of quinolizidine alkaloids from silvery lupine (*Lupinus argenteus*) ingested by domestic sheep. We collected silvery lupine pods from a native pasture in eastern Idaho near Humphrey. Seed pods were sun-air dried and finely ground to pass a 1-mm screen. The resulting seed pod mixture contained 1.6% alkaloids (anagyrine, lupanine, 5,6-dehydrolupanine, and three unknown alkaloids).

Columbia ewes (n=19) were individually penned and randomly assigned to dietary treatments to reduce or increase body condition. The diets were formulated to provide digestible energy and protein level 20% below or 20% above requirements for lactating ewes and offered to ewes being conditioned to low (LBC) or average body (ABC) condition, respectively. The mean body condition after a 9-week conditioning period was 1 for LBC and 2.5 for ewes in ABC (on a 5-point scale).

Two days before the experiment began, all ewes were fed 2.2% bw of a basal diet consisting of chopped grass and alfalfa hay. The experimental measurements began after all ewes received a single oral dose of 8.5 g kg^{-1} bw of silvery lupine pods. Ground lupine pods were diluted in 5 L of lukewarm water and gavaged using a livestock pump system. The dose of lupine used in this research resulted in a single total alkaloid dose of 139 mg kg^{-1} bw. Blood samples were drawn via jugular vein puncture at 0, 0.5, 1, 2, 3, 4, 6, 8, 12, 18, 24, 36, 48, and 60 hours after gavaging. Blood was allowed to clot for 1 hour at room temperature and was then centrifuged and serum was immediately harvested and frozen in a standard chest freezer. Serum alkaloid content was determined by gas chromatography after acid/base extraction. Absorption, distribution, elimination, and peak height were evaluated as Area Under the Curve (AUC) using the trapezoidal rule to calculate AUC.

Blood samples from the sheep contained measurable levels of alkaloids beginning 30 min after dosing and peaked by 2-3 hours. Sheep in LBC expressed higher total blood alkaloid levels and anagyrine than animals in ABC (Table 1). Other alkaloids detected in the blood did not differ among sheep in different body conditions. There was a difference in AUC between LBC and ABC for total alkaloid ($P = 0.003$) and anagyrine ($P = 0.001$), but not for other alkaloids. This difference strongly suggests that overall disposition of lupine alkaloids, and especially the teratogen anagyrine, is altered in a relatively dramatic way based on body condition (Lopez-Ortiz *et al.* 2004).

Table 1. Mean concentration ($\mu g\ mL^{-1}$ serum ± SE) of major alkaloids in serum of sheep within 60 hours after a single dose by ruminal infusion with 8.5 g kg^{-1} bw of silvery lupine (*Lupinus argenteus*). Sheep were in low (n=9) or average (n=10) body condition.

	Body Condition	
Alkaloid	Low	Average
Total	351.8 ± 31.9[a]	193.6 ± 22.2[b]
Anagyrine	128.3 ± 12.9[a]	57.6 ± 7.7[b]
Lupanine	12.7 ± 1.2[a]	11.0 ± 1.2[a]
5,6-dehydrolupanine	63.7 ± 4.7[a]	49.0 ± 4.4[a]
Unknown E	69.0 ± 7.8[a]	27.8 ± 4.5[b]

Within a row, means with different superscript letter differ statistically ($P < 0.05$).

Conclusions and Management Implications

Our research suggests that animals in low body condition are likely to eat greater quantities of chemically defended plants than animals in higher body condition. This result contrasts with contemporary intake theories that suggest animals in a low nutritional state will ingest lower amounts of PSMs because their detoxification capacity is compromised and they reach a toxic threshold and cease intake at lower consumption levels than more well-nourished animals (Illius and Jessop 1995; Provenza *et al.* 2003). It appears that the natural tendency of malnourished herbivores to increase intake overrides mechanisms related to minimizing intake of plant allelochemicals. Increased intake of chemically defended plants by animals in low body condition is particularly problematic because these animals apparently have reduced abilities to process these deleterious chemicals than animals in higher condition. These findings suggest that livestock managers should monitor body condition to mitigate consumption and effects of poisonous plants on grazing animals. Researchers are also behooved to consider body condition when designing experiments on the consumption and metabolic fate of PSMs.

References

Boyd JB and Campbell TC (1983). Impact of nutrition on detoxication. In *Biological Bases of Detoxication* (J Caldwell, and WB Jakoby, eds), pp. 287–306. Academic Press, New York.

Foley MJ, McLean S, and Cork SJ (1995). Consequences of biotransformation of plant secondary metabolites on acid-base metabolism in mammals – a final common pathway? *Journal of Chemical Ecology* 21:721–743.

Illius AW and Jessop NS (1995). Modeling metabolic costs of allelochemical ingestion by foraging herbivores. *Journal of Chemical Ecology* 21:693–719.

Lopez-Ortiz S, Panter KE, Pfister JA, and Launchbaugh KL (2004). The effect of body condition on disposition of alkaloids from silvery lupine (*Lupinus argenteus* Pursh) in sheep. *Journal of Animal Science* 82:2798–2805.

Noziere P, Attaix D, Bocquier F, and Doreau M (1999). Effects of underfeeding and refeeding on weight and cellularity of splanchnic organs in ewes. *Journal of Animal Science* 77:2279–2290.

Panter KE, James LF, and Gardner DR (1999). Lupines, poison hemlock, and *Nicotiana* spp.: Toxicity and teratogenicity in livestock. *Journal of Natural Toxins* 8:117–134.

Provenza FD, Villalba JJ, Dziba LE, Atwood SB, and Banner RE (2003). Linking herbivore experience, varied diets, and plant biochemical diversity. *Small Ruminant Research* 49:257–274.

Chapter 67

Cattle Grazing Velvet Lupine (*Lupinus leucophyllus*): influence of Associated Forages, Alkaloid Levels and Population Cycles

M.H. Ralphs [1], K.E. Panter [1], C. Gay [2], E. Motteram [2], and S.T. Lee [1]

[1]USDA-ARS Poisonous Plant Research Laboratory, Logan UT 84341 USA; [2]Field Investigative Unit, Veterinary School, Washington State University, Pullman, WA 99164, USA

Introduction

Lupines (*Lupinus* spp.) are abundant on many western rangelands. Several species contain alkaloids that can be toxic and/or teratogenic to livestock. Historically, lupines have been a major cause of sheep deaths on foothill and mountain rangelands in the western US (Kingsbury 1964). More recently, calf losses from congenital birth defects have been reported in several western states (Bagley and Panter 1996; Panter *et al.* 1997; Warnock 1997). Ingestion of lupine by the dam during the 40 to 70th day of gestation can cause a crooked calf syndrome (Shupe *et al.* 1967). The quinolizidine alkaloid, anagyrine (Keeler 1976), and some piperidine alkaloids (Keeler and Panter 1989; Panter *et al.* 1998) were shown to reduce fetal movement during this critical period of gestation (Panter *et al.* 1990), causing the spine and limbs to develop in contracted or maligned positions.

The objective of this study was to determine when cattle graze velvet lupine (*Lupinus leucophyllus* Dougl.) in the channel scabland region of eastern Washington. A second objective was to determine whether those cows that had crooked calves had a propensity to consume more lupine than those with normal calves.

Methods

Site description

The study was conducted in the Channel Scabland region of eastern Washington about 25 km southeast of Ritzville (N 46E 15.347', W 118E 13.031'), where losses of crooked calves have been catastrophic. In the spring of 1997, over 650 calves on 13 ranches were born deformed and had to be destroyed (occurrence on individual ranches ranged from 1 to 100% (C. Gay, unpublished data).

The study site was characterized by basalt plateaus with depressions and valleys carved by geologic floods from the ancient lake Missoula. The original plant community was a palouse prairie dominated by bunchgrasses but has been degraded to annual species, such as cheatgrass (*Bromus tectorum*), fiddleneck (*Amsinckia intermedia*), tansy mustard (*Descurainia pinnata*), prickly lettuce (*Lactuca serriola*) and black mustard (*Brassica nigra*). Velvet lupine occurred most frequently on the slopes and averaged 1.5 plants/m^2 in 2002 and 0.92 plants/m^2 in 2003.

Group comparison 2001

The objective was to compare lupine consumption between cows that had crooked calves and cows with normal calves. The hypothesis was that cows with crooked calves had a greater propensity to graze lupine than cows with normal calves. Five mature Hereford cows with crooked calves at their side and six cows with normal calves were obtained from a local rancher in 2001. The groups grazed together in a 100 ha pasture that was infested with lupine. The study began 10 May as lupine was beginning to grow and ended 3 August when lupine had senesced.

Diet composition was estimated by a bite count technique. Each cow was observed for 5 min periods during each major daily grazing period ranging from sun-up to 10:00 am, and from 4:00 to 9:00 pm. The number of bites of perennial grasses, annual grass, forbs and lupine were recorded and their percent of diet was calculated for each observation.

Seasonal palatability of lupine

There were relatively few crooked calves in the region in 2002 and 2003. Therefore, the study design was changed to simply define the seasonal palatability of velvet lupine. The same pasture was grazed in 2001, 2002 and 2003. Data from the cows in the 2001 study were pooled since there was no difference in lupine consumption between the groups. In 2002, ten different 3-year-old Hereford cows along with their calves grazed from 4 June to 14 August. In 2003, eight of the same cows (one cow broke her leg and another was culled), along with their new calves, grazed from 14 May to 18 August. Diets were quantified by bite counts as described above.

A bull was placed with the cows each year around 10 May and the breeding dates were observed and recorded. The susceptible period of gestation (40 to 70th day of pregnancy) was compared to the periods of lupine consumption to predict the incidence of crooked calves in the following years.

Lupine alkaloid analysis

Lupine samples were collected at roughly 2-week intervals during the study each year. Ten uniform plants were harvested each collection period from throughout the length of the pasture. They were separated into parts (leaf, stem, flower, pod, and seed), dried at 60EC for 48 hours and ground through a cyclone mill. Alkaloids were isolated following standard acid/base extraction procedures using 1% hydrochloric acid, chloroform and dilute ammonium hydroxide, and quantified by a gas chromatograph method (Gardner and Panter

1994). Deviations from the previously described procedure included the use of caffeine as an internal standard and concentration of alkaloids was quantified against a six-point anagyrine standard curve (S. Lee, unpublished method).

Results

There was no difference in lupine consumption between cows with crooked calves and cows with normal calves in 2001 ($\underline{P} = 0.17$). Lupine appeared to be equally acceptable to all cows.

Cattle selected cheatgrass while it was still green during the first part of the study (Figure 1). Crude protein content of annual grasses was 8.5% at the beginning of the study, but declined to 5% at the end. As the annual grasses matured and dried, cattle selected green forbs, such as prickly lettuce, black mustard, and tansy mustard, during the middle part of the summer (Figure 1.) Crude protein of forbs ranged from 10 to 14% during the first half of the study and fiber content (NDF) was lower than the grasses. As these forbs were depleted or matured and dried, cattle began selecting the deep-rooted lupine (Figure 1), which remained green and succulent longer than the other forbs. Crude protein content of lupine was 15% through the middle part of the study, and NDF remained below 55%. Although it was the most nutritious forage throughout the season, cattle did not graze it until the other forages were depleted or matured. We suspect the thick velvety pubescence was the reason for its lack of palatability. Lopez-Ortiz *et al.* (2005) also reported cattle grazed velvet lupine when grass and other forbs were depleted.

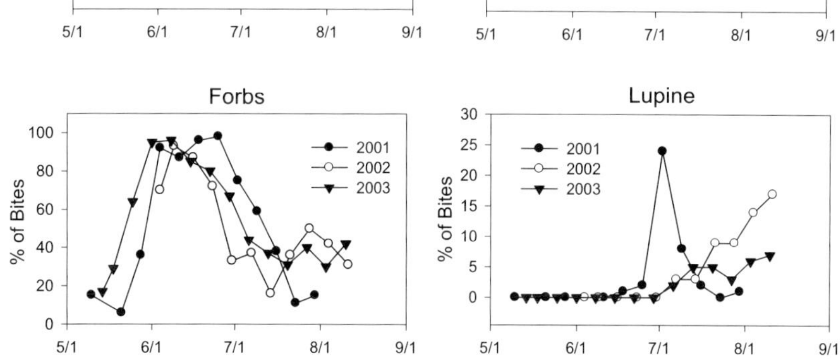

Figure1. Forage classes in cattle diets in each of the 3 years of the study. There was a difference in lupine consumption among years ($\underline{P} < 0.0001$) and the year-by-week interaction ($\underline{P} < 0.0001$) indicates the lupine consumption trends differed among years.

Lupine alkaloids

Anagyrine concentration in leaves and stems declined during the growing season. Anagyrine concentration was about 0.3% as growth began, then declined to 0.1% as the plant senesced (Figure 2). Anagyrine concentration in the floral parts averaged 0.5% and increased slightly at the beginning of pod formation, then declined after seeds shattered. Anagyrine was much higher in seeds, ranging from 1% in 2001 to 1.5% in 2002.

Discussion

The general breeding season on our study ranch was 1 May – 30 July. Therefore, the susceptible period of gestation (40–70 day) runs from 10 June – 8 Oct. Since the cows began to consume lupine during July and August when other forbs mature, ingestion of the teratogenic alkaloids will likely occur during the critical period of gestation (Table 1).

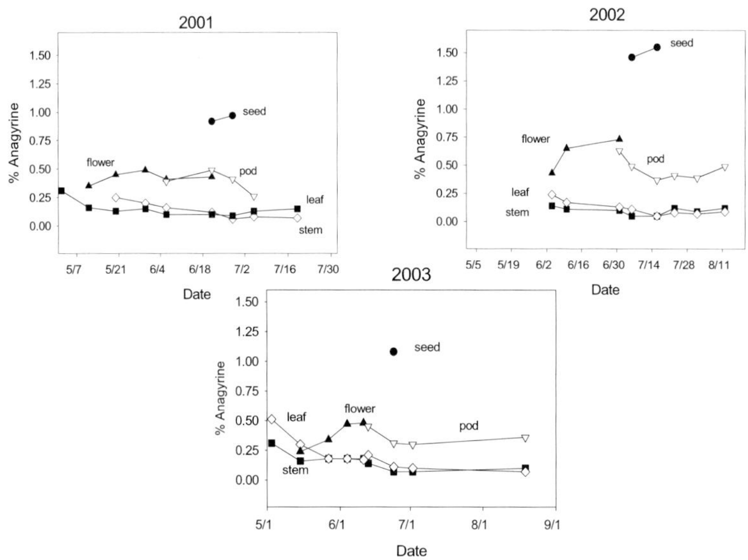

Figure 2. Concentration of the teratogenic alkaloid anagyrine in lupine plant parts.

The other part of the intoxication equation is the concentration of the toxin in the plant. Anagyrine concentration was highest in the flowers and pods, which occurred from mid-May through mid-July. Keeler *et al.* (1977) recommended restricting access to lupine when anagyrine was highest. In our study, the overlap between the susceptible period of gestation, the time cows were eating lupine, and the high concentration in the plant occurred during July. Therefore, we can refine Keeler's recommendation by denying access to lupine when anagyrine concentration is high, and when cows are eating the plant during the susceptible period.

Table 1. Overlap of lupine consumption during the susceptible period of gestation (40–70 days).

Year	Breed dates	Susceptible period	Period lupine consumed	Lupine Consumption Average[1]	Peak[2]	Max[3]
				-------------% of bites-----------		
2001	6/1–7/3	7/11–9/11	7/2–7/15	6.1 ± 0.66[a]	24	31
2002	6/18–8/30	7/28–11/2	7/7–8/15	5.9 ± 0.34[a]	17	30
2003	5/20–7/28	6/29–10/6	7/6–8/18	3.4 ± 0.30[b]	7	15

[1] Average daily amount of lupine consumed during July and August (± SE).
[2] Peak weekly lupine consumption of the entire group each year.
[3] Maximum weekly lupine consumption by individual cows.
a,b Means in the same column followed by different letters differ ($\underline{P} < 0.0001$).

The abundance of lupine is another factor influencing these catastrophic losses. Lupine density or relative abundance during the grazing seasons prior to the reported catastrophic losses listed above is not known. Density of velvet lupine plants declined from 2001 to 2003 on the study site, and throughout the Channel Scabland region (M. Ralphs, unpublished data). This decline may be a continuing trend from higher population densities that may have been present when the catastrophic losses occurred in 1997.

Management Recommendations (from General to Specific)

1. Change to fall calving to avoid the risk of the critical period of gestation during the growing season when alkaloid levels are high.
2. Delay the breeding season for 30 days. This will set the critical period back until mid-July and August when the anagyrine concentration declines following seed shatter.
3. Restrict access to lupine in July when cattle begin eating lupine and anagyrine levels are still high.
4. Avoid grazing lupine-infested pastures in July only in years when lupine is abundant. The density of lupine during the years of our study was not great enough to consume a sufficient amount of the plant to cause crooked calves.

The knowledge of plant ecology, population cycles, grazing behavior and plant toxicity can be integrated to reduce the risk of poisoning.

References

Bagley C and Panter KE (1996). Lupine poisoning: crooked calf disease. *Western Beef Producer*, April 1996, p. 6–7.

Gardner DR and Panter KE (1994). Ammodendrine and related piperidine alkaloid levels in the blood plasma of cattle, sheep and goats fed *Lupinus formosus*. *Journal of Natural Toxins* 3:107–116.

Keeler RF (1976). Lupin alkaloids from teratogenic and nonteratogenic lupins. III. Identification of anagyrine as the probable teratogen by feeding trials. *Journal of Toxicology and Environmental Health* 1:878–889.

Keeler RF and Panter KE (1989). Piperidine alkaloid composition and relation to crooked calf disease-inducing potential of *Lupinus formosus*. *Teratology* 40:423–432.

Keeler RF, James LF, Shupe JL, and Van Kampen KR (1977). Lupine-induced crooked calf disease and a management method to reduce incidence. *Journal of Range Management* 30:97–102.

Kingsbury JM (1964). *Poisonous Plants of the United States and Canada*. Prentice-Hall, Englewood Cliffs, New Jersey.

Lopez-Ortiz SL, Pfister JA, Launchbaugh KL, and Gay CC (2006). Forage availability and body condition affect intake of lupine (*Lupinus leucophyllus*) by grazing cattle. *Range Ecology Management* (in review).

Panter KE, Bunch TD, Keeler RF, Sisson DV, and Callan RJ (1990). Multiple congenital contractures (MCC) and cleft palate induced in goats by ingestion of piperidine alkaloid-containing plants: Reduction in fetal movement as the probable cause. *Clinical Toxicology* 28:69–83.

Panter KE, Gardner DR, Gay CC, James LF, Mills R, Gay JM, and Baldwin TJ (1997). Observations of *Lupinus sulphureus*–induced crooked calf disease. *Journal of Range Management* 50:587–592.

Panter KE, Gardner DR, and Molyneux RJ (1998). Teratogenic and fetotoxic effects of two piperidine alkaloid-containing lupines (*L. formosus* and *L. arbustus*) in cows. *Journal of Natural Toxins* 7:131–140.

Shupe JL, Binns W, James LF, and Keeler RF (1967). Lupine, a cause of crooked calf disease. *Journal of the American Veterinary Medical Association* 151:198–203.

Warnock D (1997). Lupine toxicity. *Western Beef Producer*, May 1997, pp. 20–22.

Chapter 68

Effects of Experience and Lactation on Lupine Consumption by Cattle

J.A. Pfister[1], S.T. Lee[1], K.E. Panter[1], E.S. Motteram[2], and C.C. Gay[2]

[1]*USDA-ARS Poisonous Plant Research Laboratory, Logan, UT 84341, USA;* [2]*Field Disease Investigation Unit, College of Veterinary Medicine, Washington State University, Pullman, WA 99164-6610, USA*

Introduction

Pregnant cattle that ingest lupine (*Lupinus* spp.) during early gestation often give birth to calves with lupine-induced arthrogryposis (i.e. crooked calf disease) (Shupe *et al.* 1967). Lupine toxicity has been especially troublesome in portions of eastern Washington state (lat N 46° long W 118°). Depending on environmental conditions, lupine density fluctuates, and when density is high, pregnant cattle can easily ingest toxic amounts of the plant. Anecdotal reports from livestock producers in Washington and elsewhere suggest that naïve cattle ingest more lupines than do experienced animals. Further, some ranchers believe that lactating cows eat more lupine than do non-lactating cows. Thus, the purpose of these trials was to determine the effects of lactation and experience on consumption of lupine (*Lupinus leucophyllus*) by cattle in the Channel Scablands of Washington state, USA.

Materials and Methods

Trial 1 2003

This trial was conducted in the Channel Scablands of eastern Washington during the summer of 2003. Six Hereford cows naïve to lupine and six Hereford cows with several years experience grazing lupine-infested rangelands in this vicinity ($\bar{x}$ weight 453 kg) were grazed together for 25 days (June 13 to July 7) in the 6.6 ha study pasture. Diets were determined by bite counts; cattle were observed for sequential 5-min periods during active grazing bouts. Typically each animal was observed for 20–30 min day^{-1}. Bite counts are not a measure of amount eaten, because bite size was not determined. Instead bite counts show relative preference for different plant species.

Forage availability (kg ha^{-1}) was measured by clipping 30, 0.25 m^2 plots each week; clipped forage was dried and weighed. Lupine density was determined by counting lupine plants in 100 1-m^2 plots randomly placed throughout the pasture.

Trial 2 2004

Six lactating and six non-lactating cows ($\bar{x}$ weight 560 kg) were grazed together for 18 days on a 2.4 ha pasture divided into two equal units. All were Hereford cows with experience grazing in lupine-infested pastures. Cattle grazed the first unit for 10 days (June 25 to July 4), and then were shifted into the second unit (July 5 to July 12) when the lupine was exhausted in the first unit. Forage availability and lupine density were measured as noted for the previous study.

Nutritional analysis of forage

Clipped forage samples were analyzed for nutritional quality and alkaloid concentration. Plants were dried at 60°C. Material was ground to pass through a 1-mm screen in a cyclone mill. Nitrogen was quantified by high-temperature combustion, followed by thermoconductometric detection (Wiles *et al.* 1998). Crude protein was calculated as N x 6.25. Alkaloids were determined by a GC-MS procedure (Lee, unpublished method).

Results and Discussion

During Trial 1 (2003) lupine density was $\leq$ 1 plant m^{-2}. Lupine plants averaged 15.2% CP in mid-June, whereas grasses and forbs averaged 7.1% and 10.6%, respectively. By early July, lupine averaged 13.9% CP, while grasses and forbs averaged 5.1% and 7.4%, respectively. Anagyrine concentration in lupines ranged from 0.17 to 0.26% during mid-June to early July. Keeler (1973) reported that 0.14% was a critical value for anagyrine concentration in lupine, above which crooked calves could be expected if consumption was sufficient.

There was no difference ($P > 0.5$) in consumption of lupine during summer 2003 by naïve and experienced cows (data not shown). Cattle ate little lupine during the first 10 days, then consumption increased to 10–14% of daily bites. Cattle began eating lupine as availability of green grass decreased. Consumption of dry grass increased markedly in the last 7 days, peaking at over 50% of the diets.

Anecdotal reports by ranchers suggested that naïve animals may eat more lupine than experienced cattle. The 2003 trial showed essentially no differences in lupine consumption between naïve and experienced cows. Experience may have a profound effect on later diet selection, as experienced animals may ingest more of a specific plant than naïve animals (Distel and Provenza 1991; Olson *et al.* 1996). Conversely, experienced animals may sample then refuse some plants based on adverse post-ingestive consequences (Provenza *et al.* 1992). Experienced sheep in South Africa refuse to eat *Senecio* spp. after 4 days exposure, even if starving (Kellerman *et al.* 1988). Cattle given a sublethal sample of *Moraea pallida* then refuse to eat the plant (Strydom and Joubert 1983). Only naïve cattle in southern Brazil are poisoned by *Baccharis coridifolia* because experienced animals learn to avoid the plant (Tokarnia *et al.* 2000).

During 2004 (Trial 2), lupine density varied from 1 to 2 plants m^{-2}. This pasture was primarily bottomland with abundant forage (> 2000 kg ha^{-1}). Lupine plants averaged 13.3% CP in late June, whereas grasses and forbs averaged 5.1% and 6.6% CP, respectively. By mid-July when the grazing trial was terminated and forage was depleted, residual lupine stems averaged 6.3% CP, while grasses and forbs averaged 4.5% and 4.3% CP, respectively. Anagyrine concentration in lupines ranged from 0.10 to 0.19% during late June to mid-July.

Cattle began eating lupine on day 4 and consumption generally increased until day 9 when lupine availability was depleted. There was no difference ($\underline{P} > 0.6$) in consumption between lactating and dry cows. Lupine consumption peaked at 10–15% of the diets. Cattle were then switched to the next paddock and grazed there for 9 days until lupine was depleted. Cattle ate mostly grasses on the first day after switching paddocks, then lupine consumption averaged 5–15% of diets during the remainder of the trial (Figure 1). Other forbs comprised the majority (60–70%) of the diets, and consumption of dry grass varied from ~10 to 20% of the diets.

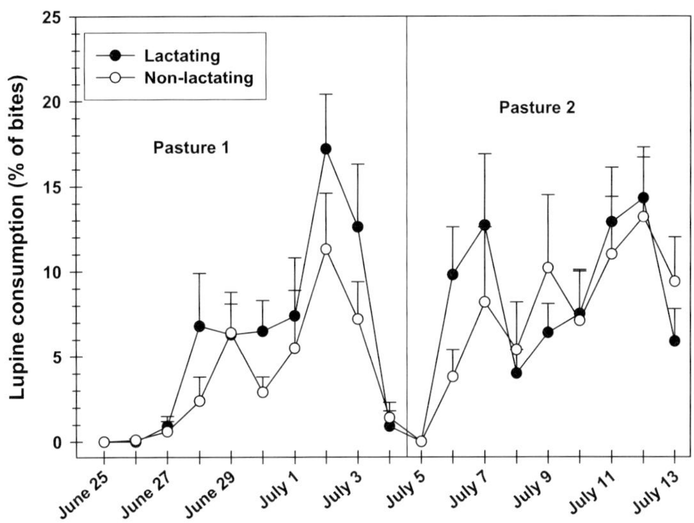

Figure 1. Lupine consumption (% of bites ± SE) by lactating and non-lactating cattle grazing lupine-infested rangeland in eastern Washington state during the summer of 2004.

Kryiazakis *et al.* (1999) suggested that animals modify their diet selection based on large physiological changes in their internal state. Milne *et al.* (1981) found that lactating sheep had greater forage intakes than did dry sheep. Lactation is nutritionally stressful for grazing cattle, particularly on ranges with mostly low quality forages. Even so we found no evidence that lactation influences cattle to eat more lupine when grazing on poor quality rangelands.

Conclusion

Scabland rangelands provided low quality nutritional choices for cattle. Lupines contained > 12% crude protein and generally were nutritionally superior to other forage choices, even green forbs. Even small amounts of lupine in cattle diets can provide necessary N for ruminal function. It is not known if supplemental N will reduce lupine consumption. During the first trial in 2003, cattle searched for green grass until this component was depleted, then ate some lupine and increasing amounts of dry grasses. During the second trial in 2004, cattle ate substantial quantities of lupine even though other forage was abundant and lupine was a relatively minor forage component. Contrary to popular wisdom, we found no evidence that naïve or lactating cattle eat more lupine than do experienced or dry cows.

References

Distel RA and Provenza FD (1991). Experience early in life affects voluntary intake of blackbrush by goats. *Journal of Chemical Ecology* 17:431–450.

Keeler RF (1973). Lupine alkaloids from teratogenic and nonteratogenic lupines. I. Correlation of crooked calf disease incidence with alkaloid distribution determined by gas chromatography. *Teratology* 7:23–30.

Kellerman TS, Coetzer JAW, and Naude TW (1988). *Plant Poisonings and Mycotoxicoses of Livestock in Southern Africa.* Oxford University Press, Cape Town, South Africa.

Kyriazakis I, Tolkamp BJ, and Emmans G (1999). Diet selection and animal state: an integrative framework. *Proceedings of the Nutrition Society* 58:765–772.

Milne JA , Maxwell TJ, and Souter WG (1981). The effect of supplementary feeding and herbage mass on the intake and performance of grazing ewes in early lactation. *Animal Production* 32:185–195.

Olson BE, Wallander RT, Thomas VM, and Kott RW (1996). Effect of previous experience on sheep grazing leafy spurge. *Applied Animal Behaviour Science* 50:161–176.

Provenza FD, Pfister JA, and Cheney CD (1992). Mechanisms of learning in diet selection with reference to phytotoxicosis in herbivores. *Journal of Range Management* 45:36–45.

Shupe JL, Binns W, James LF, and Keeler RF (1967). Lupine, a cause of crooked calf disease. *Journal of the American Veterinary Medical Association* 151:198–203.

Strydom JA and Joubert JPL (1983). The effect of predosing *Homeria pallida* Bak. to cattle to prevent poisoning. *Journal of South African Veterinary Association* 54:201–203.

Tokarnia CH, Döbereiner J, and Peixoto PV (2000). *Plantas Tóxicas do Brasil.* Editora Helianthus, Rio de Janeiro (in Portuguese).

Wiles PG, Gray IK, and Kissling RC (1998). Routine analysis of proteins by Kjeldahl and Dumas methods: review and interlaboratory study using dairy products. *Journal of the Association of Official Analytical Chemists* 81:620–632.

Chapter 69

Effect of Shade on Alkaloid Content of *Lupinus leucophyllus*

C.C. Gay[1], K.E. Panter[2], E.S. Motteram[1], J.M. Gay[1], T. Wierenga[2], and T. Platt[1]

[1] *Field Disease Investigation Unit, College of Veterinary Medicine, Washington State University, Pullman, WA 99164-6610, USA;* [2] *USDA-ARS Poisonous Plant Research Laboratory, Logan, UT 84341, USA*

Introduction

Lupine-induced arthrogryposis (crooked calf disease) is a significant problem for ranches on the semiarid Cow Creek drainage of the channeled scablands of east-central Washington state. The dominant lupine on affected ranches in this region is *Lupinus leucophyllus*. The incidence and severity of crooked calf disease varies from year to year. Anecdotal data gathered from ranchers and our own observational data indicate there have been 8 years in the 25-year period since 1980 in which one or more ranchers on the Scablands have observed high disease prevalence and severity. Our examination of this historical incidence data for the Scablands and local weather station data suggests that years of high incidence follow and are associated with years of higher than average rainfall.

It is possible that rainfall and soil moisture affect the alkaloid content and/or palatability of *L. leucophyllus*. Alternatively, it is also possible that shade from clouds passing to higher rainfall areas east of Adams County, or shade from the superior growth of other pasture species in wet years, could influence alkaloid concentrations or palatability and that this might explain patterns of abnormality seen on affected ranches. Experimental challenge has shown that fetal skeletal malformations result from exposure to teratogenic alkaloids during a later gestational age than that associated with the production of cleft palate (Panter and Keeler 1992; Panter *et al*. 1994, 1998). In herds with crooked calves we have observed variations in type and severity of deformities in a single herd. Cleft palate has occurred as a single defect or in combination with a variety of skeletal contractures. Skeletal defects can be minor and resolve spontaneously or they may be severe, preventing the calf from surviving. This indicates that lupine consumption by individual cows grazing the same pasture is not constant during its period of availability or that teratogenicity may be variable. A varying shade could explain these differences.

An effect of shade on pasture toxicity is not without precedent as the concentration of toxic tryptamine alkaloids in *Phalaris aquatica* (*tuberosa*) is higher in plants growing in the shade (Williams 1972; Ball and Hoveland 1978) and risk for toxicity in sheep is higher on cloudy days or when plant regrowth is shaded by senescent plant material. Toxicity of nitrates in plants also varies with sunlight or shade.

Materials and Methods

On a pasture with an historic high incidence of crooked calf the north-facing half of each of five plants of *L. leucophyllus* was shaded by constructing a tent with black 4 mm plastic stretching from the center of the plant clump over to its north edge. This prevented direct sunlight impacting the north-facing half of the plant but allowed indirect light to the north-facing half and full sunlight to the south-facing half. The shading was left in place for 7 days, following which the shaded and unshaded halves of the lupine plants were harvested at mid-day, and dried and stored at room temperature until analysis. Alkaloid analysis was as previously described (Gardner and Panter 1993,1994).

Differences in alkaloid concentrations between the shaded and unshaded parts of the plants were examined by two-sample *t*-tests.

Results and Discussion

Total and individual alkaloid concentrations are shown in Table 1. There was no significant ($\underline{P} > 0.1$) effect of shade on the concentration of the teratogenic alkaloid anagyrine but a significant differences were observed with 5,6-dehydrolupanine ($\underline{P} = 0.013$) and a trend to a significant difference with alkaloid unknown F ($\underline{P} = 0.052$) and total alkaloids ($\underline{P} = 0.094$). Given the small sample size, these results suggest further studies on the effects of shade on alkaloid concentration and plant palatability are warranted.

Table 1. Total and differential alkaloid concentrations in the shaded and unshaded half of each of five lupines

Lupine	Total Alkaloid	Unknown F	5,6-dehydro-lupanine	Lupa-nine	Unknown T	Unknown X	N-methyl-albine	Anagy-rine
N1	1.421	51	203	70	66	9	42	948
N2	1.801	34	97	25	55	15	80	1455
N3	1.801	149	224	32	220	25	124	981
N4	1.78	152	121	75	114	14	72	1197
N5	1.876	157	129	66	263	31	182	1007
Mean	1.7358	108.6	154.8	53.6	143.6	18.8	100	1117.6
S1	1.595	49	216	63	68	7	33	1131
S2	1.916	75	150	36	109	16	84	1401
S3	2.256	199	291	35	296	51	171	1126
S4	1.907	216	173	104	148	12	60	1142
S5	1.858	170	154	69	229	20	135	1033
Mean	1.9064	141.8	196.8	61.4	170	21.2	96.6	1166.6
	$\underline{P}$ = 0.094	**$\underline{P}$ = 0.052**	**$\underline{P}$ = 0.013**					

References

Ball DM and Hoveland CS (1978). Alkaloid levels in *Phalaris aquatica* L. as affected by environment. *Agronomy Journal* 70:977–981

Gardner DR and Panter KE (1993). Comparison of blood plasma alkaloid levels in cattle, sheep, and goats fed *Lupinus caudatus. Journal of Natural Toxins* 2:1–11.

Gardner DR and Panter KE (1994). Ammodendrine and related piperidine alkaloid levels in the blood plasma of cattle, sheep and goats fed *Lupinus formosus*. *Journal of Natural Toxins* 3:107–116.

Panter KE and Keeler RF (1992). Induction of cleft palate in goats by *Nicotiana glauca* during a narrow gestational period and the relation to reduction in fetal movement. *Journal of Natural Toxins* 1:25–32.

Panter KE, Gardner DR, and Molyneux RJ (1994). Comparison of toxic and teratogenic effects of *Lupinus formosus, L. arbustus* and *L. .caudatus* in goats. *Journal of Natural Toxins* 3:83–93.

Panter KE, Gardner DR, and Molyneux RJ (1998). Teratogenic and fetotoxic effects of two piperidine alkaloid-containing lupines (*L. formosus* and *L. arbustus*) in cows. *Journal of Natural Toxins* 7(2):131–140.

Williams JD (1972). Effects of time of day, moisture stress, and frosting on the alkaloid content of *Phalaris tuberosa*. *Australian Journal of Agricultural Research* 23(4):611–621.

Chapter 70

Year-to-year Variation in Alkaloid Concentration in *Lupinus leucophyllus* Growing on the Scablands of Central Washington

C.C. Gay[1], E.S. Motteram[1], K.E. Panter[2], and T. Wierenga [2]

[1] Field Disease investigation Unit, College of Veterinary Medicine, Washington State University, Pullman, WA 99164-7060, USA; [2] USDA-ARS Poisonous Plant Research Laboratory, Logan, UT 84341, USA

Introduction

On at-risk ranches located in the Channeled Scablands of Adams County in Washington state the severity of crooked calf disorder varies from year to year, both with respect to the number of calves born affected and to the severity of the deformities in affected calves. Calves with severe deformities cannot stand to suckle or follow the dam and are destroyed for humane reasons. Calves with cleft palate grow poorly and commonly succumb to aspiration pneumonia. Less severely affected calves (i.e. bent limbs) can stand and walk but are often not suited to range conditions.

In Adams County, ranches at risk for this disease commonly experience an incidence of moderate deformity near 5% of calves born, and have come to accept this as one of the risks of ranching this area. Periodically, there are years in which large proportions of newborn calves are affected with severe deformity. An example is the 1997 calving season when there was severe disease. Incidence data collected from 13 severely affected ranches documents 2210 calves born, of which 628 (28%) were born deformed; the majority of these were destroyed on humane grounds. Since 1980, there have been 8 years (32%) in which one or more ranchers on the Scablands have observed high disease prevalence and severity.

There are a number of possible causes of year-to-year variation in incidence and severity of crooked calf disease, including year-to-year variation in lupine, or relative availability, lupine density, lupine alkaloid content and lupine palatability. This study reports the preliminary results of an ongoing study examining year-to-year variation in lupine alkaloid content.

Materials and Methods

Monitoring transects were established at sites on four different ranches with an historic high risk for crooked calf. Site locations were recorded using geographic positioning (GPS) technology (GPS model 12XL. GARMIN International, Olathe, KS). Samples of lupine were collected during late mornings in mid-May. At this time approximately 25% of the lupines were in early flower, and bud samples were collected for analysis, dried at room temperature, and stored until analysis.

Alkaloid analysis

Analysis was via gas chromatography using modifications of published methods (Gardner and Panter 1993, 1994). Briefly, a 100 mg sample of finely ground lupine plant material was combined with 5 mL 1N HCL and 4 mL $CHCl_3$ and placed in a mechanical shaker for 15 min, then centrifuged to separate the aqueous and organic phases. The aqueous fraction was transferred to a clean tube and basified to a pH between 9 and 9.5 by drop-wise addition of concentrated NH_4OH and the basified solution was extracted twice with $CHCl_3$ (4mL, then 2mL). The combined $CHCl_3$ fractions were filtered through anhydrous Na_2SO_4, then dried under a stream of N_2 at 50°C and reconstituted in 5 mL $CHCL_3$ for analysis. Two µL of this final solution was injected into an HP 5890 gas chromatograph under conditions described by Gardner and Panter (1993) Alkaloid peaks, retention times, and peak area under the curve were compared with two known standard plant samples: *Lupinus caudatus* (USDA accession #82-7), containing quinolizidine alkaloids, and *L. formosus* (USDA accession #87-3), containing piperidine alkaloids. (Gardner and Panter 1993, 1994). Peak identifications were based on GC/MS analysis (Thermo-Finnigan, Austin, TX) and comparison of retention times with those alkaloid peaks from the standard plant samples. The interassay variation of 85 replicates over the last 6 years was less than 5%. Alkaloid concentration was determined via comparison of area under the curve of the subject alkaloid peaks with the area under the curve of known alkaloid peaks of the standard plant samples. The area under the curve was then numerically transcribed into mg alkaloid per 100 mg of dried plant or % alkaloid. The extraction method and quantitative analysis of alkaloid peaks were validated using caffeine as an internal standard, and total and individual alkaloid concentrations were quantified against a standard curve for anagyrine (S.T. Lee, unpublished data, 2003). Coefficients of variation for anagyrine in plant standards were 5.1%.

Results and Discussion

The total alkaloid in lupine over a 5-year period at each of the sites is shown in Figure 1 and the proportion of the total alkaloid represented by individual alkaloids in Figures 2–5.

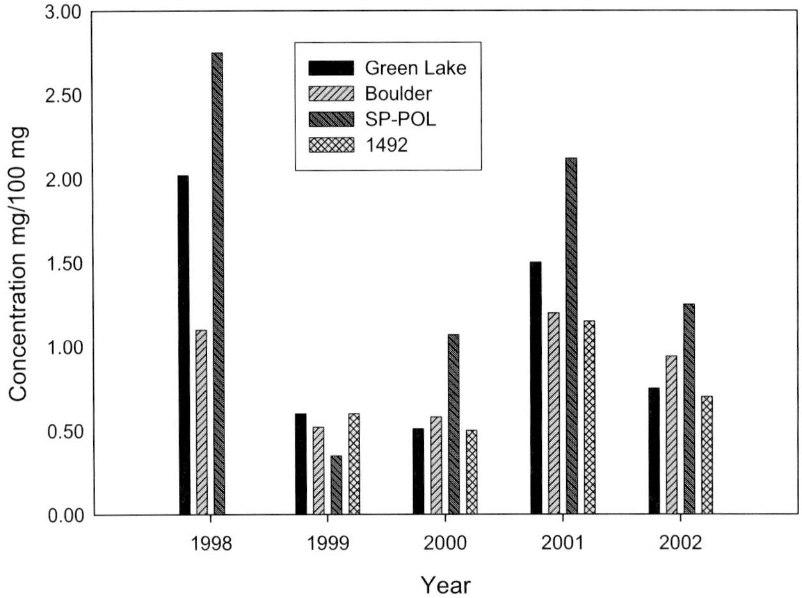

Figure 1. Year-to-year variation in total alkaloid concentrations (mg 100 mg^{-1}) in *Lupinus leucophyllus* from May sampling at all sites.

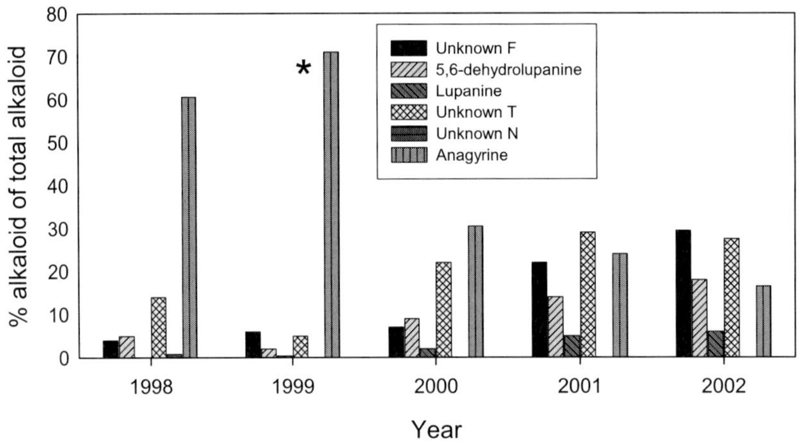

Figure 2. Year-to-year variation in individual alkaloid concentrations in *L. leucophyllus* at the Green Lake site. *about 0.4% of dry weight

There was substantial year-to-year variation in total alkaloid concentration of lupines at all sites. Total alkaloid concentration over the 5year period varied from a minimum of two-fold at one site to a maximum of eightfold at another site. In any one year the change-trend in total alkaloid concentration was the same at each site and there were years where all sites had high alkaloid concentration in lupines and years where all sites had low alkaloid concentrations. Alkaloid concentration in lupines varies with the stage of growth (Keeler *et al.* 1977; Williams and Harrison 1983; Panter and Keeler 1993). It is moderately high in the young growing leaf material, decreases in concentration in the more mature leaf growth, high in the flower bud and highest in the seed. The results reported in this study were from samples collected in mid-May when approximately 25% of the lupines are in early flower and were taken as bud samples in an attempt to minimize growth stage variability.

There was also marked year-to-year variation in the proportion of the total alkaloids represented by any one individual alkaloid and a marked variation in the proportion represented by anagyrine. Percent anagyrine in total alkaloid varied from 8 to 62%, 22 to 84%, 31 to 67% and 15 to 70% at the four sites over the 5 years. The trend to a higher or lower amount of anagyrine as a proportion of the total alkaloids was similar in any one year for all four sites.

This study was established as a long-term study to determine if there was year-to-year variation in lupine alkaloid concentrations and proportions and if this might influence the incidence and severity of crooked calf. This will require conducting the study through a high incidence year; one has not occurred since the start of this study. There was a modest increase in the incidence of crooked calf disease in this Scabland region in the year 2000, representing pastures consumed in 1999. At all sites, lupines in 1999 had the lowest concentration of total alkaloids and the highest percentage (62 to 84%) of anagyrine compared with other years.

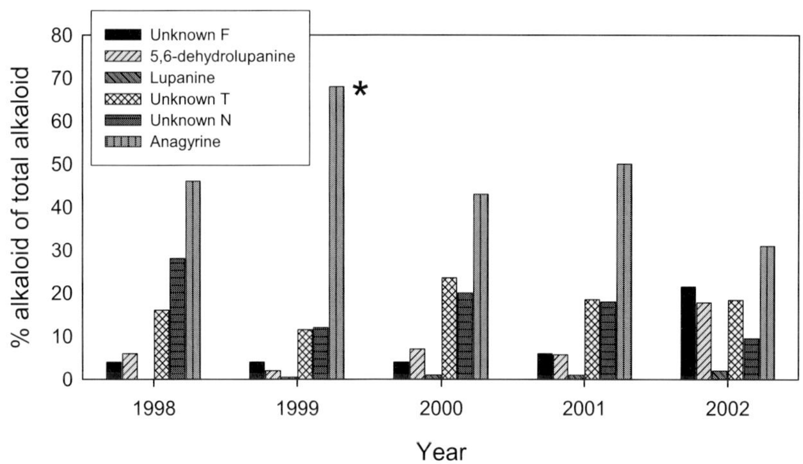

Figure 3. Year-to-year variation in individual alkaloid concentrations in *L. leucophyllus* at the Boulder site. * about 0.35% of dry weight

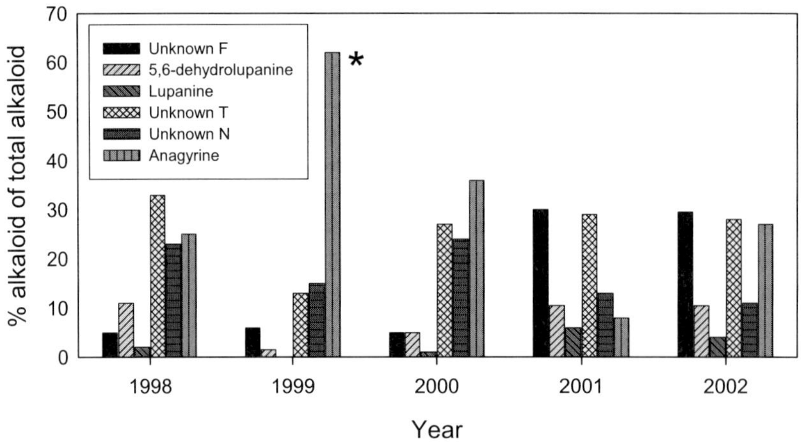

Figure 4. Year-to-year variation in individual alkaloid concentrations in *L. leucophyllus* at the SP-POL site. * about 0.25% of dry weight

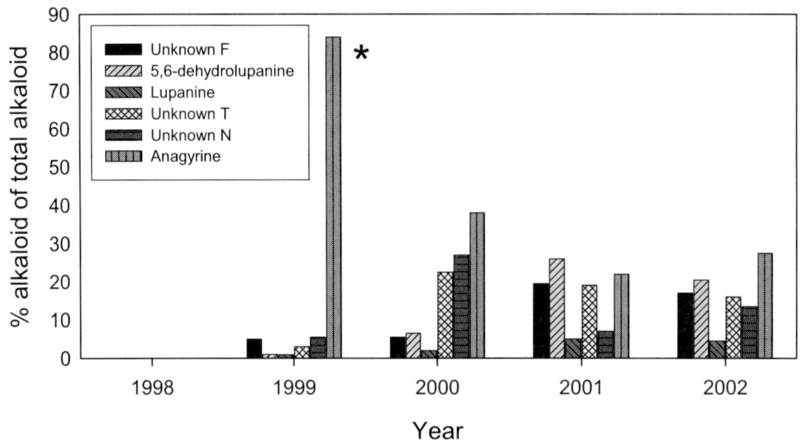

Figure 5. Year-to-year variation in individual alkaloid concentrations in *L. leucophyllus* at the 1492 site. * about 0.4% of dry weight

Whereas the production of lupine alkaloids is genetically determined, there is some indication that alkaloid concentration is influenced by environmental factors, plant nutrition, soil mineral status, and by climate (Zabolotnyi *et al.* 1973; Komarov 1980; Lovkova *et al.* 1984; Carey and Wink 1994; Muzquiz *et al.* 1994; Ralphs and Williams 1988). Our examination of the historical incidence data for the Scablands and local weather station data suggests that years of increased incidence are associated with periods of higher rainfall and years of exceptionally high incidence are precipitated by 2 preceding years of higher than normal rainfall. We believe this results in a mast year with above normal lupine flowering and seed production but it may also result in changes in alkaloid profiles to make lupines more palatable or more highly teratogenic.

This study is ongoing and will examine the influence of climatic factors on lupine alkaloid concentrations. Should these prove to be predictive for a year of serious disease it is possible that a routine May sampling and analysis might allow ranchers to take preventive strategies to avoid years of serious and economically crippling disease. It is apparent from these results that a sampling of lupine taken the year of an incident of crooked calf disease may not reflect the alkaloid exposure of the pregnant cattle grazing it in the previous spring while in early pregnancy.

References

Carey DB and Wink M (1994). Elevational variation of quinolizidine alkaloid contents in a lupine (*Lupinus argenteus*) of the Rocky Mountains. *Journal of Chemical Ecology* 20:849–857.

Gardner DR and Panter KE (1993). Comparison of blood plasma alkaloid levels in cattle, sheep, and goats fed *Lupinus caudatus*. *Journal of Natural Toxins* 2:1–11.

Gardner DR and Panter KE (1994). Ammodendrine and related piperidine alkaloid levels in the blood plasma of cattle, sheep and goats fed *Lupinus formosus*. *Journal of Natural Toxins* 3:107–116.

Keeler RF, James LF, Shupe JL, and van Kampen KR (1977). Lupine-induced crooked calf disease and a management method to reduce incidence. *Journal of Range Management* 30: 97–102.

Komarov MF (1980). Effect of different mineral nutrition conditions on alkaloid accumulation by yellow fodder lupine plants. *Sbornik Nauchnykh Trudov, Belorusskaya Sel'skokhozyaistvennaya Akademiya* 69:118–123.

Lovkova MY, Buzuk GN, and Kuz'-micheva NA (1984). Major and trace elements and accumulation of quinoline alkaloids. *Doklady Akademii Nauk SSSR* 276:250–252.

Muzquiz M, Cuadrado C, Ayet G, Cuadra C de la, Burbano C, and Osagie A (1994). Variation of alkaloid components of lupin seeds in 49 genotypes of *Lupinus albus* from different countries and locations. *Journal of Agricultural and Food Chemistry* 42:1447–1450.

Panter KE and Keeler RF (1993). Quinolizidine and piperidine alkaloid teratogen from poisonous plants and their mechanism of action in animals. *Veterinary Clinics of North America: Food Animal Practice* 9:33–40.

Ralphs MH and Williams MC (1988). Alkaloid response to defoliation of velvet lupine (*Lupinus leucophyllus*). *Weed Technology* 2:429–432.

Williams W and Harrison JEM (1983). Alkaloid concentration during development in three *Lupinus* species and the expression of genes for alkaloid biosynthesis in seedlings. *Phytochemistry* 22:85–90.

Zabolotnyi AI, Shpakova LN, Miranenka AV, and Zabalotny AI (1973). Soil conditions for growing lupine and alkaloid contents in its seeds. *Vestsi Akademii Navuk BSSR, Biyalagichnykh Navuk* 5:21–25.

Chapter 71

Conditioned Feed Aversion as a Means to Prevent Tulp (*Moraea simulans* and *Moraea pallida*) Poisoning of Cattle under Natural Grazing Conditions

L.D. Snyman, R.A. Schultz, J.P.J. Joubert, L. Labuschagne, K.M. Basson, and M.J. Molefe

Division of Toxicology, ARC-Onderstepoort Veterinary Institute, Private bag X5, Onderstepoort 0110, South Africa

Introduction

Cardiac glycoside-containing plants, such as *Moraea simulans* and *Moraea pallida*, are collectively the most important cause of plant-related poisoning of livestock in South Africa. Poisoning is characterized by posterior paresis, ruminal stasis, atrio-ventricular blocks and diarrhea (Kellerman *et al.* 1988). Except for dosing with activated charcoal, which is expensive and not successful under all circumstances, no other practical treatment is available. Conditioned feed aversion has become a practical means of teaching livestock to avoid poisonous plants (Ralphs *et al.* 2001). This has also been verified in cattle grazing on a *M. pallida* (formerly *Homeria pallida*)-infested pasture established at the Onderstepoort Veterinary Institute (Snyman *et al.* 2003), but has not yet been proved effective under natural grazing conditions.

The objective of this study was to investigate the effectiveness of conditioned feed aversion under natural grazing conditions with respect to: (1) a *M. simulans*-infested field grazing; and (2) *M. pallida*-infested harvested maize land.

Materials and Methods

Six averted and six control steers, not previously exposed to tulp (*M. simulans* or *M. pallida*), were used for each trial on tulp-infested pasture. The steers were adapted to the field grazing or harvested maize land by grazing on adjacent tulp-free sites for 2 days. The following morning one group was averted by dosing an aqueous solution of lithium chloride (120 mg kg^{-1} bw) with a stomach tube followed by administration of a mixture of epoxyscillirosidin (0.035 mg kg^{-1} bw) and a tulp hexane extract (5 g dry tulp and 15 g fresh tulp per animal), suspended in 50 mL propylene glycol, directly into the mouth. Epoxyscillirosidin, the toxin and natural aversive substance of *M. pallida,* served as a complementary aversive substance for lithium chloride, while the tulp hexane extract served as identification factor for tulp (Snyman *et al.* 2004). Averted and control steers jointly grazed the tulp-infested pasture from the next day. Steers on the *M. simulans*-infested field were kept on the field for 5 days and steers on the *M. pallida*-infested maize land for 4 days. During this time the steers were observed for clinical signs of tulp poisoning.

Results and Discussion

The number of averted and non-averted control steers poisoned on the *M. simulans*-infested field grazing and on the *M. pallida*-infested harvested maize land is shown in Table 1. Five of the control steers were poisoned on the *M. simulans*-infested field (four severely and one mildly), compared to two of the averted steers (one severely and one mildly). All six of the control steers were poisoned on the *M. pallida*-infested harvested maize land (two severely and four mildly) while none of the averted animals were poisoned. These results demonstrate that aversion treatment totally prevented steers from being poisoned on the *M. pallida*-infested harvested maize land, despite the social influence (Ralphs and Olsen 1990) of the non-averted control steers. Aversive procedures prior to exposure of steers to the *M. simulans*-infested field grazing, however, were only partially effective. This might be due to a much greater challenge presented to this group of animals as their selection for non-tulp plant material was hampered by the severe infestation of *M. simulans* in the available forage. Ingestion of *M. simulans* also would have been encouraged by social influence of the non-averted control animals. In practice, however, newly introduced steers averted to tulp will be grazed with cattle on the farm that have been naturally averted, thus influencing them to avoid tulp.

Table 1 Number of averted and non-averted control steers poisoned[a] on tulp-infested fields.

	Poisoning of steers					
	M. simulans-infested field			*M. pallida* -infested harvested maize land		
	None	Mild	Severe	None	Mild	Severe
Controls	1	1	4	0	4	2
Averted	4	1	1	6	0	0

[a] Poisoning by *M. simulans* was mainly characterized by diarrhea while steers intoxicated by *M. pallida* exhibited posterior paresis and ruminal stasis.

Conclusion

Conditioned feed aversion proved to be effective to prevent or reduce severe tulp poisoning of cattle grazing on a *M. simulans*-infested field or a *M. pallida*-infested harvested maize lands, under natural grazing conditions.

Acknowledgements

Funding was provided by the Gauteng Province (Department of Agriculture, Conservation and Environment) and the North-West Province (Department of Agriculture, Conservation, Environment and Tourism).

References

Kellerman TS, Coetzer JAW, and Naude TW (1988). *Plant Poisonings and Mycotoxicosis of Livestock in Southern Africa*, 1st ed. Oxford University Press, Cape Town.

Ralphs MH and Olsen JD (1990). Adverse influence of social facilitation and learning context in training cattle to avoid eating larkspur. *Journal of Animal Science* 68:1944–1952.

Ralphs MH, Provenza FD, Pfister, JA, Graham D, Duff GC, and Greathouse G (2001). Conditioned food aversion: from theory to practice. *Rangelands* 23:14–18.

Snyman LD, Schultz RA, Joubert JPJ, Basson KM, and Labuschagne L. (2003). Conditioned feed aversion as a means to prevent tulp (*Homeria pallida*) poisoning in cattle. *Onderstepoort Journal of Veterinary Research* 70:43–48.

Snyman LD, Kellerman TS, Schultz RA, Joubert JPJ, Basson KM, and Labuschagne L (2004). Conditioned feed aversion as a means of preventing intake of yellow-tulp (*Homeria pallida*) by livestock. In *Poisonous Plants and Related Toxins* (T Acamovic, CS Stewart, and TW Pennycott, eds), pp. 531–539. CABI Publishing, Wallingford, Oxon, UK.

Chapter 72

Poisonous Range Plants on the Fort Hall Reservation in Southeastern Idaho

D. Gunn [1], K.E. Panter[2], D.R. Gardner[2], and J.A. Pfister [2]

[1] *Fort Hall Extension, University of Idaho Fort Hall, ID 83203*, USA; [2] *USDA-ARS Poisonous Plant Research Laboratory, Logan, UT 84341, USA*

Introduction

The Fort Hall Indian Reservation in southeastern Idaho consists of approximately 122,413 ha of rangeland mainly utilized for cattle grazing by both tribal member and non-tribal member permittees. This expansive rangeland is divided into 14 range units. Each unit has unique characteristics in regard to vegetative species, soil characteristics, topography, water developments and use. Lease payments derived from grazing on reservation rangelands provide a source of income for the Shoshone–Bannock Tribes and individual tribal members. Many tribal members are also cattle producers and derive their main source of income by running cattle on reservation rangelands. Horses are utilized extensively by cattle producers as well. Although sheep aren't allowed on range units for grazing, several bands are herded through the reservation to adjacent grazing allotments. One of the main issues of concern is poisonous range plants within these range units and their effects on livestock species, including cattle, sheep and horses. Under some circumstances, poisonous range plants create and have created devastating economic losses to ranchers and landowners. An estimated 3% of cattle, sheep and horses are affected by poisonous range plants in the United States every year (Nielsen *et al.* 1988). This includes death losses of adult cattle, sheep and horses, early or late-term abortions and birth defects. Other economic losses from poisonous plants that are much more difficult to quantify include lost weight gain, long-term organ damage, impaired reproductive function, compromised immune function, increased veterinary costs, increased management costs, degradation of range condition and decreased range value.

Several poisonous range plants are found on the reservation and include, but are not limited to, the following species: bur buttercup (*Ranunculus testiculatus*), milkvetch (including *Astragalus beckwithii*, *A. mollissimus*, *A. stenophyllus*, *A. utahensis* and *A. vexilliflexux*), houndstongue (*Cynoglossum officinale*), leafy spurge (*Euphorbia escula*), broom snakeweed (*Gutierrezia sarothrae*), kochia (*Kochia scoparia*), Russian knapweed (*Centaurea repens*), horsetail (*Equisetum laevigatum*), black henbane (*Hyoscyamus niger*), poison-hemlock (*Conium maculatum*), water hemlock (*Cicuta douglasii*), death camas (including *Zigadenus paniculatus*, *Z. venenosus* and *Z. elegans*), lupine (*Lupinus arbustus, L. caudatus*, other *Lupinus* spp.), low larkspur (*Delphinium nuttallianum, D. andersonii*) and tall larkspur (*Delphinum geyeri, D. occidentale*). Those species of poisonous plants of most concern on the reservation include poison-hemlock, water hemlock, death camas, lupines and larkspurs.

Poisonous Range Plant Characteristics and Effects on Animal Health

Bur buttercup (*Ranunculus testiculatus*) is a low-growing annual which reaches heights of 5–12 cm and develops yellow flowers, which eventually develop into burrs. This plant can be found primarily in disturbed areas and in small grains, pastures, dry land and roadsides. It is quite competitive with annual and perennial seedlings and can be very toxic to sheep (Olsen *et al.* 1983). Bur buttercup contains the toxin ranunculin and causes mouth and intestinal tract irritation and death from pulmonary edema. The best management practice is to ensure that a healthy stand of perennial grasses is maintained, thus avoiding bur buttercup infestation. Depending upon the economic feasibility, management practices may also include the use of herbicides to control the plant and ensuring that sheep have other forages available for grazing.

The ***Astragalus* genus (milkvetches)** contains some 2000 species (over 350 in North America alone) and include locoweeds, selenium indicator species and nitro toxin-containing species (PPRL 2005). Milkvetches can reach up to 75 cm in height (Whitson 1999). Five milkvetches have been identified and reported on the reservation by the PPRL and the University of Idaho Extension Service. *Astragalus mollissimus*, a known locoweed, contains the loco toxin swainsonine. *Astragalus beckwithii* and *A. utahensis* have been reported by the PPRL to contain nitro toxins. Both were sampled, taxonomically identified at the Intermountain Herbarium at Utah State University, and chemically analyzed at the PPRL (Table 1). While *A. beckwithii* contained no detectable nitro toxin, *Astragalus utahensis* contained levels of 3-nitro propanol that could be considered a moderate to high risk for livestock poisoning. Although both species were found to be low in population density and probably not a significant threat at this time, populations of both species should be monitored, samples collected for further analysis, and plant density controlled to prevent future poisoning.

Table 1. 3--nitro propanol content of two vetches growing on the reservation.

Sample	mg NO_2 g^{-1} plant
A. beckwithii (Burn Area)	n.d.
A. utahensis (Spook Canyon)	8.9

Houndstongue (*Cynoglossum officinale*) is a biennial and grows up to 1.3 m tall. The leaves of this plant are rough and resemble the texture of a 'hound's tongue,' hence the name. Reddish purple flowers form fruit composed of prickly nutlets, which cling to clothing and animals (Whitson 1999). Houndstongue is toxic and contains pyrrolizidine alkaloids and is most poisonous in the rosette stage (PPRL 2005). Livestock rarely graze the plant but poisoning often occurs when calves or horses ingest it in contaminated hay. Young animals and horses are more sensitive to poisoning. Signs of poisoning include weight loss, jaundice, depression, diarrhea and photosensitivity. Sheep are more resistant to poisoning than cattle or horses. Horses are often affected through secondary photosensitization. It can be chemically controlled with broadleaf herbicides.

Leafy spurge (*Euphorbia escula*) is a perennial which reaches heights of 1 m and is listed on the Idaho Noxious Weeds list. The stems of leafy spurge are clustered with yellow, heart-shaped bracts and alternating leaves. This plant produces a latex substance which may cause dermal and ocular irritation as well as mouth and digestive tract irritation in cattle (Whitson 1999). Leafy spurge is a noxious weed found in range, pastures, roadsides and riparian areas. Its aggressive nature, lack of natural controls such as insect predators and plant disease and resistance to most herbicides has resulted in large areas of infested rangeland. Proper chemical application and/or grazing with goats are effective management tools to prevent leafy spurge from spreading.

Broom snakeweed (*Gutierrezia sarothrae*) is a perennial shrub with close-growing stems reaching heights of 60 cm. The leaves are linear and alternate and flower heads are small and yellow (Whitson 1999). The predominance of broom snakeweed can be an indicator of overgrazing or other disturbances. However, it is certainly not always an indicator of overgrazing. Broom snakeweed is alleopathic and contains saponins. It is most toxic during leaf formation and causes abortions with retained placentas in cattle, sheep and goats at any stage of gestation (PPRL 2005). It can also cause weight loss, decreased appetite, acute kidney degeneration, diarrhea, jaundice and weakened offspring if animals are forced to eat this plant (PPRL 2005). It is found in range and pastures and can be controlled with properly managed livestock grazing and herbicides such as picloram and metsulfuron (PPRL 2005). While broom snakeweed is not a significant toxic plant problem on the reservation, it is invading certain range areas, thus reducing availability of other forages.

Kochia (*Kochia scoparia*) is an annual which reaches heights of 1–2 m. The plant must not be confused with forage kochia (*Kochia prostrate*), which can serve as a dependable source of forage for livestock. The leaves are alternate and are lance-shaped on branched stems. Kochia is often used for livestock grazing, particularly during periods of drought. It can accumulate significant levels of nitrates causing nitrate poisoning in livestock (Whitson 1999; PPRL 2005). Extra caution should be practiced with pregnant cattle as the fetus is especially sensitive to the anoxic effects of nitrate poisoning (PPRL 2005). The amount of nitrate accumulation is influenced by factors such as soil nitrogen, drought, frost, herbicide use, etc. The risk of nitrate poisoning can be reduced by testing and mixing forage with low nitrate forages. Before grazing kochia, the forage should be tested for nitrate levels. Kochia is found everywhere, particularly in pastures, roadsides and some range sites. Chemical control is available but must be used in the early growth stages.

Russian knapweed (*Centaurea repens*) is a perennial that reaches heights of 40–120 cm. This plant forms considerably branched stems and monocultures of dense colonies. It develops purple flowers on cone-shaped bracts. Russian knapweed poisoning is rare but causes chewing disease (neurological degeneration) in horses. It is considered a noxious weed in Idaho and is invading fields, pastures and rangelands extensively. Russian knapweed is alleopathic and often difficult to control. Picloram can be used as an effective herbicide application and goat grazing on sensitive lands is being investigated as a potential biological control.

Black henbane (*Hyoscyamus niger*) is a member of the nightshade family and can grow as an annual or biennial, depending on conditions, and reaches heights of up to 1 m. It produces large lobed leaves and a foul, musty odor. Black henbane produces flowers which are cream-colored with purple centers and veins. It contains tropane alkaloids which are poisonous to livestock and humans. Signs of poisoning include decreased salivation and reduced intestinal motility and tachycardia. This plant is not palatable and animals will generally avoid grazing it. Black henbane can be chemically controlled (Whitson 1999; PPRL 2005).

Horsetail (*Equisetum laevigatum*) is a perennial with hollow, round and jointed stems. The plant reaches heights of 1–2 m. It reproduces via spore-producing cones. All parts of the plant are toxic. Horses are the most susceptible to poisoning but it is also poisonous to sheep and cattle. It degenerates the peripheral nerves, resulting in blindness and sudden death. Signs of poisoning include weight loss, weakness, gait abnormalities, abnormal heart rate, inability to rise and death. Poisoning is rare and generally associated with overgrazing. Most cases of poisoning that have occurred are from the plant being baled in hay which is then fed to horses (PPRL 2005).

Most Significant Infestations in Fort Hall

The most significant populations of poisonous range plants on the reservation include poison-hemlock, water hemlock, deathcamas, lupine and larkspur.

Poison-hemlock (*Conium maculatum*) is a biennial and reaches heights of 2–3 m. The plant has erect, purple mottled stems with triangular, finely divided leaves. The flowers are white and form small umbrella-shaped clusters which produce large numbers of seed. This plant contains many toxic piperidine alkaloids which affect the nervous system. The predominant alkaloids are γ-coniceine and coniine, both of which are toxic and teratogenic (PPRL 2005). Cattle and pastured swine are the livestock most often affected; however, poisoning has been reported in horses, sheep, goats and people (Panter *et al.* 1999; Knight and Walter 2004). Skeletal defects and cleft palate may be induced when plants or green seeds are ingested by pregnant cattle and pigs at susceptible stages of gestation including 30–60 and 40–100 days in swine and cattle, respectively (PPRL 2005). Poison-hemlock is found along fence lines, irrigation ditches and other moist waste places. Chemical control with broadleaf herbicides is effective with repeated application as needed when seedlings emerge (PPRL 2005).

Water hemlock (*Cicuta douglasii*) is a biennial species with erect stems that reach heights of 1–2 m. Stems are smooth and hollow with toothed leaves. The flowers are white and compound and are similar to those of poison-hemlock. The root has distinct chambers and is the most poisonous plant part, although green seed was recently reported to cause the death of cattle in Utah (Panter *et al.*, Chapter 44, this volume). It is considered one of the most poisonous range plants in North America and is often mistaken for water parsnip or other edible herbs (Whitson 1999; PPRL 2005). The toxins are a group of long-chain alcohols including cicutoxin and cicutols that are very poisonous to humans and animals and only small amounts are needed to produce poisoning and fatalities. Signs of poisoning include nervousness, tremors, collapse, grand mal seizures and sudden death. Signs develop shortly after ingestion with death generally occurring within 1–2 hours after a fatal dose (PPRL 2005). Water hemlock is found in wet areas of meadows, pastures and along streams. It can be controlled chemically and mechanically. However, care must be taken to avoid exposure of tubers as they are very palatable and highly toxic and can be easily ingested by livestock when found on the ground.

Deathcamas (*Zigadenus* spp.) is a native perennial and is one of the first plants to grow in the spring. It has white to yellowish flowers in a terminal cluster formed with three to six basal leaves and reaches heights of about 60 cm. It also produces underground bulbs (Whitson 1999). Three species have been identified on the reservation and all are toxic. The toxins are steroidal glyco-alkaloids and are considered highly toxic. All parts of the plant are poisonous at all growth stages but bulbs are especially toxic. Sheep are the most commonly affected but poisoning in cattle, horses and humans has also occurred. Excessive salivation, weakness, respiratory problems, nausea, convulsions, coma and sudden death can result after ingestion (PPRL 2005). The plant can be mistaken for edible camas lilies or wild onion and has caused illness and death in humans.

Lupines (*Lupinus* spp.) are herbaceous, perennial plants reaching heights of 1 m. The inflorescence includes a terminal raceme with flowers ranging in color from blue, purple, yellow and white, which eventually form seed pods. The stems are upright and branched with alternate, compound, palmate leaves (Whitson 1999). Most wild lupine species produce poisonous alkaloids, some of which also cause birth defects in calves. Cows ingesting lupine in early gestation (days 40 thru 100) may produce calves with skeletal defects (crooked calf disease) and cleft palates (Panter *et al.* 1999; Knight and Walter 2004). Sheep may also be poisoned but this usually occurs after heavy grazing of pods. Signs of lupine poisoning are neurological with death resulting from respiratory paralysis. While only two species have been identified on the reservation (*Lupinus caudatus*; purple and *L. arbustus*; yellow) both contain toxic and teratogenic alkaloids (Figures 1 and 2). *Lupinus caudatus* contained the quinolizidine teratogen anagyrine while *L. arbustus* contained two piperidine teratogens ammodendrine and N-methyl ammodendrine (PPRL 2005). As range surveys continue we believe other lupine species will be found. Lupine can be controlled chemically and poisoning can be managed through proper grazing methods to avoid ingestion during key gestational time periods in cattle. It is recommended that chemical analysis of all lupine species be done as taxonomy is not a reliable predictor of alkaloid content or teratogenic risk.

Tall larkspurs (*Delphinium occidentale, D. geyeri*) are perennial plants which reach heights of 1.6 m. The leaves are large and lobed. The flowers are deep purple and grow in terminal racemes. All parts of the plant are toxic to livestock, particularly to cattle. Tall larkspurs contain diterpenoid alkaloids of which the MSAL class is the most toxic (Panter *et al.* 2002). Signs of poisoning include labored breathing, muscle weakness, and neuromuscular paralysis. Chemical control methods or appropriate grazing management are available for this plant.

Low larkspurs (*Delphinium andersonii, D. nutallianum*) have been identified on the grazing units of the reservation. Large cattle losses have recently occurred (spring of 2004–05) on the reservation and are believed to have been caused by the low larkspurs, *D. andersonii and D. nutallianum* (PPRL 2005; University of Idaho Fort Hall Extension, personal communication). Analysis of the two low larkspur species supports this assumption (Table 2). The leaves are deeply parted into lobes. Blue-purple spurred flowers are produced by this plant. Low larkspur completes its growth cycle in late spring and/or early summer. It contains diterpenoid alkaloids. Low larkspur is poisonous to livestock, particularly cattle, and can result in sudden death. Death losses of approximately 1–3% occur each year due to larkspur poisoning (PPRL 2005). Signs of poisoning include nervousness, weakness, salivation, nausea, bloating, rapid heart rate and sudden death. Excitement and physical activity can intensify the effects of poisoning. Low larkspur can be controlled chemically or with timely grazing management (PPRL 2005).

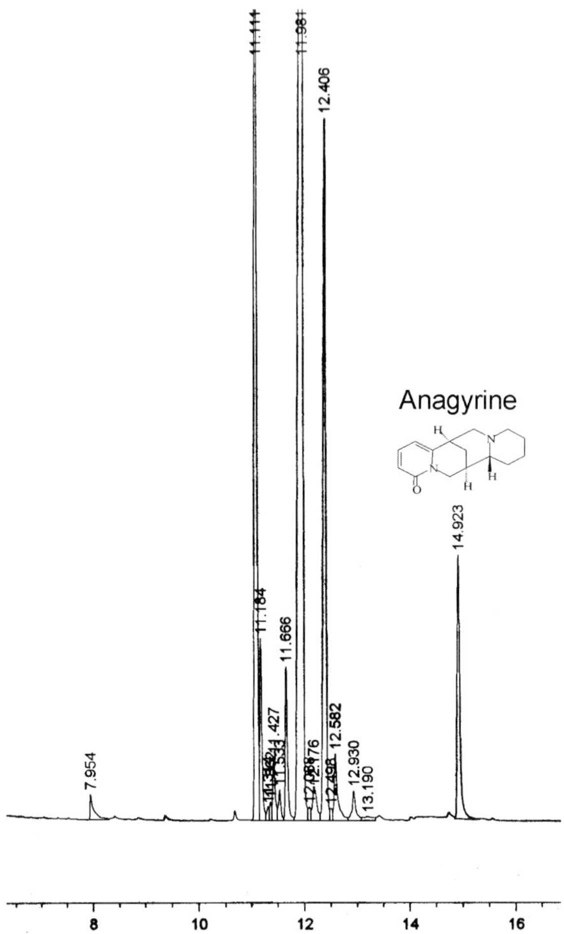

Figure 1. *Lupinus caudatus* containing quinolizidine alkaloids including the teratogen anagyrine.

Low larkspur has been particularly prevalent in certain areas on the reservation, particularly the springs of both 2004 and 2005, due to perfectly timed rains and cooler than normal spring temperatures. Low larkspur caused significant cattle death loss in the spring of 2004 (University of Idaho Fort Hall Extension, personal communication). This was due to a high concentration of toxic alkaloids (3 mg MSAL alkaloids g^{-1} dry weight is considered moderate risk; PPRL 2005). *Delphinium andersonii* is more toxic due to certain alkaloid profiles (e.g. higher nudicauline) than *D. nuttallianum* (Table 2). The following table shows alkaloid content of *Delphinum* spp. on the reservation.

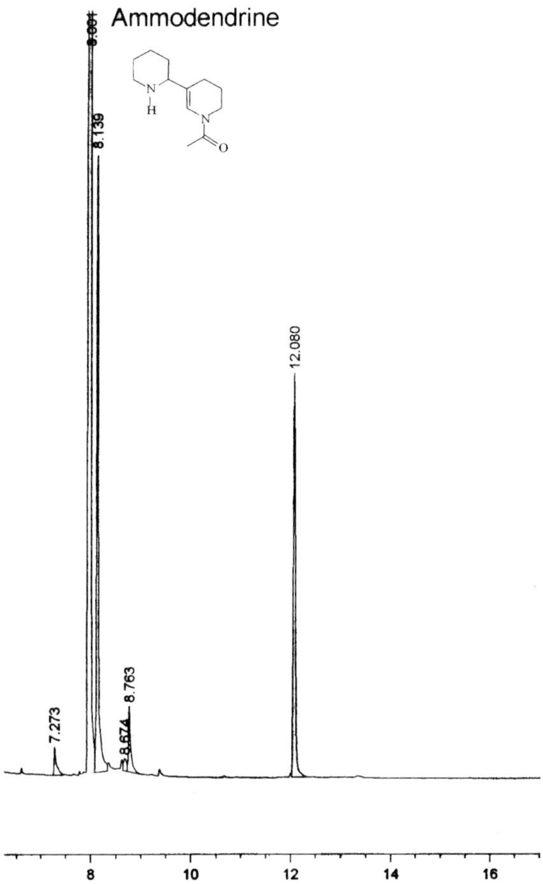

Figure 2. *Lupinus arbustus* containing the piperidine alkaloid teratogen ammodendrine.

Table 2. MSAL alkaloids (mg g^{-1}) identified in two low larkspur species on the reservation (mg alkaloid g^{-1} dry weight).

	MLA	14-DAN	BEAR	GEY	NUD	AcBEAR	TOTAL
SAMPLE 1							
D. andersonii	2.93	0.46	0.12	1.84	1.33	0.09	**6.78**
SAMPLE 2							
D. andersonii	1.92	0.37	0.04	2.31	1.30	0.06	**6.00**
SAMPLE 3							
D. nuttallianum	1.49	0.22	0.09	1.20	0.03	0.14	**3.17**
SAMPLE 4							
D. nuttallianum	1.28	0.13	0.07	1.02	0.02	0.10	**2.63**

MLA = methyllycaconitine; 14-DAN = 14-deacetylnudicauline; BEAR = bearline; GEY = geyerline; NUD = nudicauline; AcBEAR = 14-acetylbearline

Management Options

To avoid poisonous plant fatalities, birth defects, etc., the following management options can be utilized. Avoid grazing infested areas when livestock are most likely to eat poisonous plants and avoid infested areas during the plant's optimum window of toxicity, generally during the early vegetative and reproduction stages. Provide alternative feed sources to livestock when grasses or desirable forbs aren't available. Don't turn cattle out on areas where poisonous plants are the most available green feed. Feed animals well prior to turnout in the early spring season or in new grazing areas to prevent grazing of poisonous plants since many of these plants are green and more available than grasses in the early spring. Distribute salt and mineral blocks away from infested areas. Change management strategies when plant populations shift. Implementation of strict monitoring plans for poisonous plants on range units and appropriately timed turn-out dates are very beneficial to avoid livestock poisoning from poisonous plants. Producers utilizing range units should be educated as to which plants to avoid through the use of range tours, letters, meetings and workshops.

A project to inventory poisonous range plants on the reservation is currently being developed. These range plants will be collected during the growing season, identified and chemically analyzed to characterize toxin profiles at different phenological stages to determine relative grazing risk. The infested areas will be mapped utilizing GPS coordinates. Control and management strategies will be developed and implemented as well. This information will be disseminated to the Tribe, tribal members and permittees via range tours, individual ranch visits, workshops, newsletters, etc., to help reduce livestock losses. Continued development and implementation of the Fort Hall Poisonous Range Plant Project will benefit the Shoshone–Bannock Tribes and tribal members by developing improved management strategies to decrease death loss, increase awareness of the problem and improve the economic value of these rangelands.

Conclusion

In conclusion, poisonous plants can result in high morbidity, mortality and economic losses. To avoid poisonous plant effects, a management strategy should include the following options:

1. Closely monitor and observe pastures and ranges to identify potential and existing poisonous plant infestations.
2. Test existing poisonous plants for toxin levels to identify what chemical compounds are prevalent and could cause the most harm to livestock.
3. Time grazing more appropriately to avoid infested areas until other forage is available.
4. Feed and water animals well prior to turnout.
5. Mechanical, chemical and possibly grazing control methods with different livestock species can provide effective means of control.
6. As always, proper management and keen observation will help land managers and permittees avoid the possible devastating effects of poisonous range plants.

References

Knight AP and Walter RG (2004). Plants associated with congenital defects and reproductive failure. In *A Guide to Plant Poisoning of Animals in North America* (AP Knight and RG Walter, eds), pp. 1–11. Teton New Media, Jackson, Wyoming.

Nielsen DB, Rimbey NR, and James LF (1988). Economic considerations of poisonous plants on livestock. In *The Ecology and Economic Impact of Poisonous Plants on Livestock Production* (LF James, MH Ralphs, and DB Nielsen, eds), pp. 5–16. Westview Press, Boulder, Colorado.

Olsen JD, Anderson TE, Murphy JC, and Madsen G (1983). Bur buttercup poisoning of sheep. *Journal of the American Veterinary Medical Association* 183:538–543.

Panter KE, James LF, and Gardner DR (1999). Lupines, poison-hemlock, and *Nicotiana* spp: Toxicity and teratogenicity in livestock. *Journal of Natural Toxins* 8(1):117–134.

Panter KE, Manners GD, Stegelmeier BL, Lee ST, Gardner DR, Ralphs MH, Pfister JA, and James LF (2002). Larkspur poisoning: Toxicology and alkaloid structure-activity relationships. *Biochemical Systematics and Ecology* 30:113–128.

PPRL, *USDA-ARS Poisonous Plant Research Laboratory home page*, 1 July 2005. <http://www.ars.usda.gov/npa/ logan/pprl> (21 November 2005).

Whitson TD, ed (1999). *Weeds of the West*. Western Society of Weed Science in cooperation with Cooperative Extension Services, University of Wyoming. Laramie, Wyoming.

Chapter 73

Population Cycles of Poisonous Plants

M.H. Ralphs[1] and D.T. Jensen[2]

[1] *USDA-ARS Poisonous Plant Research Laboratory, Logan UT 84341, USA;* [2] *Utah State Climatologist (retired), Utah State University, Logan UT 84322, USA*

Introduction

A poisonous plant is harmless until it is consumed. Once eaten, the degree of damage depends on the amount of the plant consumed; or more specifically, the amount of toxin entering and absorbed by the body. Poisoning therefore depends on two principal variables: the toxin level in the plant and the rate of ingestion by the animal. The rate of ingestion, in turn, depends upon the availability or abundance of the plant in the ecosystem.

Weed scientists have developed the concept of an economic threshold, at which a weed's population increases to the point that it begins to suppress production of the crop. Poisonous plant populations are analogous, in that they are not a threat until their population reaches a density that an animal can consume enough of the plant, fast enough, to ingest a sufficient amount of toxin to cause harm.

The presence or abundance of specific poisonous plants in natural ecosystems is not constant, particularly in arid or semiarid rangelands. Plant populations cycle and are driven primarily by precipitation patterns and drought. The objective of this paper is to present examples of population cycles of important poisonous plants in the western US, relate them to precipitation patterns, and present management strategies to deal with fluctuating populations.

Locoweed Population Cycles

Locoweeds are the most widespread poisonous plant problem in the western US. The toxic alkaloid swainsonine causes a wasting disease. Incidence of poisoning is sporadic since many semiarid locoweed species experience extreme population cycles, increasing in wet years and dying out during drought. They germinate following autumn rains, remain green over winter, flower in spring, and may continue to grow for 1 to 2 years until the next drought occurs and the population dies back (Welsh 1989). The seed bank in the soil ranges from 600 to 4000 seeds m^2 (Ralphs and Cronin 1987; Ralphs and Bagley 1988). The seeds have hard coats and remain viable in soil for many years, thus providing an ecological advantage to exploit environmental conditions and maintain the 'boom and bust' population cycles. Poisoning problems follow these cycles, often causing catastrophic losses.

Spotted locoweed (*Astragalus lentiginosus* var. *diphysus*) causes major livestock poisoning in northern Arizona and southern Utah. However, the poisoning problems are erratic and seem to be tied to its population cycles. Outbreaks occurred in 1983–85, 1991–93, and again in 1998. Populations of spotted locoweed appear to require 2 successive wet years to establish. The first year, seeds germinate and establish, but are not very apparent. If sufficient moisture is available the second year, they grow rapidly and appear to dominate the plant community (Stan Welsh, personal communication).

Population outbreaks of Wahweap milkvetch (*Astragalus lentiginosus* var. *wahweapensis)* occurred every 6 to 8 years in the Henry Mountain area of southeast Utah (Ralphs and Bagley 1988) and were associated with above-average fall and spring precipitation. Catastrophic losses occurred in those years when Wahweap milkvetch was abundant.

Green River milkvetch (*Astragalus pubentissimus*) is a short-lived perennial forb that is occasionally abundant along the Green River corridor in southwest Wyoming, and eastern Utah, USA. It grows on lightly alkaline soils derived from shales or sandstone, and is a component of the salt-desert shrub, mixed desert shrub, and pinyon/juniper plant communities. In years of abundance, it is the principal forb in these desert communities, appearing like fields of alfalfa. Outbreaks of Green River milkvetch and resulting epidemics of poisoning occurred in 1917–18, 1957–58, and 1965–66 (James *et al.* 1968). In the 1965 outbreak in the Uinta Basin, 55% of a band of 1900 ewes died and most of the reminder of the band aborted. In another band, 45% aborted.

Crescent milkvetch (*Astragalus amphioxys*) is another short-lived perennial forb that occurs on sandy soils along the Green River corridor in eastern Utah. It is taxonomically and morphologically similar to Green River milkvetch. It contains the locoweed toxin, swainsonine, and periodically causes losses to sheep and cattle. A population outbreak occurred during the wet fall of 1994, but it died out during the dry summer and fall of 1995. A catastrophic loss of sheep occurred from grazing the remaining dead, dry stems: of 700 ewes, 340 aborted and 300 died. Another crop of Crescent milkvetch germinated in the wet spring of 1996. Crescent milkvetch density was 21.5 plants m^2 on sand dunes and 5.3 plants m^2 on sandy-gravel soil. It did not occur on the heavier clay soils. In 1997, 95% of the plants on the gravely soil died and 65% of the plants on the dunes site died. By 1998, all crescent milkvetch plants were dead. There were only widely scattered plants in 1999 and 2000, but a new crop germinated and established in the spring of 2001. Crescent milkvetch populations appear to establish during years of high fall and/or spring precipitation and die out in summer drought (Figure 1).

Woolly locoweed (*Astragalus mollissimus* var. *mollissimus*) is a principal poisonous plant in western Kansas, the panhandle of Oklahoma, eastern New Mexico and west Texas. However, it rarely persists more than 2–3 years because of reoccurring droughts and from damage inflicted by the larvae of the four-lined locoweed weevil (Pomerinke *et al.* 1995). As few as two larvae will kill the plant.

Garbancillo (*Astragalus wootoni*) is an annual in the southwest deserts and only grows during years of high winter precipitation (Sperry *et al.* 1964). It becomes a large plant that retains its succulence in this desert region, which contributes to its poisonous nature.

White locoweed (*Oxytropis sericea*) was thought to be more persistent than the *Astragalus* locoweeds. It grows on the short grass prairies of the western Great Plains and in mountain grasslands. Its preferred habitat is rocky soils where it exhibits a stress-tolerant survival strategy (Payne 1957; Ralphs *et al.* 1989). Its long tap root can access deep percolated water allowing it to survive drought, temperature and wind stress. However, its populations declined in New Mexico, Colorado and Utah during the droughts of 1996 and 2000 (Ralphs *et al.* 2001). We found negative correlations between locoweed mortality and precipitation. White locoweed died out throughout the entire western region during the widespread drought of 2001–2004. Marsh (1909) also observed that white locoweed was abundant in wet years, but nearly disappeared in dry seasons.

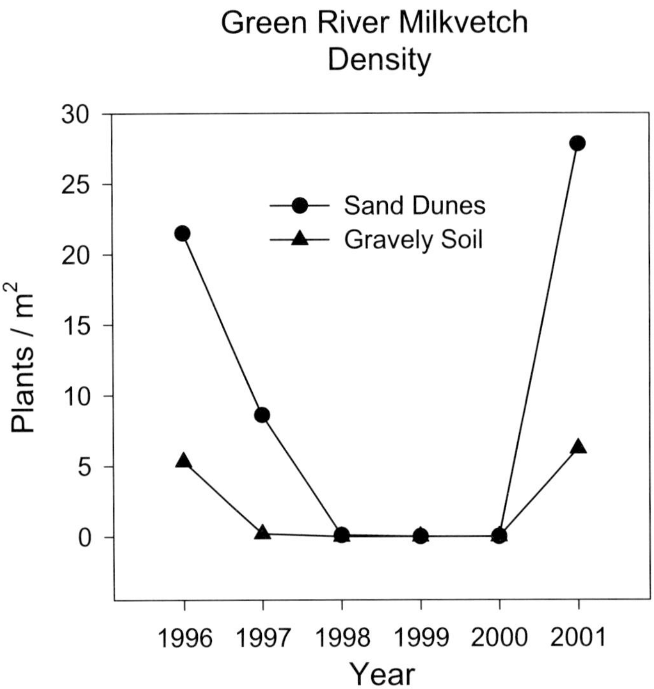

Figure 1. Density of crescent milkvetch on sand dunes and gravely soil near Green River, Utah. Large numbers of plants established following years of high fall and winter precipitation in 1995 and 2000.

Emory milkvetch (*Astragalus emoryanus* var. *emoryanus*) is a winter annual throughout much of the Rio Grande Valley of Texas and the southern half of New Mexico. It contains nitro toxins. In dry years, plants are small and scattered. When precipitation is timely and abundant, seeds germinate and plants grow profusely, often forming a veritable carpet on large areas of rangeland. Abundant precipitation fell in autumn 1975, causing unusually high rate of germination. There was adequate winter and spring moisture to support growth, and plant densities in the spring 1976 around Roswell NM had not been seen at that level for 35 years. Mortality from nitro poisoning averaged 2–3% and morbidity averaged 15–20% of all the cows in the region (Williams *et al.* 1979). Emory milkvetch has not caused poisoning problems since.

Snakeweed

Broom snakeweed (*Gutierrezia sarothrae*) is one of the most widespread range weeds in North America. It is a suffrutescent half-shrub that increases when rangelands are disturbed by overgrazing, fire and drought. It is highly competitive with desirable grasses and causes abortions in all livestock. Its populations are cyclic and appear to be related to climatic patterns. In New Mexico, broom snakeweed populations died out during droughts in 1970–71, 1984, and 1994, but rapidly re-established in the following wet winters and spring (McDaniel and Ross 2001).

We monitored broom snakeweed populations in the salt-desert shrub community of the Colorado Plateau and in crested wheatgrass (*Agropyron desertorum*) seedings typical of the Snake River Plains and Great Basin. The snakeweed population at the Colorado Plateau site at Ferron, Utah completed two cycles over the 13-year study period, dying out in the drought of 1989–90, re-establishing in the wet years of 1992–93, dying back in 1996, and finally dying out completely in 2001 (Figure 2). There were positive correlations between density of snakeweed classes and seasonal precipitation: seedlings with spring precipitation ($r = 0.63$); juveniles with winter precipitation ($r = 0.69$); and mature plants with the previous fall's precipitation ($r = 0.62$).

Only one cycle occurred at the Snake River Plains site (Figure 2). Following the snakeweed invasion into crested wheatgrass seedings in the mid-1980s, the population was at the top of its population cycle when the study began in1990, dropped back and fluctuated between 6–10% cover from 1992 to 1999, and died out in 2001. Although density of mature plants did not change much, the change in snakeweed cover was correlated with spring precipitation ($r = 0.81$) (Ralphs and Sanders 2001).

Lupine

Several lupine species contain quinolizidine and piperidine alkaloids that cause a crooked calf syndrome if the dam grazes the plant between the 40 and 70th day of gestation. The incidence of crooked calf disease is sporadic and may be tied to the plant population cycles. In the spring of 1997 in Adams County in eastern Washington, 28% of calves on 13 ranches were born deformed and had to be destroyed (Clive Gay, unpublished data). This followed abnormally high years of precipitation in 1995 and 1996. We started measuring velvet lupine (*Lupinus leucophyllus*) populations in 2002 during a series of years of below average precipitation, and found density of mature lupine plants declined from 2.1 plants m^2 in 2002 to 0.65 plants m^2 in 2004. There was low incidence of crooked calf disease during these years when lupine populations were low.

Conclusion

Poisonous plant populations cycle, particularly in arid and semiarid rangelands. Plants germinate and establish in seasons of high precipitation, and die out during drought. Management decisions to graze infested areas are based on the risk of poisoning, which is dependent on the density and availability of the poisonous plant. Ranchers should anticipate outbreaks of poisonous plants in wet years. If outbreaks materialize, they should move their livestock to other areas to prevent poisoning.

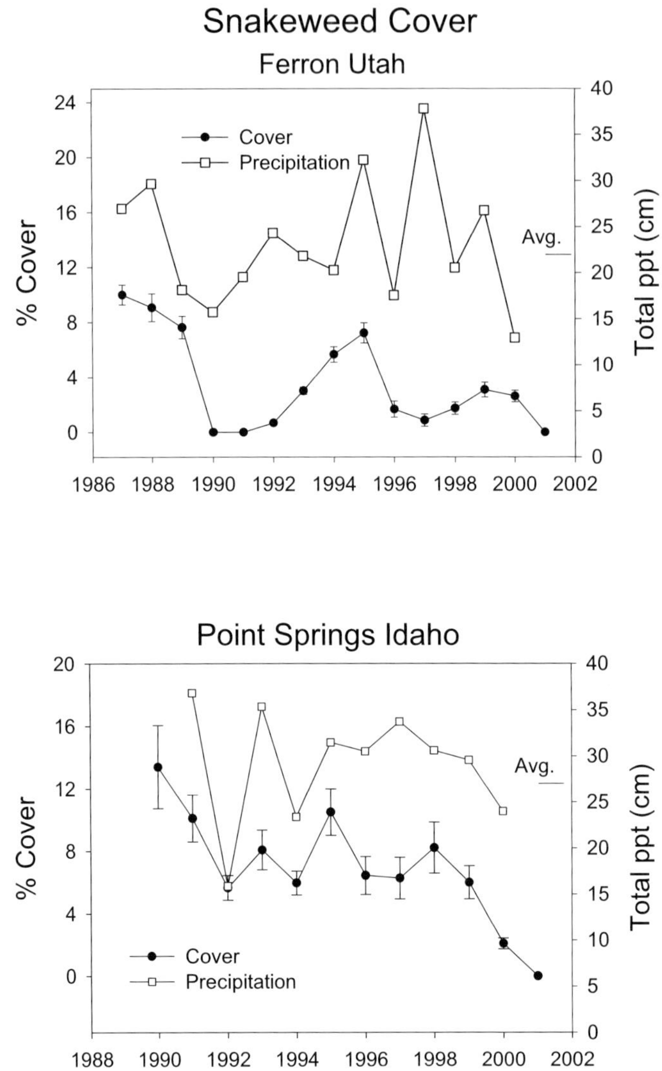

Figure 2. Broom snakeweed population cycles at Ferron Utah and Malta Idaho are related to annual precipitation (error bars are standard errors).

References

James LF, Bennett KL, Parker KG, Keeler RF, Binns W, and Lindsay B (1968). Loco plant poisoning in sheep. *Journal of Range Management* 2:360–365.

Marsh CD (1909). The locoweed disease of the plains. *USDA Animal Industries Bulletin* 112.

McDaniel KC and Ross TT (2001). Snakeweed: poisonous properties, livestock losses and management considerations. *Journal of Range Management* 55:277–284.

Payne GF (1957). Ecology and life history of the poisonous plant, white locoweed (*Oxytropis sericea* Nutt.). PhD Dissertation, Texas A&M University, College Station.

Pomerinke MA, Thompson DC, and Clason DL (1995). Bionomics of *Cleonidius trivittatus* (Coleoptera: Curculionidae): native biological control of purple locoweed (Rosales: Fabaceae). *Environmental Entomology* 24:1696–1702.

Ralphs MH and BagleyVL (1988). Population cycles of Wahweap milkvetch on the Henry Mountains and seed reserve in the soil. *Great Basin Naturalist* 48:541–547.

Ralphs MH and Cronin EH (1987). Locoweed seed in soil: density, longevity, germination and viability. *Weed Science* 35:792–795.

Ralphs MH and Sanders KD (2001). Population cycles of broom snakeweed in the Colorado Plateau and Snake River plains. *Journal of Range Management* 55:406–411.

Ralphs MH, Benson B, and Loerch JC (1989). Soil-site relationships of white locoweed on the Raft River Mountains. *Great Basin Naturalist* 49:419–424.

Ralphs MH, Gardner DR, Graham JD, Greathouse G, and Knight AP (2001). Influence of clipping and precipitation on locoweed vigor, mortality and toxicity. *Journal of Range Management* 55:394–399.

Sperry OE, Dollahite JW, Hoffman GO, and Camp BJ (1964). Texas plants poisonous to livestock. *Texas Agriculture Experiment Station Bulletin* B-1028.

Welsh SL (1989). *Astragalus* L. and *Oxytropis* DC: definitions, distributions, and ecological parameters. In *Swainsonine and Related Glycosidase Inhibitors* (LF James, AD Elbein, RJ Molyneux, and CD Warren, eds), pp. 3–13. Iowa State University Press, Ames.

Williams MC, James LF, and Bond BO (1979). Emory milkvetch (*Astragalus emoryanus* var *emoryanus*) poisoning in chicks, sheep and cattle. *American Journal of Veterinary Research* 40: 403–406.

DIAGNOSIS

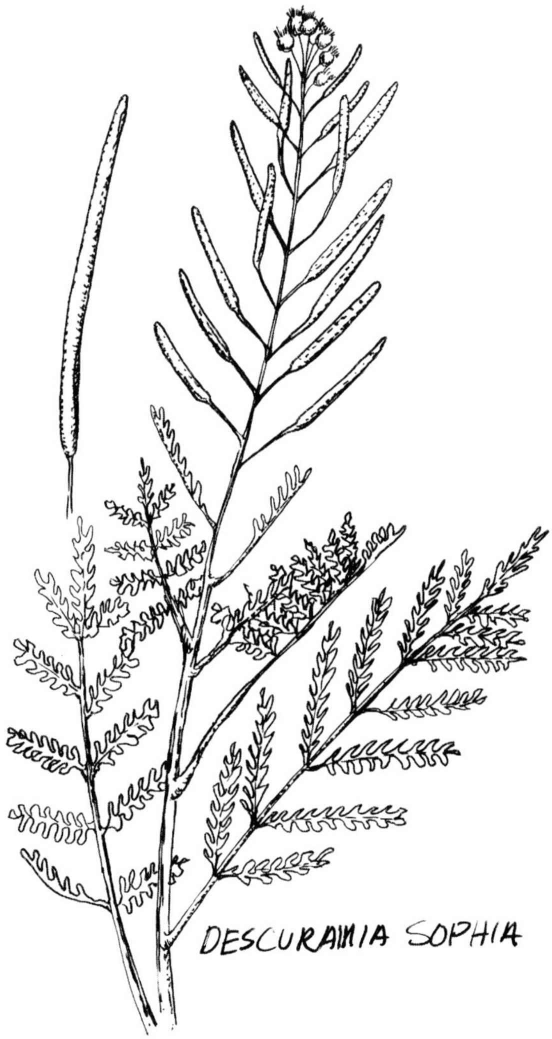

Chapter 74

Recently Encountered Poisonous Plants of Paraíba, Northeastern Brazil

F. Riet-Correa[1], R.M.T. Medeiros[1], L.A. Pimentel[1], J.C.A. Santos[1], J.A.S. de Araújo[1], O.F. de Oliveira[2], and A.F.M. Dantas[1]

[1]*CSTR, University of Campina Grande, Patos, Paraíba, Brazil, 58700-000, PB;* [2] *Escola Superior de Agricultura de Mossoró, Rio Grande do Norte, Brazil*

Introduction

Toxic plants affecting livestock in the state of Paraíba have been reviewed (Medeiros *et al.* 2004a). This chapter reports other plant intoxications observed between 2001 and 2004.

Nitrite Poisoning in Cattle

Three outbreaks of nitrite intoxication were observed in the semiarid region. One outbreak was caused by *Echinochloa polystachya* and two by *Pennisetum purpureum*. The intoxication occurred at the end of the dry season, after the first rains. In one of the outbreaks caused by *P. purpureum* part of the pasture had been fertilized with bovine feces. Five out of 11 cattle died on one farm, 27 out of 81 on another, and three out of nine on the last one. Clinical signs were respiratory distress, teeth grinding, anorexia, depression or hyperexcitability, tremors, abdominal contractions, salivation, nasal discharge, uncoordinated gait, cyanosis, and finally recumbence. Nitrates and nitrites in the blood of affected cattle and pastures were determined by the diphenylamine test. The main reason for nitrate accumulation in the grasses seemed to be the prolonged drought followed by rain. In one outbreak caused by *P. purpureum*, another reason was probably nitrogen levels from the fertilization of the grass with bovine feces (Medeiros *et al.* 2003).

Intoxication by *Ipomoea riedelli* and *Ipomoea sericophylla* in Goats

A disease of the nervous system was described in goats. Clinical signs were rough hair coat, depression, weight loss, and nervous signs including difficulties in rising, ataxia, hypermetria, wide-based stance, intention tremors, spastic paresis, nystagmus, and occasionally other signs of cranial nerve impairment. A goat with non-reversible chronic clinical signs was euthanized. The main histologic lesions observed in the cerebellum were Wallerian-type degeneration and status spongiosus of the cerebellar white matter, loss of Purkinje cells and proliferation of Bergmann glia, and few axonal spheroids in the granular layer. In visits to the farms two species of *Ipomoea* were found: *I. riedelli* in some farms and *I. sericophylla* in others. Some farms had both species.

The disease was produced experimentally in two goats ingesting fresh *I. riedelli* and in two goats ingesting fresh *I. sericophylla* as the only food. Clinical signs were observed after 20–22 days of ingestion. The intoxication was also produced in four goats ingesting the dry plants mixed with concentrated food at daily doses of 2.3 to 3 g kg^{-1} bw. The four goats (two ingesting each plant) showed clinical signs after 18-26 days of ingestion. Six goats that died naturally or were euthanized had severe vacuolization of neurons and epithelial cells of pancreas, thyroids, renal tubules and liver. On electron microscopy of Purkinje cells, numerous membrane-bordered vacuoles were observed and identified as residual or empty lysosomes (Barbosa 2003). With the lectin-histochemical analysis, cerebellar cells of goats ingesting *I. riedelli* and *I. sericophylla* gave a positive reaction to *Triticum vulgaris* and Succinil-*Triticum vulgaris*, which indicates the expression of β-D-N-acetyl-glucosaminase and acetyl-neuraminic acid, and to *Concanavalia ensiformis*, indicating the expression of α-D-mannose and α-D-glucose (Gimeno 2003, unpublished data). The chemical analysis of *I. riedelli* showed 0.14% of swainsonine and calystegines B1, B2 and C1. *Ipomoea sericophylla* contained 0.11% swainsonine but no calystegines (Molyneux and Gardner 2004, unpublished data). It was concluded that both plants cause a glycoprotein storage disease.

Intoxication by *Marsdenia* spp. (Asclepiadaceae) in Cattle, Sheep and Goats

The roots of an unknown *Marsdenia* spp. caused intoxication in cattle. The roots were removed from the soil during the removal of trees for cultivating cropland. Five cows ingested *Marsdenia* roots and became intoxicated. A few hours after ingestion most animals were in lateral recumbence with muscle tremors, mainly in the ears and head, fasciculation of many muscles, hyperesthesia, marked salivation, continuous masticatory movements, opisthotonus, clonic convulsions, and dyspnoea. Four animals died in 24 hours to 6 days. One mildly affected animal recovered. No significant lesions were observed at necropsy or histologically in the nervous system and other tissues. The roots of this plant were given to two goats at doses of 20 and 40 g kg^{-1} bw. Both animals had acute nervous signs immediately after administration and where euthanized *in extremis* after clinical manifestation periods of 30 and 50 min, respectively. No significant lesions were observed on the central nervous system and other organs (Riet-Correa *et al.* 2004a).

Another unknown *Marsdenia* spp. had been associated with sheep mortalities for nearly 10 years. During 1995, after the first rains of the wet season, 270 out of 700 sheep died after consuming leaves of this species of *Marsdenia*. Many sheep were found dead but others showed incoordination, muscular tremors, salivation and recumbency before death. This plant was given experimentally to sheep. One animal received 20g kg^{-1} bw of leaves. It had nervous signs characterized by severe incoordination, intention tremors, loss of equilibrium, wide-based stance, and falling. Clinical signs increased with movement or after the head-raising test. This sheep recovered in 62 hours. Another two sheep had similar clinical signs but recovered after ingesting 10 g kg^{-1} of the roots of this plant (Riet-Correa *et al.* 2004a).

Intoxication by *Leucaena leucocephala* in Sheep

Intoxication by *L. leucocephala* was observed in a Suffolk ram fed previously with *P. purpureum*. Because the animal refused this plant the farmer gave *L. leucocephala ad libitum* together with *P. purpureum*. Three to 4 weeks after starting the ingestion of *Leucaena* the ram showed loss of wool in the region of the neck. The administration of the plant was suspended, but 12 days later the animal lost the wool on most of the body surface. The wool started to grow again in 3–4 weeks. The plant was fed *ad libitum* to two hair sheep of the Santa Inês breed from the University experimental farm and to a crossbred native goat from another farm. Fifteen days after plant consumption, both sheep had severe signs of intoxication characterized by mild loss of hair and severe salivation. The goat did not become intoxicated after ingestion of the plant for 60 days (Riet-Correa *et al.* 2004b). The observation of one case of *L. leucocephala* and the experimental production of the intoxication in sheep demonstrated that, at least in some Brazilian regions, *Synergistes jonesii* or other similar bacteria are not present in the ruminal flora of sheep. On the other hand the resistance of the goat suggests that in other regions some ruminants have mimosine-degrading bacteria in their rumen.

Intoxication by *Centratherum brachylepis* (Asteraceae) in Cattle, Goats and Sheep

Marked anorexia, lack of rumination, ruminal atony, bloat, regurgitation, dry feces, salivation, and dyspnoea was observed in cattle grazing in areas with large amounts of *C. brachylepis*. Some cattle died in 5–8 days, and others recovered in approximately 15 days after removal from the pastures with the plant. Histologically, the stratified epithelium of the rumen had severe vacuolation and vesicle formation followed by infiltration of neutrophils with formation of intraepithelial pustules. Goats were also affected, showing ruminal atony, bloat, dry feces, salivation, and regurgitation. After 2–3 days edema of the head was also observed. Some goats were found dead, and others died in approximately 8 days. The disease occurred in the year 2004 when rains were much more frequent and in higher amounts than normal. The disease was produced by the experimental administration of *C. brachylepis* to goats and sheep. One sheep showed edema of the head and decreased ruminal movements after ingesting 25 g kg^{-1} bw of the fresh plant daily for 2 days. This animal recovered 5 days after ingestion. No clinical signs were observed within 10 days in another sheep after the ingestion of 20 g kg^{-1} bw daily of another batch of the plant. Three additional goats received 10, 20 and 30 g kg^{-1} bw daily for 10, 10 and 8 days, respectively. They had decreased ruminal movements but recovered within 2 days after the end of the plant ingestion.

Intoxication in Sheep by *Tephrosia cinerea* (Leguminoseae)

A disease known as water belly (barriga de água) characterized by chronic progressive ascites affects sheep grazing during the dry season in areas severely invaded by *T. cinerea*. Sheep are only affected after grazing for long periods in pastures where the plant is the only forage. Most animals die within a few weeks to months after first signs of the disease. Others have recovered following the first rains. At necropsy, large amounts of fluid are found in the abdominal cavity and the liver has a very irregular surface with small nodules, less than 1 mm. Histologically, liver lesions include hepatocyte degeneration, dilated lymphatic vessels, mild proliferation of bile duct cells, and periportal and subcapsular fibrosis. The disease was produced in one sheep after the administration of the seeds of the plant for 318 days.

Malformations in Goats, Sheep and Cattle Associated with *Mimosa tenuiflora* (Leguminoseae - Mimosoideae)

Bone malformations are very common in goats and sheep in the Brazilian semiarid regions (Riet-Correa *et al.* 2004c; Nóbrega *et al.* 2004). The main defects are permanent flexure of the forelimbs, cleft palate, brachygnathia, and harelip due to unilateral or bilateral aplasia or hypoplasia of the incisive bones. Other malformations are agnathia, cranioschisis, xyphosis, segmental stenosis of the colon, and meningocele. Many animals had multiple malformations, mainly in the head and legs. Malformations are also observed in cattle, but are less frequent than in small ruminants. Nevertheless, congenital flexure and other defects of the fore legs, ocular dermoids, and other eye defects have been observed in calves.

To demonstrate the cause of the malformations, experiments were performed with three plants of the semiarid region: *Prosopis juliflora*, *Aspidosperma pyrifolium* and *M. tenuiflora*. *Mimosa tenuiflora* and *P. juliflora* were given to four goats each. Fresh green *M. tenuiflora* collected daily was given *ad libitum* during the entire pregnancy; the goats also received concentrated food equivalent to 1% of body weight. Four goats received a ration with 70% *P. juliflora* and 30% hay. The goats treated with *P. juliflora* pods and four control goats delivered nine normal kids. Four goats that ingested *M. tenuiflora* during pregnancy delivered four kids, three of them with malformations. One kid had unilateral harelip due to hypoplasia of the incisive bone and unilateral corneal opacity but survived. A second kid was blind in one eye and after 3 days showed a severe abdominal distention and was euthanized. At necropsy segmental stenosis was observed in the colon. The right eye was enlarged with a cloudy appearance of the anterior chamber. Histologically, the affected gut segment was hypoplasic with complete absence of the intestinal epithelium and submucosa or only some residues of them. The eye had anterior segmental dysplasia with different alterations of the cornea, lens and uvea, mainly sinechia of the pigmented epithelium with the Descemet membrane. The other kid had unilateral harelip extending to the nostrils due to aplasia of the right incisive bone, and was completely blind, with ocular dermoids characterized by presence of hair in both eyes. Histologically, large areas of the cornea were covered by skin with epidermis, dermis, hair follicles, and sudoriparous and sebaceous glands. The lens was enlarged and occupied nearly the whole anterior chamber. These results demonstrated that *M. tenuiflora* may be responsible for the malformations observed in ruminants in the Brazilian semiarid regions (Pimentel and Riet-Correa 2005).

Experiments with *A. pyrifolium* demonstrated that this plant causes abortion in goats, but not malformations (Medeiros *et al.* 2004b).

Acknowledgements

This work was financially supported by the Programa de Apoio a Núcleos de Excelência (PRONEX-Grant Nº 001/04, CNPq, FAPESQ, MCT).

References

Barbosa RC (2003). Intoxicação por *Ipomoea riedelli* em caprinos no semi-árido da Paraíba. MS Thesis, Universidade Federal Rural de Pernambuco, Recife, 48p.

Medeiros RMT, Riet-Correa F, Tabosa IM, Silva ZA, Barbosa RC, Marques AVMS, and Nogueira FRB (2003). Intoxicação por nitratos e nitritos em bovinos por ingestão de *Echinochloa polystachya* (capim-mandante) e *Pennisetum purpureum* (capim-elefante) no sertão da Paraíba. *Pesquisa Veterinária Brasileira* 23:17–20.

Medeiros RMT, Riet-Correa F, Barbosa RC, Lima EF, Neto SAG, and Tabosa I (2004a). Plants toxic for domestic animals in the state of Paraíba, Northeastern Brazil. In *Poisonous Plants and Related Toxins* (T Acamovic, CS Stewart, and TW Pennycott, eds), pp. 151–156. CABI Publishing, Wallingford, Oxon, UK.

Medeiros RMT, Neto SAG, Riet-Correa F, Schild AL, and Sousa NL (2004b). Mortalidade embrionária e abortos em caprinos causados por *Aspidosperma pyrifolium. Pesquisa Veterinária Brasileira* 24: (Suppl.):42.

Nóbrega RS, Riet-Correa F, Nóbrega RS, Correia AP, Nóbrega JS, Aragão MC, Medeiros RM, and Tabosa IM (2004). Malformações ósseas em caprinos na região semi-árida do nordeste do Brasil. *Pesquisa Veterinária Brasileira* 24 (Suppl.): 45.

Pimentel LA and Riet-Correa F (2005). Determinação da causa de malformações em pequenos ruminantes no semi-árido da Paraíba. Relatório parcial. PIBIC, Universidade Federal de Campina Grande, 9p.

Riet-Correa F, Silva DM da, Oliveira OF, Simões SVD, Medeiros RMT, and Nóbrega RS (2004a). Intoxicação por *Marsdenia* spp. (mata-calado) em ruminantes. *Pesquisa Veterinária Brasileira* 24 (Suppl.):50–51.

Riet-Correa F, Barros SS, Araújo JAS, Vasconcelos JS, and Medeiros RM (2004b). Intoxicação por *Leucaena leucocephala* em ovinos na Paraíba. *Pesquisa Veterinária Brasileira* 24 (Suppl.):52.

Riet-Correa F, Medeiros RM, Neto SA, Tabosa IM, and Nobre VMT (2004c). Malformações ósseas em caprinos na região semi-árida do nordeste do Brasil. *Pesquisa Veterinária Brasileira* 24 (Suppl.):49–50.

Chapter 75

Heart Failure, Abortion and Neurological Signs in Cattle and Sheep Induced by *Ateleia glazioviana* and *Tetrapterys* spp. in Brazil

A.L. Schild [1], M.B. Raffi[1], G. Riet-Correa[2], C.S.L. Barros[3], and F. Riet-Correa[4]

[1]*Laboratório Regional de Diagnóstico – FV- Universidade Federal de Pelotas, Campus Universitário s/n 96010-900- Pelotas, RS, Brazil;* [2]*Departamento de Patologia –FV- Universidade Federal do Pará, Castanhal, PA, Brazil;* [3]*Departamento de Patologia, CCS, Universidade Federal de Santa Maria, Santa Maria, RS, Brazil;* [4]*Centro de Saúde e Tecnologia Rural, Universidade Federal de Campina Grande, Patos, PB, Brazil*

Introduction

Ateleia glazioviana (Leg. Papilinoideae) and *Tetrapterys* spp. (*T. acutifolia* and *T. multiglandulosa*) (Malpighiaceae) are poisonous plants that cause heart failure, abortion and neurological signs characterized by lethargy, apathy and blindness in cattle (Tokarnia *et al.* 1989; Gava and Barros 2001) and sheep (Gava *et al.* 2003). The three clinical forms can be seen together in one outbreak, but there are instances in which only a single clinical syndrome is present. Similar syndromes have been experimentally reproduced in domestic ruminants by the feeding of these plants and it was demonstrated that the amount of plant ingested determined what type of clinical syndrome the affected animal will develop (Stigger *et al.* 2002; Garcia y Santos *et al.* 2004; Raffi *et al.* 2004; Riet-Correa *et al.* 2005). The active principle of these plants is as yet unknown but experimental evidence suggests that a single substance is involved in the development of the three clinical forms observed in the field (Raffi *et al.* 2004).

Epidemiology

Ateleia glazioviana is a tree of 5–10 m in height frequently found in the southern Brazilian states. Spontaneous poisoning occurs by ingestion of the fallen leaves or the sprouts. The disease affects cattle of both sexes and several breeds over 1 year of age and occurs mainly during the fall and winter. Morbidity rates are 10–60% and lethality up to 95%. The lethargic form of the disease affects males and females alike; morbidity and lethality are variable but can be as high as 30% and 20%, respectively (Gava *et al.* 2001). *Tetrapterys acutifolia* and *T. multiglandulosa* are small shrubs occurring in large areas of southeastern Brazil. No data are available on morbidity or mortality rates associated with the toxicosis; however, problems occur in the dry months, mainly affecting cattle over 1 year of age (Tokarnia *et al.* 2000).

Heart Failure

Ingestion of these plants can cause heart failure, characterized by sudden death without previous clinical signs or by severe congestive heart failure (Tokarnia *et al.* 1989; Gava and Barros 2001). Sudden death caused by *A. glazioviana* is observed mainly in draft oxen, dairy cows and adult beef cattle at pasture. These animals are usually in good body condition and die after being handled in association with management practices (Gava *et al.* 2001). Cattle affected by congestive heart failure had subcutaneous dependent edema, engorged pulsating jugular veins and cardiac arrhythmia (Tokarnia *et al.* 1989; Gava *et al.* 2001). Cattle affected by both sudden death and congestive heart failure present sharply demarcated, firm white areas of discoloration in the myocardium. These areas are seen throughout the epicardium but are best observed in the cut surface of cross sections of the heart. The discoloration is most conspicuous around the branches of coronary artery, although in advanced cases most of the myocardium is affected. In cases of congestive heart failure, the liver is usually enlarged and has a nutmeg appearance in the cut surface and there is subcutaneous translucent gelatinous edema, ascites, hydropericardium and hydrothorax. The histopathological findings are characterized by swelling and necrosis of cardiac myofibers and interstitial fibrosis and macrophage infiltration in the cardiac muscle, as well as severe centrilobular congestion and slight fibrosis in the liver. Occasionally, scattered necrotic myocardial fibers are entrapped within fibrous tissue proliferation (Tokarnia *et al.* 1989; Gava *et al.* 2001). Experimentally the heart lesions, with clinical signs of lethargy, were reproduced in cattle by feeding the leaves of *A. glazioviana* in daily doses of 2.5, 5.0, 7.5 and 10 g kg^{-1} over long periods for a total of 120 g kg^{-1} per animal (Gava *et al.* 2001). The oral feeding of 5–20 g kg^{-1} of *Tetrapterys* spp. for 9–55 days to cattle reproduced the myocardial lesions (Tokarnia *et al.* 1989). Heart failure and nervous signs were observed in sheep receiving 180 g kg^{-1} divided in daily doses of 3 and 6 g kg^{-1} of body weight for 60 and 30 days, respectively (Riet-Correa *et al.* 2005). Hydrothorax, hydropericardium, ascites, nutmeg appearance of the liver and hard whitish irregular areas in the myocardium, mainly in the interventricular septum and left ventricle, were observed at necropsy of sheep experimentally intoxicated by *A. glazioviana* and *Tetrapterys multiglandulosa* (Stigger *et al.* 2001; Raffi *et al.* 2005; Riet-Correa *et al.* 2005). Histologically, the experimentally poisoned sheep displayed areas of myocardial fibrosis with degenerative and/or necrotic myocardial changes associated with mononuclear cell infiltration (Stigger *et al.* 2001; Riet-Correa *et al.* 2005). Mitochondrial swelling and loss of density of the mitochondrial matrix were the earliest ultrastructural changes observed (Stigger *et al.* 2001).

Neurological Signs

The neurological manifestation of the toxicosis is characterized by lethargy, apathy and occasionally blindness. Affected animals remain still for prolonged periods, have low head carriage, droopy ears and, when forced, move with an uncoordinated gait and frequently bump into objects. All pregnant cows intoxicated by *A. glazioviana* presenting this clinical syndrome aborted (Gava *et al.* 2001). Neurological signs observed in experimentally intoxicated sheep by *A. glazioviana* and *T. multiglandulosa* included depression, loss of appetite, weakness, staggering gait and prolonged periods of recumbency (Raffi *et al.* 2005; Riet-Correa *et al.* 2005). The main histopathological findings related to the neurological disease are variable degrees of spongy degeneration (*status spongiosus*) in the white matter of the brain. The lesion consisted of scattered 10–50 µm round to oblong periaxonal vacuoles, which occasionally gave a web-like appearance to the white matter. Ultrastructurally, these changes consisted of intramyelinic edema (Stigger *et al.* 2001; Riet-Correa *et al.* 2005). Frequently, cardiac lesions occurred along with brain lesions (Gava and Barros 2001; Stigger *et al.* 2001).

Abortion

Abortion caused by *A. glazioviana* and *Tetrapterys* spp. occur at any time of gestation and may be preceded by clinical signs of lethargy (Gava and Barros 2001) or brisket edema (Tokarnia *et al.* 1989). Cows that ingest these poisonous plants towards the end of their gestational periods deliver weak calves, which are unable to stand and nurse. Most of these calves die within the first few days after birth (Tokarnia *et al.* 1989; Gava and Barros 2001). The experimental feeding of these plants to pregnant cows (Garcia y Santos *et al.* 2004) and ewes (Raffi *et al.* 2004; Riet-Correa *et al.* 2005) resulted in similar diseases. Experimental trials have demonstrated that abortion and neonatal mortality in sheep induced by *A. glazioviana* and *T. multiglandulosa* occur due to the transplacental action of the toxic principle in the plant that induces cardiac and central nervous system lesions (Raffi *et al.* 2004; Riet-Correa *et al.* 2005). Pregnant sheep experimentally poisoned by *A. glazioviana* (Raffi *et al.* 2004) and *T. multiglandulosa* (Riet-Correa *et al.* 2005) and cows experimentally poisoned by *A. glazioviana* (Garcia y Santos *et al.* 2004) had several forms of reproductive failure including mummification, abortion, stillbirth and birth of weak lambs and calves which usually did not survive. Gross and histological lesions in the heart of these fetuses and newborn lambs and calves and the histopathological spongy degeneration observed in the brain were similar to the ones reported in adult cattle with the spontaneous forms of the toxicosis (Raffi *et al.* 2004; Riet-Correa *et al.* 2005).

References

Garcia y Santos MC, Schild AL, Barros SS, Riet-Correa F, Elias F, and Ramos AT (2004). Lesões perinatais em bovinos na intoxicação experimental pot *Ateleia glazioviana* (Leg. Papilionoideae). *Pesquisa Veterinária Brasileira*. 24(4):178–184.

Gava A and Barros CSL (2001). Field observations of *Ateleia glazioviana* poisoning in cattle in southern Brazil. *Veterinary and Human Toxicology* 43:37–41.

Gava A, Barros CSL, Pilati C, Barros SS, and Mori AM (2001). Intoxicação por *Ateleia glazioviana* (Leg. Papilionoideae) em bovinos. *Pesquisa Veterinária Brasileira* 21:49–59.

Gava D, Reis RN, Rocha TS, Pasquale E, and Gava A (2003). Intoxica natural por *Atelia glazioviana* (Leg. Papilionoideae) em ovinos. *Anais XI, Encontro Nacional de Patologia Veterinária,* Botucatu, SP, p. 60.

Raffi MB, Barros RR, Bragança JFM, Rech RR, Oliveira FN, and Barros CSL (2004). The pathogenesis of reproductive failure induced in sheep by the ingestion of *Ateleia glazioviana. Veterinary and Human Toxicology* 46(5):233–238.

Raffi MB, Rech RR, Sallis ESV, Oliveira FN, Barros SS, and Barros CSL (2006). Chronic cardiomyopathy and encephalic spongy changes in sheep experimentally fed *Ateleia glazioviana. Ciência Rural* 36:1860–1866.

Riet-Correa G, Terra FF, Schild AL, Riet-Correa F, and Barros SS (2005). Intoxicação experimental por *Tetrapterys multiglandulosa* em ovinos. *Pesquisa Veterinária Brasileira* 25(2):91–96.

Stigger AL, Barros CSL, Langohr IM, and Barros SS (2001). Intoxicação experimental por *Ateleia glazioviana* (Leg. Papilionoideae) em ovinos. *Pesquisa Veterinária Brasileira* 21:98–108.

Tokarnia CH, Peixoto PV, Döbereiner J, Consorte LB, and Gava A (1989). *Tetrapterys* spp. (Malpighiaceae), a causa de mortandades em bovinos caracterizadas por alterações cardíacas. *Pesquisa Veterinária Brasileira* 9:23–44.

Tokarnia CH, Döbereiner J, and Peixoto PV (2000). *Plantas Tóxicas do Brasil*. Helianthus, Rio de Janeiro.

Chapter 76

Grape and Raisin (*Vitis* spp.) Toxicosis in Dogs

C. Means and P.A. Eubig

ASPCA Animal Poison Control Center, 1717 S Philo Rd, Suite 36, Urbana, IL 61856, USA

Introduction

In 1999, veterinarians at the ASPCA Animal Poison Control Center (APCC) identified several reports where seemingly healthy dogs developed acute renal failure (ARF) after ingestion of grapes or raisins. Subsequent reports of grape/raisin toxicosis in dogs include a letter containing a brief summary of information about ten symptomatic dogs (Gwaltney-Brant *et al.* 2001), a brief report of raisin toxicosis in a Hungarian Vizsla (Penny *et al.* 2003), a letter confirming that a similar syndrome had been identified in the United Kingdom (Campbell and Bates 2003) and a case report describing the treatment of four dogs (Mazzaferro *et al.* 2004). A paper detailing histopathology in ten dogs has recently been published (Morrow *et al.* 2005). A literature search did not find evidence of a similar problem in humans.

Data Collection and Case Selection

Historical and clinical information were obtained from the AnTox™ computerized medical record database of the APCC. The APCC is a 24-hour consultation service that receives phone calls from animal owners and practicing veterinarians throughout North America concerning incidents of animal poisoning. The AnTox™ database from 1987 through 2002 was searched for cases where the APCC was contacted concerning the ingestion of grapes or raisins in dogs. Dog owners and/or veterinarians were contacted by phone or by facsimile to obtain updates on the cases and copies of the medical records.

The following criteria were met for inclusion of each case: (1) evidence of exposure (e.g. witnessed exposure, grapes/raisins in the vomitus or stool, or damaged packaging consistent with a dog ingesting the contents); (2) one or more clinical signs reported by the owner or the veterinarian; and (3) either blood urea nitrogen (BUN) or creatinine being elevated above the normal reference range subsequent to exposure. Cases were rejected if a dog coingested other substances (e.g. chocolate) or if a dog had evidence of pre-existing health problems (e.g. chronic renal failure) that might have contributed to the severity of the signs. Cases were also rejected if the dog remained asymptomatic or if azotemia did not develop.

Clinical signs

Forty-three cases met the inclusion criteria. The earliest and most consistent clinical sign in these dogs was vomiting, which occurred within the first 24 hours following ingestion in the majority of dogs. Grapes or swollen raisins were often in the vomitus. 100% of the dogs vomited. Anorexia, lethargy, and diarrhea often developed within the first 24 hours, following the onset of vomiting. In some instances grapes or swollen raisins were also present in the diarrhea. Although in some dogs signs such as decreased urine output, abdominal pain, ataxia, and weakness began as early as the day of exposure, it generally took up to 5 days for the majority of dogs to exhibit these clinical signs. Clinical signs associated with a poor outcome (death or euthanasia) include decreased urine output, ataxia, and weakness.

Laboratory

Creatinine, phosphorus, and Ca x P (calcium x phosphorus) concentrations in serum can elevate in less than 24 hours. A Ca x P concentration greater than 60 increases the risk of soft tissue mineralization. BUN and calcium may not increase for 1–3 days postexposure. Hyperkalemia or hypokalemia is possible, as well as decreased TCO_2. In some cases elevated ALT, hyperglycemia and hyperlipasemia are seen. Clinical chemistries associated with poor outcomes were hypercalcemia, Ca x P > 60, hyperkalemia, and a low TCO_2.

Toxicology

One dog had serum and urine evaluated for ethylene glycol, but no ethylene glycol was detected. Results of quantitation of 25-(OH)D_3 concentrations performed on renal and hepatic specimens revealed no elevation of calcifediol levels. Evaluation of a specimen of liver submitted for a herbicide screen revealed no detectable herbicides. No evidence of chlorinated hydrocarbon, carbamate, or organophosphorus insecticides was found upon screening one sample of raisins. No mycotoxins were detected in the same sample of raisins. Also no evidence of ochratoxin A was found in two separate samples of grapes that were evaluated.

Histopathology

The most consistent histopathologic lesion in biopsy or necropsy samples from dogs experiencing ARF following grape/raisin ingestion was renal tubular injury. Mineralization of the kidney, as well as other tissues frequently was present as well.

Treatment

Most dogs received fluid therapy. Other treatments designed to increase urine production included furosemide, dopamine and/or mannitol. Other types of medications administered included gastrointestinal protectants, antiemetics and motility modifiers, phosphate binders, antibacterial agents, corticosteroids, amlodipine and heparin. Whole blood transfusions and fresh frozen plasma transfusions were administered when necessary. Peritoneal dialysis was used in two dogs, and one recovered and one did not.

Outcome

Outcome for the 43 dogs included 23 that survived (53%), five that died (12%) and 15 that were euthanized (35). The clinical signs resolved for all 23 dogs that survived.

Discussion and Theories

The dogs in this report usually began vomiting within 24 hours of ingestion, followed by the development of anorexia, lethargy, and diarrhea in many instances. Creatinine, phosphorus, and Ca x P concentrations tended to be elevated soon after exposure, with BUN elevating 24 hours later, and then calcium values often becoming elevated within 48 to 72 hours after exposure.

Vomiting was reported in all dogs. As expected, dogs with decreased urine production were less likely to survive. Dogs exhibiting ataxia and weakness are less likely to survive. The majority of dogs in this study had elevated serum Ca x P with median values higher for dogs that died or were euthanized. Hypercalcemia impairs urinary concentrating ability as well as causes renal vasoconstriction, which leads to renal tubular injury (Kruger *et al.* 1996). Hypercalcemia also promotes soft tissue mineralization, which is most pronounced when the Ca x P is greater than 50 to 60 (Kruger *et al.* 1996). Dogs in this study with higher elevations of total calcium and/or Ca x P may have poorer outcomes because they are more likely to experience progressive renal dysfunction and to develop soft tissue mineralization. However, severe renal damage can cause increased calcium. One theory has suggested that resveratrol, an estrogenic compound and antioxidant, could interfere with calcium homeostasis by inducing renal 1-α-hydroxylase (Morrow, personal communication). Hyperkalemia in renal failure is most commonly associated with oliguria (Chew 2000).

A lower TCO_2 value, suggesting metabolic acidosis, was also associated with poor outcomes. Metabolic acidosis is often associated with oliguric renal failure in dogs (Lane *et al.* 1994).

There was often a delay in the initial evaluation of the dogs in this study as most pet owners and their veterinarians suspected that the dogs were only experiencing mild gastritis. This delay may have allowed ARF to develop.

Acute tubular necrosis, especially in the proximal tubules, was the most consistent histopathologic finding. Acute renal failure is generally caused by toxic or ischemic injury (Brown and Engelhardt 1987; Grauer 1996). An unknown chemical component of grapes and raisins is suspected of causing either direct renal tubular damage or initiating an ischemic sequence in the dogs. The toxic principle is unknown. Determining the toxic principle is difficult since not every ingestion results in ARF.

Doses of as little as 2.8 g kg^{-1} (0.1 oz kg^{-1}) of raisins and 19.6 g kg^{-1} (0.7 oz kg^{-1}) of grapes resulted in ARF. Since a dose response is lacking, an intrinsic compound that is present in varying amounts in *Vitis* spp. depending on environmental circumstances or genetic variance in the plants may be the answer. It is also possible that the toxic principle may be an extrinsic substance that is not always present on or in grapes. An idiosyncratic reaction may occur due to differences in individual dog responses to an intrinsic or extrinsic response to grapes or raisins. (Eaton and Klaassen 2001)

Mycotoxins produced by molds on grapes are a possible etiology for the clinical syndrome. The acute situation seen with the dogs differs from the naturally occurring renal damage from ochratoxin A or citrinin, generally a result of chronic exposure and a progressive clinical course (Krogh 1992).

The presence of pesticide residues on grapes and raisins has been considered. The concentrations of pesticides on the fruits often decline rapidly due to evaporation and degradation, and current evidence indicates that pesticide residues are consistently either not present or present below regulatory limits if the pesticides are applied according to good agricultural practice. (Cabras and Angioni 2000).

Treatment Recommendations

Rapid decontamination is recommended in cases of grape or raisin ingestion. However, a delay in intervention does not mean that decontamination should not be performed. The digestion of raisins and grapes into smaller particles in the stomach tends to be slow as was evidenced in some of these cases by the vomiting of swollen raisins or grapes after they had remained in the stomach overnight. Inducing emesis using standard emetics (Rosendale 2002) may recover grapes or raisins several hours after exposure. Subsequent administration of activated charcoal should be considered, especially if it is suspected that the ingested amount of grapes or raisins was not recovered by inducing emesis. Fluid diureses is typically implemented after decontamination and continued for at least 48 hours in an asymptomatic patient, and until clinical signs resolve in symptomatic patients. Clinical chemistries should be monitored in order to diagnosis early ARF.

If clinical signs and azotemia in a dog are suspected to be due to the ingestion of grapes or raisins, it is important to determine whether there are any other etiologies or contributing factors that could complicate the patient's condition. Other symptomatic and supportive care should be provided, based on clinical signs. Further investigation is needed into the mechanism(s) by which grapes and raisins are associated with renal tubular necrosis in dogs in order to determine the toxic principle.

References

Brown SA and Engelhardt JA (1987). Drug-related nephropathies. Part I. Mechanisms, diagnosis, and management. *Compendium of Continuing Education for Practicing Veterinarians* 9(2):148–160.

Cabras P and Angioni A (2000). Pesticide residues in grapes, wine, and their processing products. *Journal of Agricultural and Food Chemistry* 48(4): 967–973.

Campbell A and Bates N (2003). Raisin poisoning in dogs. *Veterinary Record* 152(12):376.

Chew DJ (2000). Fluid therapy during intrinsic renal failure. In *Fluid Therapy in Small Animal Practice*, 2nd ed. (SP DiBartola, ed), pp. 410–427. WB Saunders, Philadelphia PA.

Eaton DL and Klaassen CD (2001). Principles of toxicology. In *Casarett and Doull's Toxicology: The Basic Science of Poisons*, 6th ed. (CD Klaassen, ed), pp. 15–26. McGraw-Hill, New York.

Grauer GF (1996). Prevention of acute renal failure. *The Veterinary Clinics of North America Small Animal Practice* 26(6):1447–1459.

Gwaltney-Brant S, Holding JK, Donaldson CW, Eubig PA, and Khan SA (2001). Renal failure associated with the ingestion of grapes or raisins in dogs. *Journal of the American Veterinary Medical Association* 218(10): 1555–1556.

Krogh P (1992). Role of ochratoxin in disease causation. *Food and Chemical Toxicology* 30(3): 213–224.

Kruger JM, Osborne CA, Nachreiner RA, and Refsal KR (1996). Hypercalcemia and renal failure. *Veterinary Clinics of North America Small Animal Practice* 26(6):1417–1445.

Lane IF, Grauer GF, and Fettman MJ (1994). Acute renal failure. Part II. Diagnosis, management, and prognosis. *Compendium of Continuing Education for Practicing Veterinarians* 16(5):625–645.

Mazzaferro EM, Eubig PA, Hackett TB, Legare M, Miller C, Wingfield WE, and Wise L (2004). Acute renal failure associated with raisin or grape ingestion in 4 dogs. *Journal of Veterinary Emergency and Critical Care* 14(3):203–212.

Morrow CM, Valli VE, Volmer PE, and Eubig PA (2005). Canine renal pathology assoicated with grape or raisin ingestion: 10 cases. *Journal of Veterinary Diagnostic Investigation.* 17(3): 223–231.

Penny D, Henderson SM, and Brown PJ (2003). Raisin poisoning in a dog. *Veterinary Record* 152(10):308.

Rosendale ME (2002). Decontamination strategies. *Veterinary Clinics of North America Small Animal Practice* 32(2):311–321.

Chapter 77

Flixweed (*Descurainia sophia*): a Goitrogenic Plant

A.P. Knight[1] and B.L. Stegelmeier[2]

[1]*Department of Clinical Sciences, College of Veterinary Medicine and Biomedical Sciences, Colorado State University, Fort Collins, CO 80523, USA;* [2]*USDA-ARS Poisonous Plant Research Laboratory, Logan, UT 84341, USA*

Introduction

Flixweed (*Descurainia sophia* (L.) Webb ex Prantl), a native to Eurasia, has become a common annual weed in North America (Rollins 1993). As a member of the Brassicaceae (Cruciferae), it closely resembles the North American native tansy mustard (*Descurainia pinnata* (Walter) Britton), and is often found growing in the same habitat of dry open prairies, cultivated fields, roadsides and waste areas. Tansy mustard was first associated with a disease syndrome in cattle in which the tongue appeared paralyzed (Hershey 1945). Periodically, especially following unusual seasonal rains, tansy mustard and flixweed have been associated with a clinical syndrome in which cattle that have been grazing predominantly the pre-flowering mustard develop a staggering gait, appear blind, and have difficulty eating and drinking. In severely poisoned animals, the tongue appears affected, causing difficulty chewing and holding food in the mouth (Hershey 1945; Kingsbury 1964; Rosiere *et al.* 1975; Staley 1976; Scott 2005, personal communication). Tansy mustard and flixweed have also been associated with sporadic occurrences of hepatogenous photosensitization in cattle in Montana and Colorado, but neither of these clinical syndromes has been reproduced experimentally (Pfister *et al.* 1989, 1990; Knight 1996). Cattle readily eat rangeland diets comprising up to 31% tansy mustard without signs of intoxication, which implies the toxicity of the plant may vary considerably depending on growing conditions (Rosiere *et al.* 1975; Hakkila *et al.* 1987). Although flixweed is a member of the mustard family, and is a common weed and contaminant of hay fed to livestock, it has surprisingly not been reported as a cause of goiter in animals.

A variety of plants, especially members of the Brassicaceae, contain goitrogens that have been shown to cause goiter in humans, and various animal species including cattle, sheep, pigs, foals, goats, and poultry (Sinclair and Andrews 1958; Russel 1967; Wilson 1975; Irvine and Evans 1977; McLaren and Alexander 1979; Mawson *et al.* 1994). Most frequently goiter is associated with consumption of crop plants such as rape, or rapeseed meal (*Brassica napus*) (Russel 1967; Mallard 1981; Mawson *et al.* 1994), kale (*Brassica oleraceae*) (Paxman and Hill 1974), and other members of the mustard family (Fenwick *et al.* 1989). Other plants that have shown goitrogenic effects in animals include turnip weed (*Rapistrum rugosum*) (Stocks *et al.* 1984), lead tree (*Leucaena leukocephala*) (Hegarty *et al.* 1979), and pearl millet (*Pennisetum typhoides*) that additionally causes enterohepatonephropathy (Gadir and Adam1999). Cassava (*Mannihot esculenta*), soy beans (*Glycine max*), and linseed (*Linum* spp.) are known to contain goitrogens affecting people (McLaren and Alexander 1979).

A herd of 28 pregnant Boer goats fed grass hay containing 10-15% flixweed produced 59 kids, of which 40.7% (17 males, seven females) were either born dead, or died shortly after birth (Knight *et al.* 2005). Many of the affected kids (six male and five female) were hairless with obvious thyroid hyperplasia (Figure 1) and had abnormally large birth weights (7.0–7.5 kg). Glucosinolate analysis of the flixweed found in the hay revealed significant levels of 3-butenyl (gluconapin) and allyl (sinigrin) glucosinolate (Table 1). Subsequent removal of the flixweed from the goat's diet and providing an iodine-containing mineral prevented further occurrence of goiter in the herd.

Table 1. Glucosinolates in dried flixweed (*Descurainia sophia*) (gas chromatography analysis performed by POS Pilot Plant Corporation, Saskatoon, Saskatchewan, Canada).

Glucosinolates	Seed μmoles/g	Stems/pods μmoles/g	Fresh Seed* μmoles/g
Allyl (sinigrin)	4.30	1.88	7.81
3-butenyl (gluconapin)	27.64	10.59	65.49
4-pentenyl	0.20	0.09	0.53
2-OH-3-butenyl	<0.05	<0.05	<0.05
2-OH-4-pentenyl	<0.05	<0.05	<0.05
Methylthiobutenyl	<0.05	<0.05	<0.05
Phenylethyl	<0.05	<0.05	<0.05
Methylthiopentenyl	<0.05	<0.05	<0.05
4-OH-benzyl	<0.05	<0.05	<0.05
3-methylindolyl	<0.05	<0.05	<0.05
4-OH-3-methylindolyl	<0.05	<0.05	<0.05

*Flixweed seed collected the summer after the hay with the flixweed had been baled, indicating the decline in glucosinolates in the stored hay over the course of a year.

Congenital goiter due to iodine deficiency and/or the consumption of goitrogens has been reported in Angora, Saanen x dwarf goats and in several breeds of sheep including Merinos, and Dorsets (Rijnberk *et al.* 1977; Bath *et al.* 1979). Congenital goiter has also been reported in White and Colored German Goats associated with dietary iodine deficiency and secondary iodine deficiency due to increased dietary nitrate levels (Bostedt and Adams 2000). Angora and Boer goats and Dorset sheep appear particularly susceptible to iodine deficiency-induced goiter (Smith and Sherman 1994; Menzies and Miller 1997). Hereditary congenital goiter due to an autosomal recessive trait affecting thyroglobulin synthesis has been reported in Dutch goats (Kok *et al.* 1987), polled Dorset sheep (Davis *et al.* 1979), Merinos (Falconer 1965; Rac *et al.* 1968; de Vijlder *et al.* 1978), and other sheep breeds crossed with the Dorset breed (Davis *et al.* 1979).

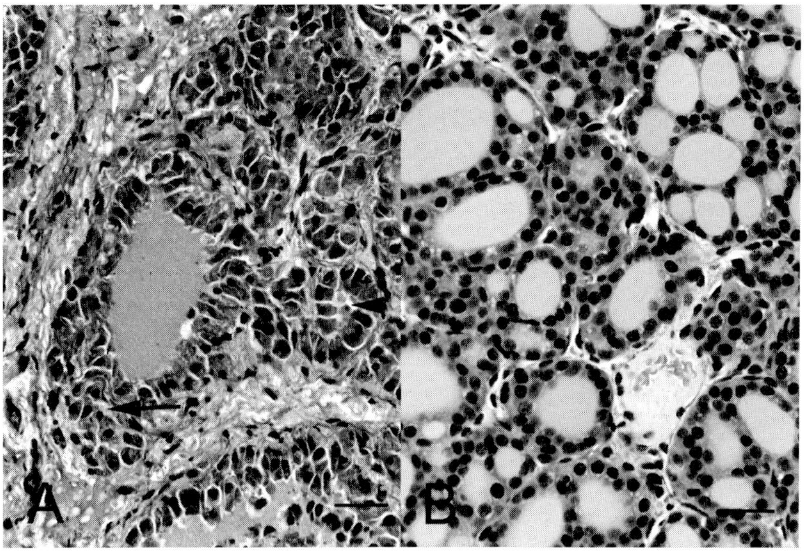

Figure 1. Photomicrographs of the thyroid glands of a goat with congenital goiter (A), and a normal goat (B). Histologically the hyperplastic goiter (A) has fewer follicles, which are often empty or contain small amounts of colloid. The follicular cells are hyperplastic with most cells forming a tall columnar or pseudostratified epithelium. The interstitial tissue is expanded with edema and wide connective tissue septae. The normal, age-matched fetal thyroid (B) is presented as a comparison. Notice the cuboidal follicular cells and relatively common colloid filled follicles. Both sections are stained with hematoxylin and eosin. The poor staining of the goiterous thyroid is due to autolysis. Bar=50 μ.

Plant Goitrogenisis

Plants that are goitrogenic contain glucosinolates, a group of sulphur-containing glucosides found throughout the plant but especially in the seeds. At least 100 glucosinolates have been identified in several plant families, with 12 having been identified in Brassica species (Stoewsand 1995). In the plant, glucosinolates are readily hydrolyzed by the plant enzyme myrosinase into isothiocyanates, thiocyanates, nitriles, sugar and sulfate ion (HSO_4) (VanEtten and Tookey 1978; Fenwick *et al.* 1989; Stoewsand 1995; Bennett *et al.* 2004). Whether isothiocyanates or nitriles are formed and in what quantity appears dependent on the specific glucosinolate in the plant. Few glucosinolates are goitrogenic and the one that is most goitrogenic is 2-hydroxy-3-butenyl glucosinolate (gluconapin) (Holst and Williamson 2004). This is the primary glucosinolate isolated from flixweed (Table 1) and is found in other Brassicas, especially rapeseed meal (Bell 1984). The glucosinolate-derived thiocyanates inhibit the uptake of iodine by the thyroid gland and at high doses over prolonged periods, decrease thyroid hormones (thyroxine T3 and T4). Thus the decrease in thyroid hormones, which normally suppress the production of thyroid stimulating hormone (TSH) from the anterior pituitary, results in excessive secretion of TSH, which stimulates increased thyroid activity and hypertrophy detected clinically as goiter (David 1976; Mawson *et al.* 1994).

Conclusions

Flixweed contains significant quantities of glucosinolates and should not be fed to pregnant goats, and other pregnant animals because of the potential for the development of fetal goiter. Animals that do not have adequate iodine in their diet or which are fed forages high in nitrate may be especially susceptible to the goitrogenic effects of flixweed and other members of the Brassicaceae.

References

Bath GF, Wentzel D, and van Tonder EM (1979). Cretinism in Angora goats. *Journal of the South African Veterinary Association* 50:237–239.

Bell JM (1984). Nutrients and toxicants in rapeseed meal: a review. *Journal of Animal Science* 58:996–1010.

Bennett RN, Mellon FA, and Kroon PA (2004). Screening crucifer seeds as sources of specific intact glucosinolates using ion-pair high-performance liquid chromatography negative ion electrospray mass spectrometry. *Journal of Agricultural & Food Chemistry* 52(3):428–438.

Bostedt SC and Adams W (2000). Juvenile goiter in a goat herd in northwest Germany. *Schweizer Archiv fur Tierheilkunde.* 142:339–347.

David JSE (1976). The effect of prolonged kale feeding on the thyroid glands of sheep. *Journal of Comparative Pathology* 86:235–241.

Davis GB, Stevenson BJ, Price MC (1979) Inherited goiter in sheep. *New Zealand Veterinary Journal* 27:126–127.

de Vijlder JJ, van Voorthuizen WF, van Dijk JE, Rijnberk A, and Tegelaers WH (1978). Hereditary congenital goiter with thyroglobulin deficiency in a breed of goats. *Endocrinology* 102:1214–1222.

Falconer IR (1965). Biochemical defect causing congenital goiter in sheep. *Nature* 205:978–980.

Fenwick GR, Heaney RK, and Mawson R (1989). Glucosinolates. In *Toxicants of Plant Origin* Vol. 2, *Glycosides* (PR Cheeke, ed), pp. 1–41. CRC Press, Boca Raton, Florida.

Gadir WSA and Adam SEI (1999). Development of goiter and enterohepatonephropathy in Nubian goats fed with pearl millet (*Pennisetum typhoides*). *The Veterinary Journal* 157:178–185.

Hakkila MD, Holechek JL, Wallace JD, Anderson DM, and Cardenas M (1987). Diet and forage intake of cattle on desert grassland range. *Journal of Range Management* 40:339–342.

Hegarty MP, Lee CP, Christie GS, Court RD, and Haydock KP (1979). The goitrogen 3-hydroxy 4(1H) pyridine, a ruminal metabolite from *Leucaena leucocephala*: effects in mice and rats. *Australian Journal of Biological Science* 32:27–40.

Hershey AL (1945). Some poisonous plant problems of New Mexico. *New Mexico Agricultural Experiment Station Bulletin* 322:20–23.

Holst B and Williamson G (2004). A critical review of the bioavailability of glucosinolates and related compounds. *Natural Product Reports* 21:425–247.

Irvine CHG and Evans MJ (1977). Hypothyroidism in foals. *New Zealand Veterinary Journal* 25:354.

Kingsbury JM (1964). *Poisonous Plants of the United States and Canada*, pp. 165–168. Prentice-Hall Inc., Englewood Cliffs, New Jersey.

Knight AP (1996). An update on plant poisoning. In *Proceedings 14th American College of Veterinary Internal Medicine Forum*, pp. 509–511.

Knight AP, Stegelmeier BL, Woods LW, and Tiwary AK (2005). Congenital goiter in goats associated with the consumption of flixweed (*Descurainia sophia*). *7th International Symposium on Poisonous Plants*, Logan, UT (abstract) p. 95.

Kok K, van Dijk JE, Sterk A, Baas F, van Ommen GJ, and de Vijlder JJ (1987). Autosomal recessive inheritance of goiter in Dutch goats. *Journal of Heredity* 78:298–300.

Mallard C (1981). Toxicity of Cruciferae and pregnancy disorders in a herd of goats *Bulletin Group Tech Vet* 3:23–25.

Mawson R, Heaney RK, Zdunczyk Z, and Kozlowska H (1994). Rapeseed meal-glucosinolates and their antinutritional effects. Part 4. Goitrogenicity and internal organs abnormalities in animals. *Nahrung* 38:178–191.

McLaren EH and Alexander WD (1979). Goitrogens. *Clinics in Endocrinology & Metabolism* 8:129–144.

Menzies PI and Miller R (1997). Abortion in sheep: diagnosis and control. In *Current Therapy in Large Animal Theriogenology* (RS Youngquist, ed), p. 534. WB Saunders, Philadelphia, Pennsylvania.

Paxman PJ and Hill R (1974). The goitrogenicity of kale and its relation to thiocyanate content. *Journal of the Science of Food and Agriculture* 25:329–237.

Pfister JA, Baker DC, Lacey JR, and Brownson R (1989). Photosensitization of cattle in Montana: is *Descurainaia pinnata* the culprit. *Veterinary & Human Toxicology* 31:225–227.

Pfister JA, Lacey JR, Baker DC, James LF, and Brownson R (1990). Is tansy mustard causing photosensitization of cattle in Montana? *Rangelands* 12:170–172.

Rac R, Hill GN, Pain RW, and Wulhearn CJ (1968). Congenital goitre in Merino sheep due to an inherited defect in the biosynthesis of thyroid hormone. *Research in Veterinary Science* 9:209–223.

Rijnberk A, De Vijlder JJ, Van Dijk JE, Jorna TJ, and Tegelaers WH (1977). Congenital defect in iodothyronine synthesis. Clinical aspects of iodine metabolism in goats with congenital goitre and hypothyroidism. *British Veterinary Journal* 133:495–503.

Rollins RC (ed) (1993). *The Cruciferae of Continental North America*, pp. 335–360. Stanford University Press, Stanford, California.

Rosiere RE, Beck RF, and Wallace JD (1975). Cattle diets on semidesert grassland: botanical composition. *Journal of Range Management* 28:89–93.

Russel AJF (1967). A note on goiter in lambs grazing rape (*Brassica napus*). *Animal Production* 9:131–133.

Sinclair DP and Andrews ED (1958). Prevention of goiter in new-born lambs from kale-fed ewes. *New Zealand Veterinary Journal* 6:87–95.

Smith MC and Sherman DM (1994). *Goat Medicine*, pp. 54–55. Lea & Febiger, Philadelphia PA.

Staley E (1976). A treatment for tansy mustard poisoning. *Oklahoma Veterinarian* 28:74.

Stocks DC, Dunster PJ, and Gibson JA (1984). Observations of the effects of *Rapistrum rugosum* on thyroid function. *Australian Veterinary Journal* 61:264–265.

Stoewsand GS (1995). Bioactive organosulfur phytochemicals in *Brassica oleracea* vegetables–a review. *Food & Chemical Toxicology* 33:537–543.

VanEtten CH and Tookey HL (1978). Glucosinolates in cruciferous plants. In *Effects of Poisonous Plants on Livestock* (RF Keeler, KR Van Kampen, and LF James, eds), pp. 507–520. Academic Press, New York.

Wilson JG (1975). Hypothyroidism in ruminants with special reference to foetal goitre. *Veterinary Record* 97:161–164.

Chapter 78

Evaluation of the Acute Toxicity of the Seeds of *Xylopia frutescens* Aubl. (Annonaceae)

J.G. Sena Filho[1], J.G.S. Melo[1], E. Araujo[1], K.A.Q. Pontual[1], I.A. Souza[2], and H.S. Xavier[1]

[1]*Laboratório de Farmacognosia, Departamento de Ciências Farmacêuticas, Universidade Federal de Pernambuco-UFPE;* [2]*Laboratório de Cancerologia Experimental, Departamento de Antibióticos, Universidade Federal de Pernambuco, Av. prof. Arthur de Sá s/n, cidade universitária, 50740-521 Recife-PE, Brazil*

Introduction

The genus *Xylopia* includes about 160 species and is a well-known source of kaurane and trachylobane diterpenoids (Waterman *et al.* 1985). *Xylopia frutescens* is a tree found in central and southern America, Africa and Asia (Rodrigues *et al.* 1981). In Brazil, where it is commonly known as embira, embira-vermelha and pau carne (Braga 1976), its seeds are used in folk medicine as bladder stimulants and as treatment for rheumatism, halitosis, and tooth decay, among other uses (Correa 1926). It is one ingredient in the preparation of 'Pau-dentro', a typical drink consumed in large quantities in the Brazilian northeast as a medicinal plant. Previous work showed the presence of sitosterol and six known kaurane diterpenes, including acutifloric acid, a diterpenic dimer, and its 7β-hydroxy-derivative, called frutoic acid (Oliveira *et al.* 1995).

Consumption of this preparation by the people in the northeast (Pernambuco, Paraíba, Rio Grande do Norte and Ceará) has increased. There is an absence of toxicological data, thus the purpose of this study was to determine the acute toxicity of the seeds of *X. frutescens*. This preliminary study determined the highest dose that killed any treated animal (D_1) as well as the dose capable of killing all animals (D_2).

We also conducted a phytochemical screening of the secondary metabolites of the seeds to determine the possible constituents that may cause toxicity in mice.

Materials and Methods

Plant material

The plant was collected in Camaragibe, Pernambuco (PE) and identified by the Laboratory of Pharmacognosy at the Pernambuco Federal University (UFPE). A voucher was deposited under number 1010-(HLF).

Ethanolic extract

Ninety grams of the dried fruits of the *X. frutescens* were triturated with 96% ethanol (Merck). After 2 days of maceration, the mixture was evaporated under vacuum with a yield of 6.99%.

Phytochemical analyses

Standard analyses were run to establish the presence of monoterpenoids, sesquiterpenoids, diterpenoids, triterpenoids, steroids (β-amirine and β-sytosterol), iridoids, saponins, sugars, flavonoids, fenylpropanoglycosides, alkaloids, coumarins, gallic acid, condensed proanthocyanidines, leuco-anthocyanidines, tannins and quinones (see Table 1).

Toxicological Assay

Toxicity trials used female Swiss albino mice (*Mus musculus*) between 45 and 60 days of age, under registration number 18 of the Brazilian Group of animal experimentation (COBEA). The animals were selected and grouped in cages with six animals each and injected ip with increasing concentrations of the extract from 250 mg kg^{-1} to 1200 mg kg^{-1}. The animals were observed for 4 hours and any adverse effects were noted (Malone 1977).

Results

At the 250 mg kg^{-1} dose, the animals were agitated with an increase in respiration and appeared pale and sleepy. A dose of 600 mg kg^{-1} resulted in the animals showing the signs described above as well as aggressive behavior and photophobia. Animals at the 900 mg kg^{-1} dose were highly agitated with conjunctive irritation and were prostrate. The 1125 mg kg^{-1} dose killed all the animals within 24 hours.

We found monoterpenoids, sesquiterpenoids, diterpenoids, and triterpenoids in the ethanolic extraction of *X. frutescens* seeds using the screening methods from the UFPE laboratory (see Table 1 for chromatography conditions). We did not find coumarins, gallic acid, condensed proanthocyanidines, iridoids, leucoanthocyanidines, tannins or quinones.

Conclusion

Pharmacological and toxicological assays of *X. frutescens* seeds will help determine if the seeds have beneficial or harmful properties when used in traditional Brazilian medicines. The 1125 mg kg^{-1} dose suggests that high doses may be toxic to people who drink seed extracts. Furthermore, the seed extracts may be harmful at lower doses. Further research will be necessary to determine the pharmacological properties of *X. frutescens*.

Table 1. Chromatography conditions of phytochemistry screening.

Compound	System of Elution[1]	Revelator	Reference
alkaloids	A	Dragendorff	Wagner & Bladt 1996
mono-, di- and sesqui-terpenoids	B	Vanillin sulphuric acid	Wagner & Bladt 1996
triterpenoids/steroids	C	Lieberman/Burchard	Harborne 1998
iridoids	A	Vanillin sulphuric acid	Wagner & Bladt 1996
saponins	A	anisaldehyde	Wagner & Bladt 1996
sugar	D	2,3,5-triphenyltetra-zolium chloride	Metz 1961
coumarins	E	2-aminoethyldiphenyl borinate	Wagner & Bladt 1996
gallic acid	A	2-aminoethyldiphenyl borinate	Stiasny 1912
flavonoids	A	2-aminoethyldiphenyl borinate	Markhan 1982; Neu 1956
fenylpropanoglyco-sides	A	2-aminoethyldiphenyl borinate	Wagner & Bladt 1996
condensed proantho-cyanidines, leuco-anthocyanidines	A	vanillin chloride	Robertson 1955
tannins	A	ferric alum, 1%	Xavier 2002 Stiasny 1912
quinones	C	potassium hydroxide, 10%	Harborne 1998

[1] A = EtOAc – HCOOH – AcOH – H_2O (100:11:11:26, v/v); B = Benzene – AcOEt (97:3, v/v); C = EtOAc – HCOOH – AcOH – H_2O (100:0.5:0.5:0.5, v/v); D = nBuOH – Me_2CO – Phosphate buffer, pH 5 (40:50:10, v/v); E = Et_2O – toluene – AcOH, 10% (1:1:1, v/v)

References

Braga R (1976). *Plantas do Nordeste Especialmente do Ceará*. Coleção Mossoroense, Brazil.

Correa MP (1926). *Dicionário das Plantas Úteis do Brasil*. Rio de Janeiro-RJ, Brazil.

Harborne JB (1998). *Phytochemical Methods*, 3rd ed. Chapman & Hall, London.

Malone RA (1977). Pharmacological approaches to natural products screening and avaliation. In *New Natural Products and Plant Drugs with Pharmacological, Biological or Therapeutical Activity* (H Warner and P Wolf, eds), pp 24–53. Springer-Verlag, Berlin.

Markhan KR (1982). *Techniques of Flavonoid Identification*, pp. 52–61. Academic Press, London.
Metz H (1961). Dünnschichtchromatographische schnellanalzse bel enyzmatischen steroid-umsetzungen. *Naturswissenschaften* 48:569–570.
Neu R(1956). A new reagent for differentiating and determining flavones of paper chromatograms. *Naturswissenschaften* 43:82.
Oliveira AB (1995). Frutoic acid, a dimeric kaurane diterpene. *Phytochemistry* 40:607–609.
Rodrigues WA (1981). Nova Iryanthera Warb. (Myristicaceae) de Anazônia. *Acta Amazônica* 11:852–854.
Stiasny E (1912). The qualitative detection and differentiation of vegetable tannins. *Collegium* pp. 483–499.
Wagner HBS and Bladt S (1996). *Plant Drug Analysis–A Thin Layer Chromatography Atlas*, 2nd ed. Springer-Verlag, Berlin.
Xavier HS, Randau KP, Barreto LCL, and Araújo E de Lemos (2002). Caracterização de taninos Hidrolizáveis II. In *53 Congresso Nacional de Botância*, Recife-Brasil.

Chapter 79

Pheasant's Eye (*Adonis aestavalis*) Toxicity in Livestock and Rodents

B.L. Stegelmeier[1], J.O. Hall[2], S.T. Lee[1], L.F. James[1], D.R. Gardner[1], K.E. Panter[1], M.H. Ralphs[1], and J.A. Pfister[1]

[1]USDA-ARS Poisonous Plant Research Laboratory, Logan, UT 84341, USA; [2] Veterinary Diagnostic Laboratory, Utah State University, Logan, UT 84322, USA

Introduction

Adonis aestavalis (pheasant's eye) is an introduced species from Eurasia that has escaped cultivation and grows as a weed in many parts of North America (Welsh *et al.* 2003). As an invading weed it often infests pastures and contaminates prepared feeds (Figure 1). Cases of poisoning have been reported in pigs and horses exposed to contaminated grain or hay (Davies and Whyte 1989; Woods *et al.* 2004). *Adonis* contains several cardenolides including adonitoxin, cymarin, K-strophanthin, and vernadigin, and historically, it has been used medicinally for its cardiac-enhancing effects (Kavalali 1973; Kopp *et al.* 1992). These cardenolides are reported to be more potent than digitoxin and they induce positive inotropic effects by inhibiting membrane–bound Na-K ATPase (Kavalali 1973). Since pheasant's eye is not highly palatable, most poisonings are caused by contaminated feed supplies. We have recently investigated *Adonis* field poisonings in both horses and cows. Four of four horses repeatedly developed anorexia, sweating and diarrhea when fed hay contaminated with *A. aestavalis*. All the animals developed biochemical changes suggestive of cardiac injury and all recovered over a period of several weeks. Four of 20 yearling steers died unexpectedly when fed *Adonis*-contaminated hay. Two steers were necropsied and the tissues were examined histologically. Both had extensive gastroenteritis with congestion and intestinal hemorrhage. Myocardial lesions were not evident histologically. The purpose of this study was to verify *Adonis* toxicity in a rodent model and better document the histologic and ultrastructural lesions of poisoning.

Methods

Adonis aestavalis collected in Cache County Utah (N41°50.75′, W111°57.51′, Intermountain Herbarium voucher #223724) was frozen, freeze dried and finely ground.

Three groups of ten Syrian hamsters were gavaged at a rate of 0, 250 and 500 mg ground plant per kg body weight four times a day for 7 days. Control animals (0 mg kg^{-1} *Adonis)* were dosed with finely ground alfalfa at similar volumes as the 500 mg kg^{-1} group. After 7 days the hamsters were euthanized; serum was collected and analyzed for biochemical and electrolyte changes. The tissues were collected and processed for histologic and ultrastructural examination. Results of serum biochemistries and electrolytes were analyzed using analysis of variance with $\underline{P} \leq 0.05$ interpreted as being significant. Where differences occurred, the means were separated using Duncan's multiple range test (Proc GLM, SAS Institute, Cary NC).

Figure 1. *Adonis aestavalis* or pheasant-eye is a glabrous, 10–20 cm tall plant with simple or branched stems. The leaves are 2-6 cm long with narrow segments and the flowers are reddish orange. It is generally found in disturbed sites or fields.

Results and Discussion

The control and low dose hamsters were clinically, biochemically and histologically unchanged. The high dose hamsters developed diarrhea, anorexia and became reluctant to move. Biochemically these animals developed significantly higher serum creatine phosphokinase activities (Table 1). They also were hyperglycemic with mild hyponatremia, hypercreatinemia and a decreased BUN/creatinine ratio. These changes are consistent with mild myocardial damage with subsequent stress and electrolyte changes. At necropsy the high dose animals all had extensive hemorrhagic gastroenteritis (Figure 2). Histologically the myocardial lesions were subtle with focal myocyte swelling (Figure 3). There was little evidence of the extensive swelling, coagulation of sarcoplasmic proteins, necrosis or fibrosis that is commonly associated with cardenolide toxicity (Galey *et al.* 1996). Ultrastructrually there were early lesions of myocardial mitochondrial swelling and myofiber disruption. The stomach, small intestine, colon and cecum had patchy hemorrhagic and necrotizing enterocolitis. The adjacent bowel was filled with blood and numerous bacteria.

Summary

These findings indicated that *A. aestavalis* is toxic; however, in many cases the myocardial lesions that have been historically associated with cardioglycoside poisoning may not be evident histologically. The lack of myocardial lesions in clinical cases may be due to the toxic *Adonis* cardioglycosides that kill the animals before morphologic lesions develop. For all animals (both the field cases and the treated hamsters) the most consistent lesion is a severe hemorrhagic gastroenteritis. The production of the classical cardioglycoside-associated myocardial lesions is variable and it is probably related to dose, duration, and recovery time. In any case *A. aestavalis* is toxic and contaminated hay should be avoided.

Table 1. Serum biochemistries of hamsters treated with 0, 250 and 500 mg kg^{-1} ground *Adonis aestavalis* four times daily for 7 days.*

	Units	0 mg kg^{-1}	250 mg kg^{-1}	500 mg kg^{-1}
Glucose	mg/dL	156±22^{A}	177±58^{AB}	378±150^{B}
BUN	mg/dL	23±3	25±6	24±3
Creatinine	mg/dL	0.35±0.10^{A}	0.38±0.05^{AB}	0.48±0.05^{B}
BUN/Creatinine	ratio	72.8±14.6^{A}	71.2±18.2^{AB}	49.8±4.2^{B}
Creatinine phosphokinase	U/L	571±69	842±91	911±142
Sodium	mequ/L	147±1^{A}	145±3^{A}	137±7^{B}
Potassium	mequ/L	9.6±0.4	9.5±0.4	10.1±0.7
Chloride	mequ/L	103±4	100±4	96±5
Carbon dioxide	mequ/L	32.8±4.6	36.0±3.3	30.2±5.9

*Data are presented as means±standard deviation. Significantly different values ($P \leq 0.05$) are indicated with different superscript letters.

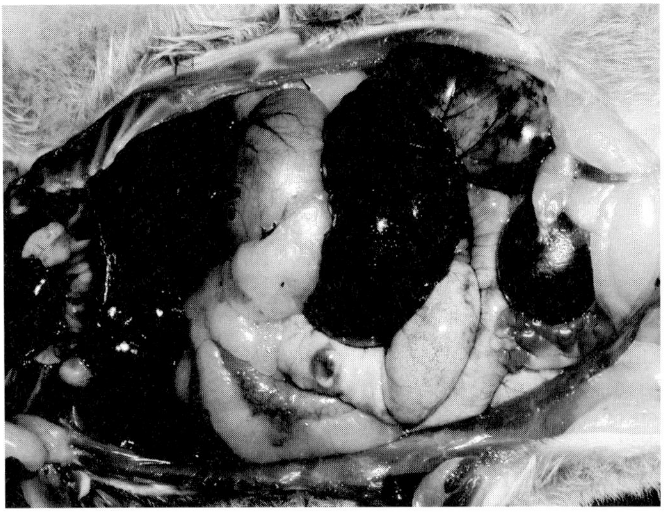

Figure 2. Photomicrograph of hemorrhagic gastroenteritis of a hamster treated with 500 mg kg^{-1} Q.I.D. for 7 days. Note the serosal gastric hemorrhage with hemorrhagic gastroenteritis. There are also hepatic congestion and hemorrhage.

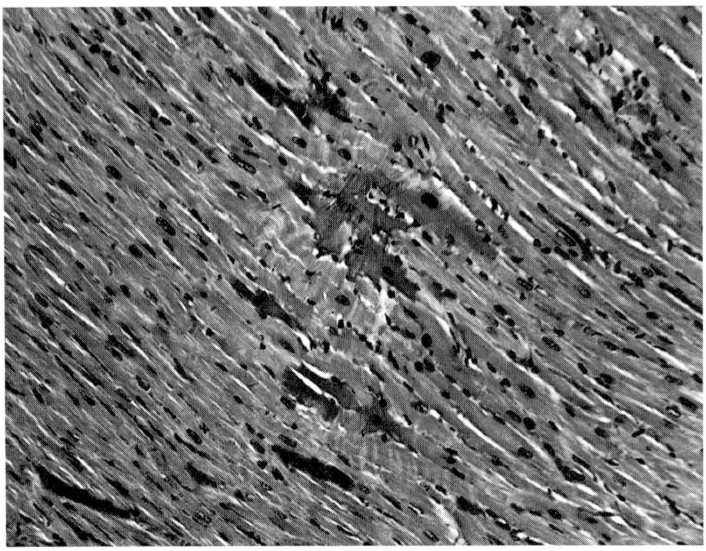

Figure 3. Photomicrograph of the heart of a hamster treated with 500 mg kg^{-1} ground *Adonis* for 7 days. Note the mild but focal myocardial swelling and degeneration.

References

Davies RL and Whyte PB (1989). *Adonis microcarpa* (pheasant's eye) toxicity in pigs fed field pea screenings. *Australian Veterinary Journal* 66:141–143.

Galey FD, Holstege DM, Plumlee KH, Tor E, Johnson B, Anderson ML, Blanchard PC, and Brown F (1996). Diagnosis of oleander poisoning in livestock. *Journal of Veterinary Diagnostic Investigation* 8:358–364.

Kavalali G (1973). Isolation of cymarin from *Adonis aestivalis*. *Lloydia* 36:426.

Kopp B, Krenn L, Kubelka E, and Kubelka W (1992). Cardenolides from *Adonis aestivalis*. *Phytochemistry* 31:3195–3198.

Welsh SL, Atwood ND, Goodrich S, and Higgins LC, eds. (2003). *A Utah Flora*, Third Ed., p. 532. Brigham Young University Press, Provo, Utah.

Woods LW, Filigenzi MS, Booth MC, Rodger LD, Arnold JS, and Puschner B (2004). Summer pheasant's eye (*Adonis aestivalis*) poisoning in three horses. *Veterinary Pathology* 41:215–220.

CHEMISTRY

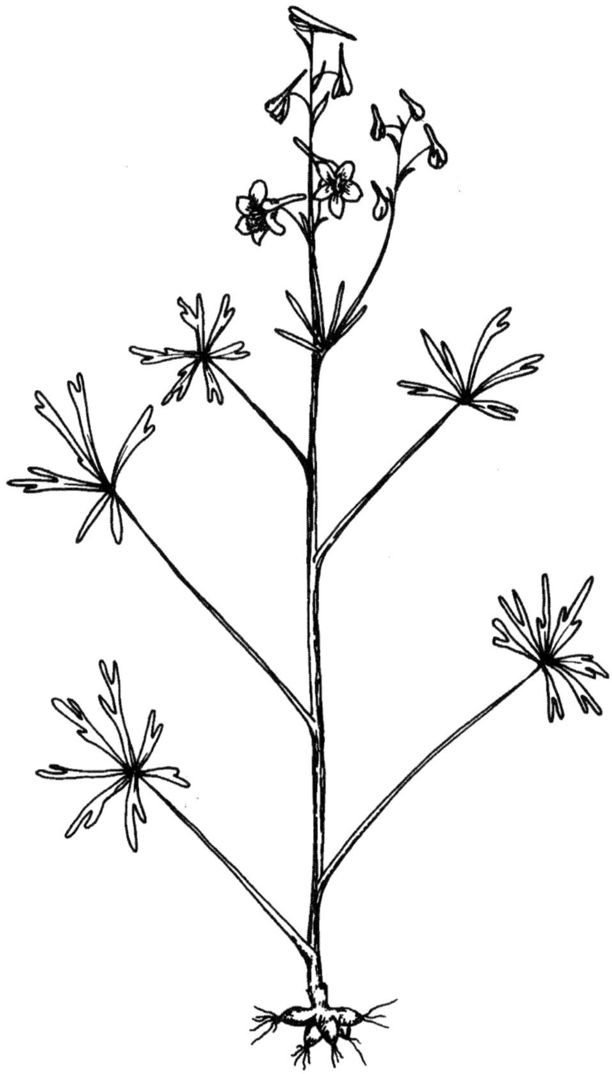

DELPHINIUM NUTTALLANIUM

Chapter 80

Separation and Isolation of Teratogenic Piperidine Enantiomers from *Nicotiana* and *Lupinus* Species

S.T. Lee[1], K.E. Panter[1], R.J. Molyneux [2], C.-W.T. Chang[3], D.R. Gardner[1], and J.A. Pfister[1]

[1]USDA-ARS Poisonous Plant Research Laboratory, Logan, UT 84341, USA; [2]USDA-ARS Western Regional Research Center, Albany, CA 94710, USA; [3]Department of Chemistry and Biochemistry, Utah State University, Logan, UT 84322-4815, USA

Introduction

Ingestion of *Nicotiana glauca* and *Lupinus formosus* by pregnant livestock at specific gestational periods can result in newborns with multiple congenital contractures (MCC) and cleft palates (Keeler *et al.* 1981, 1984; Keeler and Crowe 1984; Panter *et al.* 1990). Anabasine, a piperidine alkaloid, has been shown to be the predominant alkaloid in *N. glauca* and has been identified as the responsible teratogen (Keeler *et al.* 1984). Ammodendrine, also a piperidine alkaloid, is found in *L. formosus* and is a reported teratogen (Figure1) (Keeler and Panter 1989; Panter *et al.* 1998).

Anabasine and ammodendrine both have chiral centers. Anabasine is present in *N. glauca* (Keeler *et al.* 1984) as a racemic mixture and ammodendrine has been reported as a racemic mixture in several genera of the Leguminosae (Mears and Mabry 1971). The variability of our own measurements and of previously reported optical rotation values for ammodendrine led us to believe that lupine plants could be producing enantiomeric mixtures of ammodendrine. We converted the anabasine and ammodendrine into diastereomers. Once converted into diastereomers, the anabasine- and ammodendrine-based diastereomers were separated and isolated into milligram quantities. The separated anabasine and ammodendrine diastereomers were subsequently converted to their D and L enantiomers and their optical rotations measured. Enantiomers often have very different physiological activities (Brossi 1994), therefore the relative toxicities of the D and L anabasine and D and L ammodendrine enantiomers were measured individually in a mouse lethality bioassay.

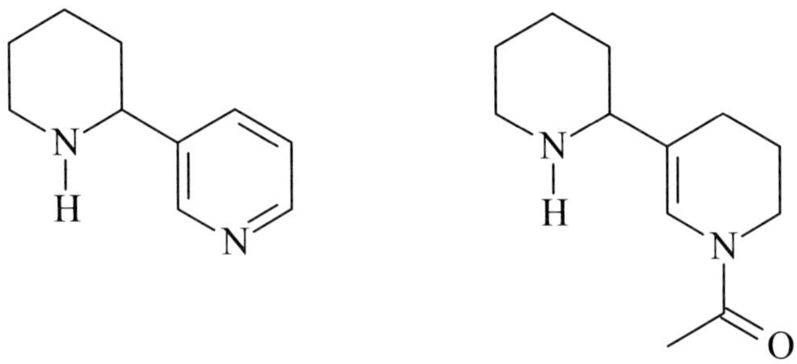

Anabasine (**1**) Ammodendrine (**2**)

Figure 1. Chemical structures of anabasine (**1**) and ammodendrine (**2**).

Materials and Methods

Plant material, extraction and piperidine isolation

Nicotiana glauca plant material was collected near Kingman, Mohave County, AZ on June 17, 2004. Lupine plant material was collected near Rio Vista Airport, Solano County, CA on June 25, 2003, and at SMIP Ranch, San Mateo County, CA on July 2, 2003. The plant material was air dried, ground in a hammer mill to pass through a 2 mm screen, and stored at ambient temperature. Anabasine and ammodendrine were extracted and isolated as previously described (Keeler *et al.* 1984; Lee *et al.* 2005).

Synthesis of anabasine- and ammodendrine-based diastereomers

We converted the anabasine enantiomers into diastereomers by a peptide coupling reaction utilizing 9-fluorenylmethoycarbonyl-L-alanine (Fmoc-L-Ala-OH) in the presence of *N*-(3-dimethylaminopropyl)-*N*'-ethylcarbodiimide hydrochloride and hydroxybenzotriazole dissolved in *N,N*-dimethyl formamide (DMF) (Figure 2). The formation of Fmoc-L-Ala-anabasine (**4**) (MH^+ 456) was monitored by electrospray ionization–mass spectrometry (ESI-MS). The conversion of **1** to **3** allowed reversed-phase HPLC analysis of the diastereomers and confirmed that the anabasine enantiomers were present in *N. glauca* plant material.

Treatment of the Fmoc-L-Ala-anabasine diastereomers (**3**) with piperidine in methylene chloride was monitored by ESI-MS for the formation of L-Ala-anabasine (**4**) (MH^+ 163). The L-Ala-anabasine diastereomers were separated and isolated in mg quantities using a semi-preparative scale 250 mm x 10 mm i.d., 5 μm, Betasil C-18 HPLC column (Thermo Hypersil-Keystone, Bellefonte, PA).

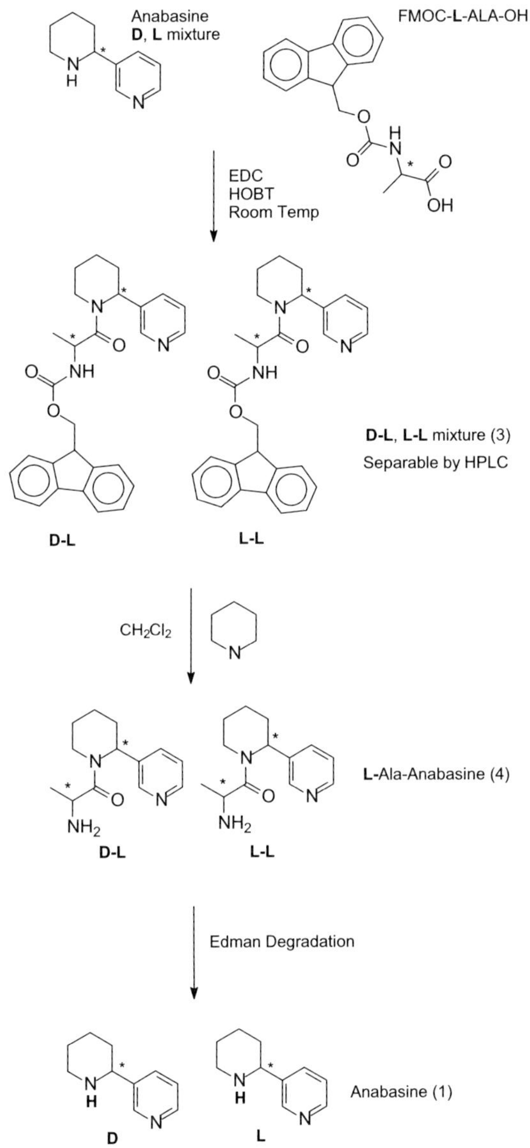

Figure 2. Conversion of anabasine **1** into diastereomers for separation and conversion of diastereomers back to enantiomerically rich isomers.

Conversion of anabasine- and ammodendrine-based diastereomers to enantiomers

Conversion of the isolated diastereomers to their respective enantiomeric forms of anabasine (**1**) was accomplished by an Edman degradation procedure to remove the L-alanine. The progress of the reaction, as monitored by ESI-MS, indicated that the major portion of the reaction residue was anabasine (**1**) (MH^+ 163). Reversed-phase HPLC chromatography of the isolated anabasine enantiomers (**1**) after conversion to the Fmoc-L-

Ala-anabasine (**3**) established the L-anabasine enantiomer to be 89% pure and the D-anabasine enantiomer to be 78% with respect to contamination from the complementary enantiomer. The optical rotation of L-anabasine has been measured as $[\alpha]_D$ −84°, −79.2°, and −80° respectively (Craig and Roy 1965; Hattori and Yamamoto 1992; Felpin *et al.* 2001). The optical rotation of our L-anabasine fraction was measured as $[\alpha]^{24}_D$ −48.1° and of our D-anabasine fraction was $[\alpha]^{22}_D$ +35.5°. Using the optical rotation measurements from our fractions and assuming the optical rotation of anabasine is $[\alpha]_D$ −80°, the purity of the L-anabasine enantiomer was calculated to be 80% and that of the D-anabasine enantiomer as 72% pure with respect to contamination from the complementary isomer.

L-Fmoc-Ala-ammodendrine diastereomers (MH^+ 502) were synthesized and then L-Fmoc-Ala-ammodendrine diastereomers (MH^+ 280) converted to L-Ala-ammodendrine diastereomers and isolated by using the methods described. The separated L-Ala-ammodendrine diastereiomers were also converted to ammodendrine enantiomers (MH^+ 209). Reverse-phase HPLC chromatography of the isolated ammodendrine enantiomers after conversion to the Fmoc-L-Ala-ammodendrine established the D-ammodendrine enantiomer to be 91% and L-ammodendrine to be 98% enantiomerically pure, respectively.

Relative toxicity of enantiomers

Known amounts of the individual alkaloids were dissolved in physiological buffered saline solution. The solutions were stored in sterile injection vials for toxicity testing. Weanling white Swiss-Webster male mice, 15 to 20 g (Simonsen Labs, Gilroy, CA) were weighed after a 12-hour fast and were dosed intravenously. Injections were performed via the tail vein in mice restrained in a plastic mouse block. Prior to injection, the mice were maintained under a heat lamp for 15 min to dilate the tail vein. The tail vein was cleaned with 70% ethanol and i.v. injections were accomplished with a tuberculin syringe equipped with a 1.27cm long 27-gauge needle. The volume injected varied depending on the dosage delivered. Time of injection, clinical effects, and time of death were noted and recorded. The LD_{50} for individual alkaloid toxicity was determined by a modified up-and-down method (Bruce 1987) and was calculated using the PROC PROBIT procedures of SAS (SAS Institute Inc., Cary, NC) on a logistic distribution of the survival data. Confidence (fiducial) intervals (95%) were also calculated using the same program.

Results and Discussion

While the optical rotation for L-anabasine has been well documented, the optical rotation values for ammodendrine that have been reported (Table 1) are inconsistent and no measures have been taken to ensure the enantiomeric purity of the isolates. Our isolates of ammodendrine were of high purity. The optical rotations of D-ammodendrine ($[\alpha]^{24}_D$ +5.4°) and L-ammodendrine ($[\alpha]^{24}_D$ −5.7°) are consistent with expected optical rotations of isolated enantiomeric pairs. The greater magnitude of the previously reported optical rotations than those reported in this study indicates that prior optical rotations for ammodendrine (**2**) are incorrect. The larger rotations are possibly due to contamination of the samples by other alkaloids with higher rotation values.

Table 1. Optical rotation data for ammodendrine (**2**).

compound	observed or reported optical rotations	plant source	references
D-Ammodendrine	$[\alpha]^{24}_{D}$ +5.4° (*c* 1.27, MeOH)	*Lupinus formosus*	This Study
L-Ammodendrine	$[\alpha]^{24}_{D}$ –5.7° (*c* 1.66, MeOH)	*L. formosus*	This Study
Ammodendrine	$[\alpha]^{24}_{D}$ +6.65° (*c* 3.9, EtOH)	*L. formosus*	Fitch *et al.* 1974
Ammodendrine	$[\alpha]^{24}_{D}$ +7.5° (*c* 0.22, EtOH)	*L. varius*	Abel-Halim *et al.* 1999
Ammodendrine	$[\alpha]^{24}_{D}$ +7.1° (*c* 0.08, MeOH)	*L. varius*	Mohamed and Hassanean 1997
Ammodendrine	$[\alpha]^{23}_{D}$ –20° (*c* 0.004, $CDCl_3$)	*Castilleja miniata*	McCoy and Stermitz 1983
Ammodendrine	$[\alpha]^{24}_{D}$ +15° (EtOH)		Dictionary of Natural Products 1999
Ammodendrine	$[\alpha]^{24}_{D}$ +11.8° (*c* 0.288, EtOH)	*L. hirsutus*	Takamatsu *et al.* 1990

Toxicity of anabasine (**1**) and ammodendrine (**2**) enantiomers was evaluated in a mouse bioassay. The LD_{50} of the L-anabasine fraction in mice was determined to be 16.2±1.0 mg kg^{-1}, n=26 and that of the D-anabasine fraction was 10.9±0.8 mg kg^{-1}, n=26. The LD_{50} of L-ammodendrine in mice was determined to be 115.0±7.0 mg kg^{-1}, *n*=20, and that of D-ammodendrine was 94.1±7.8 mg kg^{-1}, *n*=23. The anabasine enantiomers were therefore appreciably more toxic than the ammodendrine isomers. Although there was a statistical difference ($P<0.05$) in the measured toxicity between D-anabasine and L-anabasine and in the D- and L-ammodendrine pairs the observed difference in physiological activity measured in the mouse bioassay was not as large as is often the case with enantiomers (Barlow and Hamilton 1965; Aceto *et al.* 1979). It is possible that the receptor site responsible for toxicity is not chirally highly discriminatory. It is also possible that none of these compounds is the proximate toxin and that they are metabolized into compounds in which the asymmetric center is lost. This could occur by the introduction of an additional double bond in the fully saturated piperidine rings, by isomerization of the existing double bond to a position between the two piperidine rings of ammodendrine or by an oxidative process in the liver. An analogy for the latter process exists with the pyrrolizidine alkaloids, which are not toxic *per se* but are converted to toxic dehydropyrrolizidines by P450 liver enzymes (Mattocks1986)

It is recognized that the acute mouse toxicity model is not a surrogate for teratogenic crooked calf disease. Limitations in the amount of isolated enantiomeric material available only allowed us to initially investigate possible differences in toxicity based on a mouse model. However, a true measure of teratogenicity will need to be tested in a proper system such as a fetal goat model (Panter *et al.* 1990)

In this study, the similarities in toxicity between the enantiomeric pairs of anabasine (**1**) and ammodendrine (**2**) indicate that, in the absence of evidence to the contrary, both enantiomers should be taken into account in predicting toxicity of plant samples to livestock, as has been done previously.

References

Abel-Halim OB, El-Gammal AA, Abdel-Fattah H, and Takeya K (1999). Glycosidic alkaloids from *Lupinus varius*. *Phytochemistry* 51:5–9.

Aceto MD, Martin BR, Uwaydah IM, May EL, Harris LS, Izazola-Condo C, Dewey WL,

Bradshaw TJ, and Vincek WC (1979). Optically pure (+)-nicotine and biological comparisons with (-) nicotine. *Journal of Medicinal Chemistry* 22:174–180.

Barlow RB and Hamilton JT (1965). The stereospecificity of nicotine. *British Journal of Pharmacology* 25:206–212.

Brossi A (1994). Chiral drugs: Synopsis. *Medical Research Reviews*. 14:665–691.

Bruce R (1987). An up-and-down procedure for acute toxicity testing. *Fundamentals of Applied Toxicology*. 8:97–101.

Craig JC and Roy SK (1965). Optical rotatory dispersion and absolure configuration-III pyrrolidine, piperdine and tetrahydroisoquinoline alkaloids. *Tetrahedron* 21: 401–406.

Dictionary of Natural Products on CD-ROM, Version 7.2 (1999). Chapman and Hall, London.

Felpin FX, Girard S, VoThanh G, Robins RJ, Villierras J, and Lebreton J (2001). Efficient enantiomeric synthesis of pyrrolidine and piperidine alkaloids from tobacco. *Journal of Organic Chemistry* 66:6305–6312.

Fitch WL, Dollinger PM, and Djerassi CJ (1974). Alkaloid studies. LXVIII. Novel piperidyl alkaloids from *Lupinus formosus*. *Journal of Organic Chemistry* 39:2974–2979.

Hattori K and Yamamoto H. (1992). Asymmetric aza-Diels-Alder reaction mediated by chiral boron reagent. *Journal of Organic Chemistry* 57:3264–3265.

Keeler RF and Crowe MW (1984). Teratogenicity and toxicology of wild tree tobacco, *Nicotiana glauca* in sheep. *Cornell Veterinarian* 74:50–59.

Keeler RF and Panter KE (1989). Piperidine alkaloid composition and relation to crooked calf disease-inducing potential of *Lupinus formosus*. *Teratology* 40:423–432.

Keeler RF, Crowe MW, and Lambert EA (1984). Teratogenicity in swine of the tobacco alkaloid anabasine isolated from *Nicotiana glauca*. *Teratology* 30:61–69.

Keeler RF, Shupe JL, Crowe MW, Olson A, and Balls LD (1981). *Nicotiana glauca*-induced congenital deformities in calves: clinical and pathological aspects. *American Journal of Veterinary Research* 42:1231–1234.

Lee ST, Molyneux RJ, Panter KE, Chang CWT, Gardner DR, Pfister JA, and Garrossian M (2005). Ammodendrine and *N*-methylammodendrine enantiomers: Isolation, optical rotation, and toxicity. *Journal of Natural Products* 68:681–685.

Mattocks AR (1986) *Chemistry and Toxicology of Pyrrolizidine Alkaloids*. Academic Press, Orlando, Florida.

McCoy JW and Stermitz FR (1983). Alkaloids from *Castilleja miniatia* and *Penstemon whippleanus*, two host species for the plum moth, *Amblyptilia* (*Platyptilia*) *pica*. *Journal of Natural Products* 46:902–907.

Mears JA and Mabry TJ (1971). Alkaloids in the Leguminosae. In *Chemotaxonomy of the Leguminosae* (JB Harborne, D Boulter, and BL Turner, eds), pp. 73–178. Academic Press, New York.

Mohamed MH and Hassanean HA (1997). Alkaloids from seeds of *Lupinus varius* and *L. hartwegii*. *Phytochemistry* 46:365–369.

Panter KE, Bunch TD, Keeler RF, Sisson DV, and Callan RJ (1990). Multiple congenital contractures (MCC) and cleft palate induced in goats by ingestion of piperidine alkaloid-containing plants: Reduction in fetal movement as the probable cause. *Journal of Toxicology Clinical Toxicology*. 28:69–83.

Panter KE, Gardner DR, and Molyneux R J (1998). Teratogenic and fetotoxic effects of two piperidine alkaloid-containing lupines (*L. formosus* and *L. arbustus*) in cows. *Journal of Natural Toxins* 7:131–140.

Takamatsu S, Saito K, Sekine T, Ohmiya S, Kubo H, Otomasu H, and Murakoshi I (1990). Glycosidic alkaloids from *Lupinus hirsutus*. *Phytochemistry* 29:3923–3926.

Chapter 81

Heliotropium europaeum Alkaloids: a Quaternary Pyrrolizidine Alkaloid Approach to ELISA Development

S.T. Lee[1], A.M. Knill[2], A. Michalewicz[2], V. Stevens[2], and S.M. Colegate[2]

[1]*USDA-ARS Poisonous Plant Research Laboratory, Logan, UT 84341, USA;* [2]*CSIRO Livestock Industries, Australian Animal Health Laboratory, Private Bag 24, Geelong, VIC 3220, Australia*

Introduction

Heliotropium europaeum has proven to be an agricultural pest capable of killing stock and severely reducing productivity due to its pyrrolizidine alkaloid content (Everist 1981). It also has the potential to be co-harvested with edible grains and thereby contaminate the human food supply (Colegate *et al.* 1998). Existing methods for the detection of pyrrolizidine alkaloids present in *H. europaeum* (Figure 1) include an enzyme-linked immunosorbent assay (ELISA) and high pressure liquid chromatography (HPLC) or gas chromatography (GC) both linked to a mass spectrometer detector (MS) (Than *et al.* 2005). The ELISA does not detect the *N*-oxides very well, whilst the GC-MS methods do not detect *N*-oxides at all (Beales *et al.* 2004), and thus an extra step needs to be incorporated into the sample preparation protocol to reduce the *N*-oxides to their parent tertiary bases (Cavallaro *et al.* 2004). The HPLC-MS method (Colegate *et al.* 2005) is suitable for the detection of the *N*-oxides but is not suitable for rapid throughput and is expensive technology.

Previous research (Lee *et al.* 2001) in which the pyrrolizidine alkaloid riddelliine was quaternized and coupled to an immunogenic carrier protein resulted in an effective ELISA that recognized riddelliine and its *N*-oxide. This approach was adopted in an attempt to produce antibodies that could be used in an ELISA format to detect the parent pyrrolizidine alkaloids and their *N*-oxides of *H. europaeum* simultaneously.

Materials and Methods

Hapten synthesis/Protein conjugation

The pyrrolizidine alkaloids are not naturally immunogenic. Therefore the heliotrope alkaloids heliotrine, heleurine and lasiocarpine were each quaternized by reaction with Nhydroxysuccinimide (NHS) esters of brominated acids using the methods reported in Lee *et al.* 2001. The only stable quaternary compounds produced were those resulting from heleurine and lasiocarpine coupled to bromotoluic NHS ester. These haptens were subsequently linked to ovalbumin (OVA) and to fetal calf serum (FCS) protein to provide protein conjugates for ELISA plate coating and vaccination, respectively (Figure 2).

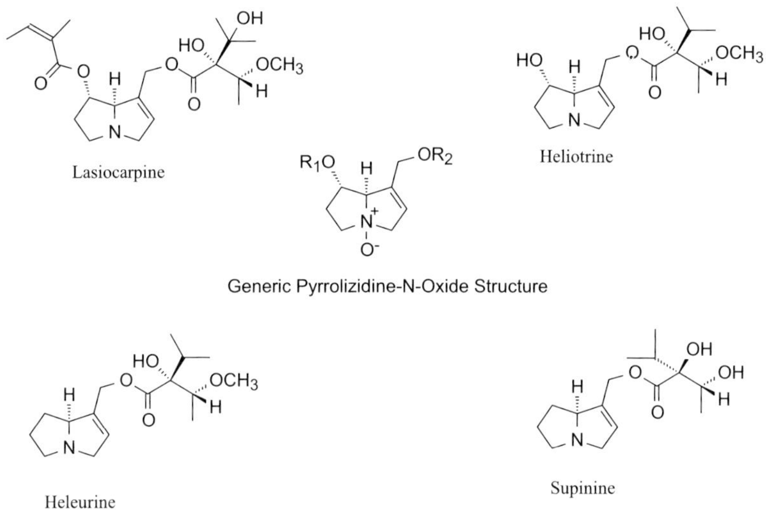

Figure 1. *Heliotropium europaeum* alkaloids.

Immunizations

For vaccination, the FCS-hapten conjugates were formulated with the CSIRO Triple Adjuvant. Two groups of four Merino rams were injected (subcutaneous on the neck) with either Lasiocarpine-Tol-FCS or Heleurine-Tol-FCS (1.0 mg conjugate in 2 mL adjuvant). At 4-weekly intervals an additional two booster vaccinations (0.5 mg conjugate in 2 mL adjuvant) were administered. Four weeks after vaccination booster three, all eight rams were injected with an equiproportional mixed antigen formulation (0.25 mg of each conjugate in 2 mL adjuvant). At 4-weekly intervals thereafter, all sheep were injected with an equiproportional mixed antigen formulation (0.5 mg of each conjugate in 2 mL adjuvant) for four additional boosters.

Antisera

Blood samples were collected from each sheep (*via* jugular vein) immediately before, and 14 days after every injection. Antibody titers in the sera were estimated using 96-well ELISA plates by determination of the binding of serial dilutions to the immobilized OVA-hapten conjugates in the ELISA plate wells.

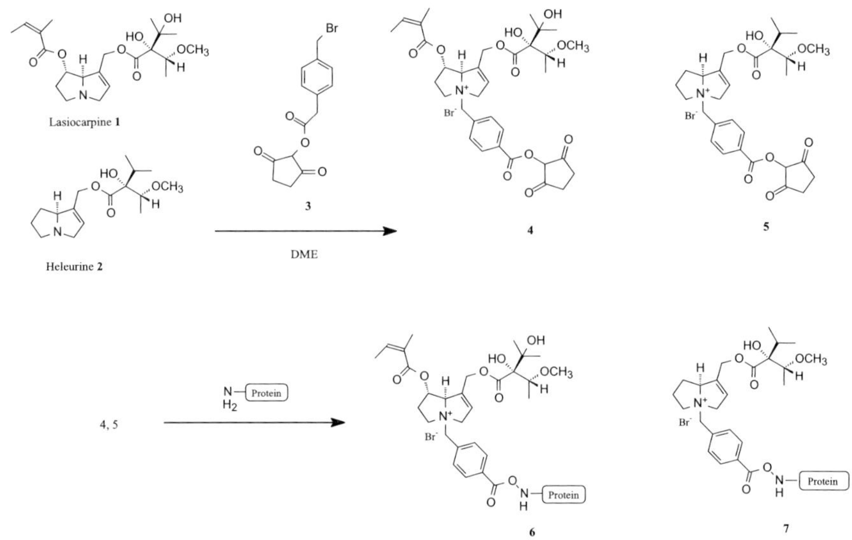

Figure 2. Lasiocarpine and heleurine protein conjugation scheme.

ELISA method and optimization

An indirect competitive ELISA format was utilized in which the soluble analyte(s) competed with surface-immobilized antigen for the antisera derived from vaccinated sheep. After washing and blocking stages in the ELISA protocol, the level of antisera binding to the surface-immobilized antigen was detected by addition of a secondary antibody and subsequent colorimetric detection.

ELISA parameters that were optimized included the dilution level of the antisera from each sheep, the concentration of the OVA-hapten conjugates used to coat the wells (*i.e.*, surface-immobilized antigen) (25 ng/well), assessment of non-specific binding to ovalbumin (none detected), blocking reagent concentration (1% skim milk powder in assay buffer for 1 hour) and the detection reagent concentrations (1/7500 of donkey antisheep antibodies conjugated to horseradish peroxidase, incubated with K-blue substrate for 25 min).

Results and Discussion

Cross-reactivity

The cross-reactivity of various pyrrolizidine alkaloids and their *N*-oxides, using the optimised ELISA conditions, are shown in Figure 3 as the % binding of the antibodies to the OVA-hapten conjugates immobilized on the surface of the ELISA plate wells. The greater the binding, the less the cross-reactivity with the pyrrolizidine alkaloids/*N*-oxides. The best response was observed when the lasiocarpine-based OVA conjugate was immobilized on the ELISA plate at 25 ng/well. The levels of analytes required to cause a 20% inhibition of binding were: heleurine-N-oxide (20 ng), heleurine (18 ng), lasiocarpine-*N*-oxide (10 ng), lasiocarpine (1.9 ng), heliotrine-*N*-oxide (60 ng) and heliotrine (20 ng) (Figure 3). It was surprising that, despite the similarity between supinine and heleurine, neither supinine-N-oxide nor supinine cross-reacted with the antisera.

The objective of increasing the relative cross-reactivity of *N*-oxides compared to their parent tertiary bases was achieved by immunizing with a quaternary hapten. By rational selection of the ELISA conditions, the major heliotrope pyrrolizidine alkaloids and their *N*-oxides are detected. However, the level of sensitivity is probably not useful in a practical sense.

Future work is required to compare this ELISA format with the previously described ELISA that recognizes the pyrrolizidine alkaloids and their tertiary bases at the same limits of detection (Lee *et al.* 2001). This comparison should include an extension of cross-reactivity studies to all heliotrope pyrrolizidine alkaloids/N-oxides and an assessment of the detection levels that might be required to monitor animal feed and the human food supply in order to prevent intoxication.

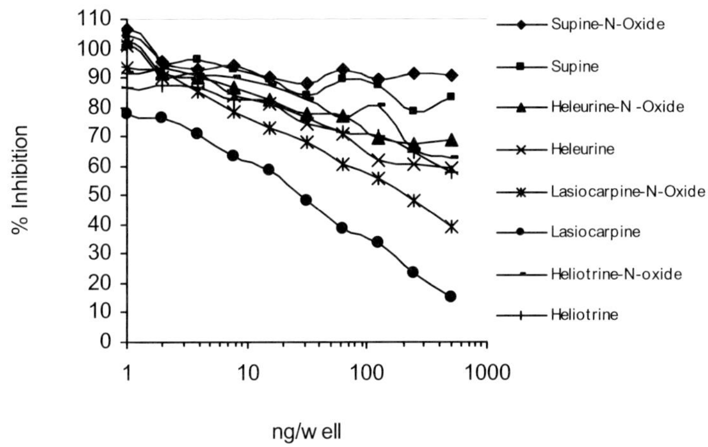

Figure 3. Competitive inhibition curves for *Heliotropium europaeum* alkaloids.

References

Beales KA, Colegate SM, and Edgar JA (2004). Experience with the quantitative trace analysis of pyrrolizidine alkaloids using GC/MS and LC/MS. In *Poisonous Plants and Related Toxins* (T Acamovic, CS Stewart, and TW Pennycott, eds), pp. 453–458. CABI Publishing, Wallingford, Oxon, UK.

Cavallaro V, Than A, Colegate SM, and Edgar JA (2004). An indirect competitive enzyme-linked immunoassay for pyrrolizidine alkaloids of *H. europaeum*. In *Poisonous Plants and Related Toxins* (T Acamovic, CS Stewart, and TW Pennycott, eds), pp. 114–119. CABI Publishing, Wallingford, Oxon, UK.

Colegate SM, Edgar JA, and Stegelmeier BL (1998). Plant-associated toxins in the human food supply. In *Environmental Toxicology: Current Developments* (J Rose, ed), pp. 317–344. Gordon and Breach Science Publishers.

Colegate SM, Edgar JA, Knill, AK, and Lee, ST (2005). Solid-phase extraction and HPLC-MS profiling of pyrrolizidine alkaloids and their N-oxides: acase study of *Echium plantagineum*. *Phytochemical Analysis* 16:108–119.

Everist SL (1981). *Poisonous Plants of Australia*, 2nd ed, pp. 119–122. Angus & Robertson Publishers, Melbourne.

Lee ST, Schoch TK, Stegelmeier BL, Gardner DR, Than KA, and Molyneux RJ (2001). Development of enzyme-linked immunosorbent assays for the hepatotoxic alkaloids riddelliine and riddelliine N-oxide. *Journal of Agriculture and Food Chemistry* 49:4144–4151.

Than KA, Stevens V, Knill A, Gallagher PF, Gaul KL, Edgar JA, and Colegate SM (2005). Plant-associated toxins in animal feed: Screening and confirmation assay development. *Animal Feed Science and Technology* 121:5–29.

Chapter 82

The Simultaneous Characterization of Pyrrolizidine Alkaloids and Their *N*-Oxides from *Echium plantagineum* using Solid Phase Extraction and LCMS

S.M. Colegate[1], J.A. Edgar[1], A.M. Knill[1], and S.T. Lee[2]

[1]*CSIRO Livestock Industries, Australian Animal Health Laboratory, Private Bag 24, Geelong, VIC 3220, Australia;* [2] *USDA-ARS Poisonous Plant Research Laboratory, Logan, UT 84341, USA*

Introduction

In addition to being liver toxins, some pyrrolizidine alkaloids and their *N*-oxides can be carcinogenic and genotoxic (Chou *et al.* 2003) and can cause pulmonary arterial hypertension, neurotoxicity and right ventricular hypertrophy (Huxtable 1989). The major, reported clinical symptoms of pyrrolizidine alkaloid intoxication in humans result from a veno-occlusive disease of the liver (Prakash *et al.* 1999). In addition to outbreaks of 'bread poisoning' when the seeds of pyrrolizidine alkaloid-producing plants were co-harvested with milling grains, pyrrolizidine alkaloids in herbs and spices have been associated with many instances of fatal hepatic disease in adult and neonatal humans. Whilst the effects of acute exposure to pyrrolizidine alkaloids on humans and animals have been well documented, the effects of long-term, low-level exposure to PAs *via* the diet (grain, milk, meat, honey, and related products) (Colegate *et al.* 1998) are unknown.

After a comprehensive risk assessment of toxic pyrrolizidine alkaloids, the German Federal Health Bureau established regulations that restrict oral exposure to pyrrolizidine alkaloids or their *N*-oxides in herbal preparations to 0.1 μg per day with the exclusion of pregnant and lactating women for which zero exposure is recommended (German Federal Health Bureau 1992). In the Netherlands, levels of pyrrolizidine alkaloids and their *N*-oxides are restricted to 0.1 μg per 100 g of food (H. van Egmond, The Netherlands National Institute for Public Health and the Environment, personal communication).

The current low levels of acceptable exposure to pyrrolizidine alkaloids in herbal products, and the co-association of the *N*-oxides, require analytical methodology that can rapidly identify both the pyrrolizidine alkaloids and their *N*-oxides. The solid phase extraction of the aqueous acid solubles of methanolic extracts of *Echium plantagineum* provided an alkaloidal fraction that was analyzed using high-pressure liquid chromatography (HPLC) coupled to ion trap, electrospray ionization mass spectrometry (ESI-MS).

Materials and Methods

Plant material and extracts

The leaves and flowers (calyx and petals, about 30–40 g) of *E. plantagineum* were collected in the vicinity of Geelong, Australia from a paddock subject to annual regrowth of the plant. Within 2 hours of collection, the fresh leaves and flowers were steeped in methanol (approximately 400 mL) at room temperature. After 48 hours standing, the methanol extracts were decanted and evaporated to dryness under reduced pressure. The resultant residues (green in color for the leaves, purple in colour for the flowers) were then extracted with 0.05M H_2SO_4 to provide the stock, aqueous acid extract of the alkaloids.

Processing of extracts

A sample (100 μL) of each of the leaf and flower extracts was applied to separate strong cation exchange (SCX) solid phase extraction (SPE) cartridges (Phenomenex Strata SCX, 500 mg/3 mL) that had been conditioned with methanol followed by 0.05M H_2SO_4. Each cartridge was then washed with water (3 mL) and then methanol (3mL) before being eluted with ammoniated methanol (methanol saturated at 0–4°C with ammonia gas and subsequently stored at room temperature, 6 mL). The ammoniated methanolic eluate was immediately evaporated to dryness under a flow of nitrogen in a heating block at 30–40°C. The residue from each sample was reconstituted into methanol (1 mL) to yield the analytical stock sample that was kept at –10°C until analyzed.

HPLC-ESI-MS analysis

An aliquot (2 μL) of an analytical stock sample was injected onto a C8 Reverse Phase column ((Phenomenex Aqua C8, 2.1 x 150 mm) protected by a guard cartridge of equivalent adsorbent. Subsequent work has also successfully utilized the C18 version of the Phenomenex 'Aqua' column for the analysis of plant samples (SM Colegate, unpublished). The plant extractives were eluted with a decreasing gradient flow (200 μL min^{-1}) of 0.1% formic acid in water (93% for 5 min before ramping down in a linear fashion to 30% over 15 min) into 0.1% formic acid in acetonitrile. Data dependent MS-MS spectra were recorded in the second scan of a two-scan sequence. Thus, nominated ions observed in the first scan (100–1200 mu) were isolated in the ion trap and then dissociated in the second scan using a generic dissociation energy of 35%.

Results and Discussion

Typical base ion chromatograms for mass to charge ratios (*m/z*) from 200 to 500 mass units (mu) are shown in Figure 1 for the solid phase extractives of the dilute acid solubles from the methanolic extracts of petals and leaves from *E. plantagineum*.

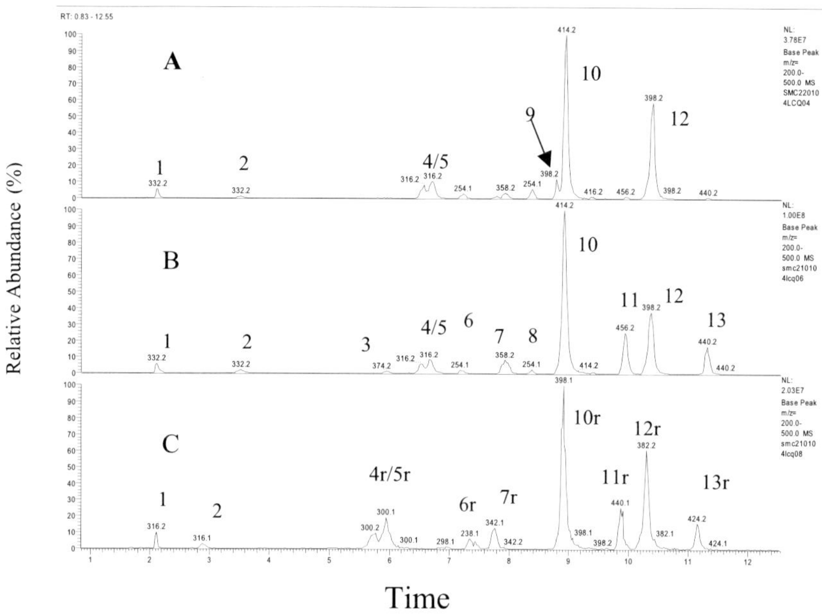

Figure 1. HPLC-ESI-MS base ion (*m/z* 200–500) chromatograms of the strong cation exchange, solid phase extractives from the dilute acid solubles of the methanolic extracts of leaves (**A**) and petals (**B**) from *Echium plantagineum*. The *N*-oxide character of the peaks in **B** was revealed after reductive treatment with an indigocarmine-based resin (**C**). The peak labels refer to compounds described in the text. 'r' indicates the reduced form of the *N*-oxides numbered in **A** and **B**.

Thirteen peaks were readily identified as pyrrolizidine-*N*-oxides in the ion chromatograms. The *N*-oxide character was implied by the significant (5–20% relative abundance) presence of a dimeric molecular ion adduct ($[2M + H]^+$) under the ESI conditions employed, and the reduction to their parent pyrrolizidine alkaloids when treated with an indigocarmine-based redox resin (Figure 1C).

Reconstructed ion chromatograms (RICs) of the full scan HPLC-ESI-MS profiles were used to selectively display ions and thereby demonstrate the presence, or absence, of minor peaks or those that co-elute with more dominant peaks. Thus, RICs for the $[M + H]^+$ ions of the tertiary base pyrrolizidine alkaloids revealed a much lower than expected relative proportion of pyrrolizidine alkaloids to their *N*-oxides (less than 1%) in the samples. However, these high proportions of *N*-oxides were consistent with the observation that senecionine-*N*-oxide is the sole product of pyrrolizidine alkaloid biosynthesis in cultured root cells of *Senecio vulgaris* (Hartmann and Toppel 1987).

Analysis of the ESI-MS data and the MS-MS data were interpreted with the benefit of previous literature reports (Culvenor 1956; Culvenor *et al.* 1981; Edgar and Smith 2000) that have identified echimidine (10r, Figure 1C), 3′-*O*-acetylechimidine (11r), echiumine (12r), 7-*O*-angelylretronecine (6r), uplandicine (3r, co-elutes with 4r/5r) and the diastereoisomers lycopsamine and intermedine (4r/5r) and their 7-*O*-acetyl derivatives (7r). Tentative structural assignments of other pyrrolizidine alkaloids and their *N*-oxides were proposed on the basis of their ESI-MS data, including selected MS^n experiments and rationalization with respect to reported and postulated MS fragments characteristic of pyrrolizidine alkaloids, and the assumption of biogenetic relationship to the major PAs present, i.e. echimidine and echiumine (Colegate *et al.* 2005).

An example of the application of MS-MS in differentiating extractives is shown in Figure 2. Peaks 9, 10r and 12 (Figure 1) each displayed a molecular ion adduct at *m/z* 398. Whilst the difference in retention time is sufficient to differentiate peak 12 (echiumine-*N*-oxide) from peaks 9 and 10r, the MS-MS data were required to differentiate peak 9 from echimidine (peak 10r), which elute very close to each other.

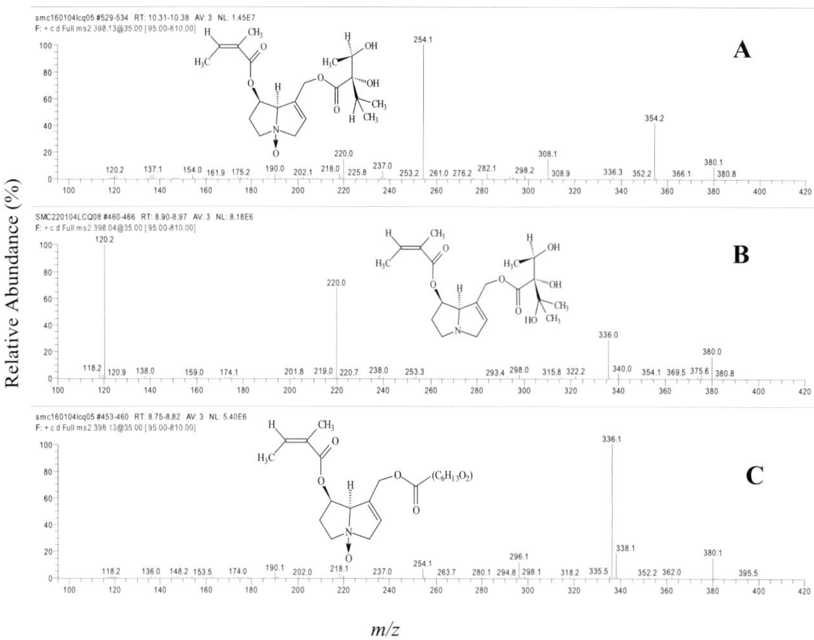

Figure 2. MS-MS differentiation of three pyrrolizidine alkaloids/*N*-oxides with the same molecular ion data (*m/z* 398) identified in *Echium plantagineum*. **A**. echiumine-*N*-oxide (peak 12, Figure 1A and B); **B**. echimidine (peak 10r, Figure 1C); and **C**. echiuplatine-*N*-oxide (peak 9, Figure 1A).

A higher proportion of acetylated pyrrolizidine-*N*-oxides was observed in the flower heads relative to the leaves. The acetylated nature of pyrrolizidine alkaloids/*N*-oxides was indicated by the observation of peaks at *m/z*180 for those acetylated at C7 and the facile loss of acetic acid ($[M+H]^+$ - 60) for those acetylated at C3′ of the esterifying acid on C9. Six pyrrolizidine alkaloids or pyrrolizidine-*N*-oxides that have not previously been reported from *E. plantagineum* were tentatively identified on the basis of mass spectra and biogenetic considerations. Three of these, 3′-O-acetylintermedine/lycopsamine (peak 7r, Figure 1C), leptanthine-*N*-oxide (peak 1 or 2) and 9-O-angelylretronecine-*N*-oxide (peak 8), have been reported elsewhere whilst three others, 3′-O-acetylechiumine-*N*-oxide (peak

which contains the binding sites for actin and ATP. Results for titin were inconclusive (Figure 3). Titin is highly susceptible to protease degradation. Regions both in the control and experimental group showed areas of breakdown.

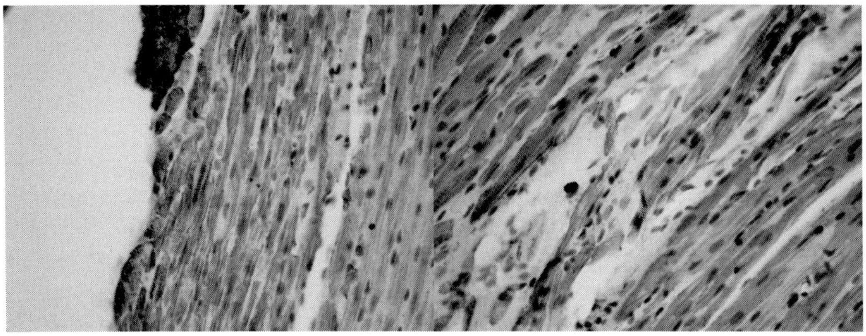

Figure 1. Actin staining of control (left) and pavetamine-treated (right) rat heart.

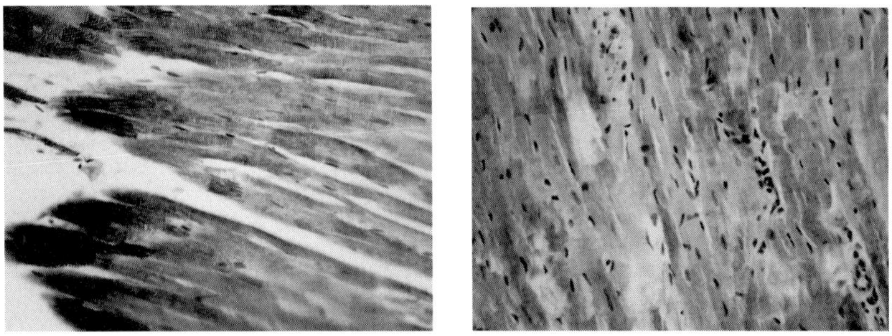

Figure 2. Myosin staining of control (left) and pavetamine-treated (right) rat heart.

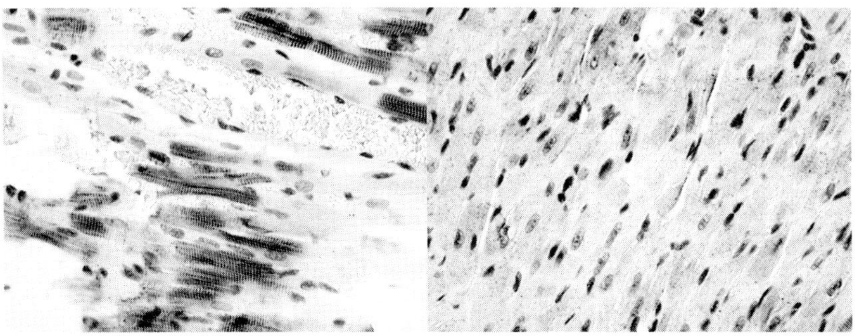

Figure 3. Titin staining of control (left) and pavetamine-treated (right) rat heart.

Microarray analysis

The up-regulated genes represent 0.22 % of the rat genome (approximately half of the genes have as yet an unknown function) and the down-regulated genes 0.05% (Table 1). Only the cardiac protein genes that were differentially expressed will be presented.

Table 1. Microarray analysis results of rat cardiac protein genes

Gene name	**U**p-regulated/**D**own-regulated
Myosin heavy chain beta isoform	U
Myosin heavy chain alpha isoform	D
Fast myosin alkali light chain isoform	U
Myosin light chain 2a isoform	D
Myosin light chain 1, atrial isoform	D
Four and a half LIM domains 1	U

The myosin heavy chain forms cross-bridges with actin and contains an enzymatic site that hydrolyzes ATP. Mammalian myocardium normally expresses the alpha isoform, but heart failure leads to down-regulation of the alpha isoform and up-regulation of the beta isoform (Sucarov *et al.* 2004). Intoxication with pavetamine also resulted in expression of the beta isoform. This transition results in slower contraction and a saving of energy. The myosin light chains have two subfamilies, the essential light chain (ELC) and the regulatory light chain (RLC). Upon phosphorylation of the RLC, contractile activity is initiated. The myosin light chain is thus the key regulatory protein in muscle contraction (Yamashita *et al.* 2003). Thus, in the case of pavetamine intoxication, the down-regulation of the myosin light chain 2 gene results in impaired contractility of the heart. The LIM domain proteins have a cysteine-rich, double-zinc finger, protein-binding motif. LIM domains mediate interaction with transcription factors and cytoskeletal proteins. The LIM domain proteins are up-regulated in hypertrophic cardiomyopathy (Lim *et al.* 2001). Intoxication with pavetamine also leads to increased expression of the four and a half LIM domain proteins.

Conclusion

At this stage it is uncertain if the expression of titin is altered. The microarray result for titin showed no change in expression between the control and treated group. It might be that the titin sequence is under-represented on the rat genome chip. Double immuno-labeling EM will clarify titin's involvement in the pathogenesis of pavetamine intoxication. Of note is the loss of the striated appearance of the myosin in the immuno-labeling of myosin. The myosin head binds to actin and although the appearance of actin seems normal, the myosin alignment is affected. Due to the complex involvement of many proteins in myofibrils, it may be that the expression of other proteins is also affected by pavetamine intoxication. Changes in the expression of cardiac myosin heavy chain and myosin light chain affect the contractile properties of the heart. Due to a decrease in the myosin light chain expression, phosphorylation by myosin light chain kinase is reduced, and the contractility is thus impaired.

In summary, some of the cardiac proteins have been identified that are involved in heart failure from pavetamine intoxication. Quantitative PCR will be done to verify the altered levels of expression of these genes.

Acknowledgements

We would like to thank Ms Leonie Labuschagne (Division of Toxicology, ARC-OVI) for technical assistance with the rat trials. We are indebted to Dr Sarah Cliff and Ms Marie Smith (Paraclinical Sciences, University of Pretoria, South Africa) for assistance during the immuno-labeling experiment. Prof. Labeit, Germany provided the antibody to titin. Funding was provided by the Gauteng Province (Department of Agriculture, Conservation and Environment) and the North-West Province (Department of Agriculture, Conservation, Environment and Tourism).

References

Diatchenko L, Lau Y-FC, Campbell AP, Chenchik A, Moqadam F, Huang, B, Lukyanov S, Yanov K, Gurskaya N, Sverdlov ED, and Siebert PD (1996). Suppression subtractive hybridization: A method for generating differentially regulated or tissue-specific cDNA probes and libraries *Proceedings of the National Academy of Sciences* 93:6025–6030.

Fourie N, Erasmus GL, Schultz RA, and Prozesky L (1995). Isolation of the toxin responsible for gousiekte, a plant-induced cardiomyopathy of ruminants in southern Africa. *Onderstepoort Journal of Veterinary Research* 62:77–87.

Gregorio CC, Trombitás K, Centner T, Kolmerer B, Stier G, Kunke K, Suzuki K, Obermayr F, Herrmann B, Granzier H, Sorimachi H, and Labeit S (1998). The NH_2 terminus of titin spans the Z-disc: Its interaction with a novel 19-kDa Ligand (T-cap) is required for sarcomere integrity. *Journal of Cell Biology* 143:1013–1027.

Kellerman TS, Coetzer JAW, and Naudé TW (1988). *Plant Poisonings and Mycotoxicoses of Livestock in Southern Africa,* 1st ed. Oxford University Press, Cape Town.

Lim D-S, Roberts R, and Marian AJ (2001). Expression profiling of cardiac genes in human hypertrophic cardiomyopathy: insight into the pathogenesis of phenotypes. *Journal of the American College of Cardiology* 38:1175–1180.

Schultz RA, Fourie N, Basson KM, Labuschagne L, and Prozesky L (2001). Effect of pavetamine on protein synthesis in rat tissue. *Onderstepoort Journal of Veterinary Research* 68:325–330.

Schutte PJ, Els HJ, Booyens J, and Pienaar JG (1984). Ultrastructure of myocardial cells in sheep with congestive heart failure induced by *Pachystigma pygmaeum*. *South African Journal of Science* 80:378–380.

Sucarov CC, Helmke SM, Langer SJ, Perryman MB, Bristow M, and Leinwand L (2004). The ku protein complex interacts with YY1, is up-regulated in human heart failure, and represses myosin heavy chain gene expression. *Molecular and Cellular Biology* 24:8705–8715.

Yamashita H, Sugiura S, Fujita H, Yasuda S-I, Nagai R, Saeki Y, Sunagawa K, and Sugi H (2003). Myosin light chain isoforms modify force-generating ability of cardiac myosin by changing the kinetics of actin-myosin interaction. *Cardiovascular Research* 60:580–588.

Chapter 84

N-Alkylated Derivatives of Swainsonine

M. Garrossian[1], S.T. Lee[1], R.J. Molyneux[2], and D.R. Gardner[1]

[1]USDA-ARS Poisonous Plant Research Laboratory, Logan, UT 84341, USA; [2]USDA-ARS Western Regional Research Laboratory, Albany, CA 94710, USA

Introduction

Locoweeds (*Astragalus* and *Oxytropis* spp.) continue to be a significant rangeland poisonous plant problem (Ralphs and James 1999). The toxin in the plant has been identified as the polyhydroxy indolizidine alkaloid named swainsonine **1** (Colegate *et al.* 1979; Molyneux and James 1982). Swainsonine inhibits cellular α-mannosidase activity (Dorling *et al.* 1978) and alters glycoprotein biosynthesis (Elbein 1989), resulting in inhibition of mannosidase II and lysosomal storage disease. Clinical signs of intoxication in animals are numerous and include depression, anorexia, weight loss, intention tremors, seizures, nervousness, dull hair coat, infertility, and abortion (Stegelmeier *et al.* 1999).

The prevention of locoweed intoxication is currently best achieved through the application of a number of management recommendations that may include rotational pasture grazing, cyclic grazing, flash grazing, delayed grazing, careful observation and removal of animals eating locoweed, culling and removal of reproductive open cows in the fall, creation of locoweed-free pasture areas through selective herbicide use, and aversion conditioning (Allison and Graham 1999). The idea of a vaccination program for use against locoweed intoxication would seem attractive; however, protective immunization against plant toxins has been of only limited success (Culvenor 1978; Payne *et al.* 1992; Edgar 1994; Than *et al.* 1998; Lee *et al.* 2003) because of the small molecular size of most plant toxins. In order to enable the low molecular weight plant toxins to provoke an immune response they must be linked to a larger carrier molecule such as a protein.

The linkage of swainsonine **1** to a large carrier molecule (protein) for the purpose of the induction of immunological activity has proven difficult because of the lack of chemical 'handles' on such a small molecule. Swainsonine has been linked via the hydroxyl groups but no swainsonine-specific antibodies were produced. It was assumed that linkage through one of the hydroxyl groups likely alters the three dimensional structure that characterizes the swainsonine-specific epitope. We sought an alternate approach by searching for a possible linkage through the nitrogen with a diverse carbon skeleton such as a 4-carbon allylic system and 4–8-carbon alkyl systems. The resulting alkyl-halide swainsonine derivatives are linked to several amino acids as a model system for future work with proteins.

Methods

Reaction scheme I. Swainsonine acetonide

Swainsonine acetonide was prepared by the reaction of swainsonine **1** with 2,2-dimethoxy propane in acetone with the pH adjusted to 3–4 by dropwise addition of methyl sulfonic acid. The reaction was refluxed for 6 hours with a Dean-Stark trap to remove the methanol formed. The reaction mixture was worked up by the removal of the solvent and extraction by base/acid/base partitioning to yield the final product, swainsonine acetonide **2** (> 90% yield).

Reaction scheme II. *N*-bromo-2-butene swainsonine and derivatives

To a 100 mL round bottom flask, equipped with a magnetic stirrer and condenser, was added 250 mg of swainsonine acetonide, 50 mL of benzene and 500 mg of 1,4-dibromo-2-butene. The reaction was refluxed for 48 hours. Diethyl ether was added, the solvent decanted, and the solid residue washed with diethyl ether/hexane, dried under vacuum to yield 395 mg of *N*-bromo-2-butene swainsonine acetonide **3**. A portion (100 mg) of the product was stirred with 2 mL water and 2 mL of 38% hydrobromic acid in acetic acid for 24 hours at 40°C. The solvent was removed by evaporation under vacuum to give a slightly yellow solid *N*-bromo-2-butene swainsonine **4**. High resolution electrospray mass spectrometry (HRESMS) $M^+ = 306.0705$; calculated for $C_{12}H_{21}BrNO_3 = 306.07048$.

Reaction scheme IIa. *N*-hydroxy-2-butene swainsonine

A small portion of **4** (15 mg) was dissolved in 1.5 mL of tetrahydrofuran and 1.5 mL water and the mixture stirred for 24 hours at room temperature and then for 3 hours at 40°C. The solvents were removed under vacuum to give 12 mg of *N*-hydroxy-2-butene swainsonine **5**.

Reaction scheme IIb. *N*-amino-2-butene swainsonine

To a small portion of **4** (50 mg) was added 1 mL of ammonium hydroxide and 1 mL of water. The mixture was stirred for 12 hours at 40°C and the solvent removed under vacuum to give 38 mg of *N*-amino-2-butene swainsonine **6**.

Reaction scheme IIc. *N*-(serine)-2-butene swainsonine

To 25 mg of **4** was added 1 mL of water and 0.5 mL of DMSO and 50 mg of serine. The reaction mixture was stirred for 48 hours, and the solvent removed under vacuum. The residue was washed twice with 2 mL of diethyl ether and 0.5 mL of hexane to remove excess serine and DMSO. The remaining residue was found to be a mixture of corresponding alcohol **5** (5%) and the serine derivative **7** (95%). If the reaction is conducted in pure water the yield of **7** is reduced to 60% and to 40% for the alcohol **5**.

Reaction scheme III. *N*-iodobutane swainsonine and derivatives

Swainsonine acetonide (200 mg) was dissolved in 30 mL of acetonitrile to which 200 mg of 1,4-diiodobutane was added. The reaction mixture was refluxed for 4 days. The solvent was mostly removed by a flow of argon, leaving a yellow residue. The residue was washed with diethyl ether and hexane (1/1) and dried under vacuum to give 50 mg of *N*-iodobutane swainsonine acetonide **8**. The acetonide group was removed by hydrolysis with iodic acid (0.5 mL 50% iodic acid) and 2 mL water upon stirring for 48 hours. The solvents were removed under vacuum to give 27 mg of *N*-iodobutane swainsonine **9**. HRESMS gave $M^+ = 356.0710$; calculated for $C_{21}H_{23}INO_3 = 356.07227$.

Reaction scheme IIIa. *N*-hydroxybutane swainsonine

To 10 mg of **9** was added 0.5 mL of DMSO, 0.5 mL water and 2 mg K_2CO_3. The reaction was stirred at 40°C for 24 hours. The reaction mixture was continuously extracted with methylene chloride for 24 hours, and the solvent removed to give *N*-hydroxybutane swainsonine **10**.

Reaction scheme IIIb. *N*-(serine)-butane swainsonine

In 0.5 mL of DMSO and 2 mL water was added 4 mg of **9**, 5 mg of serine, and 1 mg of K_2CO_3. The reaction was stirred for 48 hours at 40°C to yield *N*-(serine)-butane swainsonine **11** (50%) plus other products.

Reaction scheme IIIc. *N*-(glutathione)-butane swainsonine

To 0.5 mL of DMSO and 2 mL water was added 20 mg of **9**, 35 mg of glutathione and 3 mg of $NaHCO_3$. The mixture was stirred for 48 hours at 40°C. The solvent was removed under vacuum and the residue washed with 2 mL of dry methanol. The methanol was removed and the residue washed three times with diethyl ether and hexane (1/1) to yield a mixture of starting material **9** (5%), hydroxyl derivative **10** (5%), and *N*-glutathione-butane swainsonine **12** (90%).

Results

The overall reaction scheme is summarized in Figure 1. Initially, all reactions were carried out using swainsonine acetonide to enhance solubility in organic solvents and for protection of the hydroxyl groups. A number of *N*-alkylated derivatives of swainsonine could be produced by the reaction of swainsonine acetonide with 1,4-dibromo-2-butene and 1,4-diiodobutane. Alkyl addition on the nitrogen did occur, as evidenced by complete acetylation of the unprotected swainsonine derivatives (**4** and **9**) to give the corresponding triacetate compounds (Figure 2). Two possible isomers are produced, which are the *cis* and *trans* (relative to H-8a) adducts on the nitrogen, in a ratio of approximately 3:1. Both isomers are observed in LC/MS analysis of the product (Figure 3).

Compounds **4** and **9** were reacted to yield multiple functional groups on the end of the alkyl chain, thus providing multiple chemistry options to be used later in conjugation with the carrier molecules. In general the bromo derivatives were more reactive to a variety of functional groups than the iodo derivatives. However, we also noticed that with time the bromo reactions seemed reversible and the starting material was regenerated. Complete removal of Br anions from the reaction mixture might stop the conversion back to the alkyl bromide. The iodo derivative appeared to be more stable, but also less reactive towards OH and NH functional groups. *N*-iodobutane swainsonine did seem reactive towards SH groups as evidenced by the reaction with glutathione.

After confirmation that the above reactions were only occurring on the nitrogen atom, we recently completed the *N*-alkylation reaction directly with swainsonine, thus eliminating the use of the acetonide derivative. Identical products (**4** and **9**) were produced. Variable chain length alkyl compounds can also be used. 1,6-Diiodohexane was easily incorporated into the reaction scheme to produce the iodohexane swainsonine derivative. Future work will focus on coupling reactions to the carrier protein, injection into an animal model and testing for swainsonine-specific antibodies.

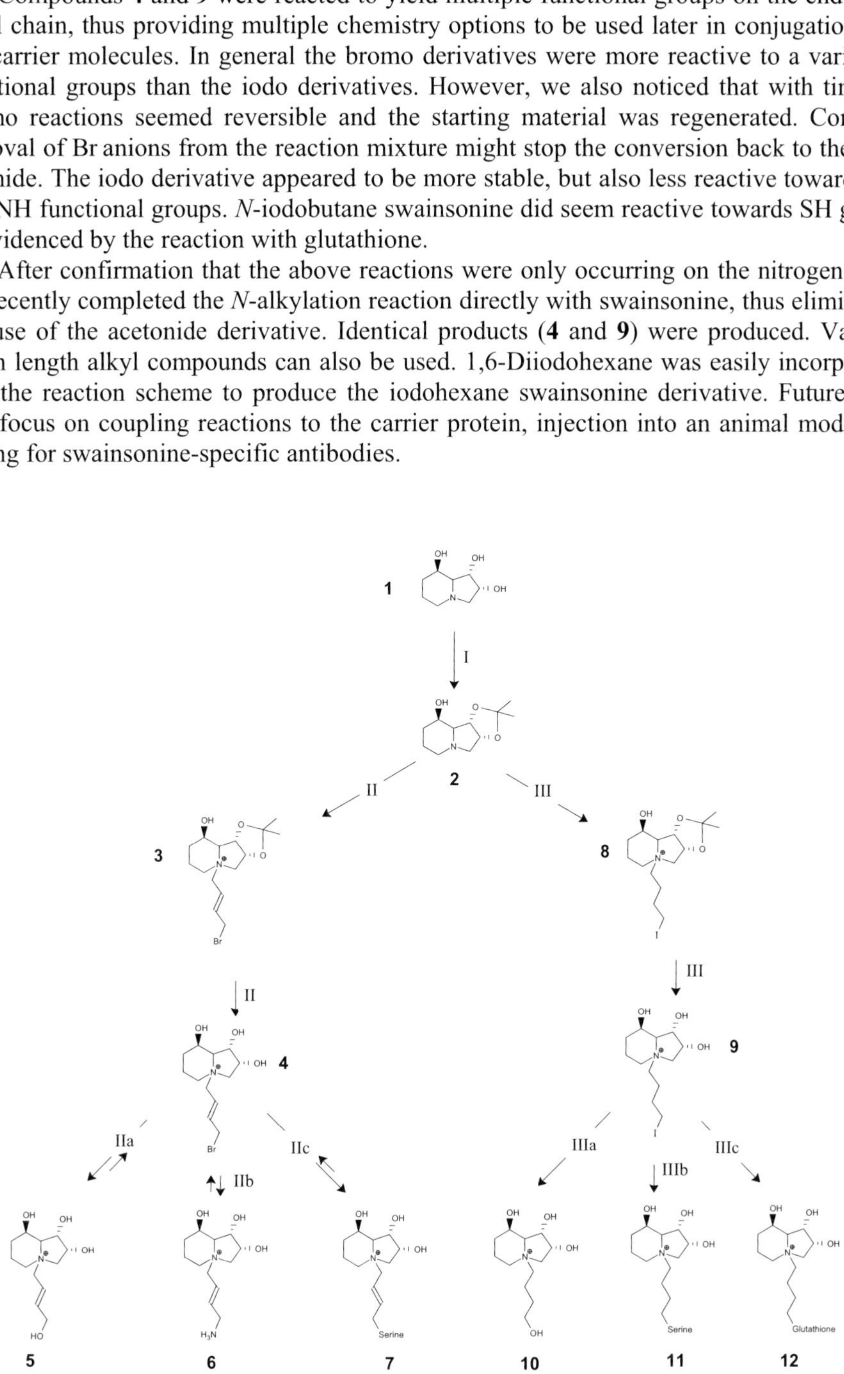

Figure 1. Overall reaction scheme for production of *N*-alkyl derivatives of swainsonine.

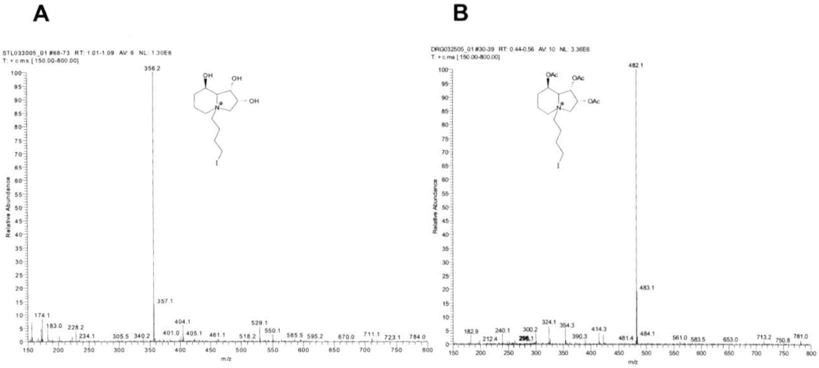

Figure 2. (A) ESMS spectrum of *N*-iodobutane swainsonine **9**; mw = 356 and (B) conversion to the triacetate, mw = 482 to prove *N*-alkylation.

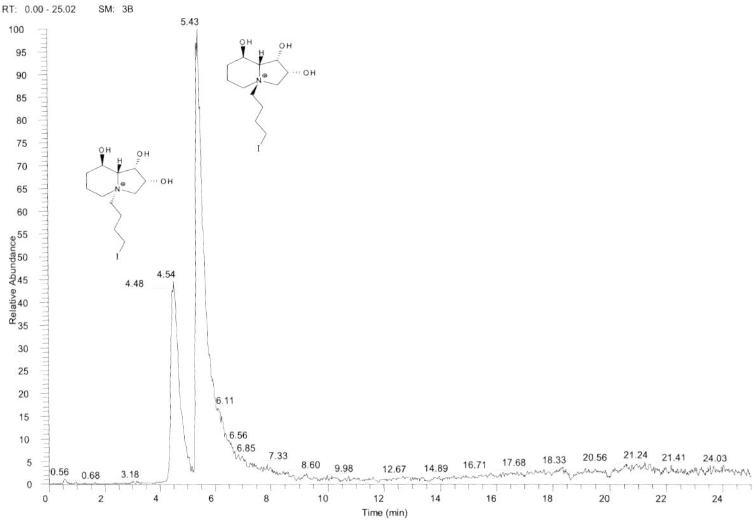

Figure 3. LC/MS separation of *cis* and *trans N*-iodobutane swainsonine.

References

Allison C and Graham JD (1999). Reducing locoism with management decisions. In *Locoweed Research Updates and Highlights* (TM Sterling and DC Thompson, eds), pp. 64–66. Research Report 730, New Mexico State University, Agricultural Experiment Station.

Colegate SM, Dorling PR, and Huxtable CR (1979). A spectroscopic investigation of swainsoine: An α-mannosidase inhibitor isolated from *Swainsona canescens*. *Australian Journal of Chemistry* 32:2257–2264.

Culvenor CCJ (1978). Prevention of pyrrolizidine alkaloid poisoning-animal adaptation or plant control. In *Effects of Poisonous Plants on Livestock* (RF Keeler, KR van Kampen, and LF James, eds), pp. 189–200. Academic Press, New York.

Dorling PR, Huxtable CR, and Vogel P (1978). Lysosomal storage in *Swainsona* spp. toxicosis: An induced mannosidosis. *Neuropathology and Applied Neurobiology* 4:285.

Edgar JA (1994). Vaccination against poisoning diseases. In *Plant-associated Toxins, Agricultural, Phytochemical and Ecological Aspects* (SM Colegate, and PR Dorling, eds), pp. 421–426. CAB International, Wallingford, UK.

Elbein AD (1989). The effects of plant indolizidine alkaloids and related compounds on glycoprotein processing. In *Swainsonine and Related Glycosidase Inhibitors* (LF James, AD Elbein, RJ Molyneux, and CD Warren, eds), pp. 155–187. Iowa State University Press, Ames, IA.

Lee ST, Stegelmeier BL, Panter KE, Pfister JA, Gardner DR, Schoch TK, and James LF (2003). Evaluation of vaccination against methyllycaconitine toxicity in mice. *Journal of Animal Science* 81:232–238.

Molyneux RJ and James LF (1982). Loco intoxication: Indolizidine alkaloids of spotted locoweed (*Astragalus lentiginosus*). *Science* 216:190–191.

Payne AL, Than KA, Stewart PL, and Edgar JA (1992). Vaccination against lupinosis. In *Poisonous Plants, Proceedings of the Third International Symposium* (LF James, RF Keeler, EM Bailey, PR Cheeke, and MP Hegarty, eds), pp. 234–238. Iowa State University Press, Ames, Iowa.

Ralphs MH and James LF (1999). Locoweed grazing. *Journal of Natural Toxins* 8(1):47–51.

Stegelmeier BL, James LF, Panter KE, Ralphs MH, Gardner DR, Molyneux RJ, and Pfister JA (1999). The pathogenesis and toxicokinetics of locoweed (*Astragalus* and *Oxytropis* spp.) poisoning in livestock. *Journal of Natural Toxins* 8(1):35–45.

Than KA, Cao Y, Michalewicz A, and Edgar JA (1998). Development of a vaccine against annual ryegrass toxicity. In *Toxic Plants and Other Natural Toxicants* (T Garland and CA Barr, eds), pp. 165–168. CABI Publishing, Wallingford, Oxon, UK.

Chapter 85

Characterization of a New Compound Eluting in an Ergovaline HPLC Assay Associated with Reproductive Problems in Dairy Cows

J.M. Duringer, C.T. Estill, and A.M. Craig

College of Veterinary Medicine, Oregon State University, Corvallis, OR 97331, USA

Introduction

Tall fescue (*Festuca arundinacea*) and perennial ryegrass (*Lolium perenne*) are perennial cool-season grasses which are infected with the endophytic fungi, *Neotyphodium coenophialum* and *N. lolii*, respectively. These endophytes have been increasingly promoted in fescue and ryegrass because they confer benefits such as pest resistance and drought tolerance to the plant (Adcock *et al.* 1997), decreasing the use of pesticides, fertilizers and irrigation. However, livestock grazing endophyte-infected (E+) grasses are negatively impacted by the production of ergot and lolitrem alkaloids from the fungi, which are responsible for a variety of insect and mammalian diseases (Joost 1995; Tor-Agbidye *et al.* 2001). Endophyte toxicosis affects cattle production by decreasing average daily weight gain and reproductive efficiency, resulting in approximately $1 billion in livestock damage each year in the United States alone (Comis 2000).

Ergot alkaloids act as α-adrenergic agonists which stimulate smooth muscle cell contraction, resulting in vasoconstriction of the extremities (Thompson and Stuedemann 1993). This can result in tissue ischemia, necrosis and sloughing of hooves, tails and ears during exposure to cold ambient temperatures (fescue foot) and reduced average daily weight gain with elevated temperatures (summer slump). Cows consuming E+ fescue straw exhibit altered reproductive hormone levels, including decreased prolactin (Emile *et al.* 2000), which is used as a diagnostic indicator of endophyte toxicosis. Additional reproductive problems include reduced pregnancy rates (Schmidt *et al.* 1983) and increased rates of early embryonic death (Schmidt *et al.* 1986).

Ergovaline is the ergot alkaloid found in highest abundance in E+ tall fescue (85–97% of total ergopeptine alkaloid content) (Lyons *et al.* 1986) and is detected along with lolitrem B in E+ perennial ryegrass (Hovermale and Craig 2001). A high-performance liquid chromatography (HPLC) method was developed to quantify ergovaline in forage samples submitted by farmers and/or veterinarians (Craig *et al.* 1994). Using this assay, threshold levels correlating concentration of ergovaline in feed to the appearance of clinical disease were established for horses, sheep and cattle (Tor-Agbidye *et al.* 2001), namely

300–500 ppb for horses (except for mares in the last 60 to 90 days of gestation when the recommended threshold is 0 ppb), 400–750 ppb for cattle and 500–800 ppb for sheep. However, it is likely that not all of the ergot alkaloids involved in endophyte toxicosis have been identified.

During sample processing for the diagnostic analysis of ergovaline, unidentified peaks occasionally elute which appear to coincide with clinical disease. In the winter of 2003-2004, such a situation occurred when numerous dairies in the Willamette Valley, Oregon encountered reproductive problems, specifically decreased cycling and conception rates. Mycotoxins, viral and bacterial pathogens were ruled out as a cause, while prolactin concentrations showed no obvious trend. Hay and silage samples (processed from the same given field for each dairy) were sent in for endophyte toxin analysis. While all samples had low concentrations of ergovaline, haylage (ensiled perennial ryegrass) samples showed the appearance of a large peak eluting at 5 min in the ergovaline HPLC chromatogram.

The objective of this research was to characterize this peak using mass spectroscopy to determine its chemical structure and identity. We believe this peak represents an ergot alkaloid toxin not previously reported in correlation with clinical signs of fescue toxicosis and should be investigated further as to its association with development of reproductive problems in livestock.

Materials and Methods

Collection and preparation of hay and silage samples

Hay and silage samples were collected from four dairies in the Willamette Valley, Oregon, from July 2003 through April 2005. Samples were dried and submitted to the College of Veterinary Medicine for ergovaline analysis. Dried samples were ground to pass through a 0.5 mm screen (Cyclotec 1093 sample mill, Tecator, Hönganäs, Sweden).

Ergovaline HPLC analysis and collection of 5-min peak

Ergovaline concentration was analyzed according to a previously reported method (Craig *et al.* 1994). Briefly, 1.0 g of ground sample was extracted over 24 hours with ergotamine tartrate (internal standard), chloroform and sodium hydroxide and then centrifuged at 2000 rpm for 5 min. Solid phase extraction (SPE) columns containing Ergosil were conditioned with chloroform before the sample supernatant was applied and drawn through the column. The column was then washed with a chloroform:acetone mixture followed by methanol. The ergot alkaloids were eluted with methanol and concentrated before analysis by HPLC. A polymeric divinyl benzene column (Jordi RP SM-500A, 5μ (150mm x 4.6mm)) was eluted with acetonitrile/2.5 mM ammonium carbonate (70/30 v/v) and run at a flow rate of 1.0 mL min^{-1}. Detection was by fluorescence with excitation and emission wavelengths of 250 and 420 nm, respectively.

A new peak, eluting at approximately 5 min in the ergovaline HPLC assay, appeared in the perennial and annual ryegrass silage samples at varying concentrations (Figure 1). To concentrate and purify this peak, samples with a large concentration of this compound were extracted using the ergovaline methodology, pooled, dried under nitrogen stream at 50ºC and reconstituted in 0.5 mL methanol. The pooled sample was then vortexed, sonicated and centrifuged for 5 min at 2000 rpm. The supernatant was injected onto the HPLC system and the peak eluting at 5 min was captured manually. This fraction was dried under a nitrogen stream at 50ºC, reconstituted with 1 mL methanol, vortexed, sonicated and centrifuged for 5 min at 2000 rpm. The final supernatant was stored in an amber vial in the freezer (−20ºC) until mass spectral analysis was performed.

Mass spectral analysis of "5-min peak"

Spectra were obtained on a LCQ Classic ion trap mass spectrometer (Thermo Finnigan) equipped with a custom-designed electrospray inlet consisting of a 30 micron i.d. steel capillary heated at 170ºC and operated at 2.7 kV. Solvent flow was controlled by a HPLC system consisting of a Waters Automated Gradient Controller with two Waters 515 HPLC pumps and a Rheodyne 8125 injector. The solvents used were water and acetonitrile each containing 0.1% acetic acid and 0.01% trifluoroacetic acid eluting under isocratic conditions in a 50:50 ratio. Samples were introduced to the spectrometer by loop injection with data acquisition in the positive ion mode. MS/MS analysis was selected for the peak with the highest relative intensity by MS.

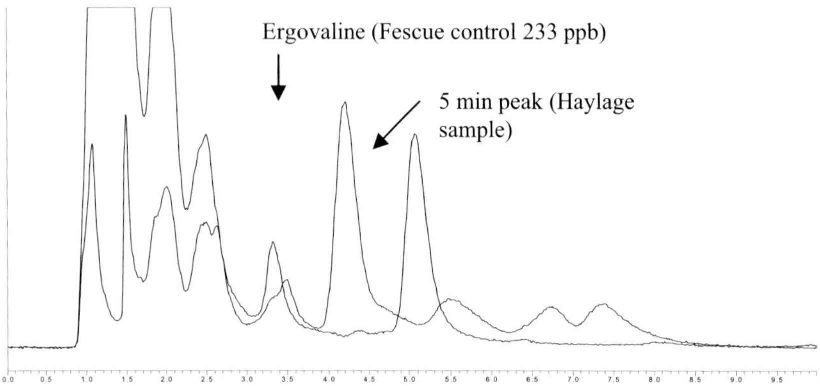

Figure 1. Chromatogram from reverse-phase HPLC of ergovaline showing the appearance of a new peak eluting around 5 min.

Results

Characterization of the '5 min peak' by LC-MS gave a pseudomolecular ion at m/z ratio 593.5 [M+H]+ (Figure 2A). LC-MS/MS analysis of ions with a m/z ratio of 593.5 produced daughter ions with a m/z ratio of 533.3 and 534.3 with little other fragmentation occurring (Figure 2B). Further mass spectrometry (MS^7) produced a scan showing peaks with a m/z ratio of 197, 223 and 268.

Conclusion

The LC-MS scan produced a peak with a m/z ratio of 593.5 which is not consistent with the molecular weight of any known ergot alkaloid. The LC-MS/MS scan produced a spectrum with two main peaks, one at 533.3 and another at 534.3. It was interesting to note the close association with ergovaline, whose molecular weight is 533.63. However, it was evident that this molecule had not completely fragmented, as the characteristic lysergic acid ring structure peaks typically present from ergot alkaloids fragmented in a mass spectrometer (*i.e.*, 197, 208, 223, 268, etc.) were absent (Lehner *et al.* 2004). Higher collision energies at MS^7 finally fragmented the molecule into many peaks, including those with a m/z ratio of 197, 223 and 268. Those at 197, 223 and 268 are consistent with molecules having the lysergic acid ring as its base; those at 235 and 342 are unknown compounds.

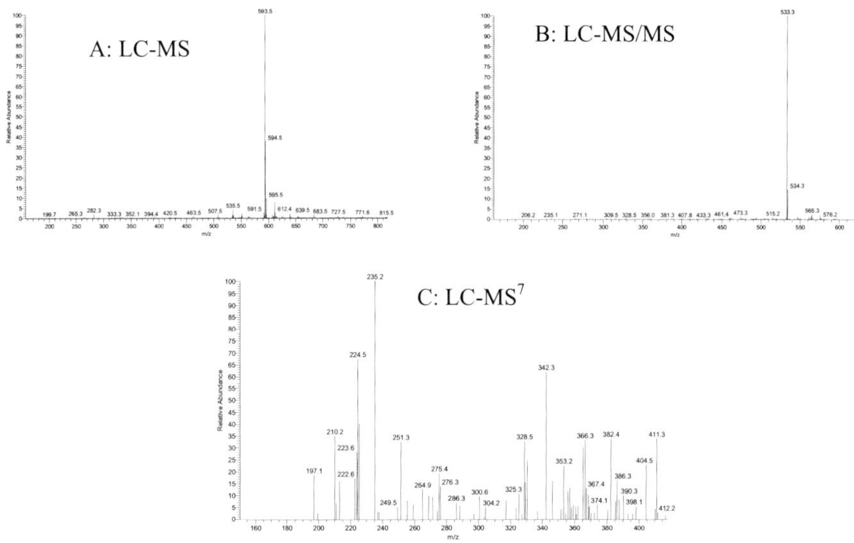

Figure 2. Mass spectroscopy of the 5-min peak by LC-MS (A), LC-MS/MS (B) and $LC\text{-}MS^7$.

Further characterization of this compound is ongoing, including analysis by LC-TOF/MS and 1H-NMR. Once this compound is identified, attempts to correlate the appearance of it in silage samples with reproductive problems in dairy herds will be made. In addition, the endophyte testing service is now aware of this unusual peak and its potential toxicity; thus, it will be able to identify samples containing this compound and caution farmers and veterinarians in using feed containing this molecule.

References

Adcock R, Hill N, Bouton J, Boerma H, and Ware G (1997). Symbiont regulation and reducing ergot alkaloid concentration by breeding endophyte-infected tall fescue. *Journal of Chemical Ecology* 23(3):691–704.

Comis D (2000). The grass farmers love to hate. *Agricultural Research.* July:4–7.

Craig AM, Bilich D, Hovermale JT, and Welty RE (1994). Improved extraction and HPLC methods for ergovaline from plant material and rumen fluid. *Journal of Veterinary Diagnostic Investigation* 6:348–352.

Emile JC, Bony S, and Ghesquiere M (2000). Influence of consumption of endophyte-infested tall fescue hay on performance of heifers and lambs. *Journal of Animal Science* 78(2):358–364.

Hovermale JT and Craig AM (2001). Correlation of ergovaline and lolitrem B levels in endophyte-infected perennial ryegrass (Lolium perenne). *Journal of Veterinary Diagnostic Investigation* 13(4):323–327.

Joost R (1995). Acremonium in fescue and ryegrass: boon or bane? A review. *Journal of Animal Science* 73:881–888.

Lehner AF, Fannin N, Bush L, Craig M, and Tobin T (2004). Fragmentation patterns of selected ergot alkaloids by electrospray tandem quadrupole mass spectrometry. *Journal of Mass Spectrometry* 39:1275–1286.

Lyons PC, Plattner RD, and Bacon CW (1986). Occurrence of peptide and clavine ergot alkaloids in tall fescue grass. *Science* 232:487–489.

Schmidt S, King CJ, Hoveland C, Clark E, Smith L, Grimes H, and Holliman J (1983). Cow-calf performance as affected by fungus infestation of Kentucky-31 tall fescue pastures. *Journal of Animal Science* 57(Suppl. 1):295.

Schmidt SP, Danilson DA, Holliman JA, Grimes HW, and Webster WB (1986). Fescue fungus suppresses growth and reproduction in replacement beef heifers. *Highlights of Agricultural Research* 33:150–151.

Thompson F and Stuedemann J (1993). Pathophysiology of fescue toxicosis. *Agriculture, Ecosystems and Environment* 44:263–281.

Tor-Agbidye J, Blythe L, and Craig A (2001). Correlation of endophyte toxins (ergovaline and lolitrem B) with clinical disease: fescue foot and perennial ryegrass staggers. *Veterinary and Human Toxicology* 43(3):140–146.

MYCOTOXINS

Chapter 86

A Study of the Bioequivalence of Corynetoxins and Tunicamycins: the Effect of these Toxins on the Viability and Morphology of Rat Hepatocytes in Tissue Culture

J.G. Allen[1], M. Retallick[2], M. Kowalski[2], S.M. Colegate[2], E.J. Speijers[1], T. Doran[2], and W. Michalski[2]

[1]*Department of Agriculture, Locked Bag No. 4, Bentley Delivery Centre WA 6983, Australia;* [2]*Plant Toxins Research Group, CSIRO Livestock Industries, Private Bag 24, Geelong VIC 3220, Australia*

Introduction

The corynetoxins are responsible for annual ryegrass toxicity (ARGT), flood plain staggers and Stewart Range syndrome, which cause high stock losses in Australia (Allen 2004). They are structurally related to the tunicamycins, having a common structural element made up of uracil, tunicamine and N-acetylglucosamine (Edgar *et al.* 1994), and differing only in regard to fatty acid side chains linked to tunicamine (Than *et al.* 2004). Both groups of compounds have high mammalian toxicity and are potent inhibitors of N-acetylglucosamine-1-phosphate transferase thereby blocking N-linked glycoprotein synthesis. They have similar biological activity at high dose rates (Stewart *et al.* 2004).

The commercially available tunicamycins have been used as a substitute for the corynetoxins in many studies of the pathogenesis, pathology and toxicology of ARGT in animals. However, interest in the effects of corynetoxins near the no observed effect level has prompted this work in which the bioequivalence of corynetoxins and tunicamycins were assessed at the cellular level.

Materials and Methods

Two experiments were conducted in parallel to study the effects of corynetoxins (av. MW 860; isolated from toxic annual ryegrass (*Lolium rigidum*)) and tunicamycins (av. MW 840; Sigma-Aldrich, St Louis, Missouri, USA). The two toxins were prepared in 0.6 mM, AnalaR grade sodium hydroxide (NaOH) (BDH Ltd., Merck, Darmstadt, Germany). Clone

9 rat hepatocytes (American Type Culture Collection, Manassas, Virginia, USA; accession number CRL-1439) were cultured in F-12K Nutrient Mixture Media (Invitrogen, Carlsbad, California, USA) with supplements. There were 10 treatments in both experiments, namely Control (F-12K media alone), NaOH Control (F-12K media with toxin-free NaOH added), 5, 50, 500 and 5000 nM corynetoxins in F-12K media, and 5, 50, 500 and 5000 nM tunicamycins in F-12K media.

Viability of toxin-exposed cells

Thirty Nunclon™ Delta Surface 24-well plates (Nunc, Roskilde, Denmark) were seeded with Clone 9 hepatocytes in toxin-free F-12K media (2.5×10^4 cells/mL). Twenty wells per plate were used. Forty-eight hours post-seeding, all of the plates were checked under low-power microscopy to ensure adequate cellular adhesion and the presence of apparently healthy cells. The media was then removed from each well and replaced with Control F-12K media, NaOH Control media or toxin-containing media, such that each of the ten dose groups were represented in two wells per plate. At 14 pre-determined time points of 0, 6, 12, 24, 30, 36, 48, 54, 60, 72, 78, 84, 96 and 108 hours following change of the media, two randomly selected plates were assessed for cell viability and proliferation rates using the trypan blue dye exclusion staining method ($n = 4$ for each dose at each time point). In this protocol, cells were trypsinised for 5 min at 37°C, mixed 1:1 with 0.2% trypan blue, then the numbers of stained (dead) and unstained (live) cells were counted using a hemocytometer. Throughout the experiment the cultures were incubated at 37°C in an atmosphere of 95% air and 5% carbon dioxide.

Morphology of toxin-exposed cells

Ten OptiCells™ (BioCrystal, Westerville, Ohio, USA) were seeded with Clone 9 hepatocytes in toxin-free F-12K media (1×10^5 cells/mL). At 48-hours post-seeding every OptiCell™ was examined under low-power microscopy to ensure adequate establishment of the cell layer. The media was then removed from the OptiCells™ and replaced with Control F-12K media, NaOH Control media or toxin-containing media. The OptiCells™ were then randomly labeled as A–J to facilitate unbiased morphological assessments. At the same 14 pre-determined time points as used in the cell viability experiment, the 'real-time' morphology of the unstained cells within the OptiCells™ was examined and recorded in photomicrographs. Throughout the experiment the cultures were incubated as for those in the cell viability study.

Statistical analysis – hepatocyte viability study

A generalized linear mixed model was fitted to the percentage of dead cells:

Fixed model: *[Control+(Toxin*Dose)]*Time*
Random model: *Plate.Time*

In the fixed model *Time* was a factor and the factor *Control* compared the Control treatment to the NaOH Control to the average of the toxin treatments. The random model allowed for a variance between plates in addition to the binomial variance expected to be seen when each cell in an aliquot is assessed as alive or dead. The data input to this model were the numbers of dead cells and the total numbers of cells counted for each well. Using Wald statistics to test fixed effects the model gave powerful tests of the effects of *Toxin* as a main effect and as an interaction with *Dose* and *Time*. The generalized linear mixed model procedure in GenStat Edition 8 was used to fit the model.

A generalized linear mixed model, similar to that described above, was fitted to the adjusted percentage of dead cells (adjusted according to no. live cells in the control treatment):

Fixed model: *Treatment /v Time*
Random model: *Plate.Time*

where *vTime* is a linear effect of time. For each treatment a response of the form

$$\log(P/(1-P)) = a + b\ \mathrm{Time}$$

is estimated, where *a* and *b* are regression coefficients, *b* is the rate of response over time and *a* is the start time of response.

Results

Viability of toxin-exposed cells

The variance between plates was significantly greater than zero and has therefore been retained in the model. There were no significant effects of Toxin ($P < 0.05$) (Table 1). However the main effect of Toxin, averaged over dose levels and times, was close to significance ($P = 0.092$). For corynetoxins 46.2% of cells were dead, whereas for tunicamycins 40.9% of cells were dead, over all times.

In most of the treatments, the proportion of dead cells was high to start with and then dropped. This was particularly so for the Control treatment: an initially high proportion dropped to close to zero after 24 hours. This was not unexpected as the tissue cultures were still growing and cells multiplying. Rather than the number of dead cells changing in this early phase it was the number of live cells that was changing. To take account of this effect the numbers of live cells in all treatments were adjusted. This was done by estimating the average number of live cells in the Control treatment from 54 hours onwards and then calculating the ratio between this value and the average number of live cells counted at each of the time points 0 to 48 hours for the Control treatment. These ratios have then been applied to counts of live cells for all treatments for the time points 0 to 48 hours. Numbers of live cells and total cells are thus adjusted upwards for these time points, and logistic regressions have been fitted to the treatments for which the number of dead cells change over time (Figure 1).

The treatments clearly fell into two groups. The first group consisted of the two Control and the 5 nM corynetoxins and tunicamycins treatments. These treatments all had a rate of response of zero. The second group consisted of the corynetoxins and tunicamycins treatments with a dose rate of ≥ 50 nM, and all these treatments had a rate of response that was greater than zero. There was no difference in the rate of response or the start of response for treatments with dose rates ≥ 50 nM (Figure 1). However there was a trend for higher dose rates to have an earlier response.

Table 1. The significance of treatment effects in the study of the effects of corynetoxin and tunicamycin on hepatocytes grown in tissue culture

Fixed term	Degrees of freedom	*P* value
Time	13	< 0.001
Control	2	< 0.001
Time Control	26	< 0.001
Dose	3	< 0.001
Time Dose	39	< 0.001
Toxin	1	0.092
Time Toxin	13	0.999
Dose Toxin	3	0.558
Time Dose Toxin	39	0.819

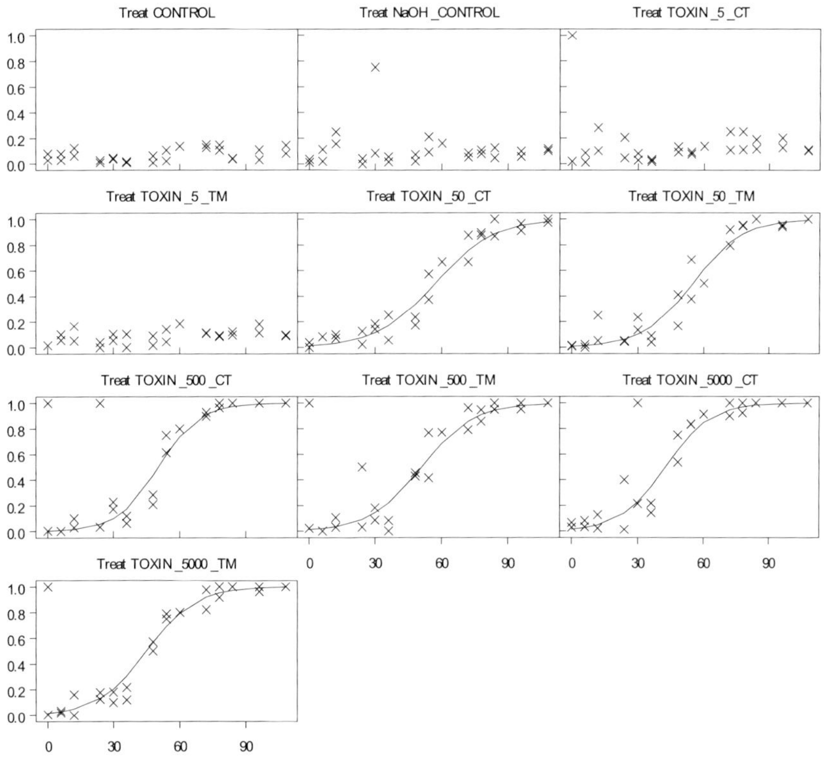

Figure 1. The adjusted proportion of dead cells at different times, and the fitted logistic regression lines, in the study of the effects of corynetoxin and tunicamycin on hepatocytes grown in tissue culture.

Morphology of toxin-exposed cells

There was no difference in the treatments up to 12 hours following the change of media. The cultures consisted of broad sheets of hepatocytes through which were scattered holes that were rapidly being filled by new cells arising via mitosis. Mitosis and normal cellular enlargement were obvious throughout this 12-hour period.

At 24 hours it became possible to distinguish between two groups of treatments. One group consisted of the two Control and the 5 nM corynetoxins and tunicamycins treatments. The other group consisted of the remaining corynetoxins and tunicamycins treatments. The distinction of these two groups remained for the rest of the experiment. It was not possible to distinguish between the treatments within the first group, but at 54 hours there appeared to be a dose rate effect in the second group. A crude scoring of the changes indicated that the severity of changes increased as the dose rate increased.

The Control and 5 nM toxin treatments showed comparatively few changes throughout the experiment. The cultures consisted of continuous sheets of normal-appearing cells that reached maximum size at about 48 hours. Mitosis was obvious throughout the experiment, and changes consistent with apoptosis first appeared at 36 hours and gradually increased in prevalence throughout the experiment. Some mild vacuolation of cytoplasm was present in some cells at 96 and 108 hours.

There were dramatic changes in all the toxin treatments of 50 nM or greater. These can be broadly divided into vacuolation of cytoplasm with other degenerate changes, separation and shrinkage of cells to form small, round bodies or cells, and disruption and disintegration of the sheets of hepatocytes due to loss of cells from the OptiCells™ surface and overall cell shrinkage. A small number of cells remained reasonably normal but were reduced in size.

Vacuolation of the cytoplasm first became apparent at 24 hours. Large numbers of cells were affected but the vacuoles were not large. With time many more cells developed vacuolation and the vacuoles were larger. The nuclei in these vacuolated cells showed a number of changes including increased density of the karyoplasms, becoming angular, margination of chromatin, nuclear membrane becoming indistinct with loss of clear nuclear definition, and disappearance of nucleoli. Vacuolation of cytoplasm with degenerate changes of the nucleus was most severe at 48 and 54 hours. The proportion of vacuolated cells attached to the OptiCell™ membrane decreased from 54 hours onwards and it was interpreted that these cells died and fell from the culture surface. Other vacuolated cells shrank and became small round bodies or cells, in which vacuoles could still be seen, often coalesced into a single vacuole.

Cell shrinkage and detachment from adjacent cells first became apparent at 30 hours. This predominantly affected non-vacuolated, or only mildly vacuolated, cells. Cells undergoing shrinkage attained their smallest size by about 48 to 54 hours, becoming small round bodies of dense cytoplasm surrounded by a thick membrane, sometimes with a small indistinct nucleus, but often with no distinct nucleus or nucleolus. The number of these cells decreased with time, and it was interpreted that these cells died and fell from the culture surface. However, many were apparently viable and therefore remained attached to the OptiCell™ membrane, and the proportion of these increased relative to vacuolated cells.

Disruption of the sheets of hepatocytes first became obvious at 36 hours when small holes in the sheet started to appear. The cell sheets became completely disrupted by 72 hours as cell loss and cell shrinkage and separation increased. From this time on the cultures consisted of a few islands of several adjacent cells, with individual round cells scattered over the remainder of the OptiCell™ membrane surface. Cell loss apparently peaked at 96 hours.

Regenerative changes first appeared at 78 hours. Some round cells extruded the dense body within their thick plasma membrane, and the empty membrane was discarded. This change became more prevalent through the remaining 30 hours of the experiment, and the extruded round bodies were seen to elongate to form small, mobile, spindle-shaped cells with a small nucleus. Some round cells developed into these spindle cells apparently without the extrusion process. The spindle cells joined up with each other and other remnant cells to start reforming sheets, and became larger with the features of an hepatocyte.

Conclusions

The effects of corynetoxins and tunicamycins on hepatocyte viability and morphology were indistinguishable in the experiments reported here. This provides further evidence that these two related toxins have equivalent bioactivity and continues to support use of the tunicamycins-based experimental model of ARGT.

In both the cell viability and the cell morphology studies, the treatments fell into two distinct groups. The first group included the two Control treatments and the 5 nM corynetoxins and tunicamycins treatments, and it was not possible to distinguish between the treatments in this group. The second group included all the toxin treatments of 50 nM or greater, and it was in this group that toxic effects of the treatments were very evident. In this group there was some evidence of a dose rate effect, but corynetoxins treatments could not be distinguished from tunicamycins treatments at each dose rate. This result indicates that the no observable effect level for corynetoxins or tunicamycins was between 5 and 50 nM in this *in vitro* study.

There was good correlation between the results of the two studies. Both identified the division of treatments into two groups, and in the group in which toxic changes were evident the time sequence of the morphological changes fitted well with the live and dead cell counts. In the morphology study cell loss commenced at about 36 hours, was extensive by 76 hours and peaked at about 96 hours. This correlates well with the dead cell counts shown in Figure 1.

References

Allen JG (2004). Annual ryegrass toxicity. In *Clinical Veterinary Toxicology* (KH Plumlee, ed), pp. 422–424. Mosby, St. Louis, Missouri.

Edgar JA, Cockrum PL, Stewart PL, Anderton NA, and Payne AL (1994). Identification of corynetoxins as the cause of poisoning associated with annual beardgrass (*Polypogon monspeliensis* (L) Desf.) and blown grass (*Agrostis avenacea* C. Gemelin). In *Plant-Associated Toxins; Agricultural, Phytochemical and Ecological Aspects* (SM Colegate and PR Dorling, eds), pp. 393–398. CABI Publishing, Wallingford, Oxon, UK.

Stewart PL, Hooper PT, Lenghaus C, Raisbeck MF, Edgar JA, and Colegate SM (2004). Effect of tunicamycins on GlcNAc-1-P transferase activity in rat tissues and toxic effects during pregnancy and lactation. In *Poisonous Plants and Related Toxins* (T Acamovic, CS Stewart and TW Pennycott, eds), pp. 448–452. CABI Publishing, Wallingford, Oxon, UK.

Than KA, Cao Y, Michalewicz A, Olsen V, Anderton N, Cockrum P, Colegate S, and Edgar JA (2004). Analysis of corynetoxins: a comparative study of an indirect competitive ELISA and HPLC. In *Poisonous Plants and Related Toxins* (T Acamovic, CS Stewart and TW Pennycott, eds), pp. 402–407. CABI Publishing, Wallingford, Oxon, UK.

Chapter 87

Dermal Toxicity of the Tropical Cyanobacterium *Cylindrospermopsis raciborskii*

I. Stewart, X. Shen, A.A. Seawright, and G.R. Shaw

National Research Centre for Environmental Toxicology, The University of Queensland, 39 Kessels Road, Coopers Plains, Queensland 4108, Australia

Introduction

Swimming and other recreational activities with cyanobacterial blooms have frequently been associated with dermatological and other symptoms (El Saadie *et al.* 1995; Pilotto *et al.* 1997). However, apart from the marine cyanobacterium *Lyngbya majuscula* which contains activators of protein kinase C (Sivonen and Jones 1999), toxic fresh water cyanobacteria were not known to contain compounds which were injurious to the skin, and the cause of skin rashes associated with exposure to these blooms was unknown.

In our recent study of epidemiological aspects of dermatological reactions possibly associated with exposure to cyanobacterial blooms in recreational water bodies consideration was given to the possibility that the latter reactions were due to delayed contact hypersensitivity, since they did not appear to be due to a direct irritant effect.

Accordingly, the mouse ear swelling test (MEST; Gad 1994) was used in a preliminary study of the capacity of a culture of *Cylidrospermopsis raciborskii* to induce delayed contact hypersensitivity. An initial step in the MEST was to apply a 75% methanolic suspension of the organism to the shaved, stripped abdominal skin of Balb/C mice for 4 days. This treatment immediately resulted in a marked inflammatory reaction, indicating that the organism contained an agent that was directly irritating to the skin. A toxin found in *C. raciborskii*, namely cylindrospermopsin (Ohtani *et al.* 1992), was isolated and when similarly applied to the mouse skin gave rise to the characteristic inflammatory reaction.

The present report describes the sequence of microscopic changes seen in the abdominal skin of Balb/C mice following a single exposure to cylindospermopsin, as well as the approximate no effect concentration of cylindrospermopsin for dermatological effect.

Materials and Methods

Time course study

Specific pathogen-free Balb/C female mice 8 to 10 weeks of age were housed in groups of six (tests) or three (controls) with free access to food and water. The abdominal skin was shaved using an electric razor and 24 hours later exposed to a 70% ethanolic solution of cylindrospermopsin of 190 μg mL^{-1}. The area of skin exposed was 2 cm^2 and the total volume of solution applied was 70 μL. The solvent was immediately evaporated from the exposed skin area with the assistance of a hair drier. Mice in groups of six were sacrificed for histopathological examination of the exposed skin at 1, 6, 12, 24, 48, 72 and 96 hours and at 5, 7 and 14 days respectively.

The skin of groups of three control mice was treated with 70 μL of 70% ethanol applied to the same skin area of 2 cm^2 and killed for pathological examination at 1, 6, 24 and 72 hours and at 7 and 14 days respectively.

Dose response study

In a subsequent dose response study, groups of six mice were treated with 70 μL of 1, ½, $^{1}/_{5}$, $^{1}/_{10}$, $^{1/}_{25}$, $^{1}/_{50}$ and $^{1}/_{100}$ of the original dose concentration and necropsied for histopathological examination of the skin at 48 hours post dosing.

All skin samples were fixed in 10% buffered neutral formalin, paraffin sections prepared and stained with hematoxylin and eosin.

Results

There were no changes either grossly or microscopically (Figure 1) to the skin of any of the control mice. In the test groups of the time course study, there was hyperemia of the skin surface visible at 24 hours and this had progressed to swelling, erosion and scab formation by 48 hours. These changes gradually subsided over succeeding days but abnormal scabiness and increased skin thickness were still visible at 7 days. By 14 days the skin had returned to normal.

Histologically, changes could be observed first at 12 hours and consisted of dermal edema with epidermal parakeratosis with foci of subcorneal polymorphonuclear infiltration. By 24 hours there was extensive neutrophilic infiltration of the epidermis as well as dermal hyperemia, swelling and moderate inflammatory cell infiltration. By 48 hours there was marked inflammatory infiltration of the dermis and subcutaneous adipose tissues, scab formation involving the surface of the epidermis and focal hyperplasia (Figure 1) with acanthosis.

By 72 hours there was a thick scab and ulceration of the epidermis with granulating surfaces of the dermis. By 4 days the latter epidermal changes had subsided although focal scabs were still present and there was marked fibrotic thickening of the dermis and acanthosis. The changes at 5 and 7 days consisted of a thickened dermis and epidermis with focal hyperkeratosis. In areas of acanthosis at this stage there were multiple apoptotic bodies as well as mitotic figures. The skin had returned to normal by 14 days.

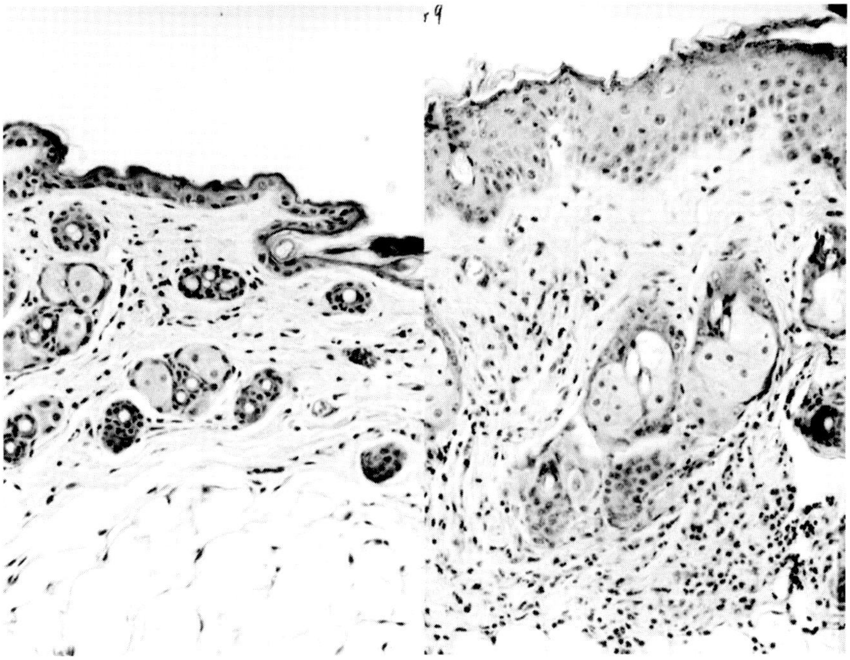

Figure 1. On the left is normal mouse abdominal skin after shaving and stripping. On the right is mouse abdominal skin 2 days after a single application of 190 µg mL^{-1} of cylindrospermopsin dissolved in 70% ethanol, photographed at the same magnification, showing hyperplasia of the epidermis and thickening with inflammatory cell infiltration of the dermis and subcutis.

In the dose response study, changes in the skin due to the single application of cylindrospermopsin at the same concentration were confirmed. Although at a dilution of $^1/_{50}$ of this dose there was little abnormal to be observed grossly on the exposed skin, histologically there was acanthosis, parakeratotic foci on the surface of the thickened epidermis and mild inflammatory cell infiltration of the dermis and subcutaneous adipose tissues. The skin was normal following exposure to $^1/_{100}$ of the concentration of the original dose. This suggests a no effect concentration of the cylindrospermopsin dissolved in 70% ethanol for skin exposure in the mouse of between 2 and 4 µg mL^{-1}.

Discussion and Conclusion

The experiment demonstrates that cylindrospermopsin is likely to be the principal agent capable of causing direct dermatological injury following exposure to blooms of *C. raciborskii*. In addition MEST studies in Balb/C mice have also shown that exposure to the toxin also causes mild delayed contact hypersensitivity (Stewart *et al.* 2005). In effect, the latter authors have also shown that patch testing for delayed hypersensitivity in a dermatology clinic in Brisbane, Queensland resulted in one patient of 39 tested reacting positively to this test.

In a recent review of possible cyanobacterial poisoning of humans in outback Australia in the early days of settlement of the continent by Europeans, Hayman (1992) speculates that the cause of this illness was due to contamination of the available surface waters with *C. raciborskii*. The illness took either of two forms, a hepatoenteric syndrome (Byth 1980) known as 'Barcoo Species' and festering sores of the exposed skin, mainly of the limbs, known as 'Barcoo Rot'. The speculation does not, at this distance in time, seem unreasonable, especially since occasional outbreaks of hepatotoxicity in cattle have been reported in these same areas (Thomas *et al.* 1998) and found to have been associated with contamination of their water supplies with *C. raciborskii*. *Cylindrospermopsis raciborskii*, it seems, has always been an environmental toxicant in tropical Australia!

References

Byth S (1980). Palm Island mystery disease. *Medical Journal of Australia* 2:40–42.

El Saadie O, Esterman AJ, Cameron S, and Roder DM (1995). Murray River water, raised cyanobactrial cell counts, and gastrointestinal and dermatological symptoms. *Medical Journal of Australia* 162:122–125.

Gad SC (1994). The mouse ear swelling test (MEST) in the 1990s. *Toxicology* 93:33–46.

Hayman J (1992). Beyond the Barcoo-probable human tropical cyanobacterial poisoning in outback Australia. *Medical Journal of Australia* 157:794–796.

Ohtani I, Moore RE, and Runnegar MTC (1992). Cylindrospermopsin, a potent hepatotoxin from the blue-green alga *Cylindrospermopsis raciborskii. Journal of the American Chemical Society* 114:7941–7942.

Pilotto LS, Burch MD, Cameron S, Beers M, Rouch GJ, Robinson P, Kirk M, Cowie C, Hardiman S, Moore C, and Attewell RG (1997). Health effects of exposure to cyanobaceria (blue-green algae) during recreational water-related activities. *Australian and New Zealand Journal of Public Health* 21:562–566.

Sivonen K and Jones G (1999). Cyanobacterial toxins. In: *Toxic Cyanobacteria in Water, a Guide to Their Public Health Consequences, Monitoring and Management* (I Chorus and J Bartram, Eds.), pp. 140–141. E & FN Spon, London.

Stewart I, Seawright AA, Robertson I, Schulter PJ, and Shaw GR (2005). Primary irritant and delayed-contact hypersensitivity reactions from the fresh water cyanobacterium *Cylindrospermopsis raciborskii* and its associated toxin cylindrospermopsin (Submitted for publication).

Thomas AD, Saker ML, Norton JH, and Olsen RD (1998). Cyanobacterium *Cylindrospermopsis raciborskii* as a probable cause of death in cattle in northern Queensland. *Australian Veterinary Journal* 76:592–594.

Chapter 88

Experimental Reproduction of *Aspergillus clavatus* Tremorgenic Neurotoxicosis in a Sheep

J.G. Allen[1], G. Doncon[1], and D.G. Wright[2]

[1]*Animal Health Laboratories, Department of Agriculture, Locked Bag No. 4, Bentley Delivery Centre WA 6983, Australia;* [2]*AGWEST Plant Laboratories, Department of Agriculture, Locked Bag No. 4, Bentley Delivery Centre WA 6983, Australia*

Introduction

In recent years the practice of feeding sprouted barley or wheat to livestock has become reasonably widespread in Australia, although not common. The sprouted cereal is germinated and grown in hydroponic systems in sheds, and is harvested for feeding after 6 to 8 days of growth.

A tremorgenic mycotoxicosis from eating sprouted grain infested with *Aspergillus clavatus* has been described in ruminants sporadically in many places in the world, and now also in Australia (El-Hage and Lancaster 2004; McKenzie *et al.* 2004). This paper reports what the authors believe to be the first experimental reproduction in Australia of the tremorgenic neurotoxicosis caused by *A. clavatus.*

History

In spring 2003 a farmer in the southwest of Western Australia lost several dairy cows after feeding them sprouted barley produced on his property using a hydroponic system in a shed. No animals were submitted for autopsy but fodder samples were cultured and *A. clavatus* isolated. The farmer closed down his hydroponic system and spelled it for 3 to 4 months. He then attempted to disinfect the shed and all equipment by spraying and scrubbing with hypochlorite solutions. In April 2004 he started to produce sprouted barley again but by day 6 noticed fungal growth in the fodder and requested testing to determine if it was toxic.

Fodder samples were collected and some submitted for culture and the remainder placed in a cold room at 4–6ºC. *Aspergillus clavatus* and a *Rhizopus* sp. were cultured. The bulk of the visible fungus on the fodder was the *Rhizopus* sp., but the *A. clavatus* could be easily detected dispersed amongst it. Yeast organisms were also present on the roots where the *Rhizopus* sp. was causing breakdown.

Culture and Identification of Fungi

Fungi present on the fodder were individually and aseptically picked off and plated onto potato dextrose agar with aureomycin hydrochloride added. The cultures were incubated at room temperature for 1 week. The fungi were then identified based on morphological characteristics. Characteristics described by Pitt and Hocking (1997) were used to confirm the identification of *A. clavatus*. A specimen of the isolate has been placed in the Mycology Collection of the Department of Agriculture Western Australia, South Perth, Western Australia (Identification Number WAC11329).

Feeding Experiment and Results

A 15-month-old Merino wether (32.0 kg) was kept in an individual pen with *ad libitum* access to water. During the experiment it was provided daily a ration composed of chaff, pellets and lupins in the approximate ratio of 6:10:1. The contaminated fodder stored in the cold room was used in this experiment.

After unsuccessful attempts to get the wether to eat the refrigerated fresh material or air-dried fodder material, it was decided to administer the air-dried fodder by stomach tube. This was achieved by hammer-milling the material to a fine powder, and dosing it as an aqueous slurry. Over a 14-day period a total of 2980 g (approximately 93 g kg^{-1}) of the material was given in ten divided doses.

The sheep maintained a full appetite for only the first 3 days of dosing. Feed intake dropped to about half of the original intake on day 4 and remained at that level until day 10. It then declined further until it ate nothing on days 14 and 15. On day 7 a rumen sample was collected just prior to and 5 hours after drenching the dried fodder to determine if it may be causing an acidosis. The pHs were 6.9 and 6.3, respectively.

Tremors were first noticed on the 7th day when the sheep was handled for drenching. The whole of the animal became mildly rigid and shook while being held. When the sheep was left alone the tremors were not apparent. Over the next week the rigidity of the body and the tremors increased in severity when the sheep was handled. It would sometimes collapse in the hindquarters while being held. When observed from a distance fine tremors over the flanks could be noticed from time to time. When taken from its pen the sheep displayed some ataxia and a stiff gait, but as the walking surface was concrete it was not sure whether this was real or compensatory for the surface. The sheep did occasionally knuckle at the fetlocks.

By day 14 the tremors were present without any stimulation, and when the animal was held for drenching they were very severe, with the whole body going rigid occasionally and the animal collapsing. It was decided not to drench any more of the dried fodder. The sheep appeared no worse the next day, but on day 16 was found to be down in lateral recumbency at the first morning inspection. Both the forelimbs and hindlimbs were shaking continuously and there were tremors over the flanks of the hindquarters. The animal could be stood up but had little strength in its legs, which continued to shake violently, and it would fall to the ground. It seemed that the hindlimbs had less capacity to hold weight. When the animal went down onto the ground it lay for a few seconds in sternal recumbency then all limbs went into extension, forcing the animal over into lateral recumbency again. The sheep was immediately euthanased and a necropsy performed.

At necropsy the liver was slightly paler, and the kidneys slightly smaller, than normal. There were no other gross changes. A full range of tissues was taken for microscopic examination, but significant microscopic changes were restricted to only the central and peripheral nervous systems. The most significant change was substantial swelling and chromatolysis of numerous neurons in the brain stem and the entire length of the spinal cord, particularly in the ventral horns. There was also Wallerian degeneration in the sciatic nerve.

Discussion

The clinical and pathological findings reported here are consistent with those reported elsewhere (McKenzie *et al.* 2004). The main exceptions are that our sheep lost its appetite, becoming completely anorectic, and had Wallerian degeneration in the sciatic nerve.

Inappetence has occurred in Western Australia in animals being fed sprouted barley contaminated with *Rhizopus* spp. (J.G. Allen, unpublished observation). As the material used in the current work was contaminated with a *Rhizopus* sp. this may account for the inappetence.

Changes in peripheral nerves have not previously been reported in this intoxication. However, this finding is not surprising when the amount of degeneration of neurons in the central nervous system is considered. Injury to a neuron cell body inevitably makes its axon unviable, and results in Wallerian-type degeneration of the axon diffusely along its length.

The optimum growth temperature for *A. clavatus* is between 20 and 35°C (Flannigan 1986; Famurewa *et al.* 1994), and accordingly it has been recommended that the temperature in sheds in which sprouted cereals are grown hydroponically should be kept as low as is economically feasible, aiming for below 20°C (McKenzie *et al.* 2004). It is noteworthy that the *A. clavatus* on our contaminated barley sprouts continued to grow prolifically at temperatures of 4–6°C.

The toxins that cause this tremorgenic neurotoxicosis are yet to be identified (McKenzie *et al.* 2004).

References

El-Hage CM and Lancaster MJ (2004). Mycotoxic nervous disease in cattle fed sprouted barley contaminated with *Aspergillus clavatus*. *Australian Veterinary Journal* 82:639–641.

Famurewa O, Oyeda MA, and Olutiola PO (1994). The influence of some physical and nutritional factors on the growth and sporulation of *Aspergillus clavatus*. *Phytopathologica and Entomologica Hungarica* 29:273–282.

Flannigan B (1986). *Aspergillus clavatus* - an allergenic, toxigenic deteriogen of cereals and cereal products. *International Biodeterioration* 22:79–89.

McKenzie RA, Kelly MA, Shivas RG, Gibson JA, Cook PJ, Widderick K, and Guilfoyle AF (2004). *Aspergillus clavatus* tremorgenic neurotoxicosis in cattle fed sprouted grains. *Australian Veterinary Journal* 82:635–638.

Pitt JL and Hocking AD (1997). *Fungi and Food Spoilage*, pp. 371–373. Blackie Academic and Professional, London, United Kingdom.

Chapter 89

Ergot Toxins and Animal Disease in Australia

A.J. Cawdell-Smith[1], W.J.K. Bakau[2], C.J. Scrivener[3], K.F.M. Reed[4], B.J. Blaney[5], and W.L. Bryden[1]

[1]*School of Animal Studies, University of Queensland, Gatton Qld 4343 Australia;* [2]*University of Vudal, East New Britain Province, Papua New Guinea;* [3] *MS and A Agriculture, Armidale NSW 2350 Australia;* [4]*Victorian Department of Primary Industries & Fisheries, Hamilton Vic 3300, Australia;* [5]*Queensland Department of Primary Industries & Fisheries, Yeerongpilly Queensland 4105, Australia*

Introduction

Most of the agriculturally important mycotoxicoses of livestock have been reported in Australia (Bryden 1998) and except for isolated outbreaks have not caused major economic losses to the animal industries. The exception is ergot alkaloids which have caused significant losses in both grazing and grain-fed animals. Within the fungal family Claviciptaceae a number of species of the fungal genus *Claviceps* and the endophyte genus *Neotyphodium* produce an array of ergot toxins. While ergot fungi invade the flowering head of grasses and grain crops, endophytic fungi grow throughout plant tissue and are transmitted in seed. The principal ergot toxins are alkaloids whose effects are very complex and in general result from actions as partial agonists or antagonists at adrenergic, dopaminergic and serotonergic receptors. The cascade of physiological events that follow result in neurohormonal, peripheral and central nervous effects (Gilman *et al.* 1985).

Claviceps spp. and Disease

In Australia, *Claviceps paspali*, *C. purpurea* and *C. africana* are the major *Claviceps* species that cause animal disease. The major alkaloid produced by *Claviceps purpurea* is ergotamine and that by *Claviceps africana* is dihydroergosine which are structurally similar. Ergotamine will cause peripheral vasoconstriction, local ischemia and gangrene in cold conditions but in warm conditions, reduced peripheral circulation results in hyperthermia. It also reduces circulating levels of prolactin (Bryden 1994). In contrast, the major toxin produced by *C. paspali* is a tremorgenic mycotoxin, paspalitrem, which produces a staggers syndrome through neurological disturbances (Cole and Dorner 1986).

Outbreaks of staggers following ingestion of paspalum or Dallis grass (*Paspalum dilatalum*) containing the ergot of *C. paspali* have been recorded in cattle, horses and sheep on the eastern seaboard of Australia for over 100 years. Cases were quite severe during the first half of the 20th century but have been much less frequent since the 1950s due to improved pasture maintenance. Prior to the summer of 1935/1936 there were sporadic problems of paspalum staggers on farms but during that summer there was a major epidemic of paspalum staggers on the coastal areas of New South Wales that extended into southern Queensland (Hurst 1942). Cattle were the species most affected, which reflected the greater number of cattle in this coastal dairying zone, where milk production was disrupted. Sheep and horses were also reported to be affected. Interestingly there are no published descriptions of paspalum staggers in horses despite it being mentioned by numerous authors in papers and text books for the last century (Mantle 1978). Recently Cawdell-Smith (2005) described a case of paspalum staggers in foals where their dams were unaffected. There are also reports of *C. paspali* infecting native grasses including water couch (*Paspalum paspalodes*, which is often mistakenly identified as *P. distichum*; Everist 1981). There is a report of cattle experiencing staggers after grazing ergotised water couch in central Queensland (Noble 1985). The indigenous ergot *Claviceps queenslandica* has been shown to infect *Paspalum orbiculare* but not *P. dilatalum*. Water couch, however, is susceptible to infection by both these ergot species (Langdon 1964). There have been no reports of animal disease resulting from indigenous ergot.

Claviceps purpurea ('rye' ergot) was recorded in Australia over 100 years ago (Barger 1931) and there is one reported case of classical gangrenous ergotism in this country. Fraser and Dorling (1983) described bilateral hind limb lameness and gangrene in two Friesian heifers that had consumed meadow hay containing ryegrass infected with the ergots of the *C. purpurea*. The outbreak occurred in Western Australia during July (winter). In the summer of 1986 (January–March) there was a severe outbreak of a disease characterized by hyperthermia, respiratory distress and loss of production that became known as bovine hyperthermia (Jessup *et al.* 1987) and was shown to be due to the ingestion of the ergots of *C. purpurea* (Burgess *et al.* 1986; Ross *et al.* 1989). The disease was characterized by high rectal temperatures (41° to 42°C), excessive salivation, inappetence, increased respiratory rate and animals seeking shade or water to reduce heat load. Wheat, oats, barley and lupins grown in the central west of New South Wales had been contaminated with ergot from annual ryegrass growing in the cereal crops. The contaminated grain was fed to cattle (some 10,000 animals) at two dairies and 12 feedlots in that region and two dairies in the coastal Illawarra region. In all cases there were no mortalities but morbidity reached 100% in some instances. The overall effect on feedlot cattle was less marked than the dairy cattle, which suffered up to 75% loss of production with early drying off. Since that time there have been a number of feedlot cases in the central west of New South Wales (Bourke *et al.* 2000) and one case in a dairy herd in the Upper Hunter region of New South Wales (Cawdell-Smith 2005). Peet *et al.* (1991) reported hyperthermia and death of feedlot cattle consuming concentrate diets contaminated with the ergots of *C. purpurea*. Lethal episodes following hyperthermia were also noted by Bourke *et al.* (2000), who suggested that the concentration of other compounds in the ergots of *C. purpurea*, possibly ergochromes, the main pigments in the ergots, may have contributed to the death of animals. There is also a report of hyperthermia in cattle grazing annual ryegrass infected with *C. purpurea* (Scrivener and Bryden 1993).

The toxicity of 'rye' ergot has been known for centuries (Barger 1931), but the toxicity of sorghum ergot (*C. africana*) was considered innocuous despite it being widespread in Africa and Asia. It was detected in Australia in 1996 and infects only sorghum and related plants like Johnson grass (Ryley *et al.* 1996). Pigs and dairy cows were poisoned on several farms about 18 months later in central Queensland (Blaney *et al.* 2000b). Sorghum ergot sclerotia have total alkaloid contents ranging from about 200 to 8000 mg kg^{-1} (0.8%), with dihydroergosine constituting >80% of the total (Blaney *et al.* 2003). The 9,10 dihydro-derivatives of rye ergot alkaloids had been considered as far less toxic than the parent compounds, and in some cases to counteract their effects. However, extensive studies of toxicity of dihydroergosine in sorghum ergot indicate a similar effect in depressing prolactin as rye ergot alkaloids (Blaney *et al.* 2000a). Lot-fed cattle are severely affected by sorghum ergot alkaloids above 1 mg dihydroergosine kg^{-1} of feed (about 0.3% ergot) (Blaney *et al.* 2001). After several feedlots in southern Queensland suffered severe production losses in 2003 from sorghum containing up to 3% ergot, the maximum concentration of sorghum ergot permitted in grain delivered to feedlots was reduced under contracts set by the National Association of Commodity Marketing Agencies to 0.1%. Government regulations specify 0.3% sorghum ergot compared to 0.02% rye ergot. While the profile of alkaloids has appeared similar in all samples of sorghum ergot so far assayed, sclerotia do vary greatly in total alkaloid content. This makes it important to relate effects on livestock to actual alkaloid concentrations. Reverse phase HPLC with fluorescence detection is an effective assay method (Blaney *et al.* 2003), and a sensitive immunoassay (ELISA) has also been developed for dihydroergotamine (Molloy *et al.* 2003).

Neotyphodium spp. and Disease

Endophyte infection of perennial ryegrass pastures in southern Australia can result in significant outbreaks of perennial ryegrass staggers or toxicosis in grazing animals: sheep, cattle, horses (Culvenor 1974) and deer (Mitchell and McCaughan 1992). *Neotyphodium lolii*, the endophyte which induces staggers, produces the tremorgenic mycotoxin, lolitrem B, which is similar in structure to paspalitrem (Cole and Dorner 1986). The endophyte also produces various ergot alkaloids including ergovaline but their role in perennial ryegrass toxicosis requires study. In North America ergovaline has been considered the major cause of syndromes (summer syndrome, fescue foot) related to the ingestion of tall fescue infected with *Neotyphodium coenophialum*. Lysergic acid amides, however, may be the main toxin (Hill 2005). Fescue foot occurred in Australia 50 years ago (Culvenor 1974) but fortunately the presence of wild *N. coenophialum* is rare and no further incidences of the disease have been reported.

Perennial ryegrass was first introduced into Australia in the early 19th century and has become the dominant improved pasture species in much of southeastern Australia. Although staggers has been associated with the grass sporadically over 100 years (Culvenor 1974), the disease was considered merely an inconvenience until 1986, when a major outbreak occurred. In that epidemic, ryegrass staggers affected 60–70% of all grazing properties in southern Victoria (Napthine and McLeod 1987) and resulted in large economic losses. Further serious losses were recorded in 1993, 2002 and 2005. In 2002 an estimated 100–200 thousand sheep and 500 cattle died (Reed *et al.* 2004). Ryegrass staggers occurs in late summer and autumn and although a large proportion in a flock or herd may show signs of intoxication (up to 80% in sheep), mortality rates are generally low. Most deaths that do occur are due to drowning or other misadventure during tetanic spasms (Foot *et al.* 1994). In epidemic years there has been catastrophic mortality in sheep flocks. High levels of ergovaline in addition to lolitrem B have been found in ryegrass and feces (Reed *et al.* 2004). In these circumstances ergot alkaloids are considered to result in heat stress which can become excessive, especially for animals bearing a high heat load. Most animals collapse in the field but some crowd into water; mass drownings of up to 400 sheep have been observed (Reed *et al.* 2005). Further, surviving ewes have recorded a high mortality of newborn lambs following the failure of lactogenesis resulting from hypoprolactinemia. In some episodes sheep growth has been reduced and fecal moisture has increased, necessitating crutching (Foot *et al.* 1994). It is likely that the production and health of dairy cows consuming wild endophyte-infected perennial ryegrass, pasture or silage is also negatively affected (Valentine *et al.* 1993; Lean 2001). Perennial ryegrass toxicosis has a most significant economic impact on grazing industries (Reed *et al.* 2005).

Conclusion

While Australia has been spared many of the mycotoxin problems that beset other countries, particularly those regions with high rainfall, this is more a matter of degree than of kind. In epidemic years the economic impact of ergot toxins is substantial but there is insufficient information to estimate the impact of the subclinical effects of these toxins (Bryden 2005) as there are sporadic cases each year. Many large areas within the 6 million hectares of naturalized perennial ryegrass containing the wild endophyte, *N. lolii*, have recently suffered major stock losses associated with mycotoxins. The recurring sub-clinical production losses associated with such grass may represent a significant annual cost to grazing industries (Reed *et al.* 2005). The various toxins produced by fungi in the family Claviciptaceae are a significant cause of animal disease in Australia. The widespread occurrence of these fungi and the variety of plant species which they can infect indicates they are likely to remain a significant problem for livestock.

References

Barger G (1931). *Ergot and Ergotism*. Guerney and Jackson, London.

Blaney BJ, Kopinski JS, Magee MH, McKenzie RA, Blight GW, Maryam R, and Downing JA (2000a). Blood prolactin depression in growing pigs fed sorghum ergot (*Claviceps africana*). *Australian Journal of Agricultural Research* 51:785–791.

Blaney BJ, McKenzie RA, Walters JR, Taylor LF, Bewg WS, Ryley MJ, and Maryam R (2000b). Sorghum ergot associated with agalactia and feed refusal in pigs and dairy cattle. *Australian Veterinary Journal* 78:102–107.

Blaney BJ, Kopinski JS, Murray, S-A, McLennan SR, Moss RJ, Downing JA, and Dingle JG (2001). Research on the toxicity of sorghum ergot and its alkaloids. In *Proceedings of the Fourth Australian Sorghum Conference* (AK Borrell and RG Henzell, eds), CD-rom format. Kooralbyn. Range Media Pty Ltd ISBN: 0-7242-2163-8

Blaney BJ, Maryam R, Murray S-A, and Ryley MJ (2003). Alkaloids of the sorghum ergot pathogen (*Claviceps africana*): assay methods for grain and feed and variation between sclerotia/sphacelia. *Australian Journal of Agricultural Research* 54:167–175.

Bourke CA, Bailey GD, and Kemp JB (2000). The case for solar radiation being more significant than ambient temperature in producing lethal hyperthermic ergotism in cattle. *Australian Veterinary Journal* 78:618–621.

Bryden WL (1994). The many guises of ergotism. In *Plant-Associated Toxins: Agricultural, Phytochemical and Ecological Aspects* (SM Colegate and PR Dorling, eds), pp. 381–386. CABI Publishing, Wallingford, Oxon, UK.

Bryden WL (1998). Mycotoxin contamination of Australian pastures and feedstuffs. In *Toxic Plants and Other Natural Toxicants* (T Garland and AC Barr, eds), pp. 464–468. CABI Publishing, Wallingford, Oxon, UK.

Bryden WL (2005). Mycotoxins and animal production: Insidious problems associated with contaminated feedstuffs. *Asian-Australasian Journal of Animal Sciences* (in press).

Burgess LW, Bryden WL, Jessup TM, Scrivener CJ, and Barrow KD (1986). Role for ergot alkaloids in bovine hyperthermia. *Proceedings of the Nutrition Society of Australia* 11:120.

Cawdell-Smith AJ (2005). Suspect mycotoxicoses in horses, cattle and dogs. *Proceedings Nutrition Society of Australia* (in press).

Cole RJ and Dorner JW (1986). Role of fungal tremorgens in animal disease. In *Mycotoxins and Phycotoxins* (PS Steyn and R Vleggaar, eds), pp. 501–511. Elsevier Scientific Publishing Co., Amsterdam, The Netherlands.

Culvenor CCJ (1974). The hazard from toxigenic fungi in Australia. *Australian Veterinary Journal* 50:69–78.

Everist SL (1981). *Poisonous Plants of Australia*, 2nd Edition. Angus & Robinson, Sydney.

Foot JZ, Woodburn OJ, Walsh JR, and Heazlewood PG (1994). Responses in grazing sheep to toxins from perennial ryegrass/endophyte associations. In *Plant-associated Toxins: Agricultural, Phytochemical and Ecological Aspects*. (SM Colegate and PR Dorling, eds), pp. 375–380. CABI Publishing, Wallingford, Oxon, UK.

Fraser DM and Dorling PR (1983). Suspected ergotism in two heifers. *Australian Veterinary Journal* 60:303–305.

Gilman AG, Goodman LS, Rall TW, and Murad F (1985). *Physiological Basis of Therapeutics*, 7:931–940. MacMillan, New York.

Hill, N (2005). Contending with fescue toxicosis in the US. In *Perennial Ryegrass Toxicosis in Australia* (KFM Reed, SW Page, and IJ Lean, eds), pp. 58–63. Meat and Livestock, Australia.

Hurst E (1942). *The Poison Plants of New South Wales*, pp.425–426. New South Wales Poison Plants Committee, Sydney.

Jessup TW, Dent CHR, Kemp JB, Christie BM, Ahrens PJ, Burgess LW, and Bryden WL (1987). Bovine idiopathic hyperthermia. *Australian Veterinary Journal* 64:353–354.

Langdon RFN (1964). Paspalum ergot in Australia. *Australian Journal of Science* 26:55.

Lean IJ (2001). Association between feeding perennial ryegrass (*Lolium perenne* cultivar Grasslands Impact) containing high concentrations of ergovaline, and health and productivity in a herd of lactating dairy cows. *Australian Veterinary Journal* 79:262–264.

Mantle PG (1978). Ergotism in horses. In *Mycotoxigenic Fungi, Mycotoxins, Mycotoxicoses* (TD Wyllie and LG Morehouse, eds), 2:185. Marcel Dekker Inc., New York.

Mitchell PJ and McCaughan CJ (1992). Perennial ryegrass staggers in fallow deer (*Dama lama*). *Australian Veterinary Journal* 69:258–259.

Molloy JB, Moore CJ, Bruyers AG, Murray SA, and Blaney BJ (2003). Determination of dihydroergosine in sorghum ergot using an immunoassay. *Journal of Agricultural and Food Chemistry* 51**:**3916–3919.

Napthine DV and McLeod IK (1987). Perennial ryegrass staggers – 1986 Victorian outbreak. *Advances in Veterinary Science 1987*, pp. 107–109. Australian Veterinary Association.

Noble JW (1985). Suspected poisoning of cattle by *Claviceps* spp on water couch. *Australian Veterinary Journal* 62:432–433.

Peet RL, McCarthy MR, and Barbetti MJ (1991). Hyperthermia and death in feedlot cattle associated with the ingestion of *Claviceps purpurea*. *Australian Veterinary Journal* 68:121.

Reed KFM, Scrivener CJ, Rainsford KA, and Walker LV (2004). *Neotyphodium* research and application in Australia. In *Neotyphodium in Cool-season Grasses* (CA Roberts, CP West and DE Spiers, eds), pp. 43–54. Blackwell Publishing Professional, Ames, Iowa.

Reed KFM, Page SW, and Lean IJ (eds) (2005). *Perennial Ryegrass Toxicosis in Australia.* Meat and Livestock Australia. North Sydney.

Ross AD, Bryden WL, Bakau W, and Burgess LW (1989). Induction of heat stress in beef cattle by feeding the ergots of *Claviceps puepurea*. *Australian Veterinary Journal* 66:247–249.

Ryley MJ, Alcorn JL, Kochman JK, Kong GA, and Thompson SM. (1996). Ergot on sorghum spp in Australia. *Australasian Plant Pathology* 25:214.

Scrivener CJ and Bryden WL (1993). Hyperthermia in cattle grazing annual ryegrass. *New Zealand Veterinary Journal* 41:215.

Valentine SC, Bartsch BD, and Carroll PD (1993). Production and composition of milk by dairy cattle grazing high and low endophyte cultivars of perennial ryegrass. In *Proceedings of 2nd International Symposium on Acremonium/Grass Interactions* (DE Hume, GC Latch, and HS Easton, eds), pp. 138–141. AgResearch, Palmerston North, New Zealand.

Chapter 90

Mycotoxins and Abnormal Embryonic Development

A.J. Cawdell-Smith[1], M.J. Edwards[2], and W.L. Bryden[1]

[1] *School of Animal Studies, University of Queensland, Gatton QLD 4343 Australia;*
[2]*Faculty of Veterinary Science, University of Sydney, Camden NSW 2006 Australia*

Introduction

Mycotoxins are secondary fungal metabolites that have become prominent over the last 40 years because of their toxicity in both man and animals especially following the discovery of the carcinogenic properties of aflatoxin. Generally, researchers have focused on the acute toxic effects of mycotoxins in relation to animal production and have explored possible carcinogenic effects when assessing health risks to human populations (Bryden 2005). The subtle effects of these toxins on embryonic development have been examined much less frequently. In this chapter the effects of mycotoxins on embryonic and fetal development, especially those toxins that are global food and feed contaminants, are reviewed. Space does not permit a complete evaluation of the literature on the teratogenicity of mycotoxins; reviews of early studies (Hayes 1981; Terao and Ohtsubo 1991) have been published.

Developmental Toxicity of Mycotoxins

There are thousands of fungal metabolites but for most, their occurrence in nature is poorly defined and the toxigenicity of relatively few has been adequately described. Table 1 lists the mycotoxins that have been screened for teratogenicity and of these aflatoxin, ochratoxin, ergot alkaloids, fumonisins, deoxynivalenol (DON) and zearalenone are of global concern.

Table 1. Mycotoxins that have been shown to be teratogenic and/or embryotoxic in different animal bioassays.

aflatoxins	ergosine	palmotoxins
alternariol	ergotamine	patulin
alternariol monmethyl ether	fumonisins	penicillic acid
chaetochromin	fusaproliferin	penitrem A
chaetoglobosin A	fusarenon - X	phomopsins
citrinin	griseofulvin	rubratoxin
cyclopiazonic acid	kojic acid	secalonic acid A, D
cytochalosin A, B, C, D	lysergic acid diethylamide	stachybotryotoxins
deoxynivalenol	mevinolin	sterigmatocystin
ergocryptine	mycophenolic acid	T-2 toxin
	ochratoxin A	tetramic acid
		viriditoxin
		zearalenone

Aflatoxins

Aflatoxins produced by *Aspergillus flavus* and *A. parasiticus* have been reported to be both teratogenic and embryotoxic. Verrett *et al.* (1964), Legator (1969) and Bassir and Adekunle (1970) all reported the occurrence of malformations in chick embryos. Administration of small doses of aflatoxin at the early stages of development was shown to cause much more significant abnormalities than larger doses at later stages of development. LeBreton *et al.* (1964) were the first group to demonstrate mammalian teratogenicity by inducing fetal growth retardation following administration of repeated small doses of aflatoxin to rats. The severity of teratogenicity in the hamster was shown to be dependent on the period of gestation during which the hamsters were treated with aflatoxin (Elis and DiPaolo 1967). Abnormalities ranged from fetal death, resorption and malformed but live fetuses when aflatoxin was administered during organogenesis. Elis and DiPaolo (1967) also showed that if aflatoxin was given when organogenesis was nearly complete, no malformed fetuses and only a few dead or resorbed fetuses were present. DiPaolo *et al.* (1967) showed in rats that higher dose rates did not cause a different range of defects but resulted in more embryos being affected; however, they were unable to produce malformations in C3H mice. Arora *et al.* (1981) were able to induce fetal anomalies in CBA mice after a single oral dose of aflatoxin B1.

Ochratoxins

The ochratoxins are produced by a number of *Aspergillus* species of which the most important is *Aspergillus ochraceus* and also *Penicillium verrucosum* (originally identified as *P. viridicatum*). Hayes *et al.* (1974) administered ochratoxin A and increased prenatal mortality and induced a variety of gross malformations including craniofacial defects and skeletal malformations in CD-1 mice. Brown *et al.* (1976) produced similar results in Sprague-Dawley rats. In another study by Hayes *et al.* (1974), a wide spectrum of lesions and fetal death and/or resorptions were induced in hamsters. Gilani *et al.* (1978) induced a range of abnormalities in chick embryos by injecting ochratoxin A into the air sac of eggs. Studies by Poppe *et al.* (1983) implicated ochratoxin A as a cause of developmental delay in the offspring of mice. Dietary protein level and composition may alter ochratoxin A-induced teratogenesis (Mayura *et al.* 1983,1984).

Ergot alkaloids

Ergot alkaloids are produced on cereal grains by *Claviceps* spp. or by endophytic fungi (*Neotyphodium* spp.) in tall fescue and perennial rye grass (Bryden 1994; Cawdell-Smith *et al*., Chapter 89, this volume). Ergot alkaloids have long been associated with reproductive failure in herbivorous mammals and man. Attempts at inducing birth defects by dosing with these toxins have produced conflicting results (Hayes 1981). Apparent teratogenic effects of the ergot alkaloid toxins have been difficult to differentiate from the impairment of embryonic nutrition as a result of impaired uterine bloodflow and resultant hypoxia (Franklin and Brent 1964; Senger *et al.* 1967). A wide range of clinical manifestations of reproductive failure have been observed in horses and cattle grazing endophyte-infected tall fescue. In the horse, most of these have been related to the effects of ergovaline on prolactin secretion and subsequent interference with normal parturition in the mare, effects on the placenta, reduced foal viability and agalactia. These effects are not suggestive of a teratogenic impact, but rather more insidious consequences of endophyte ingestion on luteal function, pregnancy rate and early embryonic deaths (Brendemuehl *et al.* 1994*)*. The increase in early embryonic loss may be due to a teratogenic effect occurring during organogenesis (Putnam *et al.* 1991).

Fusarium toxins

Fumonisins, deoxynivalenol and zearalenone are produced by *Fusarium* spp. and occur widely in nature. Fumonisins accumulate in maize and are produced predominantly by *Fusarium verticillioides* (formally *F. moniliforme*) and *F. proliferatim* (Marasas *et al.* 2001). Equine leucoencephalomalacia and porcine pulmonary edema are diseases that result from fumonisin ingestion. There is epidemiological evidence that fumonisins can cause neural tube defects in humans. This relates to a cluster of cases of anencephaly occurring among Mexican-American women who conceived in the Lower Rio Grande Valley during 1990 (Kendricks 1999), increased neural tube defects in South Africa (Viljoen *et al.* 1995), China (Moore *et al.* 1997) and in Guatemala (Voss *et al.* 2006). The experimental evidence has been less convincing. Extensive work has been carried out in an attempt to induce teratogenicity and embryotoxicity in a number of species. Javed *et al.* (1993) induced embryopathic and embryocidal effects in chick embryos by injection of fumonisin B1 into eggs. Floss *et al.* (1994a,b) and Penner *et al.* (1998) demonstrated reduced survivability of fetal hamsters in the absence of maternal toxicity. Voss *et al.* (1996) and Reddy *et al.* (1996) suggested that the developmental effects of fumonisins are mediated through maternal toxicity. Recent studies, however, have induced neural tube defects in rat embryos of mothers exposed to fumonisins without dam toxicity (Voss *et al.* 2006). The mechanism of fumonisin action will be complex (Voss *et al.* 2006) and may involve impaired folate metabolism as fumonisins inhibit folate transport (Stevens and Tang 1997). Folate deficiency of the dam is a known teratogen causing spinal tube defects in humans (Werler *et al.* 1993). Fumonisins also disrupt apoptosis (Riley 1998).

Deoxynivalenol, a trichothecene, causes feed refusal and vomiting in swine. It appears to be non-teratogenic in rats (Morrisey 1984; Morrisey and Vesonder 1985) but Khera *et al.* (1982) observed multiple teratogenic effects in mice. Zearalenone is a non-steroidal estrogen which elicits significant estrogenic activity in animals especially pigs (Hagler *et al.* 2001). Studies with pregnant rats have shown that zearalenone increases the number of resorptions and stillborns (Bailey *et al.* 1976). Ruddick *et al.* (1976) dosed pregnant rats orally with zearalenone and induced decreased fetal weights and an increased incidence of skeletal anomalies.

Implications and Future Research

Embryogenesis is a complex process of cell proliferation, differentiation, migration and organogenesis. Apoptosis is also important in embryogenesis. A toxin that disrupts any of the above processes is a potential teratogen or embryotoxic agent. Mycotoxins vary widely in their chemical structure and it would be almost impossible to speculate about their biological effects, including teratogenicity, on the basis of chemical structure alone. The biochemical lesions and cellular disruption following mycotoxin ingestion have been delineated for some toxins (Riley 1998). In many instances the effects will be unpredictable as toxicity will depend on many factors, including dosage, period of exposure, species, nutrients and feed additives and other co-occurring toxins and environmental stressors (Bryden 1982).

A number of mycotoxins have been recognized as teratogens or embryotoxic agents (Table 1). Much of the work that has been done with mycotoxins has been simple screening for birth defects. With some mycotoxins, a number of species have been used as the animal model and considerable variation between species has been observed. Other recognized teratogenic agents also show interspecies variation. Thalidomide is a teratogen in humans but not rats or mice (Lenz 1988). Some mycotoxins that have been shown to be teratogenic experimentally have required the toxin to be administered by a route other than ingestion (e.g. intraperitoneal injection) as they are poorly absorbed. The relevance of teratogenicity under these circumstances is questionable. However, it is important to consider other circumstances that may increase gut absorption and thereby cause birth defects following exposure to these agents. Studies of hyperthermia have suggested that effects of severe heat exhaustion may be the result of a break down in the physical barrier of the small intestinal wall, allowing intralumenal endotoxins to escape into the blood stream (Cronje 2005). This may also apply to some mycotoxins that are not normally well absorbed. Ergot alkaloids are known to cause hyperthermia (Bryden 1994), and hyperthermia is a teratogen (Edwards *et al.* 2003). It is possible that the two may act synergistically to cause teratogenic effects at levels which, when given separately, may not reach the teratogenic threshold. The induction of neural tube defects in a guinea pig model with hyperthermia (Cawdell-Smith *et al.* 1992) will allow the testing of interactions with other teratogens such as fumonisin, which is poorly absorbed but causes neural tube defects, and ergot alkaloids, which cause hyperthermia.

Many fungal species produce more than one toxic secondary metabolite and multiple metabolites may accumulate in a feedstuff through biosynthesis of several compounds by a single fungal species or different species. In field situations animals may therefore be subjected to low levels of a number of toxins, *viz.* a 'cocktail' of toxins. Interaction studies with ochratoxin A and aflatoxin (Wangikar *et al.* 2004) demonstrate an apparent reduction in teratogenicity but a concommitent increase in embryotoxicity. Administration of ochratoxin A with T-2 toxin increased the incidence of T-2 induced malformations (Hood *et al.* 1978). Conversely, administration of ochratoxin A with zearalenone resulted in a reduction in the number of fetotoxic and teratogenic effects (Arora *et al.* 1983). Bacon *et al.* (1995) found that co-injection of fusaric acid and fumonisin B1 resulted in a synergistic toxic response in chicken embryos. Many of these interactions may result in reduced reproductive efficiency by causing not only increased teratogenic effects but also increased embryotoxic and fetotoxic effects. These studies further highlight the difficulty of extrapolating laboratory studies with single toxins to field intoxications.

The most dramatic manifestation of teratogenesis is the birth of grossly deformed offspring but it must be recognized that any agent that causes teratogenesis has the potential to cause embryotoxicity and fetotoxicity. Embryotoxicity is much more insidious because in the field it may only present as early embryonic loss and may appear as a prolonged parturition to conception interval. Without the retrieval of the embryo it is impossible to determine the cause of embryonic loss in the early stages of pregnancy. In some laboratory species embryonic defects have only been detected by harvesting embryos early in pregnancy following the teratogenic insult as many effected embryos are resorbed as the pregnancy proceeds. A better understanding of the mechanisms of action and epidemiology of mycotoxins will help to delineate their teratogenicity and potential interactions with other teratogenic agents.

References

Arora RG, Frolen H, and Nilsson A (1981). Interference of mycotoxins with prenatal development of the mouse. *Acta Veterinaria Scandinavia* 22:534.

Arora RG, Frolen H, and Fellner-Feldegg H (1983). Inhibition of ochratoxin A teratogenesis by zearalenone and diethylstilbestrol. *Food and Chemical Toxicology* 21:779–784.

Bacon CW, Porter JK, and Norred WP (1995). Toxic interaction of fumonisin B1 and fusaric acid measured by injection into fertile chicken egg. *Mycopathologia* 129:29–35.

Bailey DE, Cox GE, Morgareidge K, and Taylor J (1976). Acute and subchronic toxicity of zearalenone in the rat. *Toxicology and Applied Pharmacology* 37:144–149.

Bassir O and Adekunle A (1970). Teratogenic action of aflatoxin B, palmotoxin B, and palmotoxin C, in the chick embryo. *Journal of Pathology* 102:49–55.

Brendemuehl JP, Boosinger TR, Pugh DG, and Shelby RA (1994). Influence of endophyte-infected tall fescue on cyclicity, pregnancy rate and early embryonic loss in the mare. *Theriogenology* 42:489–494.

Brown MH, Szczech GM, and Purmalis BP (1976). Teratogenic and toxic effects of ochratoxin A in rats. *Toxicology and Applied Pharmacology* 37:331–338.

Bryden WL (1982). Aflatoxin and animal production: an Australian perspective. *Food Technology in Australia* 34:216–223.

Bryden WL (1994). The many guises of ergotism. In *Plant-Associated Toxins: Agricultural, Phytochemical and Ecological Aspects* (SM Colegate and PR Dorling, eds), pp. 381–386. CABI Publishing, Wallingford, Oxon, UK.

Bryden WL (2005). Mycotoxins and animal production: Insidious problems associated with contaminated feedstuffs. *Asian-Australasian Journal of Animal Sciences* in press.

Cawdell-Smith J, Upfold J, Edwards M, and Smith M (1992). Neural tube and other developmental anomalies in the guinea pig following maternal hyperthermia during early neural tube development. *Teratogenesis, Carcinogenesis, Mutagenesis* 12:1–9.

Cronje PB (2005). Heat stress in livestock—the role of the gut in its aetiology and a potential role of betaine in its alleviation. *Recent Advances in Animal Nutrition in Australia* 15:107–122.

DiPaolo JA, Elis J, and Erwin H (1967). Teratogenic response by hamsters, rats and mice to aflatoxinB_1. *Nature (London)* 215:638–639.

Edwards MJ, Saunders RD, and Shiota K (2003). Effects of heat on embryos and fetuses. *International Journal of Hyperthermia* 19:295–324.

Elis J and DiPaolo JA (1967). Aflatoxin B_1 induction of malformations. *Archives of Pathology* 83:53–59.

Floss JL, Casteel SW, Johnson GC, and Rottinghaus GE (1994a). Developmental toxicity in hamsters of an aqueous extract of *Fusarium moniliforme* culture material containing known quantities of fumonisin B_1. *Veterinary and Human Toxicology* 36:5–10.

Floss JL, Casteel SW, Johnson GC, Rottinghaus GE, and Krause GF (1994b). Developmental toxicity of fumonisin in Syrian hamsters. *Mycopathologia* 128:33–38.

Franklin JB and Brent RL (1964). The effect of uterine vascular clamping on the development of rat embryos three to fourteen days old. *Journal of Morphology* 115:273–280.

Gilani SH, Bancroft T, and Reily M (1978). Teratogenicity of ochratoxin A in chick embryos. *Toxicology and Applied Pharmacology* 46:543–550.

Hagler Jr WM, Towers NR, Mirocha CJ, Eppley RM, and Bryden WL (2001). Zearalenone: Mycotoxin or mycoestrogen? In *Fusarium: Paul E. Nelson Memorial Symposium* (BA Summerell, JF Leslie, D Backhouse, WL Bryden, and LW Burgess, eds), pp. 321–331. APS Press, St Paul, Minnesota.

Hayes AW (1981). *Mycotoxin Teratogenicity and Mutagenicity*. CRC Press, Boca Raton, Florida.

Hayes AW, Hood RD, and Lee HL (1974). Teratogenic effects of ochratoxin A in mice. *Teratology* 9:93–99.

Hood RD, Kuczuk MH, and Szczech M (1978). Effects in mice of simultaneous prenatal exposure to ochratoxin A and T-2 toxin. *Teratology* 17:25–32.

Javed T, Richard JL, Bennett GA, Dombrink-Kurtzman MA, Bunte RM, Koelkebeck KW, Cote LM, Leeper RW, and Buck WB (1993). Embryopathic and embryocidal effects of purified fumonisin B1 or *Fusarium proliferatum* culture material extract on chicken embryos. *Mycopathologia* 123:185–193.

Kendricks K (1999). Fumonisins and neural tube defects in South Texas. *Epidemiology* 10:198–200.

Khera KS, Whalen C, Angers G, Vesonder RF, and Kuiper-Goodman T (1982). Embryotoxicity of 4-deoxynivalenol (vomitoxin) in mice. *Bulletin of Environmental Contamination and Toxicology* 29:487–493.

LeBreton E, Frayssinet C, Lafarge C, and deRecondo AM (1964). Aflatoxine-mecanisme de l'action. *Food Cosmetics* 2:675–680.

Legator MS (1969). Biological assay for aflatoxins. In *Aflatoxins* (LA Goldblatt, ed), pp. 107–121. Academic Press, New York.

Lenz W (1988). A short history of thalidomide embryopathy. *Teratology* 38:203–215.

Marasas WFO, Miller JD, Riley RT, and Viscontini A (2001). Fumonisins - Occurrence, toxicology, metabolism and risk assessment. In: *Fusarium: Paul E. Nelson Memorial Symposium* (BA Summerell, JF Leslie, D Backhouse, WL Bryden and LW Burgess, eds), pp. 332–359. APS Press, St Paul, Minnesota.

Mayura K, Hayes AW, and Berndt WO (1983). Effects of dietary protein on teratogenicity of ochratoxin A in rats. *Teratology* 27:147–153.

Mayura K, Parker R, Berndt WO, and Phillips TD (1984). Ochratoxin A-induced teratogenesis in rats: partial protection by phenylalanine. *Applied and Environmental Microbiology* 48:1186–1192.

Moore CA, Li S, Li Z, Hong S, Gu H, and Berry RJ (1997). Elevated rates of severe neural tube defects in a high-prevalence area in northern China. *American Journal of Medical Genetics* 73:113–118.

Morrisey RE (1984). Teratological study of Fischer rats fed a diet containing added vomitoxin. *Food and Chemical Toxicology* 22:453–460.

Morrisey RE and Vesonder RF (1985). Effect of deoxynivalenol (vomitoxin) on fertility, pregnancy and postnatal development of Sprague-Dawley rats. *Applied and Environmental Microbiology* 49:1062–1069.

Penner JD, Casteel SW, Pittman L Jr, Rottinghause GE, and Wyatt RD (1998). Developmental toxicity of purified fumonisin B1 in pregnant Syrian hamsters. *Journal of Applied Toxicology* 18:197–203.

Poppe SM, Stuckhardt JL, and Szczech GM (1983). Postnatal behavioral effects of ochratoxin A in offspring of treated mice. *Teratology* 27:293–300.

Putnam MR, Bransby DI, Schumacher J, Boosinger TR, Bush L, Shelby RA, Vaughan JT, Ball D, and Bendemuehl P (1991). Effects of the fungal endophyte *Acremonium coenophialum* in fescue on pregnant mare and foal viability. *American Journal of Veterinary Research* 52:2071–2074.

Reddy RV, Johnson G, Rottinghaus GE, Casteel SW, and Reddy CS (1996). Developmental effects of fumonisin B-1 in mice. *Myopathologia* 134:161–166.

Riley RT (1998). Mechanistic interactions of mycotoxins: Theoretical considerations. In *Mycotoxins in Agriculture and Food Safety* (KK Sinha and D Bhatnagar, eds), pp. 227–253. Marcel Dekker, Inc., New York.

Ruddick JA, Scott PM, and Harwig J (1976). Teratological evaluation of zearalenone administered orally to the rat. *Bulletin of Environmental Contamination and Toxicology* 15:678–683.

Senger PL, Lose ED, and Ulberg LC (1967). Reduced blood supply to the uterus as a cause of early embryonic death in the mouse. *Journal of Experimental Zoology* 165:335–341.

Stevens VI and Tang J (1997). Fumonisin B_1-induced sphingolipid depletion inhibits vitamin uptake via the glycosylphosphatidylinositol-anchored folate receptor. *Journal of Biological Chemistry* 272:18020–18025.

Terao K and Ohtsubo K (1991). Biological activities of mycotoxins: Field and experimental mycotoxicoses. In *Mycotoxins and Animal Foods* (JE Smith and RS Henderson, eds), pp. 227–253. CRC Press Inc., London.

Verrett JM, Marliac JP, and McLaughlin J Jr (1964). The use of chicken embryo in the assay of aflatoxin toxicity. *Journal of the Association of Official Agricultural Chemists* 47:1003–1009.

Viljoen DL, Buccimazza S, Dunne T, and Molteno C (1995). The prevalence and prevention of neural tube defects in Cape Town. *South African Medical Journal* 85:630–632.

Voss KA, Bacon CW, Norred WP, Chapin RE, Chamberlain WJ, Plattner RD, and Meredith FI (1996). Studies on the reproductive effects of *Fusarium moniliforme* culture material in rats and the biodistribution of [14C] fumonisin B_1 in rats and mice. *Natural Toxins* 4:24–33.

Voss KA, Gelineau-van Waes JB, and Riley RT (2006). Fumonisins: Current research trends in developmental toxicology. *Mycotoxin Research* 22(1):61–69.

Wangikar PB, Dwivedi P, and Sinha Neeraj (2004). Effect in rats of simultaneous prenatal exposure to ochratoxin A and aflatoxin B1. I. Maternal toxicity and fetal malformations. *Birth Defects Research* 71:343–351.

Werler MM, Shapiro SA, and Mitchell AA (1993). Periconceptional folic acid exposure risk of occurrent neural tube defects. *Journal of the American Medical Association* 269:1257–1261.

HERBAL PRODUCTS

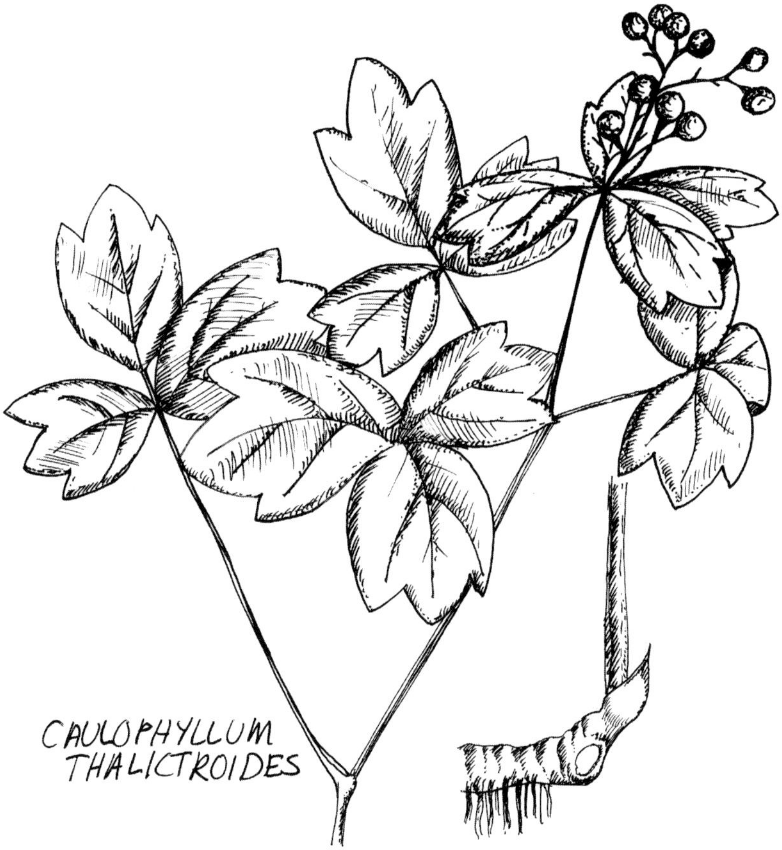

Chapter 91

Perspectives on Plant Toxicology and Public Health: an Update

J.M. Betz

Office of Dietary Supplements, U.S. National Institutes of Health, Bethesda, MD 20892, USA

Introduction

The original 'Perspectives on Plant Toxicology and Public Health' (Betz and Page 1998) presented an overview of public health issues associated with ingestion of plants by humans. At the time, the Dietary Supplement Health and Education Act (DSHEA) (United States Public Law 1994) was new, and the focus was on general interactions between people and plants. This update will focus on dietary supplements. The original categories set out in 1998 to describe adverse human/plant encounters are still largely valid, with one addition and one caveat. Adverse events are still caused by misidentification, deliberate adulteration, misuse, and inherent toxicity. Size of the 'unknown' category (e.g. idiosyncratic, reactions attributable to genetics or lack of mechanistic plausibility, etc.) is beginning to shrink as the state of our knowledge advances. Added for this update is a category that was little known in 1998, that of drug/herb interactions.

Misidentification

Mistaking one plant species for another is an old problem (Bisset and Wichtl 2001). Following consumption of a dietary supplement 'Cleansing' regimen, a young woman was admitted to the hospital because she had experienced dizziness, nausea, and vomiting. One of the products in the regimen included plantain leaf, *Plantago lanceolata* L. An investigation by the U.S. Food and Drug Administration found that the 'plantain leaf' used in the product was actually *Digitalis lanata* Ehrh. (Slifman *et al.* 1998). Other reported substitutions with potential toxicological implications have included germander (*Teucrium* spp.) for skullcap (*Scutellaria* spp.) (Gafner *et al.* 2003); *Aristolochia* spp. for *Stephania tetrandra* S. Moore or *Clematis* spp. (Vanhaelen *et al.* 1994; Lord *et al.* 1999; Nortier *et al.* 2000); and Japanese star anise (*Illicium anisatum* L.) for Chinese star anise (*I. verum* Hook. F.) (Ize-Ludlow *et al.* 2004).

Deliberate Adulteration

Some substitutions are 'historical', *i.e.*, unscrupulous suppliers take advantage of the similarity between species for economic reasons. Substitution of *Argemone mexicana* seed for Indian mustard (*Brassica juncea*) seed is a historical problem and has led to mass intoxications caused by consumption of the edible oils made from the seed (Thatte and Dahanukar 1999). Bisset and Wichtl (2001) note a number of examples of such substitutions, including *Rheum rhaponticum* L. for *Rheum officinale* Baillon root; and *Angelica archangelica* L. for *Levisticum officinale* Koch root. Fortunately, most substitutions of this sort are not toxicologically significant.

Various 'herbal' products have been found to contain undeclared pharmaceuticals. In 2002, the herbal formula PC-SPES was found to contain undeclared warfarin and diethylstilbesterol (Guns *et al.* 2002), while Sildenafil and its congeners have been found in various 'herbal viagra' products (Mitka 2003; Fleshner *et al.* 2005).

Misuse

Suicide attempts have caused several cases of poisoning. A mild intoxication by Valerian (*Valeriana officianalis* L.) was reported after a woman ingested 40–50 capsules (Willey *et al.* 1995), and intensive gastric irrigation was required to save the life of a middle-aged man who attempted suicide by ingesting aconite (*Aconitum* spp.) (Mizugaki *et al.* 1998).

Kava (*Piper methysticum* G. Forst.) is indigenous to the islands of the South Pacific, where its rhizome is used to make a beverage to induce relaxation (Weiner 1971; Nagata 1971). Pharmacological activity is attributed to arylethylene pyrones called kavalactones (Klohs *et al.* 1959). Excessive consumption causes drowsiness, motor impairment, and changes in visual and auditory perception (Garner and Klinger 1985). A Utah man stopped for driving erratically was convicted of driving under the influence after ingestion of 16 cups of kava (Swenson 1996).

Inherent Toxicity

This category is difficult to address because of problems establishing causality when adverse events are rare or delayed. The herb *Má Huáng* (*Ephedra* spp.) contains ephedrine and five structurally related alkaloids (Huang 1993). Over the years, there have been numerous adverse reactions associated with ephedra-containing products (Capwell 1995; Perrotta *et al.* 1996; Yates *et al.* 2000). In a systematic review of available information, Shekelle *et al.* (2003) reported that ephedrine and ephedra promote modest short-term weight loss in clinical trials and that ephedra or ephedrine and caffeine are associated with increased risk of psychiatric, autonomic, or gastrointestinal symptoms, and heart palpitations. A ban of the sale of ephedra for use in dietary supplements promulgated by the U.S. Food and Drug Administration in 2004 cited the Shekelle report (Department of Health and Human Services 2004).

A study by Nyska *et al.* (2005) found that an ephedra/caffeine combination caused acute hemorrhagic myocardial necrosis and a higher rate of sudden death in rats. A trial of a commercial ephedra/caffeine blend conducted in healthy volunteers by McBride *et al.* (2004) found that a single dose induced a longer maximal QTc interval and higher systolic blood pressure compared to placebo. A case series with dilated cardiomyopathy presenting to a clinic were found to have been exposed to ephedra (Peters *et al.* 2005). All patients discontinued ephedra use as advised, and all six showed improved ejection volume over the next 6 months.

Ephedra-containing products have largely been replaced by products that contain *Zhi shi* (*Citrus aurantium* L., bitter orange, Seville orange). These products are sold as supplements for the same indications as their ephedra-containing predecessors. The rationale for this substitution is the presence in bitter orange of p-synephrine, an alkaloid that is structurally similar to ephedrine. Wheaton and Stewart (1970) reported synephrine levels in a number of citrus fruit juices and plants used as foods. Penzak *et al.* (2001) reported no change in hemodynamics in a group of normotensive adults following consumption of Seville orange juice. Adverse event reports related to synephrine in food could not be found.

Nykamp *et al.* (2004) reported acute myocardial infarction in a woman who had consumed a bitter orange/caffeine supplement. The case was complicated by a smoking history, high caffeine consumption, and a pre-existing coronary artery lesion. Nasir *et al.* (2004) described a case of exercise-induced syncope associated with Qt prolongation and bitter orange/caffeine that is reminiscent of the ephedra/caffeine study by McBride *et al.* (2004). A study in healthy volunteers measured hemodynamic effects of placebo, a caffeine/synephrine product, and a synephrine/no caffeine product. The synephrine/caffeine product was associated with raised mean systolic and diastolic blood pressure, synephrine alone did not affect blood pressure, and both products caused significant increases in heart rate (Haller *et al.* 2005).

A study of effects of oral *C. aurantium* extracts on Sprague-Dawley rats found increased mortality in all treatment groups. No significant effect on blood pressure was reported, but myocardial electrical activity of all *C. aurantium* groups was altered compared to controls, with changes such as ventricular arrhythmias and enlargement of QRS complex (Calapai *et al.* 1999). These results are consistent with the Nasir *et al.* (2004) case report and point to the need for human studies such as that done by McBride *et al.* (2004).

Other

As noted, acute reactions that are rare, delayed, or that require some genetic or other predisposition in the affected individual may be difficult to detect. Establishing mechanistic plausibility in such cases becomes paramount to establishing causality. In a discussion devoted primarily to long-term safety of phytoestrogens, Sheehan (1998) pointed out that virtually all of our knowledge about potential human adverse effects of herbal products comes from reports of acute reactions to exposure rather than through well-designed experiments.

The rhizome of blue cohosh (*Caulophyllum thalictroides* (L.) Michx.) is traditionally used in pregnancy and labor. The plant and its commercial products contain the alkaloids anagyrine, N-methylcytisine, sparteine, baptifoline, and magnoflorine (Flom *et al.* 1967;

Betz *et al.* 1998). Two cases of neonatal adverse events and one of an adverse maternal effect are associated with consumption of this plant. Following maternal consumption of blue cohosh, Jones and Lawson (1998) reported a case of profound neonatal congestive heart failure, while Finkel and Zarlengo (2004) described a perinatal stroke following use to induce labor. Rao and Hoffman (2002) described a case of poisoning in a woman who had ingested a massive dose of blue cohosh in an attempt to induce an abortion. In an *in vitro* rat embryo study, Kennelly *et al.* (1999) found that the alkaloid taspine (isolated from blue cohosh) was embryotoxic but not teratogenic, while N-methylcytisine was teratogenic.

Kava was once a promising phytomedicine for treating anxiety (Pittler and Ernst 2003). Around the year 2000, reports of liver injury associated with kava extracts began to emerge. By the end of 2002, kava had been banned in Germany, Switzerland, Canada, France, Australia, and the UK. Early attempts to explain the ‘sudden’ surge in adverse events focused on ‘western’ dosage forms, which are highly processed organic solvent extracts that bear little resemblance to traditional kava preparations (Clouatre 2004). Dragull *et al.* (2003) reported the presence of the alkaloid pipermethystine in leaves and peelings of kava, and Nerurkar *et al.* (2004) reported that this compound is toxic to HepG2 human hepatoma cells *in vitro*. Leaves and peels are not used in traditional preparations. Qiu *et al.* (2005) used bioassay directed fractionation to isolate two ‘neokavalactones’ with *in vitro* toxicity to isolated human liver cells.

In retrospect, there were signals that should have caused concern. Several studies showed that chronic use of large amounts of kava are associated with adverse health effects. Dry, flaking, discolored skin and reddened eyes associated with chronic kava use was first described as ‘kawaism’ by Keller and Klohs (1963). Matthews *et al.* (1988) reported that heavy ‘traditional’ kava users were more likely to be in poor health and had greatly elevated gamma-glutamyl transferase (GGT) levels. A study of the health effects of kava in northern Australia showed dermopathy and increased levels of GGT and alkaline phosphatase in heavy users (Clough *et al.* 2003a,b). A trial by Cairney *et al.* (2003) also found elevated GGT and alkaline phosphatase levels in addition to ataxia, tremors, and sedation.

Clough *et al.* (2003b) observed that chronic, heavy users of kava consume large quantities of kavalactones in a single sitting, but that they only did this weekly. Serum GGT levels of these individuals were greatly elevated following kava consumption, but in the absence of alcoholic beverages, serum GGT levels returned to normal 1 day after discontinuation of kava. The use pattern of kava as an anxiolytic is chronic rather than acute, with product consumed several times a day for long periods of time.

One pathway by which xenobiotics are known to cause hepatotoxicity is through inhibition of hepatic cytochrome P450 (CYP450) enzymes. Matthews *et al.* (2002) investigated the ability of individual kavalactones to inhibit specific CYP450 isozymes. They found that several kavalactones caused significant inhibition of CYP2C9, CYP2C19, and CYP3A4. More importantly, they found that methysticin and dihydromethysticin form irreversible complexes with these enzymes. Chronic kava use might therefore cause a depletion of these important detoxification enzymes. Additionally, since the cytochrome P450 system is responsible for the metabolism of more than 90% of all drugs, concomitant use of western drugs or alcohol and kava may pose a significant toxicological risk (Musser 2005).

Herb–Drug Interactions

A review of herb/drug interactions was published by D'Arcy in 1993, but the subject was not the focus of much attention until a few years ago. A recent search of PubMed (http://www.pubmed.gov) using the phrase 'herb drug interactions' returned 328 articles. Fully 221 of the citations were for review articles. Original papers included ten case reports, 18 clinical trials, 14 reports of *in vitro* assays, 20 animal studies, and 17 surveys of various types (patient use patterns for herbs, physician attitudes and knowledge of interactions). A detailed discussion of herb/drug interactions could fill several large volumes and won't be attempted here. Instead, the reader is referred to recent reviews by Hu *et al.* (2005), Chavez *et al.* (2006), Venkataramanan *et al.* (2006), and Pal and Mitra (2006).

According to Goth (1974) substances influence drug effects by several mechanisms. These include effects on absorption, plasma protein binding, metabolism, action at the drug receptor site, renal excretion, and alteration of electrolyte balance. Pharmacokinetic herb/drug interactions involving hepatic enzymes such as the Cytochrome P-450 (CYP) superfamily are the focus of much attention.

Because of a number of reports of adverse effects caused by herb/drug interactions involving St. John's wort (SJW), this herb has come under intense scrutiny. Nebel *et al.* (1999) reported theophylline toxicity in a patient who discontinued SJW. Shortly after SJW had been initiated, escalation of the theophylline dose was required to offset increased theophylline clearance caused by enzyme induction. When SJW was discontinued and enzyme levels returned to normal, elevated theophylline levels caused toxicity. Acute cardiac transplant rejection followed SJW co-administration with cyclosporin (Ruschitzka *et al.* 2000) and lowered cyclosporin concentrations were recorded in a kidney transplant patient co-administered a SJW extract (Mai *et al.* 2000). A study by Mai *et al.* (2004) found that content of hyperforin in the herbal product determined the magnitude of SJW cyclosporine interactions. Yue *et al.* (2000) reported that St. John's wort may reduce the effectiveness of warfarin, while breakthrough bleeding during concomitant use of oral contraceptives and SJW was reported by Bon *et al.* (1999).

Attempts to elucidate mechanisms and quantify risk quickly followed the initial interaction reports. Using a probe drug, Markowitz *et al.* (2000) reported no CYP3A4 or CYP2D6 inhibition after 4 days of SJW administration in healthy volunteers and a trend towards CYP3A4 induction. Roby *et al.* (2000) found induction of CYP3A4 following 14 days of SJW. Piscitelli *et al.* (2000) found that 14 days of SJW reduced the area under the curve (AUC) and the trough concentrations of the anti-HIV drug indinavir in healthy volunteers, consistent with marked CYP3A4 induction. Concomitant administration of SJW and the P-gp substrate digoxin showed a trend towards inhibition of digoxin clearance after 1 day but enzyme induction and subtherapeutic digoxin levels after 10 days (Johne *et al.* 1999). Kovarik *et al.* (1999) reported that digoxin AUC increased when co-administered with a known P-gp inhibitor, valspodar. Taken together, these studies suggest that SJW/drug interactions are more pronounced than those seen in other herbs because SJW constituents may simultaneously inhibit efflux of P-gp-dependent substrates and induce CYP3A4-dependent drug clearance.

Other herbs have also been investigated. Donovan *et al.* (2004) found that valerian had little effect on CYP3A4 and no effect on CYP2D6 activity in healthy volunteers. Markowitz *et al.* (2004) reported that saw palmetto (*Serenoa repens*) does not affect these isoforms. Gurley *et al.* (2005) reported that goldenseal (*Hydrastis canadensis* L.) strongly inhibited CYP2D6 and CYP3A4/5 activity *in vivo*, whereas kava inhibited CYP2E1 and black cohosh weakly inhibited CYP2D6. Valerian was reported to be less likely to produce CYP-mediated herb/drug interactions. Gurley *et al.* (2006) reported that milk thistle (*Silybum marianum*) and black cohosh (*Actaea racemosa*) did not affect digoxin pharmacokinetics.

Potential pharmacodynamic interactions have been described. There have been several reports of bleeding events associated with garlic consumption (Burnham 1995; German *et al.* 1995). Additional case reports include bleeding events following concomitant ingestion of ginkgo (*Ginkgo biloba* L.) and aspirin (Rosenblatt and Mindel 1997), ginkgo and ergotamine/caffeine (Rowin and Lewis 1996), and Dong quai (*Angelica sinensis*) and warfarin (Page and Lawrence 1999). St. John's wort has been associated with mild serotonin syndrome when combined with antidepressants (Demott 1998; Lantz *et al.* 1999).

Effects of botanical products on drug absorption and gastrointestinal transport have also been noted. Guar gum (*Cyamopsis tetragonoloba* (L.) Taub.) prolongs gastric retention and slows absorption of digoxin, paracetemol, and bumetanide, and decreases absorption of metformin, phenoxymethylpennicillin and glibenclamide (De Smet and D'Arcy 1996). Interactions due to effects on gastric transit time (bulk-forming agents such as Psyllium, *Plantago ovata*, or stimulant laxatives such as cascara (*Rhamnus purshiana*)) are also possible (Fugh-Berman 2000).

Treasure (2000) recommends a common sense approach to interactions. He suggests that physicians keep track of all of a patient's medications and supplements, that they be alert when prescribing powerful medications and those with a narrow therapeutic window, and that patients in vulnerable groups (i.e., those subject to polypharmacy) be closely monitored. A particular area of recent emphasis that highlights these recommendations is the effect of dietary supplements on coagulation. This is such a major public health issue that the NIH recently held a conference on the topic, published in a volume of the journal *Thrombosis Research* (Marder *et al.* 2005).

Conclusions

At the moment, it appears that adverse events associated with botanical supplements are very rare. However, size and complexity of the marketplace, along with rapid introduction of new ingredients, are difficulties that public health officials face when dealing with supplement use. In North America alone, at least 700 plant species have been described as being poisonous in one way or another, and plant poisonings are often difficult to differentiate from environmental intoxications caused by pesticides and industrial chemicals and from adverse reactions to synthetic drugs (Der Marderosian and Liberti 1988). Complicating the picture is the fact that many poisoning symptoms mimic those of medical conditions not associated with a toxic exposure, and that establishing causality may therefore be difficult (Perrotta *et al.* 1995). New ingredients with known bioactive constituents such as jojoba seed meal (Booth *et al.* 1974) as a weight loss agent or the lichen X*anthoparmelia scabrosa* (Moerman *et al.* 2003) with unknown activities or poorly characterized chemistry enter the market at an accelerating pace.

Palmer *et al.* (2003) conducted an observational study of adverse event reports to US poison control centers. The authors noted that severity of events ranged from very mild to severe (including death) but that associations between events and ingredients are difficult to verify. They recommended that proactive research into hazards and risks should become a priority. As this research is pursued, adverse event cases that might otherwise have been placed in categories such as 'other' or 'unknown' may be reclassified, while other plants currently under suspicion may be found to be as harmless as proponents claim.

References

Betz JM and Page SW (1998). Perspectives on plant toxicology and public health. In *Toxic Plants and Other Natural Toxicants* (T Garland and AC Barr, eds), pp. 367–372. CABI Publishing, Wallingford, Oxon, UK.

Betz JM, Woldemariam TZ, Page SW, Obermeyer WR, and Andrzejewski D (1998). Gas chromatographic determination of quinolizidine alkaloids in blue cohosh (*Caulophyllum thalictroides* (L.) Michx. *Phytochemical Analysis* 9:232–236.

Bisset NG and Wichtl M (eds) (2001). *Herbal Drugs and Phytopharmaceuticals*, 2nd ed. CRC Press, Boca Raton, Florida.

Bon S, Hartmann K, and Kuhn M (1999). Johanniskraut: Ein Enzyminduktor? *Schweizerische Apotheker-Zeitung* 16:*535536.*

Booth AN, Elliger CA, and Waiss AC, Jr (1974). Isolation of a toxic factor from jojoba meal. *Life Sciences* 15:1115–1120.

Burnham BE (1995). Garlic as a possible risk factor for postoperative bleeding. *Plastic and Reconstructive Surgery* 95:213.

Cairney S, Maruff P, Clough AR, Collie J, and Currie BJ (2003). Saccade and cognitive impairment associated with kava intoxication. *Human Psychopharmacology—Clinical and Experimental* 18:525–533.

Calapai G, Firenzuoli F, Saitta A, Squadrito F, Arlotta MR, Costantino G, and Inferrera G (1999). Antiobesity and cardiovascular toxic effects of *Citrus aurantium* extracts in the rat: a preliminary report. *Fitoterapia* 70:586–592.

Capwell RR (1995). Ephedrine-induced mania from an herbal diet supplement. *American Journal of Psychiatry* 152:647.

Chavez ML, Jordan MA, and Chavez PI (2006). Evidence-based drug—herbal interactions. *Life Sciences* 78:2146–2157.

Clouatre DL (2004). Kava kava: examining new reports of toxicity. *Toxicology Letters* 150:85–96.

Clough AR, Jacups SP, Wang Z, Burns CB, Bailie RS, Cairney SJ, Collie A, Guyula T, McDonals SP, and Currie BJ (2003a). Health effects of kava use in an eastern Arnhem Land aboriginal community. *Internal Medicine Journal* 33:336–340.

Clough AR, Bailie RS, and Currie B (2003b). Liver function test abnormalities in users of aqueous kava extracts. *Journal of Toxicology and Clinical Toxicology* 41:821–829.

D'Arcy PF (1993). Adverse reactions and interactions with herbal medicines, Part 2: Drug Interactions. *Adverse Drug Reactions and Toxicological Reviews* 12:147–162.

Demott K (1998). St. John's wort tied to serotonin syndrome. *Clinical Psychiatry News* 26:28.

Department of Health and Human Services (2004). Final rule declaring dietary supplements containing ephedrine alkaloids adulterated because they present an unreasonable risk; final rule. *Federal Register* 69:6787–6854.

Der Marderosian AH and Liberti L (1988). *Natural Product Medicine*. George F. Stickley Co., Philadelphia.

De Smet PAGM and D'Arcy PF (1996). Drug interactions with herbals and other non-toxic remedies. In *Mechanisms of Drug Interactions* (PF D'Arcy, JC McElnay, and PG Welling, eds), pp. 327–352. Springer-Verlag, Berlin.

Donovan JL, DeVane CL, Chavin KD, Wang JS, Gibson BB, Gefroh HA, and Markowitz JS (2004). Multiple night-time doses of valerian (*Valeriana officinalis*) had minimal effects on CYP3A4 activity and no effect on CYP2D6 activity in Healthy Volunteers. *Drug Metabolism and Disposition* 32:1333–1336.

Dragull K, Yoshida WY, and Tang CS (2003). Piperidine alkaloids from *Piper methysticum*. *Phytochemistry* 63:193–198.

Finkel RS and Zarlengo KM (2004)Blue cohosh and neonatal stroke. *New England Journal of Medicine* 351:302–303.

Fleshner N, Harvey M, Adomat H, Wood C, Eberding A, Hersey K, and Guns E (2005). Evidence for contamination of herbal dysfunction products with phosphodiesterase type 5 inhibitors. *Journal of Urology* 174:636–641.

Flom MS, Doskotch RW, and Beal JL (1967). Isolation and characterization of alkaloids from *Caulophyllum thalictroides*. *Journal of Pharmaceutical Sciences* 56:1515–1517.

Fugh-Berman A (2000). Herb-drug interactions. *Lancet* 355:134–138.

Gafner S, Bergeron C, Batcha LL, Angerhofer CK, Sudberg S, Sudberg EM, Guinaudeau H, and Gauthier R (2003). Analysis of *Scutellaria lateriflora* and its adulterants *Teucrium canadense* and *Teucrium chamaedrys* by LC-UV/MS, TLC, and digital photomicroscopy. *Journal of AOAC International* 86:453–460.

Garner LF and Klinger JD (1985). Some visual effects caused by the beverage kava. *Journal of Ethnopharmacology* 13:307–311.

German K, Kumar U, and Blackford HN (1995). Garlic and the risk of TURP bleeding. *British Journal of Urology* 76:518.

Goth A (1974). *Medical Pharmacology*, 7th ed., p. 681. C.V. Mosby Co., St. Louis, Missouri.

Guns ES, Goldenberg SL, and Brown PN (2002). Mass spectral analysis of PC-SPES confirms the presence of diethylstilbestrol. *Canadian Journal of Urology* 9:1684–1688.

Gurley BJ, Gardner SF, Hubbard MA, Williams DK, Gentry WB, Khan IA, and Shah A (2005). In vivo effects of goldenseal, kava kava, black cohosh, and valerian on human cytochrome P450 1A2, 2D6, 2E1, and 3A4/5 phenotypes. *Clinical Pharmacology and Therapeutics* 77:415–426.

Gurley BJ, Barone GW, Williams DK, Carrier J, Breen P, Yates CR, Song PF, Hubbard MA, Tong T, and Cheboyina S (2006). Effect of milk thistle (*Silybum marinarum*) and black cohosh (*Cimicifuga racemosa*) supplementation on digoxin pharmacokinetics in humans. *Drug Metabolism and Disposition* 34:69–74.

Haller CA, Benowitz NL, and Jacob III P (2005). Hemodynamic effects of ephedra-free weight-loss supplements in humans. *American Journal of Medicine* 118:998–1003.

Hu Z, Yang X, Ho PC, Chan SY, Heng PW, Chan E, Duan W, Koh HL, and Zhou S (2005). Herb-drug interactions: a literature review. *Drugs* 65:1239–1282.

Huang KC (1993). *The Pharmacology of Chinese Herbs*. CRC Press, Boca Raton, Florida.

Ize-Ludlow D, Ragone S, Bernstein JN, Bruck IS, Duchowny M, and Garcia Pena BM (2004). Chemical composition of Chinese star anise (*Illicium verum*) and neurotoxicity in infants. *Journal of the American Medical Association* 291:562–563.

Johne A, Brockmoller J, Bauer S, Maurer A, Langheinrich M, and Roots I (1999). Pharmacokinetic interaction of digoxin with an herbal extract from St. John's wort (*Hypericum perforatum*). *Clinical Pharmacology and Therapeutics* 66:338–345.

Jones TK and Lawson BM (1998). Profound neonatal congestive heart failure caused by maternal consumption of blue cohosh herbal medication. *Journal of Pediatrics* 132:550–552.

Keller F and Klohs MW (1963). A review of the chemistry and pharmacology of the constituents of *Piper methysticum*. *Lloydia* 26:1–11.

Kennelly EJ, Flynn TJ, Mazzola EP, Roach JA, McCloud TG, Danford DE, and Betz JM (1999). Detecting potential teratogenic alkaloids from blue cohosh rhizomes using an in vitro rat embryo culture. *Journal of Natural Products* 62:1385–1389.

Klohs MW, Keller F, Williams RE, Toekes MI, and Cronheim GE (1959). A chemical and pharmacological investigation of *Piper methysticum* Forst. *Journal of Medicinal and Pharmaceutical Chemistry* 1:95–103.

Kovarik JM, Rigaudy L, Guerret M, Gerbeau C, and Rost KL (1999). Longitudinal assessment of a P-glycoprotein-mediated drug interaction of valspodar on digoxin. *Clinical Pharmacology and Therapeutics* 66:391–400.

Lantz MS, Buchalter E, and Giambanco V (1999). St. John's wort and antidepressant drug interactions in the elderly. *Journal of Geriatric Psychiatry and Neurology* 12:7–10.

Lord GM, Tagore R, Cook T, Gower P, and Pusey CD (1999). Nephropathy caused by Chinese herbs in the UK. *Lancet* 354:481–482.

Mai I, Kruger H, Budde K, Johne A, Brockmoller J, Neumeyer HH, and Roots I (2000). Hazardous pharmacokinetic interaction of Saint John's wort (*Hypericum perforatum*) with the immunosuppressant cyclosporin. *International Journal of Clinical Pharmacology and Therapeutics* 38:500–502.

Mai I, Bauer S, Perloff ES, Johne A, Uehleke B, Frank B, Budde K, and Roots I (2004). Hyperforin content determines the magnitude of the St. John's wort-cyclosporin drug interaction. *Clinical Pharmacology and Therapeutics* 76:330–340.

Marder VJ, Ortel TL, Grollman A, and Hasan AK (eds) (2005). Papers from the NIH conference on dietary supplements, coagulation, and antithrombotic therapies. *Thrombosis Research* 117:1–230.

Markowitz JS, DeVane CL, Boulton DW, Carson SW, Nahas Z, and Risch SC (2000). Effect of St. John's wort (*Hypericum perforatum*) on Cytochrome P-450 2D6 and 3A4 activity in healthy volunteers. *Life Sciences* 66:133–139.

Markowitz JS, Donovan JL, DeVane CL, Taylor RM, Ruan Y, Wang JS, and Chavin KD (2004). Multiple doses of saw palmetto (*Serenoa repens*) did not alter cytochrome P450 2D6 and 3A4 activity in normal volunteers. *Clinical Pharmacology and Therapeutics* 74:536–542.

Matthews JD, Riley MD, Fejo L, Munoz E, Milns NR, and Gardner ID (1988). Effects of the heavy usage of kava on physical health: summary of a pilot survey in an aboriginal community. *Medical Journal of Australia* 148:548–555.

Matthews JM, Etheridge AS, and Black SR (2002). Inhibition of human cytochrome P450 activities by kava extract and kavalactones. *Drug Metabolism and Disposition* 30:1153–1157.

McBride BF, Karapanos AK, Krudysz A, Kluger J, Coleman CI, and White CM (2004). Electrocardiographic and hemodynamic effects of a multicomponent dietary supplement containing ephedra and caffeine: a randomized controlled trial. *Journal of the American Medical Association* 291:216–221.

Mitka M (2003). FDA issues warning on 'all-natural' herbal product found to contain Viagra. *Journal of the American Medical Association* 289:2786.

Mizugaki M, Ito K, Ohyama Y, Konishi Y, Tanaka S, and Kurasawa K (1998). Quantitative analysis of *Aconitum* alkaloids in the urine and serum of a male attempting suicide by oral intake of aconite extract. *Journal of Analytical Toxicology* 22:336–340.

Moerman KL, Chai CL, and Waring P (2003). Evidence that the lichen-derived scabrosin esters target mitochondrial ATP synthase in P388D1 cells. *Toxicology and Applied Pharmacology* 90:232–240.

Musser SM (2005). Kava (*Piper methysticum*). In *Encyclopedia of Dietary Supplements* (PM Coates, MR Blackman, GM Cragg, M Levine, J Moss, and JD White, eds), pp. 373–380. Marcel Dekker, New York.

Nagata KM (1971). Hawaiian medicinal plants. *Economic Botany* 25:245–254.

Nasir JM, Durning SJ, Ferguson M, Barold HS, and Haigney MC (2004). Exercise-induced syncope associated with QT prolongation and ephedra-free xenadrine. *Mayo Clinic Proceedings* 79:1059–1062.

Nebel A, Schneider BJ, Baker RK, and Kroll DJ (1999). Potential metabolic interaction between St. John's wort and theophylline. *Annals of Pharmacotherapy* 33:502.

Nerurkar PV, Dragull K, and Tang CS (2004). In vitro toxicity of kava alkaloid, pipermethystine, in HepG2 cells compared to kavalactones. *Toxicological Sciences* 79:106–111.

Nortier JL, Martinez MC, Schmeiser HH, Arlt VM, Bieler CA, Petein M, Depierreux MF, De Pauw L, Abramowicz D, Vereerstraeten P, and Vanherweghem JL (2000). Urothelial carcinoma associated with the use of a Chinese herb (*Aristolochia fangchi*). *New England Journal of Medicine* 342:1686–1692.

Nykamp DL, Fackih MN, and Compton AL (2004). Possible association of acute lateral-wall myocardial infarction and bitter orange supplement. *Annals of Pharmacotherapy* 38:812–816.

Nyska A, Murphy E, Foley JF, Collins BJ, Petranks J, Howden R, Hanlon P, and Dunnick JK (2005). Acute hemorrhagic myocardial necrosis and sudden death of rats exposed to a combination of ephedrine and caffeine. *Toxicological Sciences* 83:388–396.

Page RL and Lawrence JD (1999). Potentiation of warfarin by dong quai. *Pharmacotherapy* 19:870–876.

Pal D and Mitra AK (2006). MDR- and CYP3A4-mediated drug-herbal interactions. *Life Sciences* 78:2131–2145.

Palmer ME, Haller C, McKinney PE, Klein-Schwartz W, Tschirgi A, Smolinske SC, Woolf A, Sprague BM, Ko R, Everson G, Nelson LS, Dodd-Butera T, Bartlett WD, and Landzberg BR (2003). Adverse events associated with dietary supplements: an observational study. *Lancet* 361:101–106.

Penzak SR, Jann MW, Cold JA, Hon YY, Desai HD, and Gurley BJ (2001). Seville (sour) orange juice: synephrine content and cardiovascular effects in normotensive adults. *Journal of Clinical Pharmacology* 41:1059–1063.

Perrotta DM, Nickey LN, Raid M, Caraccio T, Mofenson HC, Waters C, Morse D, Osorio AM, Hoshiko S, and Rutherford III GW (1995). Jimson weed poisoning-Texas, New York, and California, 1994. *Morbidity and Mortality Weekly Reports* 44:41–44.

Perrotta PM, Coody G, and Culmo C (1996). Adverse events associated with ephedrine-containing products-Texas, December 1993-September 1995. *Morbidity and Mortality Weekly Reports* 45:689–693.

Peters CM, O'Neill JO, Young JB, and Bott-Silverman C (2005). Is there an association between ephedra and heart failure? A case series. *Journal of Cardiac Failure* 11:9–11.
Piscitelli SC, Burstein AH, Chaitt D, Alfaro RM, and Falloon J (2000). Marked reduction in indinavir exposure by St. John's wort. *Lancet* 355:547–548.
Pittler MH and Ernst E (2003). Kava extract for treating anxiety. *Cochrane Database of Systematic Reviews* (1):CD003383.
Qiu SX, Liu J-H, Zhou P, Yi B-Y, and Gross ML (2005). The intrinsic hepatotoxic compounds identified from kava roots and their molecular mechanism. *American Society of Pharmacognosy*, 46th Annual Meeting, Corvallis, Oregon, 7/23–27, p. 141.
Rao RB and Hoffman RS (2002). Nicotinic toxicity from tincture of blue cohosh (*Caulophyllum thalictroides*) used as an abortifacient. *Veterinary and Human Toxicology* 44:221–222.
Roby CA, Anderson GD, Kantor E, Dryer DA, and Burstein AH (2000). St. John's wort: effect on CYP3A4 activity. *Clinical Pharmacology and Therapeutics* 67:451–457.
Rosenblatt M and Mindel J (1997). Spontaneous hyphema associated with ingestion of *Ginkgo biloba* extract [letter]. *New England Journal of Medicine* 336:1108.
Rowin J and Lewis SL (1996). Spontaneous bilateral subdural hematomas associated with chronic *Ginkgo biloba* ingestion. *Neurology* 46:1775–1776.
Ruschitzka F, Meier PJ, Turina M, Luscher TF, and Noll G (2000). Acute heart transplant rejection due to Saint John's wort. *Lancet* 355:548–549.
Sheehan DM (1998). Herbal medicines, phytoestrogens and toxicity: risk:benefit considerations. *Proceedings of the Society for Experimental Biology and Medicine* 217:379–385.
Shekelle PG, Hardy ML, Morton SC, Maglione M, Mojica WA, Suttorp MJ, Rhodes SL, Jungvig L, and Gagne J (2003). Efficacy and safety of ephedra and ephedrine for weight loss and athletic performance: a meta-analysis. *Journal of the American Medical Association* 289:1537–1545.
Slifman NR, Obermeyer WR, Aloi BK, Musser SM, Correll WA, Cichowicz SM, Betz JM, and Love LA (1998). Contamination of botanical dietary supplements by *Digitalis lanata*. *New England Journal of Medicine* 339:806–811.
Swenson J (1996). Man convicted of driving under influence of kava. *Deseret News*, Salt Lake City, Utah, 5 August.
Thatte U and Dahanukar S (1999). The Mexican poppy poisons the Indian mustard: Facts and figures. *The Journal of the Association of Physicians of India* 47:332–335.
Treasure J (2000). Herbal pharmacokinetics: A practitioner update with reference to St. John's wort (*Hypericum perforatum*) interactions. *Journal of the American Herbal Guild* 1:2–11.
United States Public Law 103-417. 103rd Congress, 25 October 1994. *Dietary Supplement Health and Education Act of 1994.*
Vanhaelen M, Vanhaelen-Fastre R, But P, and Vanherweghem JL (1994). Identification of aristolochic acid in Chinese herbs. *Lancet* 343:174.
Venkataramanan R, Komoroski B, and Strom S (2006). In vitro and in vivo assessment of herb drug interactions. *Life Sciences* 78:2105–2115.
Weiner MA (1971). Ethnomedicine in Tonga. *Economics of Botany* 25:423-450.
Wheaton TA and Stewart I (1970). The distribution of tyramine, N-methyltyramine, hordenine, octopamine, and synephrine in higher plants. *Lloydia* 33:244–254.
Willey LB, Mady SP, Cobaugh DJ, and Wax PM (1995). Valerian overdose: a case report. *Veterinary and Human Toxicology* 37:364–365.

Yates KM, O'Connor A, and Horsely CA (2000). 'Herbal Ecstasy': a case series of adverse reactions. *New Zealand Medical Journal* 113:315–317.

Yue Q-Y, Bergquist C, and Gerdén B (2000). Seven cases of decreased effect of warfarin during concomitant treatment with St. John's wort. *Lancet* 355:576–577.

Chapter 92

The Framework for Evaluating Safety of Dietary Supplements: An Overview and Potential Application to Poisonous Plant Situations

R.J. Molyneux

USDA-ARS Western Regional Research Center, Albany, CA 94710, USA

Introduction

The production and marketing of dietary supplements in the United States is exceptionally lucrative. In the year 2000, there were 29,000 different products available to consumers and 1000 new products were being introduced annually. Sales amounted to $18.7 billion per annum in the year 2002, of which $4.3 billion was estimated to be from herbs or botanicals (NBJ's Annual 2003). Safety of such products is regulated by the Dietary Supplement and Health Education Act of 1994 (DSHEA). In essence, dietary supplements are treated as analogous to foods, not drugs, so that no pre-market approval and no post-market reporting or surveillance by industry is required. The Food and Drug Administration (FDA) must therefore demonstrate significant or unreasonable risk of harm to remove or prevent product from being marketed. This presents a logistical problem because with so many products on the market only a few can be evaluated at any given time.

In 2002 a committee of the National Academies of Sciences, Institute of Medicine, was constituted to analyze the situation and develop a 'framework' which would provide a basis on which to predict potential toxicity (and by default, safety) of new dietary supplements or new formulations thereof, including consideration of interactions with drugs and xenobiotics, as well as effects on vulnerable populations. The committee adopted a science-based approach, in which information evaluated fell into four broad categories, namely: human data; animal studies; related substances; and *in vitro* experiments (Committee on the Framework for Evaluating the Safety of Dietary Supplements 2005). The approach then required that data from these categories be integrated to determine an overall level of concern for any particular product. In order to illustrate the approach, prototype monographs were developed for six diverse dietary supplements: chaparral; glucosamine; melatonin; chromium picolinate; saw palmetto; and shark cartilage. Two of these monographs revealed a potential concern for chaparral and saw palmetto, including focused reviews of liver-related risks and antiandrogenic risks for women, respectively.

Using Science-Based Principles to Establish Risk: Categories of Evidence

A number of points need to be considered with respect to the four categories of risk. Thus, for human data, a credible report of serious adverse effect raises concern, but it should be noted that historical evidence is not *prima facie* evidence of absence of harm, primarily because modern and traditional uses differ in many respects. For example, traditional use of herbal remedies was generally intermittent, as the plant itself or in the form of an aqueous infusion or tea, whereas modern use often involves long-term use of capsules or extracts, resulting in increased dose levels of active constituents over a prolonged period. Evidence of harm from animal studies should be considered as indicative of potential harm, even in the absence of information on adverse events in humans, in the same way as such criteria are applied to therapeutic drugs.

With respect to information on related substances, a taxonomic relationship to known toxic plants may be indicative of potential harm as are structural or functional similarities of specific ingredients or extracts to known compounds or modes of action. The final document published by the National Academies Press included a table of genera of concern, listing the plant family, compound classes implicated in toxicity, observed and predicted toxicities, and the toxicity class. Toxicity classes were divided into three categories: A, reports of adverse effects on major organs, teratogenicity, carcinogenesis, and/or CNS effects; B, reports of non-lethal effects, e.g. gastric upset, photosensitization, allergenicity; and C, reports where exceptional amounts were required to elicit an effect. In addition to this tabulation, an appendix was included summarizing toxicities reported, organized by plant families.

Validated *in vivo* studies can be independent indicators of risk but only if comparable exposure levels to the real-life situation are attainable and these correlate with specific adverse effects in humans and animals.

Application of the Framework: Saw Palmetto Herbal Formulations

Saw palmetto (*Serenoa repens*; Arecaceae), also known as sabal, cabbage palm, shrub palmetto, juzhong, or palmier nain is commonly recommended for treatment of benign prostatic hyperplasia (BPH) (Wilt *et al.* 1998). The general approach for treatment of BPH is to prevent conversion of testosterone, the major circulating androgen produced by interstitial cells of the testes, to the more potent androgen, 5α-dihydrotestosterone, which stimulates prostate growth, by inhibition of the enzyme 5α-reductase (Figure 1).

Saw Palmetto: Categories of Evidence and Adverse Effects

The four categories of evidence for saw palmetto can be summarized as follows. For human data there have been no adverse effects reported in clinical trials but it is important to note that pregnant women were not included. Historical use of the plant was as whole fruit or teas, not as alcoholic extracts, and generally as capsules, which are the most common preparation now consumed. In animal studies no adverse effects were noted but antiandrogenic activity has been demonstrated for fruit extract in some studies; however, effects *in utero* were not investigated.

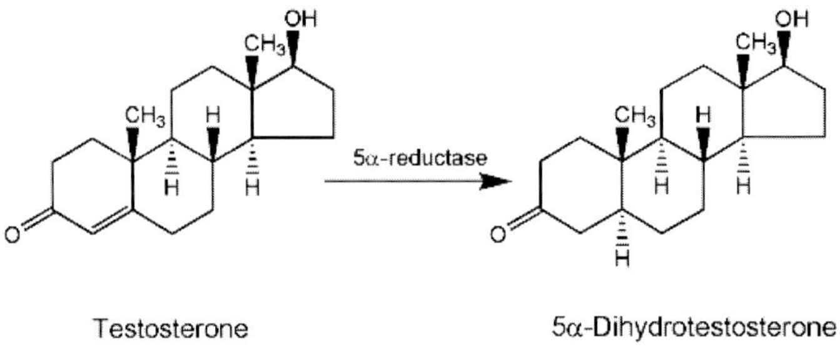

Figure 1. Reduction of testosterone to 5α-dihydrotestosterone by 5α-reductase

Information on related substances showed that the primary constituents of the fruits are fatty acids, phytosterols and flavonoids. The only taxonomic relationship of concern is that oxalates are common in family Aracaceae. Functionally related substances are steroid 5-α-reductase inhibitors such as finasteride (Proscar®). From *in vitro* experiments it was established that the fruit extract inhibits steroid 5-α-reductase conversion of testosterone to dihydrotestosterone with an IC_{50} value of 5.6 μg mL^{-1}; in comparison, finasteride has an IC_{50} of 1 μg mL^{-1} (Rhodes *et al.* 1993). It is therefore probable that the active compounds in saw palmetto are steroid 5-α-reductase inhibitors.

The major structural types identified in saw palmetto fruits are lipids, phytosterols and flavonoids. Although the sterols have not been tested for activity, daucosterol (sitosterol-*O*-β-D-glucopyranoside) has structural affinities with finasteride (Figure 2) and occurs at 0.0047% dry weight in saw palmetto, while the parent β-sitosterol occurs at 0.0033–0.22% dry weight both as the free compound and esterified with a variety of fatty acids (Hiermann, 1989). In contrast, the flavonoids kaempferol and quercitrin (Figure 2), which are also present, inhibit human prostate type 1 steroid 5α-reductase with IC_{50} values of 12 and 23 μM, respectively, compared with 0.5 μM for finasteride (Hiipakka *et al.* 2002). It may be that any therapeutic activity in saw palmetto derives from steroid 5α-reductase activity of several different structural entities.

An important observation resulting from application of the monograph is that finasteride is contraindicated in pregnant women or women who may become pregnant (GlaxoSmithKline, 2001). Low plasma dihydrotestosterone may inhibit fetal development of male external genitalia and internal reproductive organs. Animal data shows that in pregnant rats hypospadia occurs at a rate of 3.6% with a dose of 0.1 mg kg^{-1} day^{-1} together with smaller prostate and seminal vesicles and decreased anogenital distance. In pregnant rhesus monkeys defects in external genitalia of male fetuses were observed at 2 mg kg^{-1} day^{-1} p.o. These defects and the similarities in mode of action of finasteride and saw palmetto extracts and constituents all lead to a serious concern regarding the use of saw palmetto products by pregnant women. Such products are generally recommended for treatment of prostate problems in men and there is no evidence supporting any beneficial effects for women. Nevertheless, there have been some pamphlets and internet advertisements suggesting that saw palmetto could be used for treatment of urinary problems and immune system stimulation in both sexes.

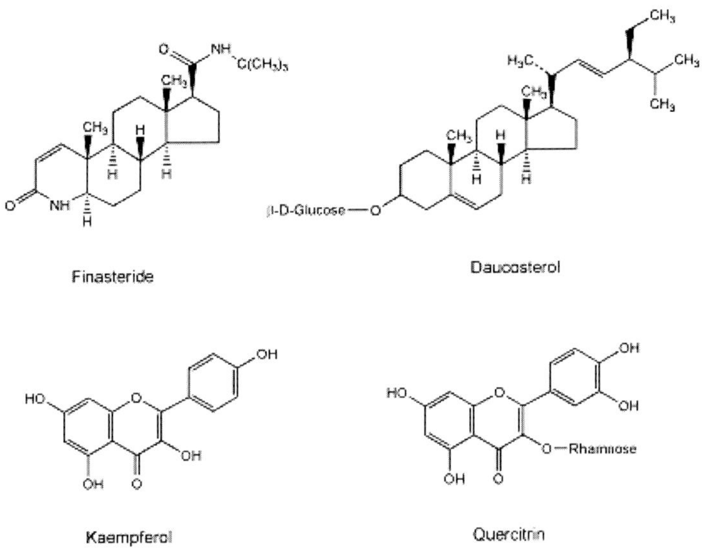

Figure 2. Structures of finasteride and the saw palmetto constituents daucosterol, kaempferol and quercitrin.

Application of the Framework to Poisonous Plant Studies: *Castanospermum australe* as Test Case.

There are a great many similarities in the application of a systematic approach to establishing the safety of herbal dietary supplements and for evaluating the toxicity of poisonous plants to livestock. The major differences are that dietary supplements are usually consumed by humans, although there is an increasing market for them for treatment of domestic pets, whereas plant toxicity primarily affects animals, with occasional episodes of poisoning in humans, particularly children. Safety and toxicity are essentially congruent but since the former can never be absolutely established, it must be regarded as an absence of provable toxicity. Given these considerations, application of the Framework to any known or suspected poisonous plant situation should yield valuable information with respect to the nature of the toxins, mode of action and prevention strategies. As a test case, a well-investigated toxic plant will be examined in a retrospective manner by application of the four categories of evidence.

Moreton Bay Chestnut or Black Bean (*Castanospermum australe*) is a tree native to northeastern Australia, producing large seeds about the same size and appearance as horse chestnuts, hence the common name. Nevertheless the plant is a member of the family Leguminosae with a typical leguminous pod which falls to the ground and splits to release the seeds when ripe. Human data are available in the form of reports that European settlers who ate the seeds suffered from vomiting, severe abdominal pain, diarrhea and dizziness. However, there is historical evidence that aborigines ate seeds after soaking and roasting (Everist 1981), suggesting that the toxic constituents are either water soluble or heat labile. Animal studies reported gastroenteritis, often fatal, in cattle and horses in the field, accompanied by emaciation, labored breathing, dull and staring eyes, and excessive urination. At autopsy there was severe inflammation of the stomach and intestinal tract. Unripe seeds (23–51 kg over 9–18 days) were found to be poisonous in feeding studies (Everist 1981). There were no reports of *in vitro* experiments and taxonomic evaluation showed that *Castanospermum* is a monotypic genus of the Leguminosae. Consideration of additional information on related substances showed the presence of saponins in the bark (Rao *et al.* 1969) and a polyhydroxy alkaloid, castanospermine, in the seeds (Hohenschutz *et al.* 1981). The latter observation was crucial in directing subsequent investigations, since castanospermine is structurally analogous to swainsonine (Figure 3), the toxin of *Swainsona* species in Australia (Colegate *et al.* 1979) and locoweeds (*Astragalus* and *Oxytropis* species) in the United States (Molyneux and James 1982).

Castanospermine

Swainsonine
Inhibits: α-mannosidase (K_i = 2μM)

Figure 3. Structural comparison of castanospermine and swainsonine.

At this point, there was no information to suggest that either the saponins or castanospermine were responsible for the episodes of livestock toxicity. However, swainsonine is a potent inhibitor of α-mannosidase (K_i = 2 μM) and this suggested that castanospermine might have similar glycosidase inhibitory properties. Subsequent *in vitro* experiments showed that *C. australe* seed extracts and castanospermine itself inhibited both α- and β-glucosidases (Saul *et al.* 1983, 1984). Bioactivity-directed fractionation has led to the isolation and identification of seven polyhydroxy alkaloids, four isomers of castanospermine with the indolizidine ring system (Molyneux *et al.* 1986, 1990, 1991) and three having a pyrrolizidine ring structure, named as australines (Molyneux *et al.* 1988; Harris *et al.* 1989; Tropea *et al.* 1989) (Figure 4). These findings have led to animal studies in rats, which showed altered glycogen distribution similar to type II glycogenesis (Saul *et al.* 1985) and vacuolar degeneration of kidney, thyroid, and liver similar to genetic glycogenesis (Pompe's disease) (Stegelmeier *et al.* 1995). All of this data is consistent with the polyhydroxy alkaloids being the toxins in the seeds because as potent inhibitors of α- and β-glucosidases they would be expected to affect digestive enzymes, leading to the observed gastrointestinal symptoms and ultimately emaciation.

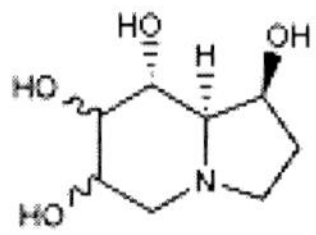

Castanospermines (4 epimers)

Inhibit: α- & β-glucosidases (K_i = 8-230 μM)

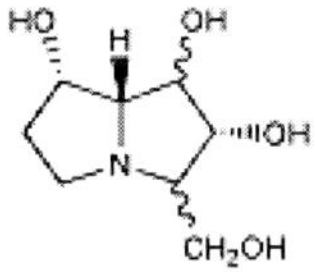

Australines (3 epimers)

Inhibit: α- & β-glucosidases (K_i = 6-21 μM)

Figure 4. Structures of indolizidine (castanospermines) and pyrrolizidine (australines) alkaloids isolated from *Castanospermum australe*.

Conclusion

The above discussion shows that the Framework for Evaluating Safety of Dietary Supplements can be applied to the purpose for which it was intended, as exemplified by the identification of a risk of genital fetal deformities in pregnant women consuming saw palmetto. Furthermore, it is a valuable systematic approach to evaluating livestock poisoning situations prior to undertaking expensive and time-consuming animal experiments, so that efforts are directed to those experiments that will yield the most information. As illustrated by the case of *C. australe*, it is probable that many experiments with horses or cattle would have been required to identify the toxin if there had not been a recognition of the structural and mode of action similarities between castanospermine and swainsonine.

References

Colegate SM, Dorling PR, and Huxtable CR (1979). A spectroscopic investigation of swainsonine: An α-mannosidase inhibitor isolated from *Swainsona canescens*. *Australian Journal of Chemistry* 32:2257–2264.

Committee on the Framework for Evaluating the Safety of Dietary Supplements (2005). *Dietary Supplements: A Framework for Evaluating Safety*. National Academies Press, Washington, DC.

Everist SL (1981). *Poisonous Plants of Australia*. Angus and Robertson Publishers, London.

GlaxoSmithKline (2001). *Approval Package. Duagen (Dutasteride) Soft Gel Capsules. Pharmacology Review(s).* Online: http://www.fda.gov/cder/foi/nda/2001/21319_Duagen.htm (accessed 29 June 2005).

Harris CM, Harris TM, Molyneux RJ, Tropea JE, and Elbein AD (1989). 1-*Epi*australine, a new pyrrolizidine alkaloid from *Castanospermum australe. Tetrahedron Letters* 30:5685–5688.

Hiermann A (1989). About contents of sabal fruits and their anti-inflammatory effect. *Archiv der Pharmazie* 322:111–114.

Hiipakka RA, Zhang HZ, Dai W, Dai Q, and Liao S (2002). Structure-activity relationships for inhibition of human 5α-reductases by polyphenols. *Biochemical Pharmacology* 63:1165–1176.

Hohenschutz LD, Bell EA, Jewess PJ, Leworthy, DP, Pryce, RJ, Arnold E, and Clardy, J (1981). Castanospermine, a 1,6,7,8-tetrahydroxyoctahydroindol-izine alkaloid, from seeds of *Castanospermum australe. Phytochemistry* 20: 811–814.

Molyneux RJ and James LF (1982). Loco intoxication: Indolizidine alkaloids of spotted locoweed (*Astragalus lentiginosus*). *Science* 216:190–191.

Molyneux RJ, Roitman JN, Dunnheim G, Szumilo T, and Elbein AD (1986). 6-*Epi*castanospermine, a novel indolizidine alkaloid that inhibits α-glucosidase. *Archives of Biochemistry and Biophysics* 251:450–457.

Molyneux RJ, Benson M, Wong RY, Tropea JE, and Elbein AD (1988). Australine, a novel pyrrolizidine alkaloid glucosidase inhibitor from *Castanospermum australe. Journal of Natural Products* 51:1198–1206.

Molyneux RJ, Tropea JE, and Elbein AD (1990). 7-Deoxy-6-*epi*castanospermine, a trihydroxyindolizidine alkaloid glycosidase inhibitor from *Castanospermum australe. Journal of Natural Products* 53:609–614.

Molyneux RJ, Pan YT, Tropea JE, Benson M, Kaushal GP, and Elbein AD (1991). 6,7-Di*epi*castanospermine, an indolizidine alkaloid inhibitor of amyloglucosidase. *Biochemistry* 30:9981–9987.

NBJ's Annual Industry Overview VIII (2003). *Nutritional Business Journal* 8: 1–9.

Rao MG, Row LR, and Rukmini C (1969). Chemistry of saponins. IV. New triterpene sapogenin, castanogenol, from the bark of *Castanospermum australe. Indian Journal of Chemistry* 7:1203–1205.

Rhodes L, Primka RL, Berman C, Vergult G, Gabriel M, Pierre-Malice M, and Gibelin B (1993). Comparison of finasteride (Proscar®) a 5 alpha reductase inhibitor, and various commercial plant extracts in *in vitro* and *in vivo* 5 alpha reductase inhibition. *Prostate* 22:43–51.

Saul R, Chambers JP, Molyneux RJ, and Elbein AD (1983). Castanospermine, a tetrahydroxylated alkaloid that inhibits ß-glucosidase and ß-glucocerebrosidase. *Archives of Biochemistry and Biophysics* 221:593–597.

Saul R, Molyneux RJ, and Elbein AD (1984). Studies on the mechanism of castanospermine inhibition of α- and ß-glucosidases. *Archives of Biochemistry and Biophysics* 230:668–675.

Saul R, Ghidoni JJ, Molyneux RJ, and Elbein AD (1985). Castanospermine inhibits α-glucosidase activity and alters glycogen distribution in animals. *Proceedings of the National Academy of Sciences USA* 82:93–97.

Stegelmeier BL, Molyneux RJ, Elbein AD, and James LF (1995). The lesions of locoweed (*Astragalus mollissimus*), swainsonine, and castanospermine in rats. *Veterinary Pathology* 31:289–298.

Tropea JE, Molyneux RJ, Kaushal GP, Pan YT, Mitchell M, and Elbein AD (1989). Australine, a pyrrolizidine alkaloid that inhibits amyloglucosidase and glycoprotein processing. *Biochemistry* 28:2027–2034.

Wilt TJ, Ishani A, Stark G, MacDonald R, Lau J, and Mulrow C (1998). Saw palmetto extracts for treatment of benign prostatic hyperplasia: A systematic review. *Journal of the American Medical Association* 280:1604–1609.

Chapter 93

Selected Herbal Toxicities in Dogs and Cats

C. Means

1717 S Philo Rd, Suite 36, Urbana IL 61802

Introduction

The sales of herbal dietary supplements in 2004 totaled $257,514,900 in mainstream retail stores, not including Wal-Mart. The top selling herbs were garlic and echinacea (Blumenthal 2005). Dogs and cats are exposed by accidental overdoses or by owner administration. Adverse effects can occur because of dosage, species differences in metabolization, and drug–herb interactions. This paper will list several popular herbs and functional foods and discuss pharmacology and clinical signs.

Sympathomimetics

Herbs that contain sympathomimetics include Ma Huang (*Ephedra sinica*), Indian common mallow (*Sida cordifolia*) and bitter orange (*Citrus aurantium*). These herbs are commonly sold as weight loss aids, decongestants, and also for illicit recreational use, as herbal ecstasy. The active components of *E. sinica* and *S. cordifolia* are alkaloids, including ephedrine and pseudoephedrine. Bitter orange contains synephrine. (DerMarderosian and Beutler 2002; Means 2002)

Pharmacologically, ephedrine and pseudoephedrine are sympathomimetic alkaloids. The alkaloids stimulate alpha- and beta-adrenergic receptors, causing the release of endogenous catecholamines at synapses in the brain and heart. This stimulation results in peripheral vasoconstriction and cardiac stimulation, resulting in tachycardia, ataxia, restlessness, tremors, and seizures (Gurley *et al.* 1998).

Underlying disease may increase an animal's susceptibility to toxicity. Some of these conditions include heart disease, diabetes, and seizure disorders. Once an animal becomes symptomatic, clinical signs may last for 36 to 48 hours. Pseudoephedrine is excreted in urine as an unchanged drug (Means 2002).

Clinical signs have been seen in dogs at 5–6 mg kg^{-1} and deaths have occurred at 10–12 mg kg^{-1}. Dogs have a narrow margin of safety compared to other species. Toxicity can be increased if other sympathomimetic substances (such as phenylpropanolomine) are taken at the same time (Ooms *et al.* 2001).

Methylxanthines

Caffeine is obtained from guarana (*Paullinia cupana*), green tea (*Camellia sinensis*), and Kola nut (*Cola nitida*), as well as other plants. Guarana may contain 3 to 5% caffeine by dry weight compared to coffee beans (1-2 % caffeine) and tea (1–4% caffeine). Theobromine and theophylline may be present in some plants, like guarana, that contain caffeine. Caffeine is used in weight loss products and for energy, as well as a diuretic.

Caffeine is a methylated xanthine. It increases cyclic AMP, releases catecholamines, and increases muscular contractility. The net effect is a positive inotropic and chronotropic effect on the heart, cerebral vasoconstriction, renal vasorelaxation, and smooth muscle relaxation in the gastrointestinal tract (Gwaltney-Brant 2001).

Caffeine is well absorbed orally. The plasma half-life in the dog is 4.5 hours. Caffeine is metabolized in the liver and undergoes enterohepatic recirculation. Caffeine is excreted through the urine (Gwaltney-Brant 2001). Pre-existing medical conditions including heart and kidney disease, and ulcers can increase toxicity of caffeine. Caffeine has caused birth defects in animals (Brinker 1998).

The LD_{50} of caffeine in dogs is reported to be 140 mg kg^{-1}. However, serious toxicity and death have been reported at doses much lower than the LD_{50}. For combinations of ma huang and guarana, the minimum dose at which clinical signs were reported is 1.3 mg kg^{-1} ma huang and 4.4 mg kg^{-1} guarana. The minimum dose at which death was reported was 5.8 mg/kg^{-1} ma huang and 19.1 mg kg^{-1} guarana. These doses were obtained by searching the ASPCA Animal Poison Control Center's Antox® database (Ooms *et al.* 2001). There are multiple drugs that can interact with caffeine. Besides pseudoephedrine and ma huang, monoamine oxidase inhibitors, aspirin, and cimetidine are commonly used medications that should not be combined with caffeine. The sedative effects of benzodiazepines may be decreased by caffeine (Brinker 1998).

Clinical signs include vomiting, restlessness and hyperactivity, polydipsia and polyuria. Tachycardia and other cardiac arrhythmias such as premature ventricular contractions (PVCs) are possible. Clinical signs progress to muscle tremors and seizures, and finally death (Gwaltney-Brant 2001; Ooms *et al.* 2001).

Miscellaneous Herbs

Griffonia simplicifolia seeds are used as a source of 5-hydroxytryptophan (5-HTP). This extract is generally used to treat depression, headaches, obesity and insomnia in humans. 5-HTP is reported to increase serotonin in the CNS. Label information may list 5-HTP, 5-hydroxytryptophan, or griffonia seed extract as an ingredient. Drug interactions with MAO inhibitors, antidepressants, and other herbs can occur.

5-HTP is rapidly and well absorbed from the gastrointestinal tract. 5-HTP readily crossed the blood–brain barrier. Once target cells are reached, 5-HTP is converted to serotonin (5-hydroxytryptamine). Serotonin is important in the regulation of sleep, cognition, behavior, temperature regulation, and other functions.

In dogs, the minimum toxic dose reported is 23.6 mg kg^{-1} and the minimum lethal dose reported in dogs is 128 mg kg^{-1}. There is not necessarily a good correlation between severity of signs and dose ingested. Signs have been reported from 10 min up to 4 hours post ingestion. Signs can last up to 36 hours (Gwaltney-Brant *et al.* 2000).

Clinical signs include seizures and tremors, depression, ataxia, and hyperesthesia. Abdominal pain, vomiting, diarrhea and drooling are common. Severe hyperthermia and blindness have been reported.

Echinacea purpurea, also known as purple coneflower, was the second most frequently purchased herb in 2004 (Blumenthal 2005). Echinacea contains essential oils as well as glyco-proteins, alkamides, and flavonoids. Echinacea is typically used as an immune stimulant and as supportive therapy for the common cold or coughs, urinary tract infections, and stomatitis (Brinker 1998; DerMaderosian and Beutler 2002)

Echinacea has a wide margin of safety. Arabinogalactan, a purified compound found in *E. purpurea*, has been dosed at 4 gm kg^{-1} IP and IV with no toxic effects. The most common clinical signs reported in dog and cat ingestions were vomiting and drooling. Rarely reported signs include lethargy, hyperactivity and erythema (Means 2002).

Chamomile refers to both German chamomile (*Matricaria recutita*) and Roman (*Anthemis nobilis*). In alternative veterinary medicine practices, chamomile may be recommended as a natural wormer, sedative, or as a treatment for aggression.

Chamomile contains essential oils, flavanoids, and hydroxycoumarins. Both fresh and dried flower heads are used. Chamomile is used as a sedative and gastrointestinal antispasmodic. It may also be taken to treat colds, bronchitis and fevers. Chamomile is contraindicated in pregnancy (McGuffin *et al.* 1997).

Bisabolol, which accounts for 50% of the essential oils found in chamomile, has an acute LD_{50} of 15 mL kg^{-1} in rats and mice. In a 4-week subacute toxicity study, 1–2 mL kg^{-1} given orally to rats produced no significant effects (DerMaderosian and Beutler 2002). Hypersensitivity and anaphylaxis have been reported.

The most common reported clinical signs include lethargy, vomiting and diarrhea. Epistaxis and hematoma development have been reported in cats (Means 2002).

Yohimbine (*Pausinystalia yohimbe*) was traditionally used to treat hypertension and angina. Today, it is used mostly as a sexual stimulant and for erectile dysfunction. Pharmacologically, yohimbine is an alpha 2-adrenergic blocking agent with a significant first pass effect. The drug is metabolized in the liver, and metabolites are eliminated in the urine. The $T_{1/2}$ in dogs is 1.5–2 hours (Plumb 2005).

Clinical signs include hyperactivity, agitation, tremors and seizures. GI signs, vomiting, diarrhea, and abdominal pain are common. An initial hypertension is usually followed by profound hypotension (Volmer *et al.* 1994).

Essential Oils

Essential oils are produced by a large number of plants. The oils are a mixture of terpenes and other chemicals. Commonly used essential oils in veterinary medicine include Melaleuca or tea tree oil (*Melaleuca alternifolia*), pennyroyal oil (*Mentha pulegium*), and D-limonene and linalool (*Citrus* spp.). In veterinary medicine, essential oils are most commonly used to treat flea infestations, hot spots or other dermatological conditions, or as wormers (Villar *et al.* 1994; Bischoff and Fessesswork 1998; Wolfe 1999).

Essential oils are rapidly absorbed both orally and dermally. They are metabolized by the liver to glucuronide and glycine conjugates. Repeated exposure can cause inductions of the hepatic enzyme systems cytochrome P-450 and UPD-glycuronyl transferase systems. Thus, pre-existing liver disease can increase the risk of toxicity. Cats appear to be more sensitive to essential oils than dogs. The acute LD_{50} varies significantly between various essential oils. Formulations mixed with an organic solvent, such as alcohol, can allow increased absorption and toxicity (Villar *et al.* 1994; Bischoff and Fessesswork 1998; Wolfe 1999).

The most common clinical signs after dermal exposures include ataxia, muscle weakness, depression, and behavioral abnormalities. Severe hypothermia and collapse as well as scrotal dermatitis have occurred in cats after d-limonene dips. A transient paresis can occur in small breed dogs when melaleuca oil is applied down the spine as a topical flea treatment. Liver failure is associated with oils, especially pennyroyal and melaleuca (Sudekum *et al.* 1992; Villar *et al.* 1994; Bischoff and Fessesswork 1998; Kaluzienski 2000).

Functional Foods

Functional foods are conventional foods consumed as part of a normal diet with the expectation that a health benefit will result, such as reducing the risk of chronic disease, or providing a specific physiologic benefit. Functional foods can be neutraceuticals, or supplements as well as part of a normal diet. Three popular functional foods include garlic, chocolate and grape juice.

Garlic (*Allium sativum*) is the most frequently used food and also the number one selling herb. The volatile oils in garlic contain the active ingredients allicin and a sulfur-containing compound. The majority of pharmaceutical activity is believed to be found in these substances. Garlic has been used to treat high cholesterol, hypertension, the common cold and diabetes (DerMaderosian and Beutler 2002)

Garlic is in the same family as onions and an overdose of garlic causes the same clinical syndrome as onions. Toxicity results in weakness, tachypnea and tachycardia. Hematological changes including hemolysis, Heinz bodies, and possibly methemoglobinemia may occur. 1.25 mL of garlic extract kg^{-1} body weight (equivalent to 5 g whole garlic kg^{-1}) for 7 days caused oxidative injury to erythrocytes (Lee *et al.* 2000).

Chocolate (*Theobroma cacao*) is used to treat asthma, premenstrual syndrome, and as a diuretic. Chocolate is a methylxanthine (see above section). Baking chocolate has 8–118 mg oz^{-1} of caffeine and 393 mg oz^{-1} of theobromine. Typically a 50-pound dog may have a mild reaction after ingesting 1 oz of baking chocolate and a moderate to severe reaction after ingesting 2 oz. Mild reactions may be seen at about 20 mg kg^{-1} of methylxanthines and moderate to severe reactions at 40–50 mg kg^{-1} methylxanthines (Gwaltney-Brant 2001).

Purple grape juice (*Vitis* spp.) is recommended to lower cholesterol and provides a protective cardiac effect. Since grapes and raisins have caused acute renal failure in dogs, grape juice can not be recommended. The toxic principle is unknown, and whether the toxin is heat stable is undetermined at this time (Mazzaferro *et al.* 2004).

Considerations

Clients who use herbal products should be advised to treat them as a medication, and keep them away from pets. Clients should be encouraged to report potential adverse reactions, or to discuss different routes of therapy if a pet's medical condition is not improving. Remember herbal products are not regulated, and amounts and contents can differ from the label. Pets can also act as sentinels for potential adverse reactions, adulterations, or other quality issues. The Veterinary Botanical Medicine Association provides veterinarians a place to report adverse effects associated with herbal treatments or overdoses. The web address is www.vbma.org.

References

Bischoff K and Fessesswork G (1998). Australian tea tree (*Melaleuca alternifolia*) oil poisoning in three purebred cats. *Journal of Veterinary Diagnostic Investigation* 10(2):208–10.

Blumenthal M (2005). Market report. *Herbalgram* 66:63.

Brinker F (1998). *Herb Contraindications and Drug Interactions,* 2nd ed. Eclectic Medical Publications, Sandy, Oregon.

DerMarderosian A and Beutler JA, eds (2002). *The Review of Natural Products.* Facts and Comparisons Publishing Group, St. Louis, Missouri.

Gurley BJ Gardner SF, White LM, and Wang PL (1998). Ephedrine pharmacokinetics after the ingestion of nutritional supplements containing *Ephedra sinica* (ma huang). *Therapeutic Drug Monitoring* 20:439–345.

Gwaltney-Brant S (2001). Chocolate intoxication. *Veterinary Medicine* 96(2):108–111.

Gwaltney-Brant SM, Albretson JC, and Khan SA (2000). 5-Hydroxytryptophan toxicosis in dogs: 21 cases (1989-1999). *Journal of the American Veterinary Medical Association* 216(12):1937–1940.

Kaluzienski M (2000). Partial paralysis and altered behavior in dogs treated with melaleuca oil. *Clinical Toxicology* 38(5):518–519.

Lee KW, Yamato O, Tajima M, Kuraoka M, Omae S, and Maede Y (2000). Hematologic changes associated with the appearance of eccentrocytes after intragastric administration of garlic extract to dogs. *American Journal of Veterinary Research* 61(11):1446–1450.

Mazzaferro EM, Eubig PA, Hackett TB, Legare M, Miller C, Wingfield WE, and Wise L (2004). Acute renal failure associated with raisin or grape ingestion in 4 dogs. *Journal of Veterinary Emergency and Critical Care* 14(3):203–212.

McGuffin M, Hobbs C, Upton R, and Goldberg A, eds (1997). *Botanical Safety Handbook.* CRC Press, New York.

Means C (2002). Selected herbal hazards. *Veterinary Clinics of North America Small Animal Practice* 32(2):367–382.

Ooms TG, Khan SA, and Means C (2001). Suspected caffeine and ephedrine toxicosis resulting from ingestion of an herbal supplement containing guarana and ma huang in dogs: 47 cases (1997-1999). *Journal of the American Veterinary Medical Association* 218(2):225–229.

Plumb DC (2005). *Veterinary Drug Handbook,* 5th ed. Iowa State University Press, Ames.

Sudekum M, Poppenga RH, Raju N, and Braselton WE (1992). Pennyroyal oil toxicosis in a dog. *Journal of the American Veterinary Medical Association* 200(6):817–818.
Villar D, Knight MJ, Hansen SR, and Buck WB (1994). Toxicity of melaleuca oil and related essential oils applied topically on dogs and cats. *Veterinary and Human Toxicology* 36(2):139–142.
Volmer PA, Tranquilli WJ, Fleming JR, and Buck WB (1994). Acute oral yohimbine toxicosis in a dog. *Canine Practice* 19(2):18–19.
Wolfe A (1999). Essential oil poisoning. *Clinical Toxicology* 37(6):721-727.

Index

Author Index